Mitochondrial Dysfunction in Neurodegenerative Disorders

Amy K. Reeve • Eve M. Simcox
Michael R. Duchen • Doug M. Turnbull
Editors

Mitochondrial Dysfunction in Neurodegenerative Disorders

Second Edition

Springer

Editors
Amy K. Reeve
Newcastle University Medical School
Wellcome Trust Centre for Mitochondrial Research
Newcastle upon Tyne
UK

Eve M. Simcox
Newcastle University Medical School
Wellcome Trust Centre for Mitochondrial Research
Newcastle upon Tyne
UK

Michael R. Duchen
Cell and Developmental Biology
University College London
London
UK

Doug M. Turnbull
Newcastle University Medical School
Wellcome Trust Centre for Mitochondrial Research
Newcastle upon Tyne
UK

ISBN 978-3-319-28635-8 ISBN 978-3-319-28637-2 (eBook)
DOI 10.1007/978-3-319-28637-2

Library of Congress Control Number: 2016941596

Printed on acid-free paper

This Springer imprint is published by Springer Nature
The registered company is Springer International Publishing AG Switzerland

Preface

The loss of neurons at any stage of life is a catastrophic event. These post-mitotic cells lack the ability to regenerate and as such their loss is an unthinkable finality. The most widely recognised diseases in which this cell loss occurs are related to advancing age, for example, Parkinson's disease and Alzheimer's disease, but there are several diseases which may occur at a much younger age, for example, multiple sclerosis and motor neuron disease. The expansion in research into roles for mitochondria in the pathogenesis of these diseases has been great in the 3 years since the first edition of *Mitochondrial Dysfunction in Neurodegenerative Disease*, and now a significant body of work is dedicated to unearthing new therapies and drugs to protect neurons against such dysfunction and loss. With ever-increasing numbers of papers being published in the field of mitochondria and neurodegenerative disease, we wanted to bring together a revised collection of articles that would update readers on current research and also introduce new research avenues, particularly into mitochondrial degradation. We hope the reader will gain insight into the significant roles that mitochondria play in the pathogenesis of these diseases, from leaders in their field. This book aims to lead the reader through the basic functions of mitochondria and what can lead to their dysfunction, to the consequences of this dysfunction on neuronal function, before finishing with the modelling of these disorders and hopes for the future. In this edition, we also include a chapter that details how mitochondria may be imaged in neurons. As an ever-expanding field of research, the last 3 years have shown us that an understanding of how dysfunction within this organelle occurs provides a crucial platform from which to launch novel interventions.

Newcastle upon Tyne, UK — Amy Katherine Reeve
Newcastle upon Tyne, UK — Eve Simcox
London, UK — Michael R. Duchen
Newcastle upon Tyne, UK — Doug M. Turnbull

Contents

Contributors

Andrey Y. Abramov, PhD, DSc Department of Molecular Neuroscience, University College London Institute of Neurology, London, UK

Marijana Aradjanski, MSc Cologne Excellence Cluster on Cellular Stress Responses in Aging-Associated Diseases (CECAD), Medical Faculty, Institute for Mitochondrial Diseases and Aging, University of Cologne, Cologne, Germany

Anna Ardissone, MD Unit of Child Neurology, IRCCS Foundation Neurological Institute "C. Besta", Milan, Italy

Kim Chisholm, BSc Department of Neuroinflammation, University College London Institute of Neurology, London, UK

Alexia Chrysostomou, BSc, MRes Wellcome Trust Centre for Mitochondrial Research, Institute of Neuroscience, Newcastle upon Tyne, UK

Michael R. Duchen, BA, MBBS, FRCP, PhD Cell and Developmental Biology, University College London, London, UK

Daniel Galtieri, BS Psychology Department of Physiology, Northwestern University, Feinberg School of Medicine, Chicago, IL, USA

Hannah Hayhurst Wellcome Trust Centre for Mitochondrial Research, Institute of Neuroscience, Newcastle University, Newcastle upon Tyne, UK

Stephen B. Helliwell, BSc, PhD Development and Molecular Pathways, Novartis Institutes for BioMedical Research, Basel, Switzerland

Gavin Hudson, BSc, PhD Institute of Genetic Medicine, International Centre for Life, Newcastle University, Newcastle upon Tyne, UK

Viktor I. Korolchuk, PhD Institute for Cell and Molecular Biosciences, Newcastle Institute for Ageing, Newcastle University, Newcastle upon Tyne, UK

Nichola Zoe Lax, PhD Wellcome Trust Centre for Mitochondrial Research, Institute of Neuroscience, Newcastle University, Newcastle upon Tyne, UK

Marthe H.R. Ludtmann, PhD Department of Molecular Neuroscience, University College London Institute of Neurology, London, UK

Diego Manni, MSc Institute for Cell and Molecular Biosciences, Newcastle Institute for Ageing, Newcastle University, Newcastle upon Tyne, UK

Meagan J. McManus, PhD Department of Pathology, Center for Mitochondrial and Epigenomic Medicine (CMEM), Children's Hospital of Philadelphia, Philadelphia, PA, USA

Arnaud Mourier, PhD Mitochondrial Biology, Max Planck Institute for Biology of Ageing, Cologne, Germany

Matthew Nguyen, BS National Institutes of Health, Medical Genetics Branch, National Human Genome Research Institute, Bethesda, MD, USA

Elsje G. Otten, MSc Institute for Cell and Molecular Biosciences, Newcastle Institute for Ageing, Newcastle University, Newcastle upon Tyne, UK

Fabian Peters, MSc Department of Neuroinflammation, University College London Institute of Neurology, London, UK

Jonathan Phillips, BSc, MRes Wellcome Trust Centre for Mitochondrial Research, Institute of Neuroscience, Newcastle University, Newcastle upon Tyne, UK

Martin Picard, PhD College of Physicians and Surgeons, Division of Behavioral Medicine, Departments of Psychiatry and Neurology, Columbia University Medical Center, New York, NY, USA

Amy Katherine Reeve, BSc(hons), PhD Newcastle University Medical School, Wellcome Trust Centre for Mitochondrial Research, Newcastle upon Tyne, UK

Mona Sadeghian, PhD Department of Experimental Neuroinflammation, University College London Institute of Neurology, London, UK

Marija Sajic, MD, PhD Department of Neuroinflammation, University College London Institute of Neurology, London, UK

Dimitra G. Schiza, MSc Department of Neuroinflammation, University College London Institute of Neurology, London, UK

Paul T. Schumacker, PhD Department of Pediatrics, Northwestern University, Feinberg School of Medicine, Chicago, IL, USA

Ellen Sidransky, MD Section of Molecular Neurogenetics, National Institutes of Health, Medical Genetics Branch, National Human Genome Research Institute, Bethesda, MD, USA

Eve Simcox, BSc(hons), PhD Newcastle University Medical School, Wellcome Trust Centre for Mitochondrial Research, Newcastle upon Tyne, UK

Kenneth J. Smith, BSc, PhD Department of Neuroinflammation, UCL Institute of Neurology, London, UK

James Surmeier, MS, PhD Department of Physiology, Northwestern University, Feinberg School of Medicine, Chicago, IL, USA

Aleksandra Trifunovic, PhD Cologne Excellence Cluster on Cellular Stress Responses in Aging-Associated Diseases (CECAD), Medical Faculty, Institute for Mitochondrial Diseases and Aging, Cologne University Clinic, University of Cologne, Cologne, Germany

Doug M. Turnbull, MD, PhD Newcastle University Medical School, Wellcome Trust Centre for Mitochondrial Research, Newcastle upon Tyne, UK

Carlo Viscomi, PhD Medical Research Council, Mitochondrial Biology Unit, Cambridge, UK

Wendy Westbroek, PhD National Institutes of Health, Medical Genetics Branch, National Human Genome Research Institute, Bethesda, MD, USA

Enrico Zampese, PhD Department of Physiology, Northwestern University, Feinberg School of Medicine, Chicago, IL, USA

Massimo Zeviani, MD, PhD Medical Research Council, Mitochondrial Biology Unit, Cambridge, UK

Part I
Mitochondria and Neurodegenerative Disease

Chapter 1
An Introduction to Mitochondria, Their Structure and Functions

Eve Michelle Simcox and Amy Katherine Reeve

Abstract Mitochondria are essential intracellular organelles whose central role in maintaining energy homeostasis places them at the heart of cell integrity, function and survival. As the reliable provision of energy is so fundamental to every aspect of cellular function, mitochondrial dysfunction inevitably has devastating implications for the cell, the tissue and the organism. This is especially critical in the nervous system, where subtle changes in signalling and function can have catastrophic global consequences. Further, as postmitotic cells are heavily dependent on oxidative phosphorylation and are morphologically enormously complex, neurons pose a unique set of challenges for the mitochondrial population that reside within them. Mitochondrial dysfunction has profound consequences for the nervous system and is implicated in a host of neurological and neurodegenerative diseases. The following chapter introduces the form and function of these fascinating organelles and introduces key concepts and vulnerabilities that may underlie their involvement in neurodegenerative conditions. This introduction lays the foundation for the following chapters, which will explore specific aspects of the roles of these organelles in a range of neurodegenerative disease.

Keywords Mitochondria • Mitochondrial function • mtDNA genetics • Neurodegeneration

---

E.M. Simcox, BSc(hons), PhD • A.K. Reeve, BSc(hons), PhD (✉)
Newcastle University Medical School, Wellcome Trust Centre for Mitochondrial Research, Newcastle upon Tyne, UK
e-mail: amy.reeve@ncl.ac.uk

A.K. Reeve et al. (eds.), *Mitochondrial Dysfunction in Neurodegenerative Disorders*, DOI 10.1007/978-3-319-28637-2_1

1.1 Introduction

1.1.1 The Mitochondrion

Mitochondria have been known to house the central metabolic pathways required for the oxidation of nutrients for well over 50 years. Their role, however, extends far beyond that of providing energy; indeed the majority of all cellular functions are reliant on the functioning of these organelles, be it directly or through an auxiliary role. Mitochondria, or the related hydrogenosomes and mitosomes found in less complex organisms, are believed to be present in all nucleated cells of the eukarya and often comprise 20 % or more of the total cell volume [1, 2], an abundance that reflects their importance.

There is a general consensus that mitochondria evolved from an alpha eubacterial ancestor which became engulfed by a host cell in a process referred to as endosymbiosis [1], although the origin of the host cell remains contentious (see [3]). This event can be considered a turning point in evolution. In gaining protection within the host cell from an ever increasingly hostile environment, the eubacterial ancestor repaid its host with a means of energy production over 15 times as powerful as before, enabling the complexity and size of organisms to increase exponentially.

This assimilation is thought to have occurred around 1.5 billion years ago [4], but the presence of mitochondria was only revealed following cytological observations during the late 1800s. Altman was the first to realise the ubiquitous nature of this intracellular organelle in 1890 and concluded that these biosomes as he termed them were 'elementary organisms' performing functions within the cell [5]. The breadth of their function and indeed importance to the cell were probably unimaginable at this time; indeed it was over 20 years later before they were first linked to respiration [6]. Eight years hence of Altmans observations, 'biosomes' underwent a change of identity when their name was refined to reflect their thread (mitos) and granular (chondrion) structure, when the term mitochondrion was coined by Benda in 1898 [7]. It was not until the electron microscopy studies of the 1950s by Palade and Sjostrand that the complex internal structure of the mitochondria began to emerge, in turn starting to reveal secrets of this remarkable organelle's functions and mechanisms of action [8].

1.1.2 Mitochondrial Structure

As independent organelles, mitochondria are typically oval in form, although through dynamic events their appearance can range from extended reticular networks through to the more typically described rod-shaped structures. The structure of the mitochondrion is dictated by its role, with every aspect of its form contributing to a highly specialised function (Fig. 1.1). The outer membrane forms the boundary between the organelle and surrounding cytosol. Its composition is similar to that of the cell membrane and, as such, allows the diffusion of lipid soluble molecules into the intermembrane space. Transport of small proteins (<5000 Da) and hydrophilic proteins is also

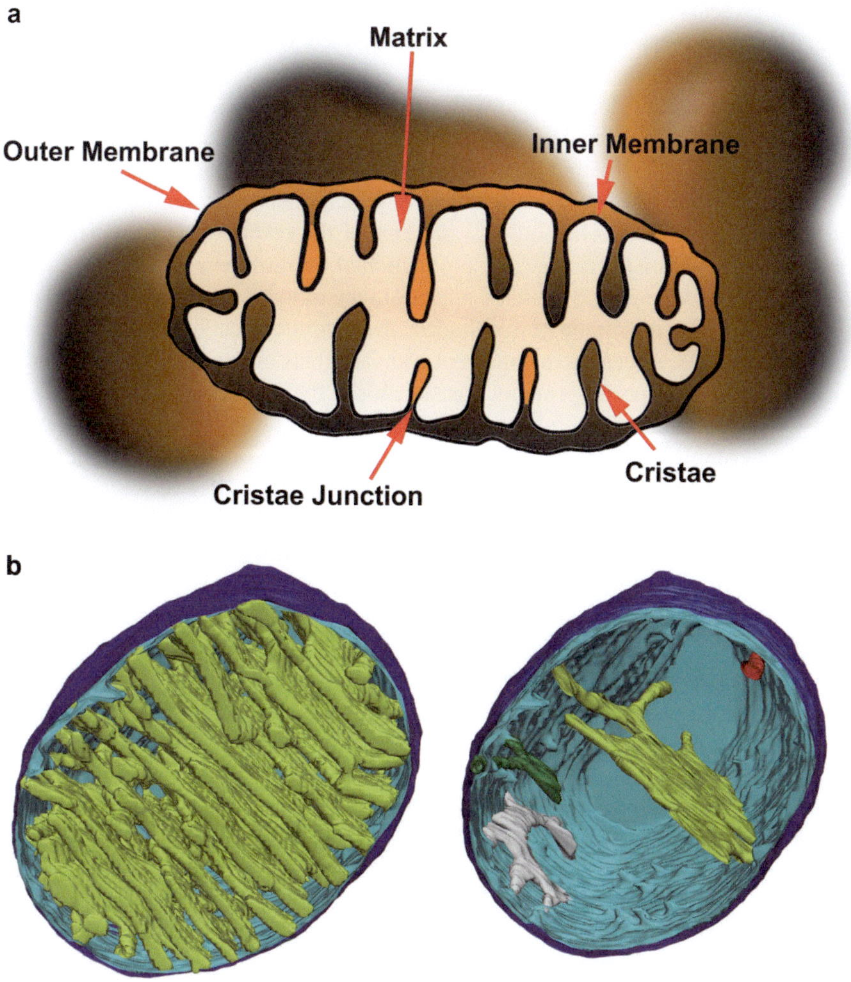

Fig. 1.1 Mitochondrial structure. (**a**) Mitochondria are double-membrane organelles, with a highly specialised structure. Here we show the main components of mitochondrial structure. (**b**) 3D electron tomography has revealed the detailed structure of the cristae within mitochondria. Each mitochondrion contains tightly packed cristae which significantly increase the surface area of the inner mitochondrial membrane. 3D tomography has enabled researchers to delve deeper into the complexities of these structures and to single out and study individual cristae (Image kindly provided by T.G. Frey, San Diego State University, and G.A. Perkins, University of California, San Diego)

made possible through porin, the voltage-dependent anion channel (VDAC), which is expressed in abundance within this outer membrane.

The internal mitochondrial membrane surrounds the central matrix of the mitochondrion and is impermeable to polar molecules and ions. This membrane is enriched in cardiolipin and is far richer in proteins than the outer membrane, containing elevated levels of proteins required for a multitude of biochemical pathways within the organelle. The impermeability of this membrane coupled with the far more permeable nature of the outer membrane results in the intermembrane space, an environment similar to that of the cytoplasm but with strict specificity for larger proteins that have a mitochondrial role. This has profound impact on mitochondrial function and is a limiting factor when targeting these organelles with drug therapies.

Although mitochondria contain their own genome, uniquely amongst organelles of animal cells, this encodes only 13 proteins of the respiratory chain. Therefore, mitochondria rely on the nucleus for production of most of their components, which must therefore be imported from the nucleus to the mitochondria. In order for larger proteins to gain access to the mitochondria, they require specific mitochondrial targeting sequences. Access to the different compartments of the organelle is governed largely by two transporter proteins, translocator of the outer membrane (TOM) and translocator of the inner membrane (TIM) (Fig. 1.2). Through a receptor protein in the TOM complex, the signal sequence is recognised and inserted alongside adjacent polypeptides. Through their proximity at close contact sites between the membranes, the polypeptide then interacts with TIM allowing the polypeptide to enter into the matrix or to move laterally into the inner membrane itself [9]. The import and folding of many intermembrane space proteins are undertaken by the mitochondrial intermembrane space assembly (MIA) complex [10, 11]. The minority of proteins that are synthesised on matrix ribosomes and require export into the inner membrane do so through the oxidase assembly machinery (OXA) [9].

A simplified schematic of the mitochondrion can be seen in Fig. 1.1a showing the double plasma membrane and convoluted internal architecture. The invagination of the inner mitochondrial membrane was long believed to give rise to the internal folds that project into the matrix, called *cristae mitochondriales*, or cristae. Electron tomography analysis, however, has now shown that this model is inaccurate with cristae actually arising from a distinct membrane connected to the inter membrane space through tubular inner membrane junctions (see Fig. 1.1b) [12]. Illustrating the highly adapted form of mitochondria in relation to their function, the cristae of the inner membrane provide a vastly enlarged surface area over which oxidative phosphorylation and maintenance of the proton gradient can occur. It is now understood that the form and complexity of these invaginations vary enormously between tissue type and even within cell type, dependent on the developmental stage or physiological condition. For example, cristae biogenesis is regulated through ATP synthase. As the ATP synthase is responsible for the majority of cellular energy produced, and mitochondrial morphology is linked to energy demand, this is a clear demonstration of the links between cellular bioenergetics and the inner membrane structure [13]. Furthermore, discrete subdomains have been identified in cristae alongside enrichment of specific mitochondrial proteins, notably OXPHOS proteins [14]. Relatively recently, cristae remodelling through ablation of the fusion protein OPA1 has been shown to impair mitochondrial respiratory chain supercomplex assembly and mitochondrial function, which in turn precipitates apoptosis [15]. The adaptive qualities of cristae and involvement in complex assembly likely mean we only partially understand the crucial role they play in mitochondrial maintenance.

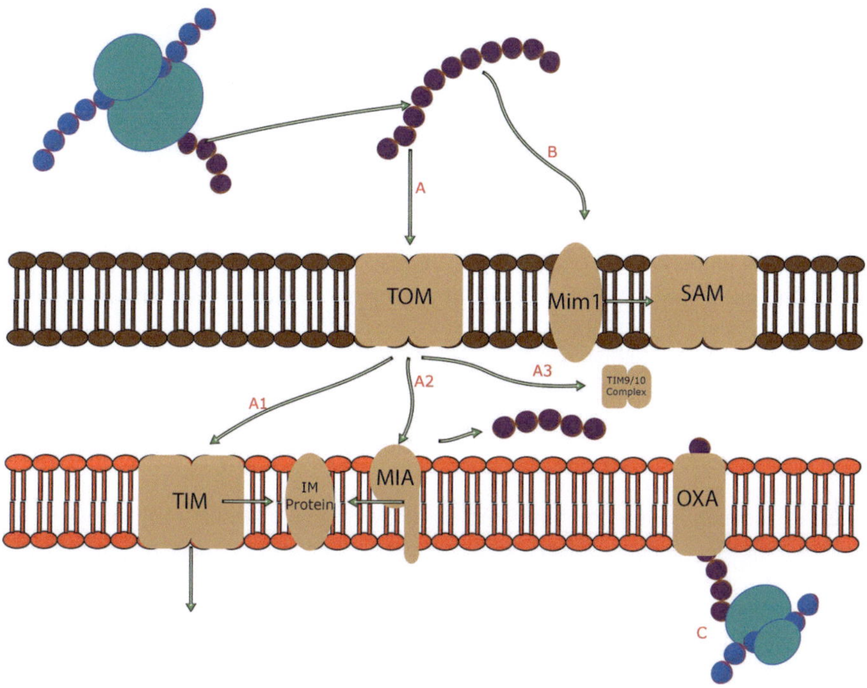

Fig. 1.2 Mitochondrial protein transport. Many of the ~1500 proteins required for mitochondrial structure and function require import into the mitochondria. The majority are transported through the outer membrane through the outer membrane translocase TOM (*A*). Once through the outer membrane, the ultimate destination of the protein dictates the next step in transportation. For proteins that continue through to the matrix, passage through the inner membrane occurs through the inner membrane translocase TIM (*A1*). Alternatively proteins that reside within the inner membrane can be transported laterally from TIM. A second route for inner membrane proteins (*IM protein*) can occur through mitochondrial intermembrane space assembly (*MIA*) machinery (*A2*). Hydrophobic precursor proteins require an inner membrane chaperone to be further transported following entry through TOM (*A3*). A proportion of proteins destined for the outer membrane bypass the TOM complex and instead may be inserted through other pathways, including mitochondrial import 1 (Mim1) and the sorting and assembly machinery (*SAM*) (*B*). Finally a few proteins are synthesised on the ribosomes of the matrix and are subsequently exported into the inner membrane by the oxidase assembly (*OXA*) machinery (*C*)

1.2 The Functions of Mitochondria

Mitochondria are required for a plethora of functions, illustrated by their presence in every nucleated eukaryotic cell. Mitochondrial functions are often over simplified; they are frequently referred to as the 'powerhouses' of the cell due to their fundamental and most well-described role in ATP production via oxidative phosphorylation; however, the most well-conserved mitochondrial function is that of iron-sulphur cluster formation. Aside from this, mitochondria are also responsible for calcium handling, apoptosis, cell signalling and ROS production.

1.2.1 Energy Provision by Oxidative Phosphorylation

The number of mitochondria per cell generally reflects the energy demands of that cell, for example, energy-demanding tissues such as muscles, cardiomyocytes and neurons tend to have more mitochondria than less energy-demanding cells. This energy transduction occurs through the transport of electrons between a chain of proteins located on the inner mitochondrial membrane, which have components encoded by both the mitochondrial and nuclear genomes. The traditional view of oxidative phosphorylation has been that it occurs through electron transport between individual complexes, randomly distributed within the mitochondrial inner membrane. However, more recent data has shown the existence of supercomplexes, with differing stoichiometries and configurations. It seems that the proton-translocating enzymes, complexes I, III and IV, can be organised into these supercomplexes which supports more efficient substrate movement and catalysis (reviewed in [16, 17]).

The first and largest complex of the electron transport chain (ETC) is composed of over 40 subunits with a combined molecular weight of 980 kDa [18] (reviewed in [19]). Of the 40, 7 subunits are encoded by the mitochondrial genome (*MT-ND1–6* and *MT-ND4L*), with the remainder encoded by nuclear DNA. Complex I (NADH/ubiquinone oxidoreductase) catalyses the oxidation of NADH, derived from the citric acid cycle, which yields two electrons that pass to a flavin mononucleotide and through a series of iron-sulphur clusters to reduce ubiquinone to ubiquinol. This transfer is coupled to the translocation of four protons across the inner mitochondrial membrane.

Succinate/ubiquinone oxidoreductase (complex II) is responsible for oxidising succinate to fumarate in the citric acid cycle. The resulting electrons are donated to the ETC through the reduction of ubiquinone. Complex II, the smallest complex of the ETC, is the only complex entirely encoded by the nuclear genome and electron flow through this complex and is not associated with proton translocation (for review see [20]).

Following the reduction of ubiquinone by complexes I and II, its re-oxidation is catalysed by complex III, ubiquinol/cytochrome *c* oxidoreductase. This third complex is composed of 11 subunits which, with the exception of cytochrome *b*, are entirely nuclear encoded. The oxidation of ubiquinol releases the two electrons transferred from complexes I and II; their transfer from ubiquinol to cytochrome *c* then occurs in two steps. First, one of the electrons is transferred to the iron-sulphur cluster and then to cytochrome *c* via cytochrome *c1*, while the second one is recycled to ubiquinol in a reaction called the Q-cycle. In this cycle two ubiquinols are oxidised, two electrons are transferred and one ubiquinone is reduced. For each electron transferred two protons are translocated across the inner mitochondrial membrane [21].

Electrons are finally transferred to complex IV (cytochrome *c* oxidase), of which three subunits are encoded by the mitochondrial genome (COXI, II, III). The substrate for this complex is cytochrome *c* which transfers electrons between complexes III and IV. This haemoprotein donates electrons one by one, through different 'states', to the intermembrane space side of complex IV. Each of these transitions results in the translocation of a proton across the membrane. As electrons are passed to molecular oxygen, a total of eight protons are translocated across the inner mitochondrial membrane, generating the proton gradient utilised by the final complex of the oxidative phosphorylation machinery to generate ATP [22].

Complex V, F_oF_1 ATP synthase, a large multi-subunit complex, contains two subunits encoded by the mitochondrial genome (ATPases 6 and 8). This complex is composed of two main domains: F_o, embedded in the inner membrane, and F_1, the catalytic domain, which forms the head and stalk of the complex and is located on the matrix side of the inner membrane. The movement of protons across the inner membrane from the intermembrane space promotes the rotation of F_1, causing the β subunits to transfer between the three conformational states. The energy harnessed during this rotation is used to synthesise ATP which can then be released on the next rotation. Three protons are required for each molecule of ATP since one proton is required for each β subunit [23, 24].

The generation of ATP through oxidative phosphorylation is an extremely efficient method of aerobic glucose metabolism, far more efficient than through the substrate level phosphorylation of anaerobic glucose metabolism merely by glycolysis. Therefore, it is easy to comprehend how any defect which affects the ability of mitochondria to carry out this essential process will greatly affect the production of ATP, which will be detrimental to cellular function. The consequences of mitochondrial dysfunction are considered in the remaining sections of this book, in particular in relation to the central nervous system.

1.2.2 Cell Death

Historically mitochondria have been mainly associated with apoptosis, which is a highly regulated active form of cell death. It is now clear that there are four forms of cell death (apoptotic, necrotic, autophagic and parthanatos) that have all been reported within the literature to be important in neurodegenerative disease and all have links to the mitochondria, although only three are ATP dependent. Of these pathways the least well defined is autophagic cell death. The exact mechanism for autophagic cell death is still to be determined, but cells lost due to this pathway do show the accumulation of autophagosomes. Recent work though has suggested that autophagy may contribute to a form of cell death mediated through the sodium/potassium ATPase [25].

Apoptosis is characterised by cell shrinkage, membrane blebbing, nuclear fragmentation and condensation of nuclear chromatin and is usually the result of intracellular signalling. Following these morphological changes, the cell is systematically dismantled and is then degraded by macrophages. There are two main apoptotic pathways which occur in mammalian cells, the intrinsic pathway and extrinsic pathway, and mitochondria play a key role in both. The intrinsic pathway is controlled by members of the Bcl-2 family, which direct death signals which target mitochondria, to facilitate the release of pro-apoptotic proteins from the intermembrane space [26]. In the extrinsic pathway activation of caspase-8, following the formation of the death-inducible signalling complex, initiates a cascade of protein interactions leading to mitochondrial outer membrane permeabilisation and cytochrome *c* release from the intermembrane space ([26] reviewed in [27]). The main effectors of apoptosis are the Bcl-2 proteins, BAX and BAK. Both these proteins interact with the mitochondrion and their loss from cells leads to resistance to a number of apoptotic stimuli. The interaction of these two key proteins facilitates the permeabilisation of the outer mitochondrial membrane and the release of cytochrome *c* which then drives apoptosis.

Necrosis tends to occur in response to extracellular stimuli and has been reported to occur following ischemia and trauma. It is a well-defined process that is characterised by cellular swelling, depletion of energy stores and a disruption of cellular membranes. Necrosis may occur in response to ischemic injury; such an injury causes a drop in pH in response to the anaerobic conditions and a depletion of ATP which causes an increase in intracellular calcium by preventing calcium uptake. These changes in calcium levels may then trigger the opening of the mitochondrial permeability transition pore (mPTP). Identified in the 1970s, the exact molecular identity of the pore remains elusive. The mPTP is believed to be a channel complex, composed of a number of different proteins, and although it's exact structure is still to be elucidated, a number of candidate proteins have been proposed including VDAC, ANT and cyclophilin D. The opening of this pore dramatically changes the permeability of the inner mitochondrial membrane, which may have irreversible effects if prolonged. This change in mitochondrial permeability depletes the mitochondrial membrane potential, disrupts OXPHOS and causes swelling of the matrix and thus disruption of the cristae structure, leading to rupture of the outer membrane and the activation of a number of hydrolytic enzymes which leads to necrosis. However, it should be noted that due to the rupture of the mitochondria, the release of pro-apoptotic proteins will also occur and so it is likely that the opening of mPTP will not only lead to necrotic cell death (reviewed in [27–30]).

Parthanatos is a form of cell death that is mediated by PARP-1 (poly(ADP-ribose) polymerase-1). This enzyme is activated by DNA strand damage, including nicks and breaks, and is a key mediator of DNA repair. PARP-1 can be activated by ischemic injury and agents damaging to DNA including ROS and ionising radiation. Parthanatos differs from other forms of cell death as it causes phosphatidylserine to be externalised onto the outer cell membrane, a loss of mitochondrial membrane potential and DNA fragmentation. There is no cellular swelling, but membrane integrity is lost. Other than the changes in mitochondrial membrane, potential PARP-1 activation also leads to the release of mitochondrial apoptosis-inducing factor. In addition, once activated PARP-1 forms poly(ADP)ribose polymers from nicotinamide adenine dinucleotide (NAD), in which it then transfers to a number of nuclear proteins. An overactivation of PARP-1 therefore leads to a depletion of NAD and subsequently ATP (reviewed in [31–33]).

The cell death that occurs as part of neurodegeneration is likely to be initiated by a number of factors including mitochondrial dysfunction. What is clear though is that changes in the functioning of this organelle will have a great impact on the survival and functioning of these cells. The relationship of the mitochondrion to neuronal cell loss and death will be considered in a number of the following chapters.

1.2.3 Calcium Handling and Energy Homeostasis

Mitochondria play a pivotal role in the buffering and maintenance of cellular calcium homeostasis. It has been known for many years that mitochondria have an enormous capacity to accumulate calcium through the expression of a uniporter in the inner mitochondrial membrane, allowing calcium to enter the mitochondrion down its electrochemical potential gradient. The molecular identity of the uniporter proved highly elusive and was only discovered in 2011 by the Rizzuto and Mootha labs. Now that the uniporter and its associated modulator proteins (MICU1/MICU2/MICU3; EMRE) are known, experimental manipulations to explore the functional consequences of the pathway will be far more precise than before [34–36] (reviewed in [37]). The efflux of calcium from the mitochondria is mediated through Na^+/Ca^{2+} exchange, although the molecular identity and complete functionality of this exchanger remain controversial [38–40]. The process of Ca^{2+} efflux ensures equilibrium is never reached, meaning in theory, mitochondria are an unlimited calcium sink. Research has also shown that calcium transport across the outer mitochondrial occurs through VDAC, the most abundant protein in the outer membrane, and the expression of this protein is directly correlated with the rapid transport of calcium into the mitochondrion [41–43].

Calcium ions are secondary messengers in many signalling pathways, and are of particular importance in neurons shaping neurotransmitter release and in driving gene expression [44]. The ability of mitochondria to take up calcium and act as spatial buffers in the cell [45–47] may have a profound impact on the spatiotemporal characteristics of the calcium signal and in shaping the activation of its downstream targets.

The transfer of calcium from cytosol to matrix is also thought to be important in energy homeostasis. Increased Ca^{2+} uptake activates the rate-limiting enzymes of the Krebs cycle through activation of pyruvate isocitrate and oxoglutarate dehydrogenases, increasing ATP production [48, 49]. This seems an elegant and simple way to couple the increase in energy demand that inevitably accompanies an increase in calcium concentration, with an increased energy supply. Calcium signals may also promote mitochondrial biogenesis, providing a longer term metabolic response to sustained activity, although it has to be said that this pathway has been characterised in muscle and is not established in neurons [50].

Calcium signalling and mitochondrial calcium uptake are also linked to apoptosis and necrosis via the mitochondrial permeability transition pore (mPTP). Ca^{2+} overload is sufficient to induce opening of this pore, which seems to serve as a key final common path to calcium-dependent necrotic cell death, as pore opening causes the collapse of the mitochondrial membrane potential and bioenergetic failure. However as pore opening can be inhibited pharmacologically by cyclosporine A, the prospect that pharmacological protection of neurons otherwise destined to die in neurodegenerative disease might be possible, remains an exciting prospect.

Calcium is also the main mediator of excitotoxicity, whereby excessive glutamatergic stimulation of neurons can lead to a large increase in calcium, leading to the activation of a number of proteins and ultimately causing neuronal death. Dopaminergic neurons of the substantia nigra express a unique Ca^{2+}-dependent pacemaking activity which may render this cellular population particularly vulnerable to dysfunctional mitochondria which are unable to fully buffer calcium. This offers intriguing links between mitochondrial (dys)function, calcium signalling and Parkinson's disease as it is these dopaminergic neurons that suffer attrition in this condition (see Chap. 5).

1.2.4 Iron Sulphur Clusters

Iron sulphur clusters are crucial inorganic cofactors for numerous biological processes including regulation of gene expression, enzyme catalysis and DNA repair. Within mitochondria themselves, Fe-S clusters function within complexes I, II and III due to their capacity as electron donors and acceptors. Dependent on function, the number of iron and sulphur atoms can vary within clusters and the maturation of these structures has been shown to be a vital function of mitochondria. Through studies within *S. cerevisiae*, two stages of Fe-S assembly have been elucidated—formation and conversion to holoenzymes. Over 20 proteins have been shown to be involved in these processes [51]. The first stage briefly consists of the conversion of cysteine to alanine, releasing sulphur by the cysteine desulphurase complex Nfs1-Isd11 [52] and then reducing to sulphide by electrons from NADH via ferredoxin reductase [53]. To enter mitochondria, iron is transported through a member of the mitochondrial solute carrier family—mitoferrin [54]. Once within the inner membrane scaffold proteins including the highly conserved IscU and Nfu [55, 56] are employed to assemble Fe-S clusters. The Fe required for the construction of these clusters is provided through frataxin. Following synthesis of the Fe-S cluster, it is released from the scaffold proteins and incorporated in apoproteins.

1.2.5 Generation of Reactive Oxygen Species

The production of reactive oxygen species (ROS) and their damaging effects on a multitude of cellular components, including proteins, lipids and DNA, have been linked to the loss of neurons in both disease and normal ageing. The brain uses approximately one-fifth of inspired oxygen, and of this 90 % is consumed by oxidative phosphorylation. Neurons are highly susceptible to the damaging effects of these radicals for many reasons including their high oxidative metabolic activity, non-replicative nature and relatively low antioxidant capacity. Mitochondria produce reactive oxygen species through the process of oxidative phosphorylation, mainly through the actions of complexes I and III. These complexes produce superoxide (O_2^-); this radical can then be converted to other reactive species including hydrogen peroxide and peroxynitrite, through reactions with SOD2 or NO, respectively (reviewed in [57]). Although damaging to cells, ROS are emerging as important signalling molecules and can alter the activity of a number of proteins, important for a variety of signalling cascades. For example, PGC1α, the master regulator of mitochondrial biogenesis, acts as a transcription factor driving synthesis of proteins required for antioxidant defences [58], while H_2O_2 can oxidise cysteine residues within phosphatases including PTEN and MAPK leading to a reduction in their ability to dephosphorylate their respective targets (reviewed in [59, 60]).

1.2.6 *Emerging Functions*

1.2.6.1 Lipid Metabolism and Transport

Although the majority of cellular lipid synthesis occurs in the endoplasmic reticulum (ER), mitochondria have been shown to contribute to this essential process. Mitochondrial integrity relies on the availability of lipids and their uptake from the ER, but mitochondria can also synthesise phosphatidylethanolamine (PE), which is then transferred to the ER, and the mitochondrial phospholipid, cardiolipin. Cardiolipin is a major component of the inner mitochondrial membrane and is required for essential mitochondrial functions and also for the maintenance of mitochondrial morphology. In yeast, it has been shown that the majority of membrane PE is synthesised within the mitochondrial inner membrane, a process dependent on ATP. As important as the transport of lipids out of the mitochondria are the transport and shuttling of lipids within the mitochondria, primarily between the inner and outer membranes. The enzymes responsible for phospholipid synthesis within the mitochondria are found within the inner mitochondrial membrane (e.g. phosphatidylserine decarboxylase (Psd1) in yeast), and a recently described protein, UPS1, has been found to be responsible for the movement of phospholipids between these two membranes, at least in yeast ([61] reviewed in [62, 63]).

1.2.6.2 Membrane Donation

As discussed in Chaps. 11 and 12, dedicated pathways exist to facilitate the degradation of dysfunctional mitochondria. A mitochondrial specific form of autophagy, mitophagy, leads to the sequestration of damaged mitochondria within a double membrane, their transport to the lysosome and ultimate destruction. Evidence from cell culture has suggested that under conditions of starvation, mitochondrial membrane is utilised in the formation of autophagosomes and that this utilisation is dependent on connections with the ER that are themselves dependent on mfn2 expression. Mitochondrial lipid content and replenishment rely on connection to the ER. The generation of autophagosomes seems to occur at discrete punctal locations along the mitochondria that then recruit LC3 and ATG5, key proteins in the facilitation of mitochondrial degradation [64].

1.2.6.3 Signalling

Aside from their damaging roles and links to cellular damage, an emerging role is appearing for ROS in cellular signalling. ROS have been proposed to be particularly important for O_2 sensing, important during hypoxia; this sensing capability relies on the production of $O_2^{\cdot-}$. $O_2^{\cdot-}$ production increases under conditions of low O_2; the $O_2^{\cdot-}$ then is converted to H_2O_2 in the mitochondrial matrix, before diffusing into the cytoplasm. Here it can stabilise hypoxia-inducible factor 1α (HIF1α), which initiates the transcription of the machinery which responds to hypoxia ([65, 66] reviewed in [67]). However the number of crucial cellular processes for which this signalling is important is increasing rapidly and now includes, amongst others, mitochondrial homeostasis, immune system regulation and oncogenic transformation (reviewed in [60, 68, 69]).

1.3 The Mitochondrial Genome

Uniquely, mitochondria contain the only nucleic acid expressed outside the nucleus, albeit a much smaller amount, totalling around 1 % of all cellular DNA. Despite the relatively small amounts of mitochondrial DNA, the genes it encodes are essential for life. The mitochondrial genome (mtDNA) is a 16,569 base pair, closed, circular, double-stranded DNA molecule (Fig. 1.3). Located within the mitochondrial matrix, mtDNA exists in multiple copies that are packaged into protein-DNA complexes known as nucleoids, which associate with the inner membrane of the mitochondria [70]. Research has suggested that each nucleoid might contain multiple copies of the genome existing alongside the protein machinery required for transcription, replication and maintenance of the mtDNA [71] (see Sect. 1.3.4). More recent research, however, has suggested that mitochondrial nucleoids may only contain one copy of the mitochondrial genome and that the main constituent of the nucleoid is the mitochondrial transcription factor (TFAM), which plays an active role in their formation [72, 73]. In line with this finding, it has been found that while an increase in mitochondrial DNA copy number increases the number of nucleoids, it does not cause a subsequent increase in their size [73].

The mitochondrial genome was first sequenced in 1981 [74] and following revision in 1999 [75] has become commonly known as the revised Cambridge Reference Sequence (rCRS). The genome encodes for a total of 37 polypeptides (including 13 proteins essential to OXPHOS), 22 transfer RNAs (tRNA) and two ribosomal RNAs (rRNA) [74]. Any further proteins required for mitochondrial function are synthesised in the cytosol and are then actively transported into the mitochondrion. Why 37 genes are retained by the mitochondrial genome when around 1500 appear to have relocated to the nuclear genome through evolutionary time is unclear but may be due to the hydrophobicity of these particular subunits which would pose a challenge for transport into the organelle after translation [76]. Alternatively, it has been proposed that the retention of these genes may enable rapid adaptation of the OXPHOS system in response to environmental changes [77].

The mitochondrial genome is composed of two strands, named heavy and light reflecting their purine to pyrimidine ratio. The heavy chain contains more guanine residues in contrast to the abundance of cytosine residues of the light chain. Only

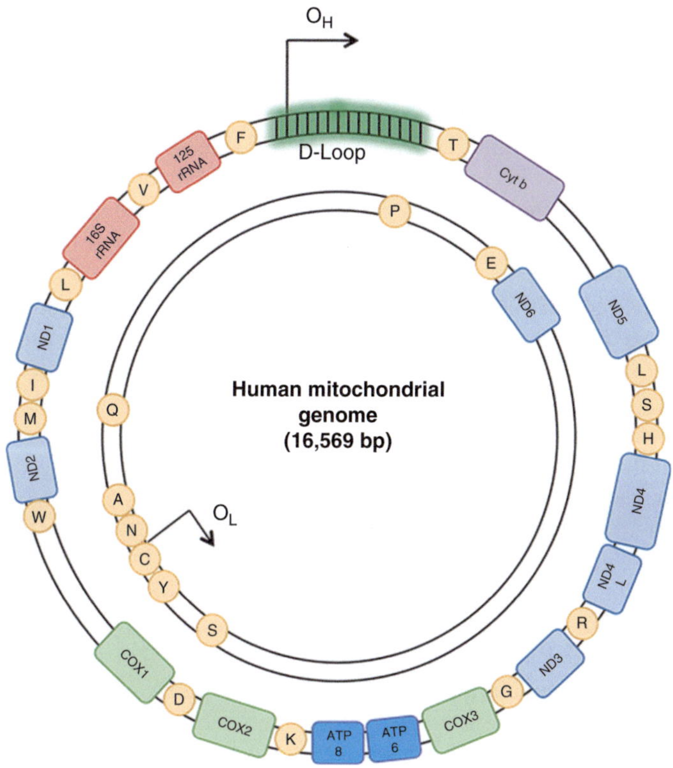

Fig. 1.3 The mitochondrial genome. The mtDNA molecule encodes 13 structural electron transport chain proteins, 7 subunits of complex I (*blue*), 1 subunit of complex III (*purple*), 3 subunits of complex IV (*green*) and 2 subunits of complex V (*turquoise*). Alongside this it encodes 2 rRNAs (*red*), all of which are punctuated by 22 tRNAs (*orange letters*). Both origins of transcription (O_H and O_L) are labelled; areas without labels represent non-coding regions (Figure reproduced with permission from Dr. C. Wilson, Washington University School of Medicine, St. Louis)

ND6 and ND8 of the mitochondrial tRNA genes (MTTs) are transcribed from the light strand; the remaining 12 protein-encoding genes, 14 MTTs and the 2 mt-rRNAs are transcribed from the heavy chain. The compact mtDNA molecules differ from the nuclear genome since they contain no introns with the only extended non-coding region being found within the displacement loop, an area that controls transcription and replication of the genome [78, 79].

1.3.1 Mitochondrial DNA Inheritance and Replication

MtDNA is exclusively inherited down the maternal lineage, although one case of paternal transmission of pathogenic mtDNA has been recorded [80]; this is a unique report and has not been repeated since. Not only is sperm mtDNA massively outnumbered in the fertilised zygote, it is highly likely that there is also an active process for degradation of male mtDNA (see Chap. 11 for details).

Replication of mitochondrial DNA occurs in a manner distinct from genomic DNA and subsequently independent of the cell cycle; however, the exact method of mtDNA replication remains a subject of controversy. Clayton and colleagues initially reported a model where mtDNA replicated in a strand-asynchronous mechanism, priming from the light strand promoter within the major non-coding region of the genome [81]. This RNA primer was then extended by the mitochondrial DNA polymerase, copying the entire heavy strand. Light strand replication was then initiated once the polymerase had passed approximately two-thirds of the entire length of the L-strand template.

The model for mtDNA replication was then revisited by Holt and colleagues, who challenged the assumption that strand-asynchronous replication was the universal method for mtDNA replication, based on the visualisation of replication intermediates by native 2D gel electrophoresis [82]. These replication intermediates were more consistent with a strand-coupled than strand-asynchronous model [83]. This work also highlighted the previously established and surprising observation that ribonucleotides are incorporated into the genome [84].

This has also led to a third replication model, RITOLS, whereby ribonucleotides are present throughout the entire lagging strand [85]. This model has been extended further to form the 'bootlace' model, where preformed L-strand RNA is incorporated in vivo at the replication fork. This RNA hybridises to the displaced H-strand and is replaced with DNA in a maturation step [86]. Despite this questions still remain and it is possible that mtDNA may be replicated by different modes at different stages of development or in different tissues. To begin addressing this, Falkenberg and colleagues have been attempting to reconstitute mtDNA replication in vitro [87]. Intriguingly, they have been able to show that the region referred to as OriL by Clayton et al. can indeed form a structure which can be primed by the mitochondrial RNA polymerase POLRMT [88]. This primer is then faithfully extended by the mitochondrial DNA polymerase gamma.

1.3.2 Mutations and Repair

For over 20 years we have known that defects in mtDNA can cause a wide variety of disorders, which often (but not solely) affect muscle and the central nervous system (reviewed in [89]; see also Chaps. 2 and 4). More recently, however, it has become apparent that many combined disorders of OXPHOS complexes are also caused by mutations in nuclear genes whose products affect mtDNA maintenance, causing multiple deletions (e.g., POLG, POLG2, ANT1, PEO1, OPA1, TYMP) or indeed mtDNA depletion (DGUOK, SUCLA2, SUCLG1, RRM2B, MPV17, TK2) (reviewed in [90]; see also Chap. 4). Mutations are found in proteins that function directly in mtDNA replication, such as the mitochondrial DNA polymerase γ (gamma), its accessory unit and the critical DNA helicase- twinkle, or in the homeostasis of mitochondrial nucleotide pools.

Mutations of the mitochondrial genome can exist as point mutations, deletions, insertions and large-scale duplications (see Chap. 3 for detail). These are all known to occur at a higher rate than those of the nuclear genome stemming from a combination of factors. As mtDNA is tethered to the inner side of the inner mitochondrial membrane, it is close to the components of the respiratory chain, the major site of reactive oxygen species (ROS) generation in the cell. Exposure to ROS is known to damage DNA and the mitochondrial genome is no exception. Secondly, the lack of non-coding DNA coupled with an absence of protective histones means deleterious mutations are far more likely to occur compared to nuclear DNA.

Increasing evidence points towards the existence of DNA repair mechanisms within the mitochondrion, despite lack of such processes being originally attributed to the high mtDNA mutation rate. It is now believed that these mechanisms are indeed present but exist at a reduced level of that seen in nDNA and may function through different mechanisms [91]. To date the most well-studied repair mechanism is short base excision repair (sBER) [92]. Less well-understood and currently incomplete pathways exist in the form of long-patch base excision repair (which functions as short BER but for multiple nucleotides), direct repair of damaged nucleotides [93] and mismatch repair (MMR).

Although the organelle does contain an array of DNA damage repair enzymes [94], it is highly likely that the toxic oxidative environment will over time overwhelm these systems leading to the eventual generation of mtDNA mutations. It has been demonstrated that accumulation of mtDNA mutations occurs over time and that the pattern of this accumulation is focal [95, 96]. This process is known as clonal expansion and has been confirmed in a variety of tissues. Although it is known that a single mutational event (be it deletion or point mutation) has the capacity to begin in a single mtDNA molecule and eventually express in an entire cell or tissue, the mechanism by which it does so is not clear. This is made more complicated by the pattern of clonal expansion differing in tissue type [97]. Several models have been proposed, as yet no definite hypothesis can be proven [98].

1.3.3 Heteroplasmy, Homoplasmy and the Threshold Effect

Akin to the fact mtDNA molecules exist in multiple copies within each mitochondrion and each cell can have numerous mitochondria, the situations of homoplasmy and heteroplasmy can occur. A cell whose mtDNA molecules are *all* wild type or *all* mutant is said to be homoplasmic. However a situation can arise whereby wild-type molecules coexist with mutated molecules giving rise to a heteroplasmic situation. This phenomenon has profound implications for mitochondrial disease. It is worth noting that although healthy cells are said to be homoplasmic, it is likely that even these have a clinically insignificant percentage of mutated mtDNA molecules (<1 %) [99, 100].

The existence of heteroplasmy within cells leads to the condition by which the ratio of mutated to wild-type molecules can (largely) dictate the clinical presentation of a deleterious mutation. Although some pathogenic homoplasmic mutations have been characterised, heteroplasmic mutants are by far more common. The threshold level for expression of a biochemical defect varies dramatically for mutation type, individual, and tissue, with reports as low as 8 % [101]. Typically, however, mutational thresholds and subsequent phenotypic expression in mitochondrial disease occur at around 70–90 % heteroplasmy. For multifactorial disorders such as neurodegenerative conditions, mutations in specific neuronal subpopulations may be sufficient to add to pathogenicity if the environment is already suitably challenged. The critical level of heteroplasmy varies between conditions and neuronal populations; these peculiarities are discussed further in Chap. 3.

1.3.4 Mitochondrial DNA Maintenance

1.3.4.1 Transcription, Translation and the Mitoribosome

Transcription

Mitochondrial transcription is unusual in the that the core transcription complex containing the mtRNA polymerase (POLRMT) and the two characterised transcription factors TFAM and TFB2 facilitate transcription of almost the entire genome from promoters borne on both strands. Detailed analysis of mitochondrial transcription and its regulation is beyond the scope of this article, but can be found in the excellent review by Larsson and colleagues [102]. Transcription of mtDNA is initiated at three sites, one on the light strand (LSP) and two on the heavy strand (HSP1 and HSP2). These large polycistronic units are essentially punctuated by mitochondrial tRNA structures, which are thought to spontaneously fold and are subsequently liberated by the mt-tRNA processing enzyme activities RNase P [103] and RNase Z. These two enzymes cleave the tRNAs at the 5′ and 3′ ends, respectively. This tRNA punctuation model appears to be correct in general, but it cannot explain how certain mRNA species are processed, as they are not surrounded by tRNA or antisense tRNA (*RNA14*, *MT-CO3*, *MT-ND5*, *MT-ND6*) [104]. All the resultant mitochondrial mRNA and tRNAs are matured, either by the 3′ terminal addition of a CCA trinucleotide extension ([CCA] nucleotidyl transferase [105]) to all mt-tRNAs or polyadenylation of mt-mRNAs of approximately 35–60 nucleotides by a mitochondrial poly(A) polymerase [106]. It must be noted, however, that at steady state a subpopulation of mt-mRNAs carries a shorter oligoadenylated extension that is not due to 3′ 5′ exonucleolytic degradation. These oligoadenylated species are the predominant form of mt-mRNAs in patients with defective mitochondrial poly(A) polymerase (mtPAP) [107] or under conditions of mtPAP depletion [106, 108] lead to the suggestion that a second poly(A) polymerase activity must be present in the mitochondrial matrix. Transcription termination is controlled by the four mitochondrial termination factors, mTERF 1–4, which act in synchrony to control and regulate the termination of this process.

Translation and the Mitoribosome

Mitoribosomes are composed of large (39S) and small (28S) subunits. The large subunit contains the 12S rRNA plus ~29 proteins, while the large subunit contains the 16S rRNA and closer to 50 proteins. Mammalian mitoribosomes contain a higher protein to RNA ratio than cytosolic and bacterial ribosomes, yielding a larger ribosome, with a dramatically altered structure and interaction with a number of mitoribosome-specific proteins.

Pioneering studies by, amongst others, O'Brien, Watanabe and Spremulli have been somewhat hampered by the absence of a faithful in vitro reconstituted system for mammalian mitochondrial protein synthesis. Consequently, although we are aware of numerous proteins that are crucial in each of the four stages of protein synthesis, initiation (If2mt, If3mt) elongation (mtEFG1, mtEFTu, mtEFTs), termination (mtRF1a, mtEFG2) and recycling (mtRRF), functional assays have had to rely on heterologous systems (for review see [109]). Unlike all other ribosomes to date, the mitoribosome is mostly comprised of protein (approx. 70 %) over RNA (30 %) where the norm is more 65 %: 35 % RNA to protein. Another striking absence from the cryoEM structure is an exit site for the mitochondrial tRNA. To date, an exit site is present in all other ribosomal structures.

The mitoribosome has an exquisite and essential interaction with the mitochondrial membrane, possibly reflecting the highly hydrophobic nature of all 13 mitochondrial gene products. This may also partly explain why it has been so difficult to reconstitute a faithful system in vitro. This critical association has been highlighted by studies of patients with mitochondrial OXPHOS disorders. Patients have recently been described who have defects in mitochondrial protein synthesis without involvement of mtDNA quality or quantity. Patients have combined OXPHOS disorders, with a wide spectrum of clinical presentation (reviewed in [110]). Patients with a form of hereditary spastic paraplegia, HSP7, have been shown to harbour mutations in the protein paraplegin [111]. This protein is a member of the heteromultimeric mAAA protease that is responsible for cleaving membrane-bound proteins. Langer and colleagues showed that the reason for the severe mitochondrial protein synthesis defect in cells lacking normal mAAA protease activity was due to impaired maturation of the mitoribosomal protein MRPL32. Further, when this protein is not cleaved, the mitoribosome is unable to associate with the inner membrane. Intriguingly, mutations in the second of the heteroligomeric components, AFG3L2, also causes severe combined OXPHOS disorder, but clinical presentation is predominantly a spinocerebellar ataxia [112].

1.3.4.2 Quality Control of Mitochondrial Protein Synthesis

The process of protein synthesis is notoriously complex, with problems often occurring early in the process leading to stalling of the ribosome and loss of the nascent peptide [113]. Stalling can also occur when mRNAs lack termination codons. Once the ribosome has stalled, there must be mechanisms to remove the peptidyl-tRNA from the complex and dissociate the ribosomal subunits. A well-described process for rescuing stalled ribosomes is found in eubacteria, where a novel RNA, tmRNA encoded by the SsrA gene, is able to access the ribosomal access site and promote its own translation, leading to the addition of a short oligopeptide to the C-terminus of the nascent protein which acts as a substrate for ubiquitination and subsequent degradation [114, 115]. This tmRNA also contains a termination codon, facilitating the hydrolysis of the complete peptidyl-tRNA and recycling of the ribosome. In the absence of a tmRNA orthologue in mitochondria, how are stalled ribosomes recognised? Human genome analysis identifies four members of a mitochondrial peptidyl-tRNA release factor family. Following the observation that the rare codons AGA and AGG promote ribosome frameshifting at the terminus of two mitochondrial open reading frames [116], we now know that only one of the release factor family, mtRF1a, is capable of terminating translation of all 13 open reading frames [117]. It is therefore highly likely that other members of this family will be intimately involved in rescuing stalled ribosomes. Indeed, one of these members, ICT1, has been shown to have become integrated into the mitochondrial ribosome [118]. Although its exact function is currently unknown, it is highly likely to be involved in ribosome rescue, particularly as its peptidyl-tRNA hydrolase activity is codon independent, potentially allowing it to release peptides stalled at any or indeed no A site codon. The roles of the remaining two factors are unknown although defects of one member, C12orf65, have been shown to cause mitochondrial disease [119]. Although tmRNA is a mechanism used widely by eubacteria for rescuing stalled ribosomes, it is not essential for viability and recently other proteins have been also shown to promote such rescue, one of which, YaeJ, is believed to be the orthologue of ICT1 [120]. Such observations make it highly likely that the other members of this family will also be important in quality control of the translation system in mammalian mitochondria.

1.3.5 Mitochondrial Dynamics, Degradation and Movement

1.3.5.1 Mitochondrial Dynamics

The notion of mitochondria as lone ellipsoid organelles has long passed. Mitochondria can exist in a multitude of forms, shown to be dynamic and dependent on numerous factors. The transition between elongated reticular networks and punctate structures is dependent on two antagonistic processes, fission and fusion, both of which are vital for cell survival as they directly affect organelle number/shape and location. The interaction between individual organelles is vital for maintaining respiratory function and mixing mitochondrial content which benefits mtDNA stability and is capable of rescuing damaged phenotypes [121]. Changes in both these processes have been shown to cause certain neurodegenerative conditions and are implicated in the pathogenic pathway of many others. These processes are reviewed in Chap. 6.

The Fusion Machinery

Fusion events are mediated through the mitofusins of the outer mitochondrial membrane and OPA1 of the inner membrane. The mitofusins in mammals consist of the two homologues Mfn1 and Mfn2; both are large GTPases containing hydrophobic heptad repeats, with long transmembrane repeats that contain charged residues [122]. The N and C terminal of both protrude into the cytosol [123]. The ability to form a U shape within the membrane is thought possible through the unusually long transmembrane domain and the charged residues within it. Ablation of either Mfn1 or Mfn2 in cells causes greatly reduced levels of fusion, removing both, and halts all mitochondrial fusion [124]. OPA1 is situated in the inner membrane and functions within the intermembrane space [125]. It is a dynamin family GTPase and was first identified through its mutation in dominant optic atrophy [126]. Eight protein isoforms are generated through mRNA splicing and possibly more through post-translational processing [127]. Removal of OPA1 causes loss of mitochondrial fusion as with the mitofusins but also causes severe modification in mitochondrial cristae structure [125]. Why mutations in this protein should especially cause retinal ganglion cell degeneration remains obscure.

The Fission Machinery

Mitochondrial fission generates individual smaller organelles that are necessary for transportation purposes as well as selecting unwanted mitochondria for programmed removal by mitophagy. FIS1 and DRP1 are heavily involved in fission events. Dynamin-related protein 1 (Drp1) is predominantly found within the cytosol although a subset has been shown to localise to mitochondrial tubules. Furthermore, a proportion of these puncta have gone on to be fission sites [128]. As a dynamin family GTPase, Drp1 contains a characteristic GTPase domain, a central domain and a GTPase effector domain. Its removal increases mitochondrial length and tubule complexity [129]. FIS1 is a small outer membrane protein that contains a single C-terminal transmembrane domain and within the cytosolic domain a helical bundle [130]. Knockdown, as with DRP1, causes elongation of mitochondria, while overexpression causes fragmentation of the mitochondrial network [131].

1.3.5.2 Mitophagy

If mitochondrial damage is irrecoverable through fusion, the option to remove individual organelles through fission and subsequent mitophagy exists (reviewed in Chap. 11). Removal of damaged or unnecessary mitochondria is a highly regulated event. Current understanding highlights PINK1 accumulation on the mitochondrial outer membrane and the recruitment of the ubiquitin ligase, PARKIN, as key facilitators of this process. It is worth noting however that several factors in this relatively recent discovery remain unclear and it is likely that other components of the pathway or indeed novel pathways are yet to be uncovered. To highlight just one example, recent work has shown that Fbxo7 participates in mitochondrial maintenance through direct interaction with PINK1 and Parkin and acts in Parkin-mediated mitophagy. Mutations in Fbxo7 (encoded by PARK15) also cause early-onset autosomal recessive Parkinson's disease [132]. PINK1 and PARKIN therefore likely function alongside, and may be dependent on, various factors yet to be revealed. Any inhibition of the mitophagy pathway leads to accumulation of dysfunctional mitochondria that can then harm other cellular components [133, 134]. Furthermore, useful components that would ordinarily be recycled are confined in these accumulations and the control of apoptotic signals is lost.

1.3.5.3 Mitochondrial Motility

Mitochondria not only show dynamic movement within the network but are also highly motile throughout cells providing a mechanism to position them where they are most needed. For example, it has been estimated that around 35 % of neuronal mitochondria are estimated to be moving at any given time [135–137]. Trafficking becomes especially important in cells with axons that may be a metre or more in length. Sophisticated trafficking machinery exists to enable mitochondria to be readily transported to areas of increased energy demand or (perhaps) returned for degradation, although direct evidence for this scenario is limited. Mitochondria typically have set patterns of movement. These can be fairly uniform over long distances or show more 'stop-start-reconsider' characteristics, where movements are shorter and often punctuated with a pause that can precede a change in direction. The significance of these patterns remains obscure.

Within neurons mitochondria are largely moved along microtubules (Fig. 1.4), although movement using actin filaments has been reported for smaller distances and within dendritic spines and growth cones [138]. The 'docking' and 'shipping' of mitochondria onto these cellular tracks are facilitated through motor proteins and a plethora of adaptor proteins such as Milton, Miro, myosin and dynactin. Microtubules display uniform polarity in axons with all positive poles aligning at the axonal terminal; the tracking of mitochondria along these is mediated via different motors dependent on direction. For antegrade movement (towards the axon terminal), kinesin motors are employed; conversely, retrograde movement utilises dynamin motors [139]. Disruption of mitochondrial motility can have ramifications for the cellular environment, including lack of energy where required and an inability to return mitochondria to be recycled meaning they are left in often 'at-risk' areas.

1.4 The Cause and Effects of Mitochondrial Dysfunction

Mitochondrial dysfunction may be caused by a number of factors. Mutations of mitochondrial DNA as well as of essential mitochondrial maintenance genes cause mitochondrial disease. However, mtDNA deletions are also prominent within the

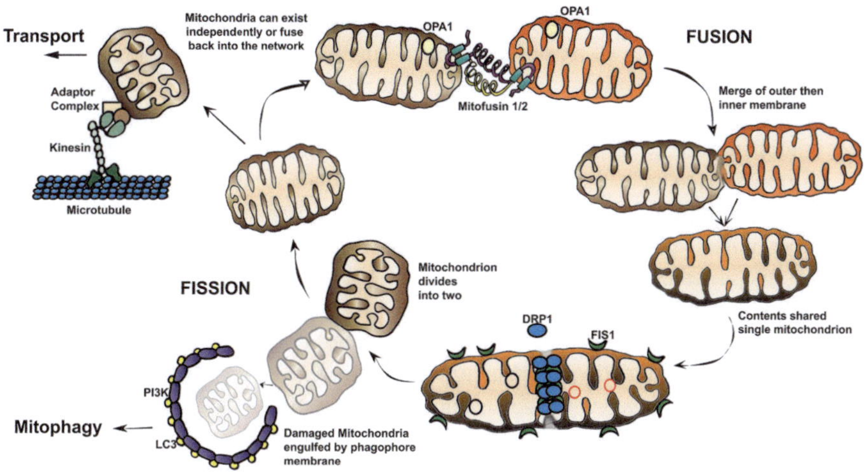

Fig. 1.4 Mitochondrial dynamics. Mitochondria are highly dynamic organelles that through the processes of fission and fusion can maintain a reticular network within cells. Changes in this network are often seen in disease and in the presence of mitochondrial defects. Mitochondrial fusion allows the joining of two mitochondria so that their outer and inner membranes fuse and their internal contents, including mtDNA, mix. Upon damage to the mitochondria or loss of membrane potential, mitochondria may separate through fission. Damaged mitochondria which fission from the network may then be degraded by the process of mitophagy

ageing brain and in diseases such as Parkinson's disease; in fact early-onset forms of Parkinson's disease can be caused by mutations within essential mitochondrial proteins including Pink1, Parkin and DJ-1. The cause of and expansion to high levels of mtDNA deletions with ageing, however, are still heavily debated, and with all the proposed causes of mitochondrial dysfunction, researchers struggle to disentangle cause from effect as many of the supposed causes of mitochondrial dysfunction are exacerbated by changes within the mitochondria. For example, mounting evidence shows that alpha-synuclein, the protein which aggregates to form Lewy bodies in Parkinson's disease, interacts with mitochondria, causing their dysfunction, through inhibition of complex I [140, 141]. However, it is the aggregating forms of the protein which have these effects and in order to aggregate alpha-synuclein must undergo a conformational change, which may be caused by damage by reactive oxygen species (see Chap. 10). The damage of proteins and DNA caused by ROS is also likely to cause mitochondrial dysfunction particularly with advancing age. Oxidative damage may contribute to the aggregation of protein into structures such as Lewy bodies, amyloid plaques and neurofibrillary tangles, which form the pathological hallmarks of a number of neurodegenerative diseases. ROS are also likely to contribute to the accumulation of mitochondrial DNA mutations, damage to mitochondrial membranes and damage to key mitochondrial proteins, including those of the respiratory chain.

As previously mentioned mitochondria are also a significant source of reactive oxygen species within the cell. Their ability to assemble iron-sulphur clusters ensures that mitochondria also contain a high concentration of iron which through the Fenton reaction can increase the production of hydroxy radicals. Mitochondrial dysfunction has long been proposed to lead to an increase in the production of ROS which contribute not only to the damage of a number of cellular components but also cause further mitochondrial dysfunction. It is feasible that dysfunction within the mitochondrion could also lead to a reduction in iron-sulphur cluster formation, which may lead to an accumulation of iron within them which would again exacerbate production of ROS. A disruption in iron metabolism has been associated with a number of neurodegenerative disorders including Parkinson's disease, Alzheimer's disease and neurodegeneration with brain iron accumulation (NBIA), which has now been linked to mutations in a number of mitochondrial genes (see Chap. 4).

In this edition we have invited leaders in their fields to delve into the consequences of mitochondrial dysfunction, from a number of different perspectives with the aim of understanding how these mechanisms contribute to neuronal death and dysfunctions in neurodegenerative disease. Measuring the consequences of mitochondrial dysfunction particularly within human tissue is challenging, highlighting the need for good animal models which faithfully recapitulate mitochondrial defects (see Chap. 13).

References

1. Margulis L, editor. Origin of eukaryotic cells. New Haven: Yale University Press; 1970.
2. Martin W, Mentel M. The origin of mitochondria. Natl Educ. 2010;3(9):58.
3. Embley TM, Martin W. Eukaryotic evolution, changes and challenges. Nature. 2006;440(7084):623–30.
4. Gray MW. The endosymbiont hypothesis revisited. Int Rev Cytol. 1992;141:233–357.
5. Altman R. Die Elementarorganismen Und Ihre Beziehungen Zu Den Zellen. Leipzing: Zeit & Comp; 1890. p. 145.
6. Battelli F, Stern L. Die Oxydationsfermente. Ergebn Physiol Biol Chem Exp Pharmakol. 1912;15:96–268.
7. Benda C. Ueber die Spermatogenese der Vertebraten und höherer Evertebraten, II. Theil: Die Histiogenese der Spermien. Arch Anat Physiol. 1898;73:393–8.
8. Palade GE. The fine structure of mitochondria. Anat Rec. 1952;114(3):427–51.
9. Neupert W, Herrmann JM. Translocation of proteins into mitochondria. Ann Rev Biochem. 2007;76:723–49.
10. Chacinska A, Pfannschmidt S, Wiedemann N, Kozjak V, Sanjuan Szklarz LK, Schulze-Specking A, et al. Essential role of Mia40 in import and assembly of mitochondrial intermembrane space proteins. EMBO J. 2004;23(19):3735–46.
11. Naoe M, Ohwa Y, Ishikawa D, Ohshima C, Nishikawa S, Yamamoto H, et al. Identification of Tim40 that mediates protein sorting to the mitochondrial intermembrane space. J Biol Chem. 2004;279(46):47815–21.
12. Frey TG, Mannella CA. The internal structure of mitochondria. Trends Biochem Sci. 2000;25(7):319–24.
13. Mannella CA. Structure and dynamics of the mitochondrial inner membrane cristae. Biochim Biophys Acta. 2006;1763(5–6):542–8.
14. Vogel F, Bornhovd C, Neupert W, Reichert AS. Dynamic subcompartmentalization of the mitochondrial inner membrane. J Cell Biol. 2006;175(2):237–47.
15. Cogliati S, Frezza C, Soriano ME, Varanita T, Quintana-Cabrera R, Corrado M, et al. Mitochondrial cristae shape determines respiratory chain supercomplexes assembly and respiratory efficiency. Cell. 2013;155(1):160–71.

16. Acin-Perez R, Enriquez JA. The function of the respiratory supercomplexes: the plasticity model. Biochim Biophys Acta. 2014;1837(4):444–50.
17. Genova ML, Lenaz G. Functional role of mitochondrial respiratory supercomplexes. Biochim Biophys Acta. 2014;1837(4):427–43.
18. Koopman WJ, Nijtmans LG, Dieteren CE, Roestenberg P, Valsecchi F, Smeitink JA, et al. Mammalian mitochondrial complex I: biogenesis, regulation, and reactive oxygen species generation. Antioxid Redox Signal. 2010;12(12):1431–70.
19. Efremov RG, Baradaran R, Sazanov LA. The architecture of respiratory complex I. Nature. 2010;465(7297):441–5.
20. Cecchini G. Function and structure of complex II of the respiratory chain. Ann Rev Biochem. 2003;72:77–109.
21. Iwata S, Lee JW, Okada K, Lee JK, Iwata M, Rasmussen B, et al. Complete structure of the 11-subunit bovine mitochondrial cytochrome bc1 complex. Science (New York, NY). 1998;281(5373):64–71.
22. Tsukihara T, Aoyama H, Yamashita E, Tomizaki T, Yamaguchi H, Shinzawa-Itoh K, et al. The whole structure of the 13-subunit oxidized cytochrome c oxidase at 2.8 A. Science (New York, NY). 1996;272(5265):1136–44.
23. Ferguson SJ. ATP synthase: what dictates the size of a ring? Curr Biol. 2000;10(21): R804–8.
24. Yoshida M, Muneyuki E, Hisabori T. ATP synthase – a marvellous rotary engine of the cell. Nat Rev Mol Cell Biol. 2001;2(9):669–77.
25. Liu Y and Levine B. Autosis and autophagic cell death: the dark side of autophagy. Cell Death and Differentiation. 2015;22:367–73.
26. Pradelli LA, Beneteau M, Ricci JE. Mitochondrial control of caspase-dependent and -independent cell death. Cell Mol Life Sci. 2010;67(10):1589–97.
27. Orrenius S, Gogvadze V, Zhivotovsky B. Calcium and mitochondria in the regulation of cell death. Biochem Biophys Res Commun. 2015;460(1):72–81.
28. Morciano G, Giorgi C, Bonora M, Punzetti S, Pavasini R, Wieckowski MR, et al. Molecular identity of the mitochondrial permeability transition pore and its role in ischemia-reperfusion injury. J Mol Cell Cardiol. 2015;78:142–53.
29. Morciano G, Giorgi C, Balestra D, Marchi S, Perrone D, Pinotti M, et al. Mcl-1 involvement in mitochondrial dynamics is associated with apoptotic cell death. Mol Biol Cell. 2016;1:20–34.
30. Bonora M, Pinton P. The mitochondrial permeability transition pore and cancer: molecular mechanisms involved in cell death. Front Oncol. 2014;4:302.
31. Fatokun AA, Dawson VL, Dawson TM. Parthanatos: mitochondrial-linked mechanisms and therapeutic opportunities. Br J Pharmacol. 2014;171(8):2000–16.
32. Virag L, Robaszkiewicz A, Rodriguez-Vargas JM, Oliver FJ. Poly(ADP-ribose) signaling in cell death. Mol Aspects Med. 2013;34(6):1153–67.
33. Andrabi SA, Dawson TM, Dawson VL. Mitochondrial and nuclear cross talk in cell death: parthanatos. Ann N Y Acad Sci. 2008;1147:233–41.
34. Patron M, Raffaello A, Granatiero V, Tosatto A, Merli G, De Stefani D, et al. The mitochondrial calcium uniporter (MCU): molecular identity and physiological roles. J Biol Chem. 2013;288(15):10750–8.
35. Kamer KJ, Mootha VK. MICU1 and MICU2 play nonredundant roles in the regulation of the mitochondrial calcium uniporter. EMBO Rep. 2014;15(3):299–307.
36. Sancak Y, Markhard AL, Kitami T, Kovacs-Bogdan E, Kamer KJ, Udeshi ND, et al. EMRE is an essential component of the mitochondrial calcium uniporter complex. Science (New York, NY). 2013;342(6164):1379–82.
37. Kamer KJ, Mootha VK. The molecular era of the mitochondrial calcium uniporter. Nat Rev Mol Cell Biol. 2015;16(9):545–53.
38. Boyman L, Williams GS, Khananshvili D, Sekler I, Lederer WJ. NCLX: the mitochondrial sodium calcium exchanger. J Mol Cell Cardiol. 2013;59:205–13.

39. Palty R, Hershfinkel M, Sekler I. Molecular identity and functional properties of the mitochondrial Na+/Ca2+ exchanger. J Biol Chem. 2012;287(38):31650–7.
40. Palty R, Sekler I. The mitochondrial Na(+)/Ca(2+) exchanger. Cell Calcium. 2012;52(1):9–15.
41. Gincel D, Zaid H, Shoshan-Barmatz V. Calcium binding and translocation by the voltage-dependent anion channel: a possible regulatory mechanism in mitochondrial function. Biochem J. 2001;358(Pt 1):147–55.
42. Keinan N, Pahima H, Ben-Hail D, Shoshan-Barmatz V. The role of calcium in VDAC1 oligomerization and mitochondria-mediated apoptosis. Biochim Biophys Acta. 2013;1833(7):1745–54.
43. Weisthal S, Keinan N, Ben-Hail D, Arif T, Shoshan-Barmatz V. Ca(2+)-mediated regulation of VDAC1 expression levels is associated with cell death induction. Biochim Biophys Acta. 2014;1843(10):2270–81.
44. Clapham DE. Calcium signaling. Cell. 2007;131(6):1047–58.
45. Boitier E, Rea R, Duchen MR. Mitochondria exert a negative feedback on the propagation of intracellular Ca2+ waves in rat cortical astrocytes. J Cell Biol. 1999;145(4):795–808.
46. Tinel H, Cancela JM, Mogami H, Gerasimenko JV, Gerasimenko OV, Tepikin AV, et al. Active mitochondria surrounding the pancreatic acinar granule region prevent spreading of inositol trisphosphate-evoked local cytosolic Ca(2+) signals. EMBO J. 1999; 18(18):4999–5008.
47. Babcock DF, Hille B. Mitochondrial oversight of cellular Ca2+ signaling. Curr Opin Neurobiol. 1998;8(3):398–404.
48. Jouaville LS, Pinton P, Bastianutto C, Rutter GA, Rizzuto R. Regulation of mitochondrial ATP synthesis by calcium: evidence for a long-term metabolic priming. Proc Natl Acad Sci U S A. 1999;96(24):13807–12.
49. McCormack JG, Halestrap AP, Denton RM. Role of calcium ions in regulation of mammalian intramitochondrial metabolism. Physiol Rev. 1990;70(2):391–425.
50. Bruton JD, Aydin J, Yamada T, Shabalina IG, Ivarsson N, Zhang SJ, et al. Increased fatigue resistance linked to Ca2+-stimulated mitochondrial biogenesis in muscle fibres of cold-acclimated mice. J Physiol. 2010;588(Pt 21):4275–88.
51. Lill R. Function and biogenesis of iron-sulphur proteins. Nature. 2009;460(7257):831–8.
52. Adam AC, Bornhovd C, Prokisch H, Neupert W, Hell K. The Nfs1 interacting protein Isd11 has an essential role in Fe/S cluster biogenesis in mitochondria. EMBO J. 2006;25(1):174–83.
53. Lange H, Kaut A, Kispal G, Lill R. A mitochondrial ferredoxin is essential for biogenesis of cellular iron-sulfur proteins. Proc Natl Acad Sci U S A. 2000;97(3):1050–5.
54. Paradkar PN, Zumbrennen KB, Paw BH, Ward DM, Kaplan J. Regulation of mitochondrial iron import through differential turnover of mitoferrin 1 and mitoferrin 2. Mol Cell Biol. 2009;29(4):1007–16.
55. Tong WH, Rouault T. Distinct iron-sulfur cluster assembly complexes exist in the cytosol and mitochondria of human cells. EMBO J. 2000;19(21):5692–700.
56. Tong W-H, Jameson GNL, Huynh BH, Rouault TA. Subcellular compartmentalization of human Nfu, an iron-sulfur cluster scaffold protein, and its ability to assemble a [4Fe-4S] cluster. Proc Natl Acad Sci U S A. 2003;100(17):9762–7.
57. Murphy MP. How mitochondria produce reactive oxygen species. Biochem J. 2009;417(1):1–13.
58. St-Pierre J, Drori S, Uldry M, Silvaggi JM, Rhee J, Jager S, et al. Suppression of reactive oxygen species and neurodegeneration by the PGC-1 transcriptional coactivators. Cell. 2006;127(2):397–408.
59. Giorgio M, Trinei M, Migliaccio E, Pelicci PG. Hydrogen peroxide: a metabolic by-product or a common mediator of ageing signals? Nat Rev Mol Cell Biol. 2007;8(9):722–8.
60. Ray PD, Huang BW, Tsuji Y. Reactive oxygen species (ROS) homeostasis and redox regulation in cellular signaling. Cell Signal. 2012;24(5):981–90.

61. Connerth M, Tatsuta T, Haag M, Klecker T, Westermann B, Langer T. Intramitochondrial transport of phosphatidic acid in yeast by a lipid transfer protein. Science (New York, NY). 2012;338(6108):815–8.
62. Tatsuta T, Scharwey M, Langer T. Mitochondrial lipid trafficking. Trends Cell Biol. 2014;24(1):44–52.
63. Gohil VM, Greenberg ML. Mitochondrial membrane biogenesis: phospholipids and proteins go hand in hand. J Cell Biol. 2009;184(4):469–72.
64. Hailey DW, Rambold AS, Satpute-Krishnan P, Mitra K, Sougrat R, Kim PK, et al. Mitochondria supply membranes for autophagosome biogenesis during starvation. Cell. 2010;141(4):656–67.
65. Bonello S, Zahringer C, BelAiba RS, Djordjevic T, Hess J, Michiels C, et al. Reactive oxygen species activate the HIF-1alpha promoter via a functional NFkappaB site. Arterioscler Thromb Vasc Biol. 2007;27(4):755–61.
66. Chandel NS, McClintock DS, Feliciano CE, Wood TM, Melendez JA, Rodriguez AM, et al. Reactive oxygen species generated at mitochondrial complex III stabilize hypoxia-inducible factor-1alpha during hypoxia: a mechanism of O2 sensing. J Biol Chem. 2000;275(33):25130–8.
67. Galanis A, Pappa A, Giannakakis A, Lanitis E, Dangaj D, Sandaltzopoulos R. Reactive oxygen species and HIF-1 signalling in cancer. Cancer Lett. 2008;266(1):12–20.
68. Hamanaka RB, Chandel NS. Mitochondrial reactive oxygen species regulate cellular signaling and dictate biological outcomes. Trends Biochem Sci. 2010;35(9):505–13.
69. D'Autreaux B, Toledano MB. ROS as signalling molecules: mechanisms that generate specificity in ROS homeostasis. Nat Rev Mol Cell Biol. 2007;8(10):813–24.
70. Albring M, Griffith J, Attardi G. Association of a protein structure of probable membrane derivation with HeLa cell mitochondrial DNA near its origin of replication. Proc Natl Acad Sci U S A. 1977;74(4):1348–52.
71. Holt IJ, He J, Mao C-C, Boyd-Kirkup JD, Martinsson P, Sembongi H, et al. Mammalian mitochondrial nucleoids: organizing an independently minded genome. Mitochondrion. 2007;7(5):311–21.
72. Kukat C, Wurm CA, Spahr H, Falkenberg M, Larsson NG, Jakobs S. Super-resolution microscopy reveals that mammalian mitochondrial nucleoids have a uniform size and frequently contain a single copy of mtDNA. Proc Natl Acad Sci U S A. 2011;108(33):13534–9.
73. Kukat C, Davies KM, Wurm CA, Spahr H, Bonekamp NA, Kuhl I, et al. Cross-strand binding of TFAM to a single mtDNA molecule forms the mitochondrial nucleoid. Proc Natl Acad Sci U S A. 2015;112(36):11288–93.
74. Anderson S, Bankier AT, Barrell BG, de Bruijn MH, Coulson AR, Drouin J, et al. Sequence and organization of the human mitochondrial genome. Nature. 1981;290(5806):457–65.
75. Andrews RM, Kubacka I, Chinnery PF, Lightowlers RN, Turnbull DM, Howell N. Reanalysis and revision of the Cambridge reference sequence for human mitochondrial DNA. Nat Genet. 1999;23(2):147.
76. Claros MG, Perea J, Shu Y, Samatey FA, Popot JL, Jacq C. Limitations to in vivo import of hydrophobic proteins into yeast mitochondria. The case of a cytoplasmically synthesized apocytochrome b. Eur J Biochem/FEBS. 1995;228(3):762–71.
77. Wallace DC. Why do we still have a maternally inherited mitochondrial DNA? Insights from evolutionary medicine. Annu Rev Biochem. 2007;76:781–821.
78. Clayton DA. Replication of animal mitochondrial DNA. Cell. 1982;28(4):693–705.
79. Clayton DA. Mitochondrial DNA, replication: what we know. IUBMB Life. 2003;55(4–5):213–7.
80. Schwartz M, Vissing J. Paternal inheritance of mitochondrial DNA. N Engl J Med. 2002;347(8):576–80.
81. Clayton DA. Replication and transcription of vertebrate mitochondrial DNA. Annu Rev Cell Biol. 1991;7:453–78.

82. Brewer BJ, Fangman WL. A replication fork barrier at the 3′ end of yeast ribosomal RNA genes. Cell. 1988;55(4):637–43.
83. Holt IJ, Lorimer HE, Jacobs HT. Coupled leading- and lagging-strand synthesis of mammalian mitochondrial DNA. Cell. 2000;100(5):515–24.
84. Grossman LI, Watson R, Vinograd J. The presence of ribonucleotides in mature closed-circular mitochondrial DNA. Proc Natl Acad Sci U S A. 1973;70(12):3339–43.
85. Yasukawa T, Reyes A, Cluett TJ, Yang MY, Bowmaker M, Jacobs HT, et al. Replication of vertebrate mitochondrial DNA entails transient ribonucleotide incorporation throughout the lagging strand. EMBO J. 2006;25(22):5358–71.
86. Reyes A, Kazak L, Wood SR, Yasukawa T, Jacobs HT, Holt IJ. Mitochondrial DNA replication proceeds via a 'bootlace' mechanism involving the incorporation of processed transcripts. Nucleic Acids Res. 2013;41(11):5837–50.
87. Korhonen JA, Pham XH, Pellegrini M, Falkenberg M. Reconstitution of a minimal mtDNA replisome in vitro. EMBO J. 2004;23(12):2423–9.
88. Fuste JM, Wanrooij S, Jemt E, Granycome CE, Cluett TJ, Shi Y, et al. Mitochondrial RNA polymerase is needed for activation of the origin of light-strand DNA replication. Mol Cell. 2010;37(1):67–78.
89. Tuppen HA, Blakely EL, Turnbull DM, Taylor RW. Mitochondrial DNA mutations and human disease. Biochim Biophys Acta. 2010;1797(2):113–28.
90. Copeland WC. Inherited mitochondrial diseases of DNA replication. Annu Rev Med. 2008;59:131–46.
91. Gredilla R, Bohr VA, Stevnsner T. Mitochondrial DNA repair and association with aging – an update. Exp Gerontol. 2010;45(7–8):478–88.
92. Bohr VA. Repair of oxidative DNA damage in nuclear and mitochondrial DNA, and some changes with aging in mammalian cells. Free Radic Biol Med. 2002;32(9):804–12.
93. Satoh MS, Huh NH, Rajewsky MF, Kuroki T. Enzymatic removal of O-6-ethylguanine from mitochondrial-DNA in rat-tissues exposed to N-ethyl-N-nitrosourea in vivo. J Biol Chem. 1988;263(14):6854–6.
94. Boesch P, Weber-Lotfi F, Ibrahim N, Tarasenko V, Cosset A, Paulus F, et al. DNA repair in organelles: pathways, organization, regulation, relevance in disease and aging. Biochim Biophys Acta. 2011;1813(1):186–200.
95. Muller-Hocker J, Seibel P, Schneiderbanger K, Kadenbach B. Different in situ hybridization patterns of mitochondrial DNA in cytochrome c oxidase-deficient extraocular muscle fibres in the elderly. Virchows Arch A Pathol Anat Histopathol. 1993;422(1):7–15.
96. Schwarze SR, Lee CM, Chung SS, Roecker EB, Weindruch R, Aiken JM. High-levels of mitochondrial-DNA deletions in skeletal-muscle of old rhesus-monkeys. Mech Ageing Dev. 1995;83(2):91–101.
97. Liu VW, Zhang C, Nagley P. Mutations in mitochondrial DNA accumulate differentially in three different human tissues during ageing. Nucleic Acid Res. 1998;26(5):1268–75.
98. Elson JL, Samuels DC, Turnbull DM, Chinnery PF. Random intracellular drift explains the clonal expansion of mitochondrial DNA mutations with age. Am J Hum Genet. 2001;68:802–6.
99. Lightowlers RN, Chinnery PF, Turnbull DM, Howell N. Mammalian mitochondrial genetics: heredity, heteroplasmy and disease. Trends Genet. 1997;13(11):450–5.
100. He YP, Wu J, Dressman DC, Iacobuzio-Donahue C, Markowitz SD, Velculescu VE, et al. Heteroplasmic mitochondrial DNA mutations in normal and tumour cells. Nature. 2010;464(7288):610–U175.
101. Sacconi S, Salviati L, Nishigaki Y, Walker WF, Hernandez-Rosa E, Trevisson E, et al. A functionally dominant mitochondrial DNA mutation. Hum Mol Genet. 2008;17(12):1814–20.
102. Falkenberg M, Larsson NG, Gustafsson CM. DNA replication and transcription in mammalian mitochondria. Ann Rev Biochem. 2007;76:679–99.
103. Holzmann J, Frank P, Loffler E, Bennett KL, Gerner C, Rossmanith W. RNase P without RNA: identification and functional reconstitution of the human mitochondrial tRNA processing enzyme. Cell. 2008;135(3):462–74.

104. Temperley RJ, Wydro M, Lightowlers RN, Chrzanowska-Lightowlers ZM. Human mitochondrial mRNAs – like members of all families, similar but different. Biochim Biophys Acta. 2010;1797(6–7):1081–5.
105. Nagaike T, Suzuki T, Tomari Y, Takemoto-Hori C, Negayama F, Watanabe K, et al. Identification and characterization of mammalian mitochondrial tRNA nucleotidyltransferases. J Biol Chem. 2001;276(43):40041–9.
106. Tomecki R, Dmochowska A, Gewartowski K, Dziembowski A, Stepien PP. Identification of a novel human nuclear-encoded mitochondrial poly(A) polymerase. Nucleic Acids Res. 2004;32(20):6001–14.
107. Crosby AH, Patel H, Chioza BA, Proukakis C, Gurtz K, Patton MA, et al. Defective mitochondrial mRNA maturation is associated with spastic ataxia. Am J Hum Genet. 2010;87(5):655–60.
108. Slomovic S, Schuster G. Stable PNPase RNAi silencing: its effect on the processing and adenylation of human mitochondrial RNA. RNA. 2008;14(2):310–23.
109. Koc E, Haque MD, Spremulli LL. Current views of the structure of the mammalian mitochondrial ribosome. Isr J Chem. 2010;50(1):45–59.
110. Chrzanowska-Lightowlers ZM, Horvath R, Lightowlers RN. 175th ENMC international workshop: mitochondrial protein synthesis in health and disease, 25–27th June 2010, Naarden, The Netherlands. Neuromuscul Disord. 2011;21(2):142–7.
111. Nolden M, Ehses S, Koppen M, Bernacchia A, Rugarli EI, Langer T. The m-AAA protease defective in hereditary spastic paraplegia controls ribosome assembly in mitochondria. Cell. 2005;123(2):277–89.
112. Di Bella D, Lazzaro F, Brusco A, Plumari M, Battaglia G, Pastore A, et al. Mutations in the mitochondrial protease gene AFG3L2 cause dominant hereditary ataxia SCA28. Nat Genet. 2010;42(4):313–21.
113. Das G, Varshney U. Peptidyl-tRNA hydrolase and its critical role in protein biosynthesis. Microbiology. 2006;152(Pt 8):2191–5.
114. Keiler KC, Waller PR, Sauer RT. Role of a peptide tagging system in degradation of proteins synthesized from damaged messenger RNA. Science (New York, NY). 1996;271(5251):990–3.
115. Muto A, Ushida C, Himeno H. A bacterial RNA that functions as both a tRNA and an mRNA. Trends Biochem Sci. 1998;23(1):25–9.
116. Temperley R, Richter R, Dennerlein S, Lightowlers RN, Chrzanowska-Lightowlers ZM. Hungry codons promote frameshifting in human mitochondrial ribosomes. Science (New York, NY). 2010;327(5963):301.
117. Soleimanpour-Lichaei HR, Kuhl I, Gaisne M, Passos JF, Wydro M, Rorbach J, et al. mtRF1a is a human mitochondrial translation release factor decoding the major termination codons UAA and UAG. Mol Cell. 2007;27(5):745–57.
118. Richter R, Rorbach J, Pajak A, Smith PM, Wessels HJ, Huynen MA, et al. A functional peptidyl-tRNA hydrolase, ICT1, has been recruited into the human mitochondrial ribosome. EMBO J. 2010;29(6):1116–25.
119. Antonicka H, Ostergaard E, Sasarman F, Weraarpachai W, Wibrand F, Pedersen AM, et al. Mutations in C12orf65 in patients with encephalomyopathy and a mitochondrial translation defect. Am J Hum Genet. 2010;87(1):115–22.
120. Handa Y, Inaho N, Nameki N. YaeJ is a novel ribosome-associated protein in Escherichia coli that can hydrolyze peptidyl-tRNA on stalled ribosomes. Nucleic Acids Res. 2011;39(5):1739–48.
121. Youle RJ, van der Bliek AM. Mitochondrial fission, fusion, and stress. Science (New York, NY). 2012;337(6098):1062–5.
122. Chan DC. Mitochondrial fusion and fission in mammals. Annu Rev Cell Dev Biol. 2006;22:79–99.
123. Rojo JV, Merino AM, Gonzalez LO, Vizoso F. Expression and clinical significance of pepsinogen C in epithelial ovarian carcinomas. Eur J Obstet Gynaecol Reprod Biol. 2002;104(1):58–63.

124. Chen H, Detmer SA, Ewald AJ, Griffin EE, Fraser SE, Chan DC. Mitofusins Mfn1 and Mfn2 coordinately regulate mitochondrial fusion and are essential for embryonic development. J Cell Biol. 2003;160(2):189–200.
125. Griparic L, van der Wel NN, Orozco IJ, Peters PJ, van der Bliek AM. Loss of the intermembrane space protein Mgm1/OPA1 induces swelling and localized constrictions along the lengths of mitochondria. J Biol Chem. 2004;279(18):18792–8.
126. Delettre C, Lenaers G, Griffoin JM, Gigarel N, Lorenzo C, Belenguer P, et al. Nuclear gene OPA1, encoding a mitochondrial dynamin-related protein, is mutated in dominant optic atrophy. Nat Genet. 2000;26(2):207–10.
127. Westermann B. Mitochondrial fusion and fission in cell life and death. Nat Rev Mol Cell Biol. 2010;11(12):872–84.
128. Schmid SL, Frolov VA. Dynamin: functional design of a membrane fission catalyst. Annu Rev Cell Dev Biol. 2011;27:79–105.
129. Smirnova E, Griparic L, Shurland DL, van der Bliek AM. Dynamin-related protein Drp1 is required for mitochondrial division in mammalian cells. Mol Biol Cell. 2001;12(8):2245–56.
130. Dohm JA, Lee SJ, Hardwick JM, Hill RB, Gittis AG. Cytosolic domain of the human mitochondrial fission protein Fis1 adopts a TPR fold. Proteins. 2004;54(1):153–6.
131. Lee YJ, Jeong SY, Karbowski M, Smith CL, Youle RJ. Roles of the mammalian mitochondrial fission and fusion mediators Fis1, Drp1, and Opa1 in apoptosis. Mol Biol Cell. 2004;15(11):5001–11.
132. Burchell VS, Nelson DE, Sanchez-Martinez A, Delgado-Camprubi M, Ivatt RM, Pogson JH, et al. The Parkinson's disease-linked proteins Fbxo7 and Parkin interact to mediate mitophagy. Nat Neurosci. 2013;16(9):1257–U135.
133. Osellame LD, Rahim AA, Hargreaves IP, Gegg ME, Richard-Londt A, Brandner S, et al. Mitochondria and quality control defects in a mouse model of Gaucher disease – links to Parkinson's disease. Cell Metab. 2013;17(6):941–53.
134. Osellame LD, Duchen MR. Quality control gone wrong: mitochondria, lysosomal storage disorders and neurodegeneration. Br J Pharmacol. 2014;171(8):1958–72.
135. Chen S, Owens GC, Crossin KL, Edelman DB. Serotonin stimulates mitochondrial transport in hippocampal neurons. Mol Cell Neurosci. 2007;36(4):472–83.
136. Overly CC, Rieff HI, Hollenbeck PJ. Organelle motility and metabolism in axons vs dendrites of cultured hippocampal neurons. J Cell Sci. 1996;109(Pt 5):971–80.
137. Waters J, Smith SJ. Mitochondria and release at hippocampal synapses. Pflugers Arch. 2003;447(3):363–70.
138. Saxton WM, Hollenbeck PJ. The axonal transport of mitochondria. J Cell Sci. 2012;125(Pt 9):2095–104.
139. Baas PW, Black MM, Banker GA. Changes in microtubule polarity orientation during the development of hippocampal neurons in culture. J Cell Biol. 1989;109(6 Pt 1):3085–94.
140. Devi L, Raghavendran V, Prabhu BM, Avadhani NG, Anandatheerthavarada HK. Mitochondrial import and accumulation of alpha-synuclein impair complex I in human dopaminergic neuronal cultures and Parkinson disease brain. J Biol Chem. 2008;283(14):9089–100.
141. Reeve AK, Ludtmann MH, Angelova PR, Simcox EM, Horrocks MH, Klenerman D, et al. Aggregated alpha-synuclein and complex I deficiency: exploration of their relationship in differentiated neurons. Cell Death Dis. 2015;6:e1820.

Part II
Mitochondrial Dysfunction: Causes and Effects?

Chapter 2
Neurodegeneration in Mitochondrial Disorders

Jonathan Phillips, Hannah Hayhurst, and Nichola Zoe Lax

Abstract Mitochondria are critically responsible for the generation of energy in the form of adenine triphosphate (ATP) through the mitochondrial respiratory chain (see Chap. 1). The central nervous system (CNS) performs highly energy-intensive tasks and is therefore particularly dependent on ATP. Defects residing within the complexes of the respiratory chain can affect the synthesis of ATP and consequently severely compromise neuronal function. It is unsurprising then that mitochondrial DNA (mtDNA) defects are an important cause of neurological disease. The clinical presentation is often heterogeneous in terms of age of onset and different neurological signs and symptoms which might include ataxia, seizures, cognitive decline, blindness and stroke-like episodes. The clinical course can vary considerably, but in many patients there are progressive neurological decline and marked neurodegeneration. Our understanding of the mechanisms underpinning neurodegenerative changes due to mitochondrial DNA defects is limited due to the availability of appropriate animal models of disease. However, studies on human post-mortem CNS tissues have provided an invaluable insight into the distribution and severity of neuronal degeneration in patients harbouring mitochondrial DNA defects. In this chapter, we describe the neuropathological changes occurring in the CNS associated with different mutations of the mitochondrial genome and discuss the mechanisms which might contribute to neural dysfunction and cell death.

Keywords Mitochondria • Mitochondrial DNA • Respiratory chain deficiency • Neurodegeneration • Neuropathology

2.1 Introduction

Mitochondrial disease was first described in 1962, when Luft and colleagues described a patient with non-thyroidal hypermetabolism [1]. Since this discovery, major advances in our understanding of mitochondrial biology and genetics have

---

J. Phillips, BSc, MRes • H. Hayhurst • N.Z. Lax, PhD (✉)
Wellcome Trust Centre for Mitochondrial Research, Institute of Neuroscience, Newcastle University, Newcastle upon Tyne, UK
e-mail: nichola.lax@ncl.ac.uk

A.K. Reeve et al. (eds.), *Mitochondrial Dysfunction in Neurodegenerative Disorders*, DOI 10.1007/978-3-319-28637-2_2

permitted the recognition of a number of mitochondrial disorders [2]. Mitochondrial diseases are heterogeneous disorders that arise from dysfunction of the mitochondrial respiratory chain due to mutations in either the mitochondrial or nuclear genome. Mitochondrial diseases might be described as primary disorders arising from mtDNA defects, either in the form of point mutations or rearrangements, or secondary disorders due to intergenomic signalling failure resulting in accumulation of mtDNA deletions or mtDNA depletion. MtDNA defects and respiratory chain abnormalities are increasingly linked with the pathogenesis of other neurodegenerative disorders, such as Parkinson's disease. As discussed in Chap. 1, mitochondria contain their own circular double-stranded DNA encoding for 13 polypeptides of the mitochondrial respiratory chain, 22 transfer RNAs (tRNA) and 2 ribosomal RNAs (rRNA) [3]. This genome is under dual genetic control, with many mtDNA maintenance genes encoded by the nuclear genome.

Mitochondrial DNA diseases present an important social and economic burden with an estimated prevalence of 1 in 10,000 people with a clinical manifestation of the disease and a further 1 in 6000 at risk of developing mitochondrial disease [4].

Neurological deficits are consistently reported in patients with mitochondrial disease and are often the most disabling [5]. Neurological symptoms are wide ranging and may include seizures, dementia, peripheral neuropathy, sensorineural deafness, stroke-like episodes and cerebellar ataxia. It is important to recognise that patients with mitochondrial disease do not always conform to the exact clinical criteria/presentation for a specific syndrome and many only manifest with a few features (Table 2.1).

There are a number of reasons why the central nervous system (CNS) is particularly vulnerable to mitochondrial dysfunction. Firstly, the brain is highly metabolically active and therefore particularly susceptible to bioenergetic failure [6]. Secondly, it has considerably fewer antioxidant defences than other tissues, so it is less able to protect against excessive reactive oxygen species (ROS) production [7]. Thirdly, with the exception of the subventricular zone, olfactory epithelium and the hippocampus, most neurons within the brain are post-mitotic and therefore irreplaceable [8]. Consequently, any particular neuronal insult will prove fatal to the cell if not alleviated in some way. In addition, there is increasing evidence linking mitochondrial dysfunction to neuronal cell loss in age-related neurodegenerative disorders, including Alzheimer's disease and Parkinson's disease.

2.2 Neurodegeneration in Mitochondrial Diseases

2.2.1 Brain Atrophy

Post-mortem brain atrophy is a common observation in patients with mitochondrial disease though quantitative data are sparse. In Newcastle, a series of 18 brains dissected at post-mortem with genetically and clinically diverse mitochondrial disease showed evidence of atrophy in 16. The average fresh brain weight was reduced by

Table 2.1 Clinical features of mitochondrial syndromes

Disorder	Primary feature	Causative mutations
Alpers' syndrome	Childhood onset characterised by severe developmental delay, intractable epilepsy and liver failure	The most common mutations are autosomal recessive mutations to the gene, *POLG*, which encodes the mitochondrial DNA polymerase
Chronic progressive external ophthalmoplegia (CPEO)	Bilateral ptosis and external ophthalmoplegia	CPEO can be caused by point mutations in the *MT-TL1* gene or mtDNA rearrangements resulting from mutations in nuclear-encoded mitochondrial maintenance genes such as *POLG*
Leigh syndrome (LS)	Presentation of nystagmus, hypotonia and respiratory dysfunction due brainstem dysfunction in infancy	The mtDNA point mutation, m.8993T>G, accounts for 10 % of affected individuals. Approximately 30 % of patients have a mutation to the nuclear-encoded gene *SURF1* which is involved in the assembly and maintenance of complex IV
Leber hereditary optic neuropathy (LHON)	Degeneration of the retinal ganglion cell layer and optic nerve leading to bilateral subacute loss of central vision with a predilection to males	95 % of LHON cases are due to the following mtDNA point mutations: m.3460G>A (*MT-ND1*), m.11778G>A (*MT-ND4*) and m.14484T>C (*MT-ND6*)
Kearns-Sayre syndrome (KSS)	Presentation of retinitis pigmentosa and progressive external ophthalmoplegia before the age of 20 years	KSS results from a sporadic large-scale deletion in the mtDNA. The most common deletion size is 4779 bp
Mitochondrial encephalomyopathy with lactic acidosis and stroke-like episodes (MELAS)	The phenotype is characterised by lactic acidosis, stroke-like episodes and seizures	80 % of MELAS cases are caused by the point mutation m.3243A>G (*MT-TL*)
Myoclonic epilepsy with ragged red fibres (MERRF)	Presentation of cerebellar ataxia, myoclonic epilepsy, myoclonus and myopathy in childhood and adolescence	Over 80 % of MERRF cases are caused by the point mutation, m.8344A>G (*MT-TK*)
Polymerase gamma encephalopathies	The typical syndromes associated with *POLG* mutations are ataxia neuropathy spectrum (ANS) and myoclonic epilepsy myopathy sensory ataxia (MEMSA). Characteristic symptoms associated with these syndromes are ataxia, myoclonic seizures, neuropathy and myopathy	Depletion, multiple deletion and multiple point mutations to mtDNA due to either autosomal dominant or autosomal recessive mutations to the gene, *POLG*, which encodes the mitochondrial DNA polymerase

about 16 % (range: 2.4–36.7 %; Table 2.2) compared to control individuals matched for age and gender. Of the 16 patients with atrophic brains, over half of the patients were aged 20–60 years and harboured a m.3243A>G mutation, three patients were aged between 18 and 55 years and harboured autosomal recessive polymerase gamma (*POLG*) mutations, two patients aged 42 and 58 harboured the m.8344A>G mutation, one patient harboured the m.13094T>C mutation associated with a MELAS (mitochondrial encephalomyopathy, lactic acidosis and stroke-like episodes)/Leigh overlap phenotype, one patient aged 50 harboured the m.14709T>C mutation and one patient aged 40 years harboured a single large-scale mtDNA deletion and Kearns-Sayre syndrome (KSS). Brain weights were within the normal limits in two patients aged 59 and 79 years harbouring autosomal recessive *POLG* mutations.

2.2.2 Neuroradiological Imaging

Neuroradiological imaging in patients with mitochondrial diseases often demonstrates focal lesions in cerebral cortex, white matter, basal ganglia and brainstem and particularly in patients with longer disease duration, generalised cerebral and cerebellar atrophy, all suggestive of CNS neuron loss. Enlargement of the fourth ventricle is often profound in younger patients and could be a predictor of cerebellar atrophy [9, 10].

Most of the imaging findings are diverse and nonspecific showing little correlation with genotype or biochemical phenotype, but in some of the mitochondrial diseases, lesions show a tendency for particular anatomical areas of the brain. For instance, in patients with MELAS, there is often prominent dilation of occipital horns with focal stroke-like cortical lesions in a non-vascular distribution generally in the posterior of the brain consistent with stroke-like episodes. At early stages of the disease, the lesions show features of vasogenic oedema, rather than ischaemic-like lesions [11], and often exhibit rapid resolution associated with clinical improvement. However, presence of cerebral atrophy in long-standing disease suggests association with gradual attrition of neuronal cells [9, 10]. In patients with MERRF (myoclonic epilepsy with ragged red fibres) due to the m.8344A>G mutation, neuroradiological imaging most often shows atrophy in the cerebral cortex, cerebellum, superior cerebellar peduncle and brainstem [12, 13]. More recently, MRI showed evidence of thalamic demyelination in a young male harbouring m.8344A>G but without the typical clinical presentation [14]. In patients with Leigh syndrome necrotic-like lesions distributed along the brainstem, thalamus and basal ganglia are considered characteristic [9]. Patients harbouring *POLG* mutations demonstrate a combination of thalamic and cortical lesions without basal ganglia involvement but with lesions in deep cerebellar nuclei and inferior olivary nuclei which can be indicative of mitochondrial spinocerebellar ataxia and epilepsy (MSCAE). In contrast, the main finding in patients with *POLG* mutations and Alpers' syndrome with paediatric onset is stroke-like lesions in occipital cortex [15], although thalamic changes have also been described, particularly in patients presenting as young adults [16, 17].

Table 2.2 Post-mortem brain weights

Patients with mitochondrial disease						Normal cases (Dekaban, Sadowsky. *Ann Neurol*. 1978;4:345–356)		
Age (years)	Sex	Genetic defect	Wt (g)	Wt (% normal)	Wt (% reduction)	Age range (years)	Sex	Wt (g) mean ± SD
47	M	m.3243A>G	1355	94.1 %	5.9 %	41–50	M	1440 ± 20
45	M	m.3243A > G	1200	83.3 %	16.7 %	41–50	M	1440 ± 20
20	F	m.3243A>G	829	63.3 %	36.7 %	19–22	F	1310 ± 10
30	M	m.3243A>G	1174	81.5 %	18.5 %	22–30	M	1440 ± 20
60	F	m.3243A>G	984	78.7 %	21.3 %	56–60	F	1250 ± 20
37	F	m.3243A > G	930	72	28 %	31–40	F	1290 ± 30
42	F	m.3243A>G	1070	82.9 %	17.1 %	41–50	F	1290 ± 20
57	F	m.3243A>G	1150	92 %	8 %	56–60	F	1250 ± 20
42	F	m.8344A>G	1075	83.3 %	16.7 %	41–50	F	1290 ± 30
58	M	m.8344A>G	1121	79.5 %	20.5 %	56–60	M	1410 ± 10
34	F	m.13094T>C	1260	97.6 %	2.4 %	31–40	F	1290 ± 30
55	M	m.14709T>C	1325	94 %	6 %	56–60	M	1410 ± 10
40	F	Single mtDNA deletion	1040	80.6 %	19.4 %	31–40	F	1290 ± 30
18	F	*POLG*	1098	81.9 %	18.1 %	16–18	F	1340 ± 30
24	F	*POLG*	1081	83.2 %	16.8 %	22–30	F	1300 ± 10
55	M	*POLG*	1292	91.6 %	8.4 %	56–60	M	1410 ± 10
59	M	*POLG*	1480	105 %	0 %	56–60	M	1410 ± 10
79	M	*POLG*	1421	105 %	0 %	76–81	M	1350 ± 10

In some mitochondrial diseases imaging shows predominantly white matter lesions. In patients with KSS symmetrical lesions in the white matter of the cerebrum, cerebellum, globus pallidus, dorsal midbrain and thalamus are consistently reported [9]. In mitochondrial neurogastrointestinal encephalopathy (MNGIE) patients, there are diffuse white matter lesions in cerebral hemispheres, brainstem and cerebellar peduncle in addition to demyelinating peripheral neuropathy [18].

Intracerebral calcification has been commonly reported in MELAS, KSS and Leigh syndrome. A neuroimaging study from 1998 revealed that basal ganglia (BG) calcification, involving the caudate, putamen, thalamus and global pallidus, was the most common finding in 54 % of patients harbouring the m.3243A>G mutation [10]. A similar distribution pattern is seen in ageing suggesting that BG calcification in mitochondrial disease, and particularly in MELAS, may occur due to accelerated ageing [19]. This particular feature is also seen in a range of insults to the brain including anoxia [20], exposure to radiation [21], and parathyroid [22] and hypothyroid disorders [23], suggesting that this brain region responds to damage by laying down minerals. Although BG calcification in mitochondrial diseases is often severe, it has not been reported in association with BG atrophy. It is also unclear if there are any functional symptoms related to this pathology in the patients with mitochondrial diseases, and while there is evidence of BG calcification in other diseases, there does appear to be a lack of correlation between the degree of pathology and emergence of clinical symptoms [24, 25]. It is intriguing that in the very old and also in younger patients with psychiatric illnesses, psychotic symptoms are strongly associated with basal ganglia calcification [19].

2.3 Neurological Features, Genetics and Neuropathology of Mitochondrial Diseases

Microscopic analyses of post-mortem brain tissue from patients with mitochondrial disease typically show profound neuronal cell loss, evidence of grey matter cortical lesions and accompanying astrogliosis and microgliosis. There is also evidence of mitochondrial respiratory chain impairments, predominantly affecting complexes I and IV in remaining cells, which can be defined using either a sequential cytochrome *c* oxidase and succinate dehydrogenase (COX/SDH) assay or the use of monoclonal antibodies raised against various subunits of the respiratory chain complexes.

2.3.1 Adult-Onset Disorders

2.3.1.1 Kearns-Sayre Syndrome

Symptoms and Cause

Kearns-Sayre syndrome (KSS) is characterised by triad of symptoms including retinitis pigmentosa, progressive external ophthalmoplegia and an onset before the age of 20 years [26]. Other neurological features include deafness, cerebellar ataxia, raised cerebrospinal fluid (CSF) protein levels, subclinical neuropathy and cognitive impairments (impairments in visuospatial attention and executive function) that are commonly associated with KSS. As KSS is a multisystem disorder, symptoms are not restricted to the nervous system; patients also often present with endocrinopathies, short stature, complete heart block, proximal myopathy and dysphagia [27].

KSS is caused by a single large-scale deletion or complex rearrangements of mtDNA that typically arise sporadically. MtDNA deletions range in size from 2.0 to 7.0 kb with about one third of patients harbouring a 'common deletion' of 4977 bp [28]. These deletions are typically located in the major arc between the between the two proposed origins of replication (O_H and O_L).

Neuropathology

Neuropathologically, KSS is characterised by spongiform encephalopathy in the white matter tracts of the cerebrum, cerebellum (Fig. 2.1a), basal ganglia, thalamus and spinal cord [26, 29, 30]. The spongiform encephalopathy varies in severity from myelin splitting to vacuolation of the tissue [30]. The susceptibility of the white matter in KSS is confirmed by the observation of preferential loss of myelin-associated glycoprotein (MAG), and 2′,3′-cyclic nucleotide phosphodiesterase (CNPase) the reduction in oligodendrocyte lineage cells (see Fig. 2.1b, c) and respiratory chain deficiency in mature oligodendrocytes (see Fig. 2.1d, e) which contribute to distal-dying back of oligodendrocytes in the cerebellum [31]. The pathology is not only restricted to the white matter, with neuronal cell loss observed in the cerebellum and brainstem. In the cerebellum, there is moderate loss of Purkinje cells accompanied by spongiform degeneration, capillary proliferation and reduced mitochondrial respiratory chain protein expression. The brainstem reveals neuronal cell loss and gliosis plus iron deposits, specifically in the globus pallidus [26].

2.3.1.2 Mitochondrial Encephalomyopathy Lactic Acidosis and Stroke-Like Episodes

Symptoms and Cause

The most common mutation associated with mitochondrial encephalomyopathy lactic acidosis and stroke-like episodes (MELAS) is the m.3243A>G mutation in the $tRNA^{Leu}$ gene. Central to the phenotype of MELAS are a triad of symptoms: lactic acidosis, stroke-like episodes and seizures. In addition, individuals can also suffer with hemianopia, hemiparesis, maternally inherited diabetes, deafness and ataxia. m.3243A>G is not the sole mutation to cause the MELAS phenotype; point mutations at positions m.3251A>G [32], m.3271T>C [33] and m.3291T>C [34] within the $tRNA^{Leu}$ gene can also produce a MELAS phenotype. Additionally to the $tRNA^{Leu}$ gene mutations, point mutations within $tRNA^{Val}$ (m.1642G>A) [35] and subunit III of cytochrome *c* oxidase (m.9957T>C) have been shown to be causative of a MELAS phenotype [36].

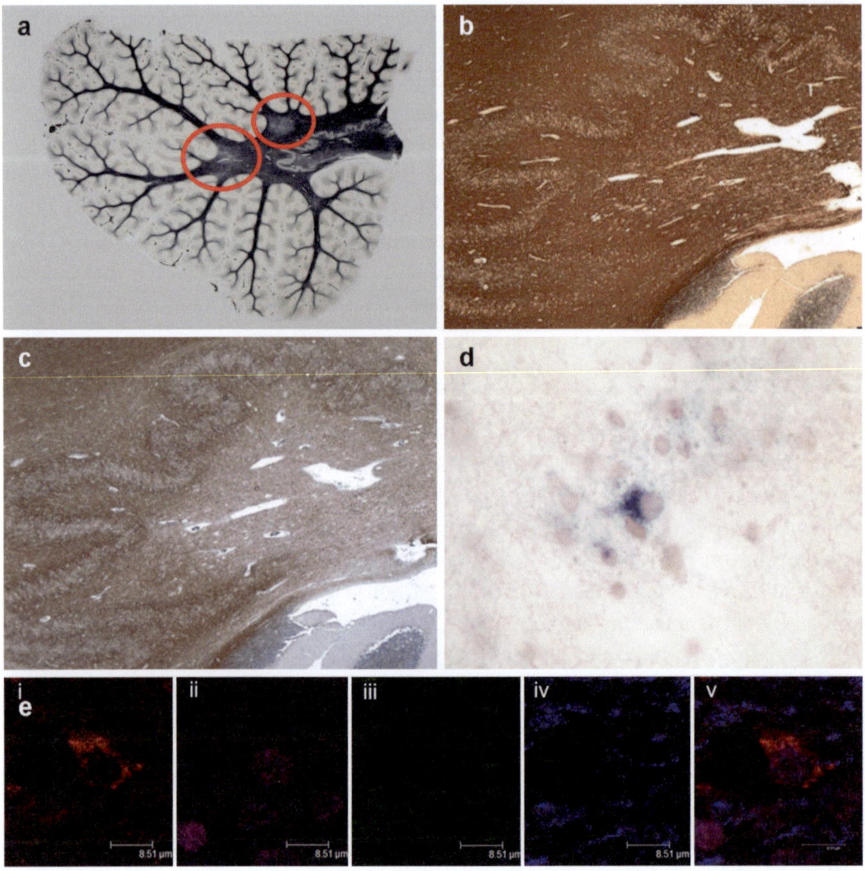

Fig. 2.1 White matter pathology in patients harbouring a single large-scale mtDNA deletion and KSS. Multiple foci of mild spongiform degeneration and myelin pallor are observed in the cerebellar deep white matter of a patient with KSS harbouring a single large-scale mtDNA deletion (**a** *circled*, Loyez myelin stain). High magnification of the white matter neighbouring the dentate nucleus reveals high and uniform expression of myelin basic protein (**b** anti-myelin basic protein immunohistochemistry), while there is a selective loss of CNPase expression in the corresponding region indicating a loss of oligodendrocytes in the dentate nucleus (**c** anti-CNPase immunohistochemistry). Sequential COX/SDH histochemistry reveals COX-deficient cells in the white matter (**d** COX/SDH), while immunofluorescent studies confirm the presence of oligodendrocytes and their nuclei (**e**i and ii anti-Olig2, *red*, and nuclei, *magenta*) which demonstrate a loss of complex IV expression (**e**iii anti-COX1 – *green*) despite high density of mitochondria (**e**iv anti-porin – *green*) which can be seen as a lack of colocalisation in the merged image (**e**v). A loss of complex IV expression is indicative of respiratory chain deficiency in those cells

Neuropathology

The neuropathological hallmarks of MELAS are defined by the presence of 'microinfarct-like' lesions in the occipital, parietal and temporal lobes and cerebellar cortex (Fig. 2.2a, b). The lesions may feature laminar dehiscence and become cystic;

they are associated with necrosis, increased astrogliosis and capillary proliferation. While these lesions are not associated with any major vascular territories [3], there is evidence of respiratory chain impairment in the cells comprising the cerebrovasculature in addition to neurons [37, 38]. Blood vessel wall calcification in the basal ganglia, thalamus and globus pallidus is particularly prominent in these patients; however there is no neuronal involvement in these regions [10]. The cerebellum is almost always affected in these patients with evidence of widespread atrophy of the molecular and granular cell layers accompanied by profound Purkinje cell loss (see Fig. 2.2c, d) and loss of neurons from dentate nucleus. Surviving Purkinje cells undergo a number of morphological changes with the appearance of axonal swellings known as axonal torpedoes in the granular cell layer (see Fig. 2.2g), loss of presynaptic terminals, increased dendritic arborisation and trapped mitochondria within thickened dendritic processes [37, 39]. Respiratory chain deficiencies involving complexes I and IV are particularly prominent in Purkinje cells (see Fig. 2.2e, f) [37, 38] and this has recently been documented in interneurons within the temporal, occipital and frontal lobes [40].

2.3.1.3 Myoclonic Epilepsy with Ragged Red Fibres

Symptoms and Cause

Originally described in 1980, myoclonic epilepsy with ragged red fibres (MERRF) occurs in childhood and early adulthood and is clinically characterised by deafness, seizures, myoclonus, myopathy, ataxia and the presence of ragged red fibres (RRF) in muscle [41]. The RRF may be observed following Gomori trichrome staining to reveal the presence of accumulated mitochondria below the plasma membrane of muscle fibres which might affect the contour of the fibre giving it an irregular or 'ragged' appearance. The typical point mutation associated with MERRF is the m.8344A>G mutation within $tRNA^{lys}$, though a point mutation in the same tRNA, m.8356T>C, produces a phenotype similar to that of MERRF [42].

Neuropathology

The neuropathology of MERRF is characterised by neuronal loss, demyelination and astrocytosis located preferentially within the dentate nucleus and Purkinje cells of the cerebellum and inferior olive, red nuclei and substantia nigra of the brainstem [43]. Pathology is also observed in the spinal cord, with a significant reduction in neurons of Clarke's nucleus as well as mild neuron loss in the anterior and posterior horn [30]. Immunohistochemical analysis of the mitochondrial respiratory chain proteins shows reduced expression of subunits comprising complexes I and IV within the remaining neurons of the cerebellar dentate nucleus, inferior olivary nuclei and frontal cortex [37, 44]. Quantitative immunofluorescent analysis of substantia nigra neurons in a patient with m.8344A>G revealed that 22 % and 23 % of dopaminergic neurons did not contain subunits of complexes I and IV, respectively, despite preserved mitochondrial mass [45]. Another study observed reduced

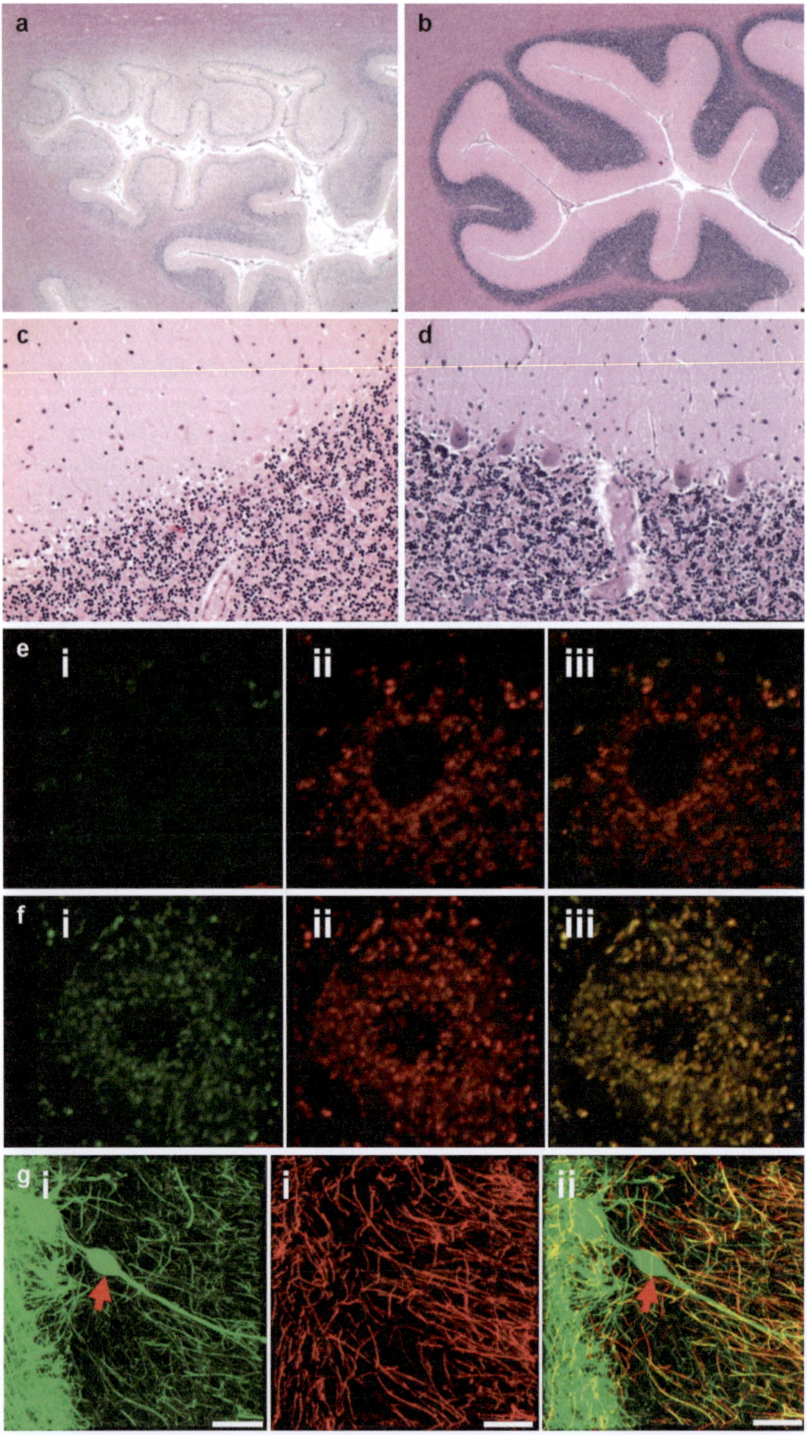
a
b
c
d
e
i
ii
iii
f
i
ii
iii
g
i
i
ii

GABAergic interneuron density in the frontal, temporal and occipital cortices of two patients harbouring the m.8344A>G mutation with remaining interneurons displaying a significant reduction in complex I and complex IV expression [40]. A study looking into the susceptibility of specific neuronal types harbouring the m.8344A>G mutation found that there was no correlation between heteroplasmy level and selective vulnerability of the neurons [46]. This study shows that the pathogenesis of MERRF is likely multifactorial, and neuronal vulnerability is not exclusively determined by high levels of mutated mtDNA.

2.3.1.4 Leber Hereditary Optic Neuropathy

Symptoms and Cause

First described by Theodore Leber in 1871, Leber hereditary optic neuropathy (LHON) is clinically characterised by bilateral subacute loss of central vision due to the focal degeneration of the retinal ganglion cell layer and optic nerve [47]. The disease typically manifests between 20 and 40 years of age, with males being more commonly afflicted. The primary cause of 95 % of LHON cases can be attributed one of three mtDNA point mutations, including m.3460G>A, m.11778G>A and m.14484T>C [48]. Each mutation is present in genes that encode for subunits of complex I of the respiratory chain including *MT-ND1*, *MT-ND4* and *MT-ND6*, respectively.

Neuropathology

The neuropathology of LHON is characterised by progressive degeneration of the retinal ganglion cells and axons affecting the optic nerve. In the early stages of the disease, it is the small calibre axons of the papillomacular bundle that are lost before the rest of the fibres leading to optic atrophy [49]. In a number of LHON cases, the

Fig. 2.2 Cerebellar pathology in patients harbouring the m.3243A>G mutation. Infarct-like lesion involving the molecular layer, Purkinje cells, granular cell layer and the white matter of the cerebellar cortex of a patient (**a** haematoxylin and eosin stain; scale bar 100 μm) in contrast to intact morphology of the cerebellar cortex in a control individual (**b** haematoxylin and eosin stain; scale bar 100 μm). Severe Purkinje cell loss and reduced granular cell density in the cerebellar cortex of a patient (**c** haematoxylin and eosin stain; scale bar 100 μm), while preserved Purkinje cell density is seen in the cerebellar cortex of a control individual (**d** haematoxylin and eosin stain; scale bar 100 μm). Purkinje cells from patients typically reveal prominent loss of complex I (**ei** NUDUFA13 subunit) expression despite high expression of complex IV (COX4 subunit **eii**) which can be seen in the colocalised image (**eiii**). While Purkinje cells from control individuals reveal high expression of both complexes I (**fi**) and IV (**fii**) and can be seen in the colocalised image (**fiii**). Scale bar 100 μm. The main constituent of axonal torpedoes (*red arrow*) is phosphorylated neurofilament H (**gi**). There is a lack of myelin (myelin basic protein, **gii**) around the axonal torpedo as seen in the colocalised image (**gii**). Scale bar 100 μm

involvement of the spinal cord is observed with demyelination, axon loss and macrophage activation in the gracile fasciculus of the posterior column of the spinal cord [50]. As well as the observation of necrotic lesions and gliosis in the brainstem [51].

2.3.1.5 Polymerase Gamma Encephalopathies

Symptoms and Cause

The mitochondrial genome is replicated by mitochondrial polymerase gamma (POLG). The human mtDNA polymerase is a 195 kDa heterotrimer consisting of a 140 kDa catalytic subunit (PolγA) and two identical 55 kDa accessory units (PolγB). The C-terminus of the catalytic subunit of PolγA is responsible for the polymerase function, while the N-terminus is responsible for exonuclease activity and proofreading of mtDNA. The linker mediates a focal contact with the dimeric accessory subunit. PolγA is encoded by the nuclear *POLG* gene. Primary mutations in the *POLG* gene represent one of the mechanisms that cause secondary mutations in the mtDNA, including depletion of mtDNA, multiple mtDNA deletions or multiple point mutations in mtDNA [52].

Recently, numerous mutations have been described in *POLG* in association with a spectrum of clinical/neurodegenerative phenotypes, including autosomal dominant and autosomal recessive progressive external ophthalmoplegia (PEO) [53–55], myoclonic epilepsy myopathy sensory ataxia (MEMSA) [56], ataxia neuropathy spectrum (ANS) [57], parkinsonism [58] and Alpers' syndrome [59].

Neuropathology

The first neuropathological description of a patient harbouring two recessive heterozygous *POLG* mutations, p.Ala467Thr and a novel p.X1240Gln and secondary multiple mtDNA deletions, was in 2000 [60]. In this patient, the inferior medullary olives were the most severely affected region, showing massive neuronal loss. Moderate to extensive neuronal loss was observed throughout the cerebellum and to a lesser extent in the dentate nucleus and the red nuclei, while the pons remained unaffected. This pattern is indicative of cerebello-olivary atrophy and probably underlies the cerebellar ataxia observed in patients with multiple deletions. A later neuropathological study reported that the spinal cord tissues demonstrated evidence of severe myelin and axonal loss within the posterior columns and in dorsal spinal roots. Severe neuronal loss and respiratory chain abnormalities were also present in dorsal root ganglia and in paraspinal sympathetic ganglia (Fig. 2.3). Ventral and lateral spinal tracts, motor roots and motor neurons were intact [61]. There was also severe depletion of neurons from the substantia nigra without Lewy body formation. Throughout the brain of this patient, there was lack of correlation between the distribution of COX-deficient neurons and the degree of neuropathological damage, similar to the findings in patients with MELAS and MERRF.

The neuropathology of a patient harbouring multiple deletions due to two heterozygous *POLG* mutations revealed the loss of pigmented neurons in the substantia nigra (SN) with alpha-synuclein-reactive Lewy body formation. In addition, unlike in patients with typical Lewy body-Parkinson's disease (PD) but similar to patients with some mitochondrial diseases, there were loss of Purkinje cells, loss of neurons from the cerebellar dentate nucleus, profound loss of myelin and axons from posterior columns of spinal cord and loss of neurons from gracile nucleus. In the patient with the *POLG* mutations, the remaining SN neurons had high levels of mtDNA deletions (~65 %) associated with COX deficiency in 21 % of the neurons; these levels show a manifold increase compared to <3 % of COX-deficient SN neurons observed in PD patients and <1 % of SN neurons observed in ageing controls [62]. High levels of mtDNA deletions in the substantia nigra have recently been shown in

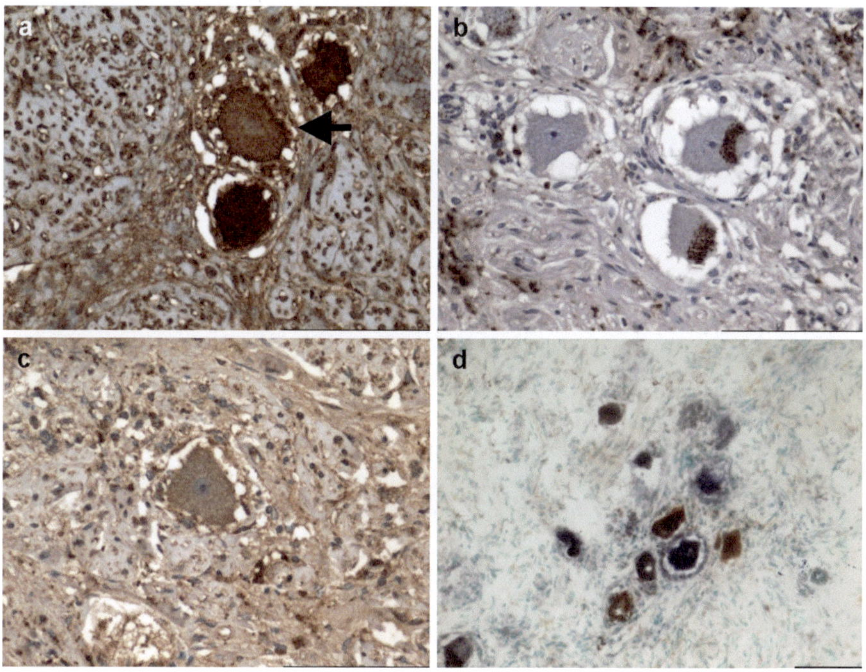

Fig. 2.3 Evidence of prominent deficiencies of mitochondrial respiratory chain proteins within the sensory neurons of the dorsal root ganglia and myelin loss from the dorsal columns of the spinal cord in patients harbouring *POLG* mutations and multiple mtDNA deletions. Unusually reduced mitochondrial density in one neuron (*arrow*) and uneven cellular distribution within sensory neurons (**a** anti-porin immunohistochemistry) which also demonstrate profound neuronal deficiency of the NDUFB4 subunit of complex I with focal cytoplasmic clumping of immunoreactivity (**b** anti-NDUFB4 immunohistochemistry) and reduced expression of the SDHA subunit of complex II within the neurons of the dorsal root ganglia (**c** anti-SDHA immunohistochemistry), similar to the distribution observed with porin. Dual COX/SDH histochemistry reveals a high level of complex IV (COX) deficiency in some neurons (**d** *blue cells*), while in other neurons the COX activity is relatively unaffected (**d** *brown cells*). Further mitochondrial abnormalities can be observed with markedly reduced expression and focal cytoplasmic clumping of subunits I and IV of complex IV in all neurons (**e** anti-COX1 immunohistochemistry and (**f**) anti-COX4 immunohistochemistry – *arrow*). Loss of myelin from the dorsal columns of the spinal cord can be seen in patients (**g** haematoxylin and eosin staining and (**h**) Loyez myelin staining). Scale bar represents 100 μm

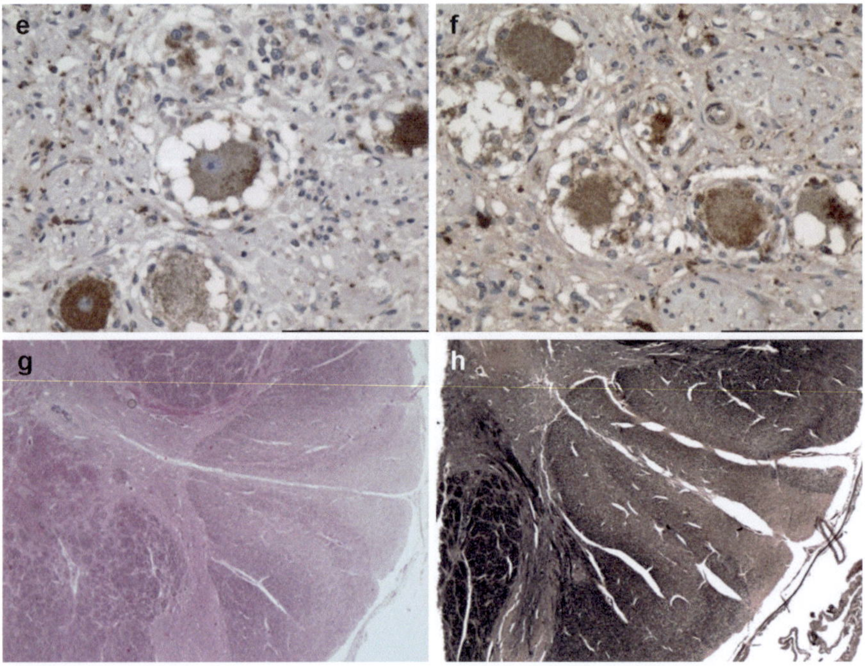

Fig. 2.3 (continued)

a larger cohort of patients with *POLG* mutations. In these patients, severe and time-dependent neuronal cell loss was observed in the cerebellum (affecting both Purkinje cells and dentate nucleus), inferior olivary nucleus and substantia nigra in conjunction with a high percentage of remaining neurons with complex I deficiency. They described foci of necrotic lesions (similar to those observed in patients with the m.3243A>G mutation) in the neocortex, hippocampus, thalamus and cerebellar cortex which they define as 'focal energy-dependent neuronal necrosis' [63].

2.3.2 Childhood-Onset Disorders

2.3.2.1 Leigh Syndrome

Symptoms and Cause

A progressive neurodegenerative disorder has an onset in infants usually before the age of 2 [64]. Leigh syndrome is clinically defined by brainstem dysfunction that results in nystagmus, hypotonia, motor and intellectual retardation and respiratory dysfunctions [65]. In addition to brainstem dysfunction, neurological symptoms including ataxia, dystonia and optic atrophy are common. The symptoms arise due to necrotic lesions to the brainstem, thalamus and basal ganglia, the cause of which has been attributed to a possible failure in oxidative phosphorylation due to either a mtDNA or nuclear DNA mutation.

Neuropathology

Leigh syndrome can be identified by symmetrical hypointensities on CT scan or hyperintense lesions on T2-weighted MRI in the anterior basal ganglia, medial thalami, periaqueductal region of the midbrain and pons and cerebellar hemispheres [66]. Macroscopically these lesions present with variable vacuolation, astrogliosis, neuron loss, vascular proliferation and demyelination [43]. While in the substantia nigra, necrosis, macrophage infiltration and axonal swelling had also been observed [67]. The lesions that appear in Leigh syndrome are very heterogeneous, likely due to the fact that the mutations associated with the disease are in a multitude of nuclear and mtDNA genes encoding mitochondrial enzymes, including pyruvate dehydrogenase, and respiratory complexes I, II, IV and V; therefore a different mutation will result in a differing phenotype.

2.3.2.2 Alpers' Syndrome

Symptoms and Cause

Alpers' syndrome, also known as Alpers-Huttenlocher syndrome, is an early-onset neurodegenerative disorder characterised by a clinical triad of severe developmental delay, intractable epilepsy and liver failure. Presentation is often sudden with uncontrollable seizures in a previously healthy child. The onset is bimodal: it most commonly affects infants between the ages of 2 and 4 years (range, 3 months to 8 years), but has a second peak of onset between the ages of 17 and 24 years (range, 10–27 years) [68]. Other symptoms include ataxia, hypotonia, cortical blindness, spasticity and dementia. The disorder is progressive and often leads to death from hepatic failure or status epilepticus before age 3 years [69].

Alpers' syndrome is caused by autosomal recessive mutations in *POLG*. Defects in *POLG* compromise replication and repair of mtDNA and can therefore result in mtDNA depletion or accumulation of mtDNA deletions, though depletion is more frequently reported in these patients. This affects mitochondrial respiratory chain function through impaired oxidative phosphorylation, lowered ATP generation leading to cellular dysfunction and death in affected tissues [70].

Neuropathology

In 1931, Bernard Alpers first used the term 'diffuse progressive degeneration of the grey matter of the cerebrum', describing neuropathological findings such as neuronal loss and gliosis [71]. Subsequent studies have also shown degeneration of the cerebral grey matter with loss of neurons, astrogliosis and capillary proliferation

[59, 69]. On macroscopic neuropathological examination, symmetrical atrophy with laminar necrosis and neuronal degeneration is seen in the cerebral cortex (with predominant involvement of the temporal and occipital lobes), hippocampi, olfactory bulbs and cerebellum [72]. The presence of focal ischaemic or necrotic lesions is variable. Neuronal respiratory chain deficiencies, affecting complex I, and lowered mtDNA content correlating with the severity of neurodegeneration have recently been described in a large cohort of patients with *POLG* mutations [63].

2.4 Mechanisms of Neurodegeneration in Mitochondrial Diseases

Understanding the mechanisms contributing to neurodegeneration in patients with mitochondrial disease is complicated, and though post-mortem studies are crucial for documenting the neuropathological changes, it can be difficult to disentangle cause and effect and therefore precisely delineate mechanisms contributing to neuronal cell dysfunction and death.

2.4.1 Mitochondrial Homeostasis

Neurons have a high and constant demand for ATP generated via mitochondrial metabolism which is reflected by high mitochondrial mass within the neuronal cell body, axon, presynaptic terminals and dendritic branches [73]. To support this metabolic demand, mitochondria are highly dynamic and continuously move, fuse and divide. Although mitochondria are essential for ATP generation, mitochondria also play essential roles in the production of iron sulphur clusters and in calcium handling, apoptosis and ROS signalling (see Chap. 1). To some extent, all of these processes have been implicated in neurodegenerative diseases.

2.4.2 Mitochondria and Calcium Handling

Mitochondria play a pivotal role in the tight regulation of intracellular calcium (Ca^{2+}) levels and achieve this via uptake through a membrane potential-driven carrier, known as the mitochondrial calcium uniporter (MCU) [74]. Ca^{2+} homeostasis is particularly important for neuronal function where Ca^{2+} is intricately linked to neurotransmitter vesicle release, synaptic plasticity and mitochondrial transport [75]. Neuronal mitochondria transition between mobile and stationary states in response to intracellular [Ca^{2+}] [76]. It is thought that this mechanism allows mitochondrial retention in presynaptic terminals and postsynaptic dendritic spines, where calcium influx is dynamic and requires tight regulation to maintain neurotransmission. While there have been a limited number of studies investigating the effect of mtDNA mutations on Ca^{2+} homeostasis, these have often focused on non-neuronal cell lines where physiological differences may be manifold. Fibroblasts derived from patients harbouring the m.3243A>G mutation revealed high basal Ca^{2+} levels and sustained elevation of calcium in response to depolarisation following stimulation with potassium [77]. Furthermore ρ^o osteosarcoma cells derived from a patient with the m.3243A>G mutation found a perturbed intra-mitochondrial calcium homeostasis [78]. A recent study of mature neurons derived from embryonic stem (ES) cells harbouring mtDNA mutations displayed a normal Ca^{2+} baseline, but following repeated glutamate stimulation, there was a progressive deficit in Ca^{2+} transients [79]. These studies show that mtDNA mutations not only impair ATP generation but also the ability of mitochondria to regulate Ca^{2+} homeostasis which could contribute to neurodegeneration since an increase in neuronal calcium concentration is a mechanism of excitotoxicity [80].

2.4.3 *Mitochondria and Cell Death*

There are three defined modes of cell death which might become induced in energetically compromised neurons: necrosis, apoptosis and autophagy. Mitochondria are intricately involved in the process of apoptosis since disruption of mitochondrial transmembrane potential can cause the release of cytochrome *c* and apoptosis-inducing factor from the mitochondria into the cytoplasm. Although the mechanism of neuronal cell death in patients with mitochondrial disease is not known, there are a few studies which examine apoptosis in muscle fibres in conjunction with respiratory chain deficiency; however the results from these reports are inconsistent and contradictory. One study found increased expression of factors associated with apoptosis (Fas, caspase-3 and p75) and increased TUNEL labelling (indicating double-strand breaks in DNA) in skeletal muscle biopsies from patients with point mutations or mtDNA deletions [81]. However in a different study, TUNEL labelling was not significantly increased in muscle biopsies from patients with mitochondrial disease; however the expression of apoptotic makers (Bax and Apaf-1) and cytochrome *c* release from mitochondria were observed in ragged red fibres, indicating initiation of apoptosis [82]. Another study found that a peptide known as humanin was upregulated in patients with MELAS and MERRF and this was inhibiting apoptosis in muscle fibres [83]. While a study investigating neuropathology of patients with *POLG* encephalopathy observed no neurons that were positive for caspase-3 immunohistochemistry, two patients with Alpers' syndrome demonstrated some TUNEL staining in neurons [63]. The timeframe for capturing neurons undergoing cell death is likely to be narrow, and so it is challenging to identify the exact mode of cell death in post-mortem brain tissue.

2.4.4 *Oxidative Stress*

Mitochondrial metabolism is also a major source of reactive oxygen species (ROS) production in the cell [84]. Unpaired electrons may leak from complexes I and III and react with oxygen to produce damaging superoxide. The excessive production and release of ROS have been linked with neurodegeneration as ROS can react with and oxidise molecules such as proteins, DNA and lipids resulting in oxidative stress. Oxidative stress occurs when there is an imbalance in the creation and detoxification of reactive oxygen species leading to the oxidisation and damage of proteins, DNA and lipids. However at normal levels, ROS can act as intracellular signals. Dysfunction within mitochondria leading to increased ROS production has been postulated as a pathological mechanism for both disease and the typical ageing phenotype [85]. In a mouse model of LHON where mice harbour a mutation in the *MT-ND6* gene (the human equivalent of the m.14600G>A mutation), decreased complex I activity and increased ROS production were observed in synaptosomes, while ATP homeostasis was preserved which lead the authors to speculate that a chronic increase in ROS production and oxidative damage was the pathological driving force for LHON [86]. Increased ROS production was also observed in primary skin and muscle fibroblasts from a mouse where complex I accessory subunit (*NDUFS4*) was knocked out [87]. However, fibroblasts derived from a French Canadian variant of Leigh syndrome due to a mutation in the *LRPPRC* gene, which encodes for a protein involved in mitochondrial mRNA stability, presented with mitochondrial abnormalities but no increase in ROS production [88]. This was also observed in a different mouse model of mitochondrial disease, where the *POLG* gene is mutated and there was no reported increase in ROS production [89]. As shown by the conflicting results from these studies, it is unlikely that the increase in ROS is solely responsible for neurodegeneration but might act in conjunction with dysfunction of other important mitochondrial pathways to cause neural dysfunction and degeneration.

2.5 Modelling Neuropathology of Mitochondrial Disease

Recent developments using cell culture models have allowed a more mechanistic approach to understand the impact of mitochondrial dysfunction on neural function. Differentiated neurons derived from mouse embryonic stem cells harbouring mtDNA defects have been generated including wild type parental cells, polymorphic variant cells, mild complex IV mutant cells and severe complex I mutant cells. Those neurons harbouring severe complex I mutations revealed the most dramatic alterations; with reduced ability to undergo neuronal differentiation and delayed development [90]. Despite this, differentiated neurons showed maintenance of high mitochondrial membrane potential with reversal of the ATP synthase and marked ROS production [91] while demonstrating an impaired response to Ca^{2+} transients following repetitive neuronal stimulation [79].

A more promising approach to understand mechanisms of neural dysfunction and degeneration derives from the reprogramming of patient fibroblasts into induced pluripotent stem cells (IPSCs). These cells contain the original nuclear and mitochondrial DNA from patients and are capable of differentiation into any cell type, including post-mitotic neurons. A recent study using neurons derived from patient fibroblasts harbouring the m.3243A>G mutation revealed bimodal segregation of mutated mtDNA with cells containing either 100 % wild type and 100 % mutated mtDNA and prominent complex I deficiency recapitulating the respiratory chain abnormalities detected in post-mortem tissues [92, 93]. Patient-derived IPSCs will not only improve understanding of disease mechanisms but may also play a role in regenerative medicine since cellular homoplasmy for the m.8993T>G mutation was genetically corrected using somatic cell nuclear transfer [94].

The generation of mouse models has helped towards understanding the pathogenesis of mitochondrial disease. The majority of models manipulate nuclear DNA, due to technical challenges associated with modification of the mitochondrial genome and the complexity of the mitochondrial bottleneck, and in recent years many groups have begun to take advantage of Cre-Lox recombination technology which permits the excision of specific genes that are flanked by two LoxP sites through the activity of an enzyme Cre recombinase. Though mouse models are discussed elsewhere in this book (see Chap. 14), there are recent developments using mice where complex I (e.g., *NDUFS4* [95]), complex III (e.g., *UQCRFS1* [96]), complex IV (e.g., *COX10* [96, 97]) and mtDNA maintenance (e.g., *TFAM* [98–100]) genes are floxed by two LoxP sites which induce knockout of the gene when crossed with mice expressing a Cre recombinase under control of neuronal-specific promoter. In these mice, the effects of the gene knockout on neural function and degeneration can be monitored and the mechanisms contributing to neurodegeneration can be further delineated.

2.6 Conclusion

Deciphering the mechanisms of neurodegeneration in patients with mtDNA diseases poses a difficult challenge to neuroscientists because of the diversity in clinical presentations, different types of mtDNA mutations and variable pattern of brain involvement. However, increased understanding of the processes underlying CNS degeneration in patients with mtDNA diseases may be informative for other neurodegenerative disorders where a mitochondrial aetiology is suspected.

There is some evidence that mutations within the mitochondrial genome or mitochondrial dysfunction play a role in a number of neurodegenerative disorders including Parkinson's disease (PD), Huntington's disease (HD), Alzheimer's disease (AD) and multiple sclerosis [101–104]. While the role of these reported mitochondrial abnormalities, whether causative for the onset of neurodegeneration or a consequence of ongoing neurodegenerative processes, it is clear that mitochondrial dysfunction would act to exacerbate disease pathogenesis. However, much like the mitochondrial encephalomyopathies, our understanding of the pathophysiological mechanisms involved in these diseases remains incomplete.

Acknowledgements This work was supported by the Wellcome Trust (074454/Z/04/Z) and UK NIHR Biomedical Research Centre for Ageing and Age-related disease award to the Newcastle upon Tyne Hospitals NHS Foundation Trust.

Glossary

Anopia A defect to the visual field.

Ataxia Progressive loss of coordination and balance in hands, arms and legs often due to dysfunction of the cerebellum.

Axonal torpedoes Fusiform swellings of the Purkinje cell axons.

Bradykinesia An abnormal slowness of movement, a characteristic clinical symptom seen in Parkinson's disease.

Cardiomyopathy The deterioration of heart muscle resulting in compromised heart function.

Chronic progressive external ophthalmoplegia Slowly progressive paralysis of the extraocular muscle which results in bilateral, symmetrical, progressive ptosis (drooping of the eyelids) followed by ophthalmoparesis (paralysis) months to years later.

Cortical blindness Total loss or partial loss of visual field with normal pupillary responses resulting from damage to the occipital lobe.

Dysarthria A speech disorder characterised by poor articulation. Often present in patients suffering from cerebellar ataxia.

Dysphagia Swallowing problems often present in conjunction dysarthria in patients suffering from cerebellar ataxia.

Dystonia Dystonia is characterised by involuntary and uncontrollable muscle spasms which can force affected parts of the body into abnormal, sometimes painful, movements or postures.

Encephalopathy Any disease that results in the alteration of the structure and function of the brain.

Endocrinopathy A disorder in the function of the endocrine gland which can often result in hormone imbalances.

Epilepsy Recurrent seizures (often described as fits).

Extrapyramidal features May be defined as an inability to initiate movements or an inability to remain motionless.

Hypotonia A state of low muscle tone resulting in reduced muscle strength.

Laminar dehiscence Splitting of the six cortical layers.

Lipoma A benign tumour consisting of brown fat cells that forms under the skin.

Myoclonus Brief involuntary twitching of a muscle or a group of muscles.

Nystagmus Involuntary eye movements that may result in reduced or limited vision.

Oxidative stress The imbalance of reactive oxygen species creation and detoxification leading to the oxidisation of proteins, DNA and lipids.

Peripheral neuropathy Damage to the nerves comprising the peripheral nervous system which can result in a combination of weakness, autonomic dysfunction and sensory abnormalities.

Proximal myopathy Weakness of those muscles located closely (proximal) to the body.

Ptosis A drooping of the upper eyelids.

Retinitis pigmentosa Retinitis pigmentosa is a disorder of the retina which is characterised by dysfunction of the photoreceptors resulting in incurable blindness or tunnel vision.

Stroke-like episodes Neurological deficits often described as resembling stroke however do not conform to a vascular territory. They are often accompanied by cortical blindness, hemianopsia, or hemiparesis.

Wolff-Parkinson-White syndrome This syndrome is a heart condition caused by extra electrical activity and can lead to increased heart rate.

Vasogenic oedema The leakage of fluids from the vascular network into the extracellular space.

References

1. Luft R, Ikkos D, Palmieri G, Ernster L, Afzelius B. A case of severe hypermetabolism of nonthyroid origin with a defect in the maintenance of mitochondrial respiratory control: a correlated clinical, biochemical, and morphological study. J Clin Invest. 1962;41:1776–804.
2. Taylor RW, Pyle A, Griffin H, Blakely EL, Duff J, He L, et al. Use of whole-exome sequencing to determine the genetic basis of multiple mitochondrial respiratory chain complex deficiencies. JAMA. 2014;312(1):68–77.
3. Anderson S, Bankier AT, Barrell BG, de Bruijn MH, Coulson AR, Drouin J, et al. Sequence and organization of the human mitochondrial genome. Nature. 1981;290(5806):457–65.
4. Schaefer AM, McFarland R, Blakely EL, He L, Whittaker RG, Taylor RW, et al. Prevalence of mitochondrial DNA disease in adults. Ann Neurol. 2008;63(1):35–9.
5. McFarland R, Taylor RW, Turnbull DM. A neurological perspective on mitochondrial disease. Lancet Neurol. 2010;9(8):829–40.
6. Attwell D, Laughlin SB. An energy budget for signaling in the grey matter of the brain. J Cereb Blood Flow Metab. 2001;21(10):1133–45.
7. Coyle JT, Puttfarcken P. Oxidative stress, glutamate, and neurodegenerative disorders. Science. 1993;262(5134):689–95.
8. Gould E. How widespread is adult neurogenesis in mammals? Nat Rev Neurosci. 2007;8(6):481–8.
9. Haas R, Dietrich R. Neuroimaging of mitochondrial disorders. Mitochondrion. 2004;4(5–6):471–90.
10. Sue CM, Crimmins DS, Soo YS, Pamphlett R, Presgrave CM, Kotsimbos N, et al. Neuroradiological features of six kindreds with MELAS tRNA(Leu) A2343G point mutation: implications for pathogenesis. J Neurol Neurosurg Psychiatry. 1998;65(2):233–40.
11. Kolb SJ, Costello F, Lee AG, White M, Wong S, Schwartz ED, et al. Distinguishing ischemic stroke from the stroke-like lesions of MELAS using apparent diffusion coefficient mapping. J Neurol Sci. 2003;216(1):11–5.

12. Huang CC, Wai YY, Chu NS, Liou CW, Pang CY, Shih KD, et al. Mitochondrial encephalomyopathies: CT and MRI findings and correlations with clinical features. Eur Neurol. 1995;35(4):199–205.
13. Ito S, Shirai W, Asahina M, Hattori T. Clinical and brain MR imaging features focusing on the brain stem and cerebellum in patients with myoclonic epilepsy with ragged-red fibers due to mitochondrial A8344G mutation. Am J Neuroradiol. 2008;29(2):392–5.
14. Erol I, Alehan F, Horvath R, Schneiderat P, Talim B. Demyelinating disease of central and peripheral nervous systems associated with a A8344G mutation in tRNALys. Neuromuscul Disord. 2009;19(4):275–8.
15. Harding AE, Holt IJ, Cooper JM, Schapira AH, Sweeney M, Clark JB, et al. Mitochondrial myopathies: genetic defects. Biochem Soc Trans. 1990;18(4):519–22.
16. Harding BN, Alsanjari N, Smith SJ, Wiles CM, Thrush D, Miller DH, et al. Progressive neuronal degeneration of childhood with liver disease (Alpers' disease) presenting in young adults. J Neurol Neurosurg Psychiatry. 1995;58(3):320–5.
17. Wolf NI, Rahman S, Schmitt B, Taanman JW, Duncan AJ, Harting I, et al. Status epilepticus in children with Alpers' disease caused by POLG1 mutations: EEG and MRI features. Epilepsia. 2009;50(6):1596–607.
18. Gamez J, Minoves T. Abnormal brainstem auditory evoked responses in mitochondrial neurogastrointestinal encephalomyopathy (MNGIE): evidence of delayed central conduction time. Clin Neurophysiol. 2006;117(11):2385–91.
19. Ostling S, Andreasson LA, Skoog I. Basal ganglia calcification and psychotic symptoms in the very old. Int J Geriatr Psychiatry. 2003;18(11):983–7.
20. Midroni G, Willinsky R. Rapid postanoxic calcification of the basal ganglia. Neurology. 1992;42(11):2144–6.
21. Numaguchi Y, Hoffman Jr JC, Sones Jr PJ. Basal ganglia calcification as a late radiation effect. Am J Roentgenol Radium Ther Nucl Med. 1975;123(1):27–30.
22. Illum F, Dupont E. Prevalences of CT-detected calcification in the basal ganglia in idiopathic hypoparathyroidism and pseudohypoparathyroidism. Neuroradiology. 1985;27(1):32–7.
23. Halpern JP, Boyages SC, Maberly GF, Collins JK, Eastman CJ, Morris JG. The neurology of endemic cretinism. A study of two endemias. Brain. 1991;114(Pt 2):825–41.
24. Wider C, Dickson DW, Schweitzer KJ, Broderick DF, Wszolek ZK. Familial idiopathic basal ganglia calcification: a challenging clinical-pathological correlation. J Neurol. 2009;256(5):839–42.
25. Brannan TS, Burger AA, Chaudhary MY. Bilateral basal ganglia calcifications visualised on CT scan. J Neurol Neurosurg Psychiatry. 1980;43(5):403–6.
26. Oldfors A, Fyhr IM, Holme E, Larsson NG, Tulinius M. Neuropathology in Kearns-Sayre syndrome. Acta Neuropathol. 1990;80(5):541–6.
27. Maceluch JA, Niedziela M. The clinical diagnosis and molecular genetics of kearns-sayre syndrome: a complex mitochondrial encephalomyopathy. Pediatr Endocrinol Rev. 2005;4(2):117–37.
28. Zeviani M, Moraes CT, DiMauro S, Nakase H, Bonilla E, Schon EA, et al. Deletions of mitochondrial DNA in Kearns-Sayre syndrome. Neurology. 1998;51(6):1525. a.
29. Tanji K, DiMauro S, Bonilla E. Disconnection of cerebellar Purkinje cells in Kearns-Sayre syndrome. J Neurol Sci. 1999;166(1):64–70.
30. Sparaco M, Bonilla E, DiMauro S, Powers JM. Neuropathology of mitochondrial encephalomyopathies due to mitochondrial DNA defects. J Neuropathol Exp Neurol. 1993;52(1):1–10.
31. Lax NZ, Campbell GR, Reeve AK, Ohno N, Zambonin J, Blakely EL, et al. Loss of myelin-associated glycoprotein in kearns-sayre syndrome. Arch Neurol. 2012;69(4):490–9.
32. Morten KJ, Cooper JM, Brown GK, Lake BD, Pike D, Poulton J. A new point mutation associated with mitochondrial encephalomyopathy. Hum Mol Genet. 1993;2(12):2081–7.
33. Goto Y, Horai S, Matsuoka T, Koga Y, Nihei K, Kobayashi M, et al. Mitochondrial myopathy, encephalopathy, lactic acidosis, and stroke-like episodes (MELAS): a correlative study of the clinical features and mitochondrial DNA mutation. Neurology. 1992;42(3):545.

34. Hess JF, Parisi MA, Bennett JL, Clayton DA. Impairment of mitochondrial transcription termination by a point mutation associated with the MELAS subgroup of mitochondrial encephalomyopathies. Nature. 1991;351(6323):236–9.
35. Taylor RW, Chinnery PF, Haldane F, Morris AAM, Bindoff LA, Turnbull DM, et al. MELAS associated with a mutation in the valine transfer RNA gene of mitochondrial DNA. Ann Neurol. 1996;40(3):459–62.
36. Manfredi G, Schon EA, Moraes CT, Bonilla E, Berry GT, Sladky JT, et al. A new mutation associated with MELAS is located in a mitochondrial DNA polypeptide-coding gene. Neuromuscul Disord. 1995;5(5):391–8.
37. Lax NZ, Hepplewhite PD, Reeve AK, Nesbitt V, McFarland R, Jaros E, et al. Cerebellar ataxia in patients with mitochondrial DNA disease: a molecular clinicopathological study. J Neuropathol Exp Neurol. 2012;71(2):148–61.
38. Lax NZ, Pienaar IS, Reeve AK, Hepplewhite PD, Jaros E, Taylor RW, et al. Microangiopathy in the cerebellum of patients with mitochondrial DNA disease. Brain. 2012;135(6):1736–50.
39. Tanahashi C, Nakayama A, Yoshida M, Ito M, Mori N, Hashizume Y. MELAS with the mitochondrial DNA 3243 point mutation: a neuropathological study. Acta Neuropathol. 2000;99(1):31–8.
40. Lax NZ, Grady J, Laude A, Chan F, Hepplewhite PD, Gorman G, et al. Extensive respiratory chain defects in inhibitory interneurons in patients with mitochondrial disease. Neuropathol Appl Neurobiol. 2015. doi:10.1111/nan.12238 [Epub ahead of print].
41. Mancuso M, Orsucci D, Angelini C, Bertini E, Carelli V, Comi GP, et al. Phenotypic heterogeneity of the 8344A> G mtDNA "MERRF" mutation. Neurology. 2013;80(22):2049–54.
42. Silvestri G, Moraes CT, Shanske S, Oh SJ, DiMauro S. A new mtDNA mutation in the tRNA(Lys) gene associated with myoclonic epilepsy and ragged-red fibers (MERRF). Am J Hum Genet. 1992;51(6):1213–7.
43. Tanji K, Kunimatsu T, Vu TH, Bonilla E. Neuropathological features of mitochondrial disorders. Semin Cell Dev Biol. 2001;12(6):429–39. See comment in PubMed Commons below.
44. Sparaco M, Schon EA, DiMauro S, Bonilla E. Myoclonic epilepsy with ragged-red fibers (MERRF): an immunohistochemical study of the brain. Brain Pathol. 1995;5(2):125–33.
45. Grünewald A, Lax NZ, Rocha MC, Reeve AK, Hepplewhite PD,A, Rygiel K, et al. Quantitative quadruple-label immunofluorescence of mitochondrial and cytoplasmic proteins in single neurons from human midbrain tissue. J Neurosci Methods. 2014;232:143–9.
46. Zhou L, Chomyn A, Attardi G, Miller CA. Myoclonic epilepsy and ragged red fibers (MERRF) syndrome: selective vulnerability of CNS neurons does not correlate with the level of mitochondrial tRNAlys mutation in individual neuronal isolates. J Neurosci. 1997;17(20):7746–53.
47. Man PYW, Turnbull DM, Chinnery PF. Leber hereditary optic neuropathy. J Med Genet. 2002;39(3):162–9.
48. Mackey DA, Oostra RJ, Rosenberg T, Nikoskelainen E, Bronte-Stewart J, Poulton J, et al. Primary pathogenic mtDNA mutations in multigeneration pedigrees with Leber hereditary optic neuropathy. Am J Hum Genet. 1996;59(2):481–5.
49. Milesi J, Rocca M, Bianchi-Marzoli S, Petrolini M, Pagani E, Falini A, et al. Patterns of white matter diffusivity abnormalities in Leber's hereditary optic neuropathy: a tract-based spatial statistics study. J Neurol. 2012;259(9):1801–7.
50. Jaros E, Mahad DJ, Hudson G, Birchall D, Sawcer SJ, Griffiths PG, et al. Primary spinal cord neurodegeneration in leber hereditary optic neuropathy. Neurology. 2007;69(2):214–6.
51. Sadun AA, Win PH, Ross-Cisneros FN, Walker SO, Carelli V. Leber's hereditary optic neuropathy differentially affects smaller axons in the optic nerve. Trans Am Ophthalmol Soc. 2000;98:223.
52. Hudson G, Chinnery PF. Mitochondrial DNA polymerase-gamma and human disease. Hum Mol Genet. 2006;15(Spec No 2):R244–52.
53. Lamantea E, Tiranti V, Bordoni A, Toscano A, Bono F, Servidei S, et al. Mutations of mitochondrial DNA polymerase gammaA are a frequent cause of autosomal dominant or recessive progressive external ophthalmoplegia. Ann Neurol. 2002;52(2):211–9.

54. Van Goethem G, Dermaut B, Lofgren A, Martin JJ, Van Broeckhoven C. Mutation of POLG is associated with progressive external ophthalmoplegia characterized by mtDNA deletions. Nat Genet. 2001;28(3):211–2.
55. Van Goethem G, Schwartz M, Lofgren A, Dermaut B, Van Broeckhoven C, Vissing J. Novel POLG mutations in progressive external ophthalmoplegia mimicking mitochondrial neurogastrointestinal encephalomyopathy. Eur J Hum Genet. 2003;11(7):547–9.
56. Van Goethem G, Martin JJ, Dermaut B, Lofgren A, Wibail A, Ververken D, et al. Recessive POLG mutations presenting with sensory and ataxic neuropathy in compound heterozygote patients with progressive external ophthalmoplegia. Neuromuscul Disord. 2003;13(2):133–42.
57. Hakonen AH, Goffart S, Marjavaara S, Paetau A, Cooper H, Mattila K, et al. Infantile-onset spinocerebellar ataxia and mitochondrial recessive ataxia syndrome are associated with neuronal complex I defect and mtDNA depletion. Hum Mol Genet. 2008;17(23):3822–35.
58. Davidzon G, Greene P, Mancuso M, Klos KJ, Ahlskog JE, Hirano M, et al. Early-onset familial parkinsonism due to POLG mutations. Ann Neurol. 2006;59(5):859–62.
59. Huttenlocher PR, Solitare GB, Adams G. Infantile diffuse cerebral degeneration with hepatic cirrhosis. Arch Neurol. 1976;33(3):186–92.
60. Cottrell DA, Ince PG, Blakely EL, Johnson MA, Chinnery PF, Hanna M, et al. Neuropathological and histochemical changes in a multiple mitochondrial DNA deletion disorder. J Neuropathol Exp Neurol. 2000;59(7):621–7.
61. Lax NZ, Whittaker RG, Hepplewhite PD, Reeve AK, Blakely EL, Jaros E, et al. Sensory neuronopathy in patients harbouring recessive polymerase gamma mutations. Brain. 2012;135(Pt 1):62–71.
62. Betts-Henderson J, Jaros E, Krishnan KJ, Perry RH, Reeve AK, Schaefer AM, et al. Alpha-synuclein pathology and Parkinsonism associated with POLG1 mutations and multiple mitochondrial DNA deletions. Neuropathol Appl Neurobiol. 2009;35(1):120–4.
63. Tzoulis C, Tran GT, Coxhead J, Bertelsen B, Lilleng PK, Balafkan N, et al. Molecular pathogenesis of polymerase gamma-related neurodegeneration. Ann Neurol. 2014;76(1):66–81.
64. Sofou K, De Coo IF, Isohanni P, Ostergaard E, Naess K, De Meirleir L, et al. A multicenter study on Leigh syndrome: disease course and predictors of survival. Orphanet J Rare Dis. 2014;9:52.
65. Finsterer J. Leigh and Leigh-like syndrome in children and adults. Pediatr Neurol. 2008;39(4):223–35.
66. McKelvie P, Infeld B, Marotta R, Chin J, Thorburn D, Collins S. Late-adult onset Leigh syndrome. J Clin Neurosci. 2012;19(2):195–202.
67. Wick R, Scott G, Byard RW. Mechanisms of unexpected death and autopsy findings in Leigh syndrome (subacute necrotising encephalomyelopathy). J Forensic Legal Med. 2007;14(1):42–5.
68. Saneto RP, Cohen BH, Copeland WC, Naviaux RK. Alpers-Huttenlocher syndrome. Pediatr Neurol. 2013;48(3):167–78.
69. Harding BN. Review article: progressive neuronal degeneration of childhood with liver disease (Alpers-Huttenlocher syndrome): a personal review. J Child Neurol. 1990;5(4):273–87.
70. Naviaux RK, Nguyen KV. POLG mutations associated with Alpers' syndrome and mitochondrial DNA depletion. Ann Neurol. 2004;55(5):706–12.
71. Alpers BJ. Diffuse progressive degeneration of the gray matter of the cerebrum. Arch Neurol Psychiatry. 1931;25(3):469–505.
72. Khan A, Trevenen C, Wei X-C, Sarnat HB, Payne E, Kirton A. Alpers syndrome: the natural history of a case highlighting neuroimaging, neuropathology, and fat metabolism. J Child Neurol. 2012;27(5):636–40.
73. Sheng ZH, Cai Q. Mitochondrial transport in neurons: impact on synaptic homeostasis and neurodegeneration. Nat Rev Neurosci. 2012;13(2):77–93.
74. Kirichok Y, Krapivinsky G, Clapham DE. The mitochondrial calcium uniporter is a highly selective ion channel. Nature. 2004;427(6972):360–4.
75. MacAskill AF, Rinholm JE, Twelvetrees AE, Arancibia-Carcamo IL, Muir J, Fransson A, et al. Miro1 is a calcium sensor for glutamate receptor-dependent localization of mitochondria at synapses. Neuron. 2009;61(4):541–55.

76. Yi M, Weaver D, Hajnóczky G. Control of mitochondrial motility and distribution by the calcium signal. J Cell Biol. 2004;167(4):661–72.
77. Moudy AM, Handran SD, Goldberg MP, Ruffin N, Karl I, Kranz-Eble P, et al. Abnormal calcium homeostasis and mitochondrial polarization in a human encephalomyopathy. Proc Natl Acad Sci U S A. 1995;92(3):729–33.
78. Brini M, Pinton P, King MP, Davidson M, Schon EA, Rizzuto R. A calcium signaling defect in the pathogenesis of a mitochondrial DNA inherited oxidative phosphorylation deficiency. Nat Med. 1999;5(8):951–4.
79. Trevelyan AJ, Kirby DM, Smulders-Srinivasan TK, Nooteboom M, Acin-Perez R, Enriquez JA, et al. Mitochondrial DNA mutations affect calcium handling in differentiated neurons. Brain. 2010;133(Pt 3):787–96.
80. Dong X-x, Wang Y, Z-h Q. Molecular mechanisms of excitotoxicity and their relevance to pathogenesis of neurodegenerative diseases. Acta Pharmacol Sin. 2009;30(4):379–87.
81. Mirabella M, Di Giovanni S, Silvestri G, Tonali P, Servidei S. Apoptosis in mitochondrial encephalomyopathies with mitochondrial DNA mutations: a potential pathogenic mechanism. Brain. 2000;123(Pt 1):93–104.
82. Ikezoe K, Nakagawa M, Yan C, Kira J-I, Goto Y-I, Nonaka I. Apoptosis is suspended in muscle of mitochondrial encephalomyopathies. Acta Neuropathol. 2002;103(6):531–40.
83. Kariya S, Hirano M, Furiya Y, Ueno S. Effect of humanin on decreased ATP levels of human lymphocytes harboring A3243G mutant mitochondrial DNA. Neuropeptides. 2005;39(2):97–101.
84. Chen Q, Vazquez EJ, Moghaddas S, Hoppel CL. Production of reactive oxygen species by mitochondria central role of complex III. J Biol Chem. 2003;278(38):36027–31.
85. Loeb LA, Wallace DC, Martin GM. The mitochondrial theory of aging and its relationship to reactive oxygen species damage and somatic mtDNA mutations. Proc Natl Acad Sci U S A. 2005;102(52):18769–70.
86. Lin CS, Sharpley MS, Fan W, Waymire KG, Sadun AA, Carelli V, et al. Mouse mtDNA mutant model of Leber hereditary optic neuropathy. Proc Natl Acad Sci U S A. 2012;109(49):20065–70.
87. Valsecchi F, Grefte S, Roestenberg P, Joosten-Wagenaars J, Smeitink JAM, Willems PHGM, et al. Primary fibroblasts of NDUFS4–/– mice display increased ROS levels and aberrant mitochondrial morphology. Mitochondrion. 2013;13(5):436–43.
88. Burelle Y, Bemeur C, Rivard M-E, Thompson Legault J, Boucher G, Morin C, et al. Mitochondrial vulnerability and increased susceptibility to nutrient-induced cytotoxicity in fibroblasts from Leigh syndrome French Canadian patients. PLoS ONE. 2015;10(4):e0120767.
89. Trifunovic A, Hansson A, Wredenberg A, Rovio AT, Dufour E, Khvorostov I, et al. Somatic mtDNA mutations cause aging phenotypes without affecting reactive oxygen species production. Proc Natl Acad Sci U S A. 2005;102(50):17993–8.
90. Kirby D, Smulders-Srinivasan T, Rennie K, Acin-Perez R, Whittington M, Enriquez J, et al. Transmitochondrial embryonic stem cells containing pathogenic mtDNA mutations are compromised in neuronal differentiation. Cell Prolif. 2009;42(4):413–24.
91. Abramov AY, Smulders-Srinivasan TK, Kirby DM, Acin-Perez R, Enriquez JA, Lightowlers RN, et al. Mechanism of neurodegeneration of neurons with mitochondrial DNA mutations. Brain. 2010;133(Pt 3):797–807.
92. Kodaira M, Hatakeyama H, Yuasa S, Seki T, Egashira T, Tohyama S, et al. Impaired respiratory function in MELAS-induced pluripotent stem cells with high heteroplasmy levels. FEBS Open Biol. 2015;5:219–25.
93. Hamalainen RH, Manninen T, Koivumaki H, Kislin M, Otonkoski T, Suomalainen A. Tissue- and cell-type-specific manifestations of heteroplasmic mtDNA 3243A> G mutation in human induced pluripotent stem cell-derived disease model. Proc Natl Acad Sci U S A. 2013;110(38):E3622–30.
94. Ma H, Folmes CD, Wu J, Morey R, Mora-Castilla S, Ocampo A, et al. Metabolic rescue in pluripotent cells from patients with mtDNA disease. Nature. 2015;524(7564):234–8.

95. Quintana A, Kruse SE, Kapur RP, Sanz E, Palmiter RD. Complex I deficiency due to loss of Ndufs4 in the brain results in progressive encephalopathy resembling Leigh syndrome. Proc Natl Acad Sci U S A. 2010;107(24):10996–1001.
96. Diaz F, Garcia S, Padgett KR, Moraes CT. A defect in the mitochondrial complex III, but not complex IV, triggers early ROS-dependent damage in defined brain regions. Hum Mol Genet. 2012;21(23):5066–77.
97. Fukui H, Diaz F, Garcia S, Moraes CT. Cytochrome c oxidase deficiency in neurons decreases both oxidative stress and amyloid formation in a mouse model of Alzheimer's disease. Proc Natl Acad Sci U S A. 2007;104(35):14163–8.
98. Sorensen L, Ekstrand M, Silva JP, Lindqvist E, Xu B, Rustin P, et al. Late-onset corticohippocampal neurodepletion attributable to catastrophic failure of oxidative phosphorylation in MILON mice. J Neurosci. 2001;21(20):8082–90.
99. Ekstrand MI, Terzioglu M, Galter D, Zhu S, Hofstetter C, Lindqvist E, et al. Progressive parkinsonism in mice with respiratory-chain-deficient dopamine neurons. Proc Natl Acad Sci U S A. 2007;104(4):1325–30.
100. Good CH, Hoffman AF, Hoffer BJ, Chefer VI, Shippenberg TS, Backman CM, et al. Impaired nigrostriatal function precedes behavioral deficits in a genetic mitochondrial model of Parkinson's disease. FASEB J. 2011;25(4):1333–44.
101. Bender A, Krishnan KJ, Morris CM, Taylor GA, Reeve AK, Perry RH, et al. High levels of mitochondrial DNA deletions in substantia nigra neurons in aging and Parkinson disease. Nat Genet. 2006;38(5):515–7.
102. Campbell G, Ziabreva I, Reeve AK, Krishnan KJ, Reynolds R, Howell O et al. Mitochondrial DNA deletions and neurodegeneration in multiple sclerosis. Ann Neurol. 2011;69(3): 481–92.
103. Horton TM, Graham BH, Corral-Debrinski M, Shoffner JM, Kaufman AE, Beal MF, et al. Marked increase in mitochondrial DNA deletion levels in the cerebral cortex of Huntington's disease patients. Neurology. 1995;45(10):1879–83.
104. Lin MT, Simon DK, Ahn CH, Kim LM, Beal MF. High aggregate burden of somatic mtDNA point mutations in aging and Alzheimer's disease brain. Hum Mol Genet. 2002;11(2):133–45.

Chapter 3
The Ageing Brain, Mitochondria and Neurodegeneration

Gavin Hudson

Abstract The brain is a complex and energy-demanding organ, which like any organ is subject to the ravages of time. Mitochondria are synonymous with their role in energy production, which is particularly critical to high-energy-demanding cells such as neurons.

Here we discuss ageing of the brain, initially setting the scene by introducing the core concepts associated with brain ageing; discussing the physiological, genetic and cognitive changes which occur over time; and subsequently introducing the roles that mitochondria play in the 'normal' brain ageing process.

The final section of the chapter discusses the role of both inherited and somatic mitochondrial DNA variation in neurodegeneration, initially in the context of primary mitochondrial disorders (such as Leber's hereditary optic neuropathy, myoclonic epilepsy and ragged-red fibres and mitochondrial encephalomyopathy lactic acidosis and stroke-like episodes) and subsequently in the context of common, but more complex, neurodegenerative diseases such as Parkinson's disease, Alzheimer's disease, Huntington's disease, amyotrophic lateral sclerosis, Friedreich's ataxia, hereditary spastic paraplegia and multiple sclerosis.

Keywords Brain ageing • Mitochondrial DNA (mtDNA) • mtDNA mutation • Neurodegeneration • Ageing

3.1 The 'Normal' Ageing Brain

Like any organ, the human brain is susceptible to the endless march of time with the effects of ageing often manifesting as diverse pathologies. To understand the role of mitochondrial DNA (mtDNA) variation in brain ageing, we must first understand what brain ageing entails.

G. Hudson, BSc, PhD
Institute of Genetic Medicine, International Centre for Life, Newcastle University, Newcastle upon Tyne, UK
e-mail: gavin.hudson@ncl.ac.uk

A.K. Reeve et al. (eds.), *Mitochondrial Dysfunction in Neurodegenerative Disorders*, DOI 10.1007/978-3-319-28637-2_3

3.1.1 Physiological Changes

At the organ level, brain weight and volume decline at a rate of around 5 % per decade after 40 years of age [1], with rates increasing in individuals over 70 years [2]. Volume loss is associated with an increase in ventricular volume and cerebrospinal fluid (CSF) spaces. Longitudinal magnetic resonance imaging (MRI) and comparative cross-sectional studies have shown that age-related brain changes are regional [2, 3] with the hippocampus and frontal lobes most affected. Estimates indicate that, over an age range of 30–90 years, the cerebral cortex, hippocampus and cerebral white matter are subjected to volumetric losses of 14 %, 35 % and 26 %, respectively [4, 5]. The occipital cortex appears least affected.

The exact mechanism of reduction is up for debate, with studies indicating that grey matter shrinkage [6] and changing neuronal morphology [7], rather than the dogmatic theory of neuronal loss, are the principle components contributing to decreasing brain volume. Additional ageing hallmarks include dendritic sprouting [8, 9], a compensatory mechanism to maintain synapse number and cell number [10], a loss of dendritic arbour [11] and a decline in white matter [12].

The effects of 'normal' ageing are also seen at the cellular level. Lipofuscin, the pigmented residues of lysosomal digestion, accumulates in neurons; it is a hallmark of ageing and characteristic of a failure to eliminate the products of peroxidation-induced cell damage [13]. Neurofibrillary tangles and plaques, hallmarks of Alzheimer's disease (AD), are identifiable in hippocampal, amygdala and entorhinal cortex neurons in healthy aged brains [14] although at subclinical levels. Amyloid β-peptide (Aβ)-containing plaques, another hallmark of AD, form in the grey matter of 'normal' aged brains, although present only in small numbers compared to the diseased brain. Advanced age increases the frequency of vacuoles, immunoreactive for neurofibrillary tangles (155K and 210K neurofilament polypeptides) [15], in the cell bodies of pyramidal and hippocampal cells leading to granulovacuolar degeneration [16]. Large (30 μm long) rod-shaped, paracrystalline structures, termed Hirano bodies, accumulate in hippocampal pyramidal cells [17]. Similar to granulovacuoles, these bodies appear to be composed of cytoskeletal proteins (including actin) [18] as well as the constituents of neurofibrillary tangles [19]. As we age, a reduction of oxygen through a reduction in blood supply causes cerebral microinfarcts (cellular necrosis). Typically associated with dementia, these microinfarcts, difficult to detect by conventional means (MRI), are surprisingly common in aged brains (typically >80 years) [20].

3.1.2 Genetic and Transcriptomic Changes

Genetic factors, such as apolipoprotein E haplotype (*APOE*), have been identified largely through the investigation of pathological ageing such as in AD [21]. However, these studies are mixed, with some indicating age-related defects in elderly ε4-haplotype carriers [22] and others showing no significant effect [23]. Studies in mice indicate that genetic loci (Hipp1a, Ch1 and Hipp5a, Ch5) can modulate hippocampal structure and volume during normal ageing (potentially mediated through the genes: retinoid X receptor gamma, *Rxrg*, and fibroblast growth factor receptor 3, *Fgfr3*, respectively) supporting the role of genetics in brain ageing [24].

At the transcriptomic level, work in mice indicates that the gene expression landscape of the brain, particularly genes involved in stress and inflammatory response, changes dramatically over time and can be modulated by dietary restriction [25]. This supports earlier work investigating the role of ageing in mice at the protein level [26]. In humans, a reduction in synaptic plasticity, modulated by a reduction in α-amino-3-hydroxy-5-methyl-4-isoxazolepropionic acid (*AMPA*) receptor expression, is a hallmark of ageing [27]. In addition reductions of key brain transcripts, including N-methyl-D-aspartate (NMDA) receptors (required for learning and memory) [28], GABA-A receptor subunits (required for neurotransmission) [29] and genes involved in long-term potentiation (e.g. *CALM1*, calmodulin 1, required for calcium signalling) [30], begin at around 40 years of age.

3.1.3 Cognitive Changes

Cognitive performance declines with age, although the differentiation between 'normal ageing,' 'non-pathological ageing' and neurodegenerative disease confounds quantitative measurement. The most observable phenotype associated with 'normal' brain ageing is a loss of memory, with no significant loss of visuospatial, cognitive function or language [31]. Memory is classified as four distinct classes: episodic (memory of situation, e.g. a memory of a birthday party), semantic (memory of facts, e.g. Washington, DC, is the capital of the USA), procedural (memory of motor skills, e.g. how to walk) and working (transient memory, for new and stored information) [32]. Studies have investigated the effects of ageing on memory using neuropsychological testing and neuroimaging, demonstrating that brain ageing has a profound effect on both episodic and semantic memory.

Episodic memory, typically affected in AD, is thought to decline from middle age onwards, affecting both recall and recognition in elderly life [33]. Semantic memory also decreases from middle age and then declines rapidly during elderly life [33], driven in part by slower cognitive reaction times, lower attention levels and slower processing speeds in aged individuals [6, 34]. In addition to memory, ageing affects an individual's orientation and attention. Orientation, or the awareness of one's surroundings, has been shown to decline mildly with advancing age [35], although this is inconsistent between studies, possibly reflecting the interaction of other variables [36]. Attention, or our ability to focus our brain's resources on a task, declines with age; moreover secondary phenotypes associated with advancing age [37], i.e. impaired hearing or vision, compound valid attention testing. Indeed, it is worth noting that differentiating these separate cognitive functions, which are intrinsically linked, makes data interpretation difficult [38],

3.1.4 *The Drivers of Brain Ageing*

Dopamine, a neurotransmitter involved in motor control, arousal, cognition and reward, declines from early adulthood at a rate of around 10 % per decade, subsequently affecting both cognitive and motor function in aged individuals [33]. Serotonin (a neurotransmitter) and brain-derived neurotrophic factor (BNDF, a neuronal growth factor) are both expressed to maintain synaptic plasticity and both decline with age. We also know that sex hormones, particularly oestrogen, can affect dopaminergic efficacy – purportedly relaying a protective effect in neurodegenerative disorders such as AD [39]. As we age, cerebrovascular efficiency falls, which in turn impairs our ability to metabolise glucose [33], although subsequent data indicates that impaired glucose metabolism may be an effect of cellular atrophy, rather than a cause [40]. Finally, it is becoming clearer that mitochondria play an integral role in brain ageing, from modulating calcium influx and homeostasis [41] to direct dysfunction and the increased proliferation of reactive oxygen species (ROS) [42].

3.2 Mitochondrial Function in the 'Normal' Ageing Brain

Even in the absence of a primary pathogenic defect, changes in mitochondrial function are intrinsically linked to 'normal' brain ageing, attributed principally to increased mitochondrial dysfunction, a direct loss of cellular energy or an increase in ROS production, which increases mitochondrial decay [43]. It is worth noting that, from a bioenergetic perspective, the brain is an extremely heterogeneous and dynamic organ; for example, there is a reciprocal relationship between neurons, which have aerobic terminals, and astrocytes, which are more dependent on glycolysis [44]. This complexity, which makes energetic interpretations difficult, must be considered when investigating mitochondrial function in the brain.

Aged animal brains show a significant decline in mitochondrial function, characterised as a loss of phosphorylation capacity, decreased mitochondrial membrane potential [45], decreased respiration, ATP synthesis and activation of the mitochondrial permeability transition pore. The aged brain shows a characteristic reduction in two key respiratory chain (RC) enzymes: NADH/ubiquinone oxidoreductase (complex I, CI) and cytochrome *c* oxidase (complex IV, CIV) [46]; however, the extent and localisation of the RC deficit are variable [47]. This reduction in RC activity appears to trigger a corresponding proliferation of mitochondria with age, presumably an attempt to attenuate a loss of cellular energy [48] and neuronal mtDNA [49].

The brain, a high-energy organ, is particularly vulnerable to oxidative damage, in part due to a lack of antioxidant enzymes and the increased abundance of polyunsaturated fats, transition metals and the high rate of oxygen consumption [50]. Markers of oxidative stress, carbonylation (protein oxidation), lipid oxidation and the oxidation of the mitochondrial genome [51] are increased with age; however, whether increased oxidative stress markers are a result of a loss of antioxidant enzyme activity or an increase in pro-stress factors is contentious [52]. The coexistence of mitochondrial dysfunction and increased oxidative stress in the brain likely acts as a trigger for apoptotic activation, yet the role of apoptosis in the 'normal' ageing brain is controversial, with studies indicating that mitochondrially mediated apoptosis is not a cardinal element of 'normal' brain ageing [53] and is rather restricted to disease [54].

3.2.1 Mitochondrial DNA in the 'Normal' Ageing Brain

The risk of mtDNA mutation is high, primarily due to its localisation near to the site of oxidative stress [55], and is not restricted by tissue type or disease [56]. The result is a spectrum of mtDNA variation, ranging from the 'inherited' population-level polymorphisms which classify mtDNA haplogroups [57] to the acquired or somatic variation that accumulates with age (either large-scale deletions or single base-pair substitutions).

Despite strong links to mitochondrial functionality [58, 59], there is little evidence for the role of inherited mtDNA variation in 'normal' brain ageing. However, there are several reports linking inherited variation to the progression of neurodegenerative disorders such as Parkinson's and Alzheimer's disease (discussed later).

Conversely, there is mounting evidence indicating that acquired mtDNA variation accumulates in the brain during 'normal' ageing and much work has focused on the aptly named 'common deletion' (a 5 Kb deletion with known disease associations [60]), with reports showing that age-related oxidative damage increases the levels of the common deletion in several regions of the 'healthy' brain, particularly in the cortex, cerebellum, putamen [61, 62] and more recently substantia nigra and basal ganglia [60], and appears to correlate with regional differences in energy demand [63]. In addition to the common deletion, several studies have reported an increase in random and variable mtDNA deletions, typically smaller in size (>50 bp), which can be differentially distributed amongst brain regions [62, 64]. Studies of acquired single nucleotide variation in the brain are sparse; however studies have shown that healthy individuals harbour an increase in the frequency of low-level heteroplasmic variation in cytochrome *c* oxidase subunit I (*MTCO1*), the core catalytic subunit of complex IV of the RC [65] and the non-coding, regulatory, D-loop region [66].

3.3 Mitochondrial DNA Mutation and Neurodegeneration

As discussed, mitochondria are the cellular powerhouses, supplying critical ATP through oxidative phosphorylation. Secondary roles in apoptosis, responses to cellular stress and modulation of reactive oxygen species make mitochondrial function a central component of almost all cellular survival (see Chap. 1). The central nervous system (CNS), a highly energy-demanding system, is dependent on efficient mitochondrial function. A disruption of this function, through mitochondrial DNA mutation, is in many cases associated with neurodegenerative disease (Fig. 3.1).

3.3.1 Mitochondrial DNA Mutations

Mitochondrial DNA, a circular 16.5 Kb molecule encoding the 13 core RC polypeptides, has a significantly higher mutation rate compared to its nuclear counterpart (around 3× or 2.7×10^{-5} base pairs per generation) [67], due in part to an absence of histone and its proximity to the inner mitochondrial membrane – the site of ROS production. However, mitochondria do contain antioxidant and DNA repair enzymes, including oxoguanine glycosylase (*OGG1*, responsible for 8-oxoguanine base excision) and (*MUTYH*, involved in oxidative DNA damage repair);

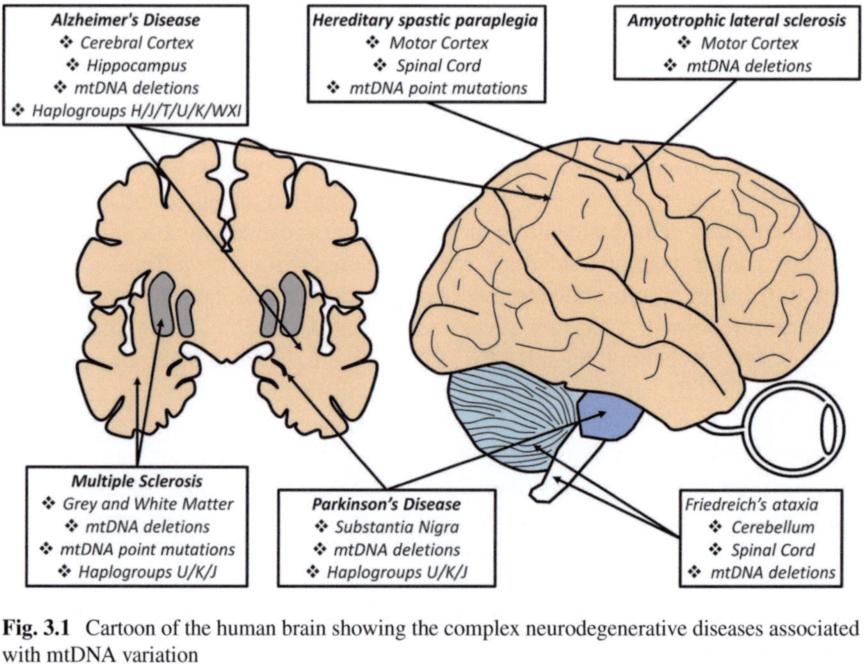

Fig. 3.1 Cartoon of the human brain showing the complex neurodegenerative diseases associated with mtDNA variation

furthermore mitochondrial gene expression is adaptive and isolated from the nucleus, allowing the post-translational machinery to generate individual gene products in response to ROS through retrograde signalling [68]. This retrograde signalling stimulates an adaptive nuclear response to mtDNA impairment, which modulates the expression of over 40 nuclear genes [69].

This high mutation rate and lack of effective mtDNA repair mechanisms lead to the formation of two classes of mtDNA variants: (1) inherited mtDNA variants, typically single base-pair exchanges that are present at varying population frequencies, and (2) acquired or somatic mtDNA variants, which can be either single base-pair exchanges or deletions of mtDNA.

3.3.2 Primary Mitochondrial Disorders and Neurodegeneration

Since the 1980s mitochondrial dysfunction has been a hallmark of metabolic disease [70] and subsequently over 300 mtDNA mutations have been associated with disease [71]. More recently, the phenotypic spectrum has broadened to include neurodegenerative disease [72], typically as a function of a primary mtDNA defect.

Leber's hereditary optic neuropathy (LHON) is the commonest cause of maternally inherited blindness and a cardinal example of how mtDNA mutations cause neurodegeneration [73]. MtDNA mutations (m.3460G>A, m.11778G>A and m.14484T>C in ~90 % of cases) in NADH-dehydrogenase subunits (complex I) reduce cellular energy and increase oxidative stress and ROS production to cause a selective loss of retinal ganglion cells (RGC) in the retina whilst sparing the retinal pigmented epithelium and photoreceptor layer [74]. Atrophy of neuronal cell bodies and demyelination of the optic nerve are associated with impaired activity of the EAAT1 (excitatory amino acid transporter) glutamate transporter and increased mtDNA mutation-mediated ROS production triggering the apoptotic cascade [75]. RGC selectivity is attributed to high energy demand; however, atypical individuals may experience secondary neuronal loss.

Myoclonic epilepsy with ragged-red fibres (MERRF [76]) is typically attributed to m.8344A>G in the mitochondrial tRNA lysine gene (*MT-TK*) and presents as a progressive degeneration of the olivocerebellar pathway [77, 78], with severe neuronal loss in the inferior olivary complex, Purkinje cells and dentate nucleus. Biochemical studies indicate that m.8344A>G significantly impairs mitochondrial protein synthesis, hampering oxidative phosphorylation, decreasing ATP production and mitochondrial membrane potential and finally culminating in apoptotic cell death [79].

In addition, mtDNA mutations are associated with cerebellar or sensory ataxia [80], where a secondary loss of Purkinje or other cerebellar cells is seen in syndromes typically associated with a primary mtDNA defect, including Kearns-Sayre syndrome (KSS [81]), mitochondrial encephalopathy and stroke-like episodes (MELAS [82]) and neuropathy, ataxia and retinitis pigmentosa (NARP [83]).

3.3.3 Mitochondrial DNA Mutation in Complex Neurodegeneration

There is increasing evidence that an increase in oxidative stress and apoptosis is linked to the aetiopathogenesis of several age-related neurodegenerative diseases. Much research has focused on the role of mitochondrial impairment and oxidative stress in the onset and progression of the major ageing brain disorders, particularly Parkinson's disease (PD) and AD; however, research has also linked mitochondrial mutations to Huntington's disease (HD) and amyotrophic lateral sclerosis (ALS) (see Fig. 3.1).

3.3.3.1 Parkinson's Disease

Parkinson's disease is a degenerative disorder of the central nervous system, primarily affecting the motor system through a selective loss of dopaminergic neurons in the substantia nigra pars compacta. Early disease is typically movement related, manifesting as tremor, rigidity, slowness of movement initiation and gait instability, progressing to cognitive impairment in advanced stages.

The link between mitochondrial dysfunction and postmortem PD tissue is well recognised [84]. The substantia nigra pars compacta (SNpc), the most vulnerable tissue in PD patients, suffers an age-related reduction of complex I activity [85] and exposure to potent inhibitors of respiratory chain activity such as rotenone [86], 1-methyl-4-phenyl-1,2,3,6-tetrahydropyridine (MPTP) [87] and certain pesticides [88] manifest with neurological abnormalities similar to PD. The combined reduction in cellular energy availability and an increase in ROS may lower the threshold for neuronal apoptosis killing the energy-demanding dopaminergic neurons in the SNpc.

Despite a consistent link between mitochondrial proteomics and PD, the underlying genetic mechanisms remain complex and conflicting, due in part to the complex interactions of both inherited and acquired mtDNA mutations.

The role of inherited mtDNA variation, mediated primarily through mtDNA haplogroups, in the aetiopathology of PD is well studied, although the results are often conflicting and contentious.

Evolutionarily, inherited mtDNA mutations have created stable population subgroups termed 'mitochondrial haplogroups,' groups of phylogenetically similar mtDNA molecules, separated by a collection of stable and frequent polymorphic variants [57]. Many of these subdivisions occurred over 50,000 years ago, developing as humans migrated into new geographical regions. Over 95 % of Europeans belong to one of ten major haplogroups, H, J, T, U, K (a subgroup of U), M, I, V, W and X, with each haplogroup defined by specific mtDNA sequence variants within the population. The role of inherited mtDNA variants and their defined haplogroups has evolved; once thought to be benign population variants, mtDNA haplogroups are now known to directly affect mitochondrial function

[58, 89] and have become associated with a broad spectrum of human disease [90–92]. In addition, there is evidence to suggest that inherited mtDNA may influence neuronal cytosolic pH and calcium regulation, which over time could influence neuronal function and viability [93].

In early work, the common European haplogroup J, driven largely by the presence of a single nucleotide substitution in MT-ND3 (m.10398G>A), was shown to be underrepresented in PD, suggestive of a protective effect, and is interesting given the purported effects of haplogroup J on 'normal' ageing, where it is overrepresented in healthy centenarians [94]. This is supported by a moderately powered study which found that a phylogenetically linked cluster containing haplogroup J (JTUK) reduced PD risk [95]. Smaller studies have reported a contradictory increase in PD risk with haplogroup J [96], but sample number and genetic heterogeneity, critically misunderstood in the highly mutable mtDNA [97], may account for this disparity. More recently, a much larger study and meta-analysis confirmed the role of haplogroup J/K in reducing the risk of PD [98], an effect which was replicated in a follow-up study later [99]. A meta-analysis of all available mtDNA-PD data was able to identify that the phylogenetically linked H/V haplogroups increased the risk of PD [99].

It is intriguing that ostensibly common, and once thought benign, variation in the mitochondrial genome could convey a direct effect on the pathoaetiology of PD. More likely, haplogroup J/K/H/V are genetic tags of further mtDNA variation not directly investigated in these studies. For example, haplogroup K contains the lowest frequency of nonsynonymous complex I gene variants (when compared to the other haplogroups), raising the possibility that natural selection against genetic variation of complex I reduces the risk of PD in haplogroup K individuals. Similarly, an association with haplogroup H, which is linked to infection survival, raises the possibility that natural selection has led to the emergence of variants which predispose to age-related neurodegenerative disease through antagonistic pleiotropy. Similar to HD, where an increased CAG(n) repeat length in the Huntingtin gene causes a neurodegenerative disorder, but is associated with a lower incidence of cancer, recent mutations in mtDNA may be tolerated in humans because they increase the chance of surviving early-life insults such as sepsis, but are pathogenic in later life through the increased generation of reactive oxygen species. In both of these instances, advances in genotype and sequencing technology will allow researchers to understand the role of inherited mtDNA variants, by enabling more and more sequences to be collected and analysed.

More speculative is the exact role of acquired or age-related somatic, mtDNA variants. Much early work focused on the role of mtDNA deletions, large and small lesions in the genome. These deletions typically remove large proportions of the heavily coded mtDNA molecule, dramatically affecting mitochondrial protein synthesis and replication, which due to their obvious pathogenic nature appear heteroplasmic (a mixed population of mtDNA molecules within a cell or tissue).

Early studies failed to find a direct association between mtDNA deletion levels in the brains of PD patients, showing no significant increase when compared to age-matched healthy control tissue [100–102]. Advances in mtDNA isolation technology, enabling capture from single cells, and improvements in the PD diagnostic efficacy improved deletion detection. Studies taking advantage of these technologies show that the frequency of mtDNA deletions in individual neurons in the SNpc of PD patients clonally accumulates with age, increasing after 65 years old [103], and is associated with a decrease in cytochrome *c* oxidase activity [104]. Both of these reports suggest that the SNpc is particularly vulnerable to free radical-mediated mtDNA damage and that mtDNA deletions may play a role in the pathoaetiology of PD.

The identification of mtDNA deletions in non-PD-related disease [105] and healthy individuals [106] questions the exact contribution of these lesions to the development of PD. In addition, the observation of mtDNA deletions in newborn brains [107] raises the question of whether the clonal expansion mtDNA deletions are the cause of, or are driven by, SNpc failure.

Studies of acquired mtDNA point mutations, typically heteroplasmic, in PD are extremely limited and are often focused on idiopathic cases with no evidence of mitochondrial dysfunction. Studies have identified heteroplasmic variation in PD cases [108, 109]; however, without supporting biochemical or functional data, these variants and their role in the development of PD are difficult to interpret.

3.3.3.2 Alzheimer's Disease

Alzheimer's disease is a progressive loss of neurons and synapses in the cerebral cortex and subcortical regions and is the commonest cause of dementia in the elderly [110]. The disease is typically associated with plaques or neurofibrillary tangles (aggregates of hyper-phosphorylated tau protein).

Like PD, there are several lines of evidence linking mitochondrial dysfunction and oxidative damage in the aetiopathogenesis of AD. Animal experiments using mice which overexpress precursor amyloid protein (APP) demonstrated mitochondrial dysfunction and ATP production through a reduction in mitochondrial protein synthesis [111] or an interaction with mitochondrial matrix proteins [112]. Mitochondrial dysfunction, primarily a reduction in neuronal cell energy, promotes tau phosphorylation, an indication that mitochondrial vitality and oxidative stress are linked to the neurofibrillary tangles seen in AD [113].

Unlike PD, the mitochondrial genetics of AD remain nebulous. Studies have shown that inherited mtDNA variants, again mediated through mtDNA haplogroups, may affect the pathoaetiology of AD; however, like PD these studies are conflicting. Reports show that haplogroups H, U, K, T, I, W and X (seven of the ten typical European haplogroups) associate with an increased risk of developing AD, with effect sizes ranging from OR>2 [114–120]. If correct, the only haplogroups not associated with AD are haplogroups J and V, which is seemingly unlikely, especially given the reciprocal relationship between mtDNA haplogroups.

It is possible that geographical variation could affect these results, with the relative contribution of specific mtDNA variants varying in different ethnic groups; however, this would imply that the overall effect of inherited mtDNA on AD is small, or restricted to very rare pathogenic variants, and would certainly require significantly larger studies to fully investigate. A more recent and comprehensive study supports this hypothesis, investigating the role of inherited mtDNA variants in a large AD cohort concluding that there is no clear role for inherited mtDNA variation in the pathoaetiology of AD [121]. Further, when mtDNA from AD patients is transferred to Rho0 cell lines (cells devoid of native mtDNA), a biochemical defect is observed, suggesting an effect rare inheritable mtDNA abnormalities [122]; however, to date, studies have failed to find an over-representation of inherited mtDNA mutations in AD patients [123].

Despite strong evidence that amyloid beta (Aβ) enters mitochondria and increases ROS production [124], likely increasing mtDNA mutation formation, few studies have estimated the contribution of acquired mtDNA variation in AD. The role of the 'common' mtDNA deletion, a recurring pathogenic ~5 Kb mtDNA deletion, [125] was assessed in AD, but no significant association was found [126]. The same authors did identify a significantly increased acquired mtDNA mutational burden, two to threefold higher than controls, in the parietal gyrus, hippocampus and cerebellum of AD patient brains – consistent with the hypothesis of oxidative-induced cell death seen in AD [49] and supported by a more recent study [65]. Additionally, it has been hypothesised that somatic mutations in the mtDNA control region (displacement or D-loop) impair mitochondrial transcription and translation in AD brains [127].

3.3.3.3 Huntington's Disease

Huntington's disease is a comparatively early-onset neurodegenerative disorder (physical symptoms can begin at any age from infancy to old age, but usually begin between 35 and 44 years of age) and is caused by autosomal dominant repeat expansions in the Huntington gene. Early pathology begins in the striatum but rapidly progresses to other brain regions.

Mitochondrial dysfunction appears to play an important role in the progression of HD, characterised by metabolic deficits and a decrease in core mitochondrial enzymes [128]. In addition, mutant huntingtin protein (Htt) interacts with the mitochondrial membrane, detrimentally affecting calcium metabolism [129], and Htt cytotoxicity triggers mitochondrial transcriptional changes and initiates a cascade of energy failure ultimately leading to cell death [130]. At the mtDNA level, increased oxidative stress resulting from Htt toxicity induces mtDNA lesion formation and a reduction in mtDNA copy number, which in turn triggers the vicious circle of ROS-mediated mtDNA damage [130]; however, this appears more as a consequence rather than a cause of disease progression. HD patients have higher frequencies of mtDNA deletions when compared to age-matched controls; however, much of this is based on measurements in peripheral blood [131, 132], with only one study investigating cortical tissue [133].

Studies of inherited mtDNA variation are conflicted. Early work investigating the role of mtDNA haplogroups in HD concluded that they do not affect disease development or progression [134]; however, this study was relatively small compared to typical genetic association studies. Contrary to this, a more recent study demonstrated that HD patients with a haplogroup H (the commonest European haplogroup, ~40–45 % population) background showed significantly higher ATP concentrations than non-H HD patients [135]. This is puzzling, given the synonymous nature of the haplogroup H defining mtDNA variant (m.7028C, Ala375Ala in *MTCOI*), and likely reflects either a statistical power inadequacy or haplogroup tagging or rare mtDNA variants, which only analysis of whole mtDNA genome sequencing could identify.

3.3.3.4 Amyotrophic Lateral Sclerosis

Amyotrophic lateral sclerosis (ALS, often termed Lou Gehrig's disease in the USA, Charcot disease or motor-neurone disease in the UK) is a typically sporadic disease (~95 % of cases), causing progressive loss of neurons in the motor cortex and spinal cord. Clinical hallmarks of ALS include muscle stiffness and twitching, gradually progressing to atrophy [136].

Although the underlying mechanism is not fully understood, reports link nDNA mutations in superoxide dismutase 1 (*SOD1*), one of three superoxide dismutases responsible for destroying free superoxide radicals, to mitochondrial dysfunction in ALS patients [137]. Mutant SOD1 inhibits voltage-dependent anion selective protein channel 1 (*VDAC1*) [138], reducing energy production and driving an increase in oxidative damage in transgenic mice motor neurons [139]. ALS patients were found to harbour significantly higher levels (>30×) of the mtDNA common deletion, compared to matched controls, in the motor cortex [140], and spinal cord [141]. It is likely, given the reported *SOD1*-mediated reduction in mtDNA repair enzymes in ALS patients [142], that mtDNA lesions and point mutations are proliferating; although this has yet to be fully investigated.

3.3.3.5 Friedreich's Ataxia

Friedreich's ataxia (FA) is the commonest form of hereditary ataxia, pathophysiologically characterised as a reduction of the mitochondrial protein frataxin in the peripheral sensory nerves, dorsal root ganglia, posterior columns, spinocerebellar and corticospinal tracts of the spinal cord, gracile and cuneate nuclei, dorsal nuclei of Clarke and dentate nucleus [143]. This deficiency causes an accumulation of iron in mitochondria and a reciprocal depletion in neuronal cytosol, which in turn leads to respiratory chain dysfunction and an increase in oxidative stress [144]. At the mtDNA level, frataxin-deficient yeast hybrid cell lines show a progressive loss of mtDNA and a reciprocal rise in mtDNA lesion formation [145, 146].

3.3.3.6 Hereditary Spastic Paraplegia

Hereditary spastic paraplegia (HSP) is not a single entity, rather a group of neurodegenerative disorders phenotypically characterised by progressive lower extremity weakness and spasticity (progressive stiffness and contraction) [147]. Mitochondrial dysfunction is linked to two forms of HSP: SPG13, the mitochondrial chaperonin Hsp60 [148], and SPG7, a nuclear-encoded mitochondrial metalloprotease protein which increases ROS [149].

Investigations into the effects of inherited mtDNA mutation on HSP are limited. An mtDNA mutation in *MTATP6* (m.9176T>C) was identified in a large HSP pedigree [150] and a mutation in the mitochondrial 12 s rRNA (m.1432A>G) was identified in an Amish family who demonstrated abnormal mRNA maturation [151]. A single study investigating the role of mtDNA haplogroups appears negative, but is largely inconclusive due to low sample size and genetic heterogeneity [152].

3.3.3.7 Multiple Sclerosis

Multiple sclerosis (MS) is one of the commonest neuroinflammatory diseases in the world [153] and is phenotypically characterised by autonomic, visual, motor and sensory deficits [154]. Until relatively recently, MS pathology was restricted to the formation of lesions or plaques and the destruction of myelin in the spinal cord; however, improvements in imaging technology confirm that neurodegeneration begins in the grey and white matter of MS patients [155, 156].

Mitochondrial dysfunction appears a common component of MS. N-Acetylaspartate and N-acetylaspartylglutamate (NAA), which correspond to MS relapses, are markers for neuronal integrity produced by mitochondria and indicate mitochondrial functionality [157]. RC chain function, particularly complex I, is reduced in MS tissue and complexes I and III are reduced in non-lesional MS motor cortex tissue [158, 159]. Further, there are established links between inherited mtDNA mutation and MS, with female primary LHON (discussed previously) mutation carriers presenting with an MS-like phenotype – termed 'Harding's disease' [160]. In support, m.3460G>A (a primary LHON mutation) was identified in a small cohort of neuromyelitis optica cases, where spinal nerve degeneration is a predominant feature [161], indicating that inherited mtDNA variants could modulate MS pathogenesis. However, subsequent studies focusing on the role of LHON mutations (m.3460G>A, m.11778G>A and m.14484T>C) refute a link to MS [162, 163].

Early attempts to link mtDNA haplogroups to MS indicate that European super-haplogroup 'UK' increases MS risk, with a reciprocal risk reduction for individuals on a super-haplogroup 'JT' background [164, 165]. These associations were confirmed in a much larger investigation into inherited mtDNA mutations in late-onset disease, with haplogroup U increasing MS risk and haplogroup J/K reducing MS risk [99].

Attempts to identify somatic or acquired mtDNA mutations, focusing on the identification of mtDNA deletions in normal-appearing grey and white matter or paraspinal muscle, failed to find a significant difference between MS cases and controls [166, 167].

3.4 Conclusion

The modulation of mitochondrial function is a key component to neurodegeneration, both during 'normal' ageing and disease (Fig. 3.2). Changes in mitochondrial function as we age and when disease manifests are often mediated by mitochondrial DNA variation, which can be either inherited (typically single base-pair exchanges) or acquired as we age (either single base-pair exchanges or mtDNA deletions).

Despite strong evidence of mitochondrial involvement in several neurodegenerative diseases, the ubiquity of mitochondria makes identifying disease-causing mtDNA mutations, and how they interact with disease-specific pathologies, difficult. Nevertheless, if we are to fully understand if this phenomenon is the cause or correlation, it is important that studies of mtDNA in neurodegenerative disease continue.

As the evidence demonstrates, mitochondrial function, mediated through genomic flux, can play a role in neuronal loss, making mitochondria a useful target for therapeutic strategies in an increasingly aged population.

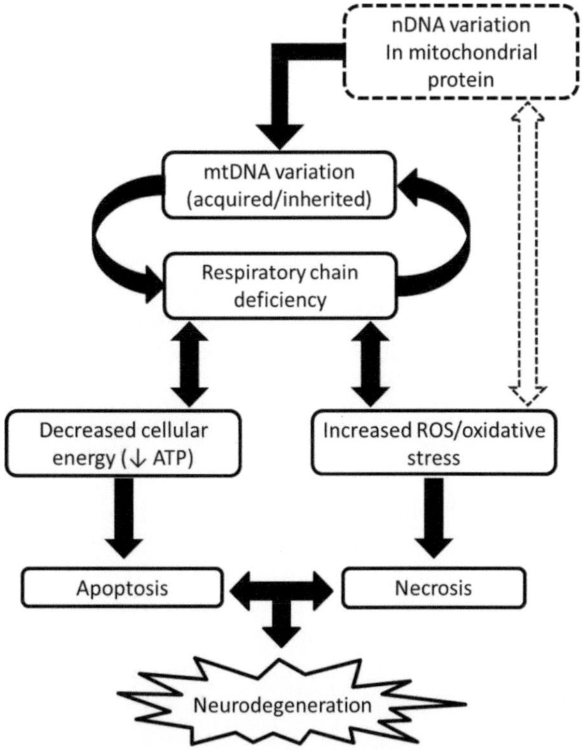

Fig. 3.2 Summary of the complex interaction between mitochondrial dysfunction, mtDNA mutation and neurodegeneration

References

1. Svennerholm L, Bostrom K, Jungbjer B. Changes in weight and compositions of major membrane components of human brain during the span of adult human life of Swedes. Acta Neuropathol. 1997;94(4):345–52.
2. Scahill RI, Frost C, Jenkins R, Whitwell JL, Rossor MN, Fox NC. A longitudinal study of brain volume changes in normal aging using serial registered magnetic resonance imaging. Arch Neurol Chicago. 2003;60(7):989–94.
3. Trollor JN, Valenzuela MJ. Brain ageing in the new millennium. Aust N Z J Psychiatry. 2001;35(6):788–805.
4. Esiri MM, Wilcock GK, Morris JH. Neuropathological assessment of the lesions of significance in vascular dementia. J Neurol Neurosurg Psychiatry. 1997;63(6):749–53.
5. Jernigan TL, Archibald SL, Fennema-Notestine C, Gamst AC, Stout JC, Bonner J, et al. Effects of age on tissues and regions of the cerebrum and cerebellum. Neurobiol Aging. 2001;22(4):581–94.
6. Raz N. The ageing brain: structural changes and their implications for cognitive ageing. In: Dixon R, Bäckman L, Nilssonn L, editors. New frontiers in cognitive ageing. New York: Oxford University Press; 2004.
7. Burke SN, Barnes CA. Neural plasticity in the ageing brain. Nat Rev Neurosci. 2006;7(1):30–40.
8. Kolb B, Gibb R, Robinson TE. Brain plasticity and behavior. Curr Dir Psychol Sci. 2003;12(1):1–5.
9. Kolb B, Whishaw IQ. Brain plasticity and behavior. Annu Rev Psychol. 1998;49:43–64.
10. Anderton BH. Ageing of the brain. Mech Ageing Dev. 2002;123(7):811–7.
11. Flood DG, Buell SJ, Horwitz GJ, Coleman PD. Dendritic extent in human dentate gyrus granule cells in normal aging and senile dementia. Brain Res. 1987;402(2):205–16.
12. Head D, Buckner RL, Shimony JS, Williams LE, Akbudak E, Conturo TE, et al. Differential vulnerability of anterior white matter in nondemented aging with minimal acceleration in dementia of the Alzheimer type: evidence from diffusion tensor imaging. Cereb Cortex. 2004;14(4):410–23.
13. Gray DA, Woulfe J. Lipofuscin and aging: a matter of toxic waste. Sci Aging Knowledge Environ. 2005;2005(5):re1.
14. Braak H, Braak E. Neuropathological stageing of Alzheimer-related changes. Acta Neuropathol. 1991;82(4):239–59.
15. Kahn J, Anderton BH, Probst A, Ulrich J, Esiri MM. Immunohistological study of granulovacuolar degeneration using monoclonal-antibodies to neurofilaments. J Neurol Neurosurg Psychiatry. 1985;48(9):924–6.
16. Xu M, Shibayama H, Kobayashi H, Yamada K, Ishihara R, Zhao P, et al. Granulovacuolar degeneration in the hippocampal cortex of aging and demented patients – a quantitative study. Acta Neuropathol. 1992;85(1):1–9.
17. Gibson PH, Tomlinson BE. Numbers of Hirano bodies in the hippocampus of normal and demented people with Alzheimer's disease. J Neurol Sci. 1977;33(1–2):199–206.
18. Galloway PG, Perry G, Gambetti P. Hirano body filaments contain actin and actin-associated proteins. J Neuropathol Exp Neurol. 1987;46(2):185–99.
19. Dickson DW, Liu WK, Kress Y, Ku J, DeJesus O, Yen SH. Phosphorylated tau immunoreactivity of granulovacuolar bodies (GVB) of Alzheimer's disease: localization of two amino terminal tau epitopes in GVB. Acta Neuropathol. 1993;85(5):463–70.
20. Smith EE, Schneider JA, Wardlaw JM, Greenberg SM. Cerebral microinfarcts: the invisible lesions. Lancet Neurol. 2012;11(3):272–82.
21. Saunders AM, Strittmatter WJ, Schmechel D, George-Hyslop PH, Pericak-Vance MA, Joo SH, et al. Association of apolipoprotein E allele epsilon 4 with late-onset familial and sporadic Alzheimer's disease. Neurology. 1993;43(8):1467–72.

22. Deary IJ, Whiteman MC, Pattie A, Starr JM, Hayward C, Wright AF, et al. Apolipoprotein E gene variability and cognitive functions at age 79: a follow-up of the Scottish Mental Survey of 1932. Psychol Aging. 2004;19(2):367–71.
23. Jorm AF, Mather KA, Butterworth P, Anstey KJ, Christensen H, Easteal S. APOE genotype and cognitive functioning in a large age-stratified population sample. Neuropsychology. 2007;21(1):1–8.
24. Lu L, Airey DC, Williams RW. Complex trait analysis of the hippocampus: mapping and biometric analysis of two novel gene loci with specific effects on hippocampal structure in mice. J Neurosci. 2001;21(10):3503–14.
25. Lee CK, Weindruch R, Prolla TA. Gene-expression profile of the ageing brain in mice. Nat Genet. 2000;25(3):294–7.
26. Tsugita A, Kawakami T, Uchida T, Sakai T, Kamo M, Matsui T, et al. Proteome analysis of mouse brain: two-dimensional electrophoresis profiles of tissue proteins during the course of aging. Electrophoresis. 2000;21(9):1853–71.
27. Henley JM, Wilkinson KA. AMPA receptor trafficking and the mechanisms underlying synaptic plasticity and cognitive aging. Dialogues Clin Neurosci. 2013;15(1):11–27.
28. Magnusson KR, Brim BL, Das SR. Selective vulnerabilities of N-methyl-D-aspartate (NMDA) receptors during brain aging. Front Aging Neurosci. 2010;2:11.
29. Limon A, Reyes-Ruiz JM, Miledi R. Loss of functional GABA(A) receptors in the Alzheimer diseased brain. Proc Natl Acad Sci U S A. 2012;109(25):10071–6.
30. Lu T, Pan Y, Kao SY, Li C, Kohane I, Chan J, et al. Gene regulation and DNA damage in the ageing human brain. Nature. 2004;429(6994):883–91.
31. Mayeux R, Small SA, Tang MX, Tycko B, Stern Y. Memory performance in healthy elderly without Alzheimer's disease: effects of time and apolipoprotein-E. Neurobiol Aging. 2001;22(4):683–9.
32. Parkin A. Human-memory and amnesia – Cermak, Ls. Q J Exp Psychol A. 1983;35(Aug):544–5.
33. Nyberg L, Bäckman L. Cognitive ageing: a view from brain imaging. In: Dixon R, Bäckman L, Nilsson L, editors. New frontiers in cognitive ageing. New York: Oxford University Press; 2004.
34. Lustig C, Buckner RL. Preserved neural correlates of priming in old age and dementia. Neuron. 2004;42(5):865–75.
35. Sweet JJ, Suchy Y, Leahy B, Abramowitz C, Nowinski CJ. Normative clinical relationships between orientation and memory: age as an important moderator variable. Clin Neuropsychol. 1999;13(4):495–508.
36. Hopp GA, Dixon RA, Grut M, Bäckman L. Longitudinal and psychometric profiles of two cognitive status tests in very old adults. J Clin Psychol. 1997;53(7):673–86.
37. Kensinger EA. Cognition in aging and age related disease. In: Hof PR, Mobbs CV, editors. Handbook of the neuroscience of aging. London: Elsevier; 2009. p. 249–56.
38. Cabeza R. Cognitive neuroscience of aging: contributions of functional neuroimaging. Scand J Psychol. 2001;42(3):277–86.
39. Compton J, van Amelsvoort T, Murphy D. HRT and its effect on normal ageing of the brain and dementia. Br J Clin Pharmacol. 2001;52(6):647–53.
40. Ibanez V, Pietrini P, Furey ML, Alexander GE, Millet P, Bokde AL, et al. Resting state brain glucose metabolism is not reduced in normotensive healthy men during aging, after correction for brain atrophy. Brain Res Bull. 2004;63(2):147–54.
41. Toescu EC, Verkhratsky A, Landfield PW. Ca2+ regulation and gene expression in normal brain aging. Trends Neurosci. 2004;27(10):614–20.
42. Melov S. Modeling mitochondrial function in aging neurons. Trends Neurosci. 2004; 27(10):601–6.
43. Balaban RS, Nemoto S, Finkel T. Mitochondria, oxidants, and aging. Cell. 2005;120(4): 483–95.
44. Magistretti PJ, Pellerin L. Cellular mechanisms of brain energy metabolism and their relevance to functional brain imaging. Philos Trans R Soc B. 1999;354(1387):1155–63.

45. LaFrance R, Brustovetsky N, Sherburne C, Delong D, Dubinsky JM. Age-related changes in regional brain mitochondria from Fischer 344 rats. Aging Cell. 2005;4(3):139–45.
46. Navarro A, Boveris A. The mitochondrial energy transduction system and the aging process. Am J Physiol Cell Physiol. 2007;292(2):C670–86.
47. Navarro A. Mitochondrial enzyme activities as biochemical markers of aging. Mol Aspects Med. 2004;25(1–2):37–48.
48. Barrientos A, Casademont J, Cardellach F, Estivill X, Urbano-Marquez A, Nunes V. Reduced steady-state levels of mitochondrial RNA and increased mitochondrial DNA amount in human brain with aging. Mol Brain Res. 1997;52(2):284–9.
49. Hirai K, Aliev G, Nunomura A, Fujioka H, Russell RL, Atwood CS, et al. Mitochondrial abnormalities in Alzheimer's disease. J Neurosci. 2001;21(9):3017–23.
50. Halliwell B, Gutteridge JMC. Free radicals in biology and medicine. New York: Oxford University Press; 1999.
51. Navarro A, Boveris A. Brain mitochondrial dysfunction in aging, neurodegeneration, and Parkinson's disease. Front Aging Neurosci. 2010;2:34.
52. Ceballospicot I, Nicole A, Clement M, Bourre JM, Sinet PM. Age-related-changes in antioxidant enzymes and lipid-peroxidation in brains of control and transgenic mice overexpressing copper-zinc superoxide-dismutase. Mutat Res. 1992;275(3–6):281–93.
53. Pollack M, Leeuwenburgh C. Apoptosis and aging: role of the mitochondria. J Gerontol A Biol Sci Med Sci. 2001;56(11):B475–82.
54. Swerdlow RH. Mitochondria and cell bioenergetics: increasingly recognized components and a possible etiologic cause of Alzheimer's disease. Antioxid Redox Signal. 2012;16(12):1434–55.
55. Swerdlow RH. Brain aging, Alzheimer's disease, and mitochondria. BBA Mol Basis Dis. 2011;1812(12):1630–9.
56. Payne BAI, Wilson IJ, Yu-Wai-Man P, Coxhead J, Deehan D, Horvath R, et al. Universal heteroplasmy of human mitochondrial DNA. Hum Mol Genet. 2013;22(2):384–90.
57. Torroni A, Huoponen K, Francalacci P, Petrozzi M, Morelli L, Scozzari R, et al. Classification of European mtDNAs from an analysis of three European populations. Genetics. 1996;144(4):1835–50.
58. Gomez-Duran A, Pacheu-Grau D, Martinez-Romero I, López-Gallardo E, López-Pérez MJ, Montoya J, et al. Oxidative phosphorylation differences between mitochondrial DNA haplogroups modify the risk of Leber's hereditary optic neuropathy. Biochim Biophys Acta. 2012;1822(8):1216–22.
59. Pello R, Martin MA, Carelli V, Nijtmans LG, Achilli A, Pala M, et al. Mitochondrial DNA background modulates the assembly kinetics of OXPHOS complexes in a cellular model of mitochondrial disease. Hum Mol Genet. 2008;17(24):4001–11.
60. Meissner C, Bruse P, Mohamed SA, Schulz A, Warnk H, Storm T, et al. The 4977 bp deletion of mitochondrial DNA in human skeletal muscle, heart and different areas of the brain: a useful biomarker or more? Exp Gerontol. 2008;43(7):645–52.
61. Cortopassi GA, Arnheim N. Detection of a specific mitochondrial-DNA deletion in tissues of older humans. Nucleic Acids Res. 1990;18(23):6927–33.
62. Corraldebrinski M, Horton T, Lott MT, Shoffner JM, Beal MF, Wallace DC. Mitochondrial-DNA deletions in human brain – regional variability and increase with advanced age. Nat Genet. 1992;2(4):324–9.
63. Pickrell AM, Fukui H, Wang X, Pinto M, Moraes CT. The striatum is highly susceptible to mitochondrial oxidative phosphorylation dysfunctions. J Neurosci. 2011;31(27):9895–904.
64. Fukui H, Moraes CT. Mechanisms of formation and accumulation of mitochondrial DNA deletions in aging neurons. Hum Mol Genet. 2009;18(6):1028–36.
65. Lin MT, Simon DK, Ahn CH, Kim LM, Beal MF. High aggregate burden of somatic mtDNA point mutations in aging and Alzheimer's disease brain. Hum Mol Genet. 2002;11(2):133–45.
66. Williams SL, Mash DC, Zuchner S, Moraes CT. Somatic mtDNA mutation spectra in the aging human putamen. Plos Genet. 2013;9(12):e1003990.

67. Schneider S, Excoffier L. Estimation of past demographic parameters from the distribution of pairwise differences when the mutation rates vary among sites: application to human mitochondrial DNA. Genetics. 1999;152(3):1079–89.
68. Mercer TR, Neph S, Dinger ME, Crawford J, Smith MA, Shearwood AM, et al. The human mitochondrial transcriptome. Cell. 2011;146(4):645–58.
69. Epstein CB, Waddle JA, Hale 4th W, Davé V, Thornton J, Macatee TL, et al. Genome-wide responses to mitochondrial dysfunction. Mol Biol Cell. 2001;12(2):297–308.
70. Scholte HR. The biochemical basis of mitochondrial diseases. J Bioenerg Biomembr. 1988;20(2):161–91.
71. Brandon MC, Lott MT, Nguyen KC, Spolim S, Navathe SB, Baldi P, et al. MITOMAP: a human mitochondrial genome database – 2004 update. Nucleic Acids Res. 2005;33:D611–3.
72. Morais VA, De Strooper B. Mitochondria dysfunction and neurodegenerative disorders: cause or consequence. J Alzheimers Dis. 2010;20:S255–63.
73. Kirches E. LHON: mitochondrial mutations and more. Curr Genomics. 2011;12(1):44–54.
74. Carelli V, Ross-Cisneros FN, Sadun AA. Mitochondrial dysfunction as a cause of optic neuropathies. Prog Retin Eye Res. 2004;23(1):53–89.
75. Beretta S, Mattavelli L, Sala G, Tremolizzo L, Schapira AH, Martinuzzi A, et al. Leber hereditary optic neuropathy mtDNA mutations disrupt glutamate transport in cybrid cell lines. Brain. 2004;127:2183–92.
76. Fukuhara N, Tokiguchi S, Shirakawa K, Tsubaki T. Myoclonus epilepsy associated with ragged-red fibres (mitochondrial abnormalities): disease entity or a syndrome? Light-and electron-microscopic studies of two cases and review of literature. J Neurol Sci. 1980;47(1):117–33.
77. Shoffner JM, Wallace DC. A mitochondrial transfer rnalys mutation causes myoclonic epilepsy and ragged-red fiber disease. Prog Neuropathol. 1991;7:161–7.
78. Wallace DC, Zheng XX, Lott MT, Shoffner JM, Hodge JA, Kelley RI, et al. Familial mitochondrial encephalomyopathy (MERRF): genetic, pathophysiological, and biochemical characterization of a mitochondrial DNA disease. Cell. 1988;55(4):601–10.
79. Rommelaere G, Michel S, Malaisse J, Charlier S, Arnould T, Renard P. Hypersensitivity of A8344G MERRF mutated cybrid cells to staurosporine-induced cell death is mediated by calcium-dependent activation of calpains. Int J Biochem Cell B. 2012;44(1):139–49.
80. Zeviani M, Simonati A, Bindoff LA. Ataxia in mitochondrial disorders. Handb Clin Neurol. 2012;103:359–72.
81. Kearns TP. External ophthalmoplegia, pigmentary degeneration of the retina, and cardiomyopathy: a newly recognized syndrome. Trans Am Ophthalmol Soc. 1965;63:559–625.
82. Goto Y, Horai S, Matsuoka T, Koga Y, Nihei K, Kobayashi M, et al. Mitochondrial myopathy, encephalopathy, lactic acidosis, and stroke-like episodes (MELAS): a correlative study of the clinical features and mitochondrial DNA mutation. Neurology. 1992;42(3 Pt 1):545–50.
83. Holt IJ, Harding AE, Petty RK, Morgan-Hughes JA. A new mitochondrial disease associated with mitochondrial DNA heteroplasmy. Am J Hum Genet. 1990;46(3):428–33.
84. Winklhofer KF, Haass C. Mitochondrial dysfunction in Parkinson's disease. BBA Mol Basis Dis. 2010;1802(1):29–44.
85. Schapira AH, Cooper JM, Dexter D, Jenner P, Clark JB, Marsden CD. Mitochondrial complex I deficiency in Parkinson's disease. Lancet. 1989;1(8649):1269.
86. Sherer TB, Betarbet R, Testa CM, Seo BB, Richardson JR, Kim JH, et al. Mechanism of toxicity in rotenone models of Parkinson's disease. J Neurosci. 2003;23(34):10756–64.
87. Forno LS, Delanney LE, Irwin I, Langston JW. Similarities and differences between Mptp-induced parkinsonism and Parkinson's disease. Neuropathologic considerations. Adv Neurol. 1993;60:600–8.
88. Betarbet R, Sherer TB, MacKenzie G, Garcia-Osuna M, Panov AV, Greenamyre JT. Chronic systemic pesticide exposure reproduces features of Parkinson's disease. Nat Neurosci. 2000;3(12):1301–6.

89. Gomez-Duran A, Pacheu-Grau D, Lopez-Gallardo E, Díez-Sánchez C, Montoya J, López-Pérez MJ, et al. Unmasking the causes of multifactorial disorders: OXPHOS differences between mitochondrial haplogroups. Hum Mol Genet. 2010;19(17):3343–53.
90. Hudson G, Carelli V, Spruijt L, Gerards M, Mowbray C, Achilli A, et al. Clinical expression of Leber hereditary optic neuropathy is affected by the mitochondrial DNA-haplogroup background. Am J Hum Genet. 2007;81(2):228–33.
91. Fuku N, Park KS, Yamada Y, Nishigaki Y, Cho YM, Matsuo H, et al. Mitochondrial haplogroup N9a confers resistance against type 2 diabetes in Asians. Am J Hum Genet. 2007;80(3):407–15.
92. Chinnery PF, Elliott HR, Syed A, Rothwell PM, Oxford Vascular Study. Mitochondrial DNA haplogroups and risk of transient ischaemic attack and ischaemic stroke: a genetic association study. Lancet Neurol. 2007;9(5):498–503.
93. Kazuno AA, Munakata K, Nagai T, Shimozono S, Tanaka M, Yoneda M, et al. Identification of mitochondrial DNA polymorphisms that alter mitochondrial matrix pH and intracellular calcium dynamics. PLoS Genet. 2006;2(8):1167–77.
94. Rose G, Passarino G, Carrieri G, Altomare K, Greco V, Bertolini S, et al. Paradoxes in longevity: sequence analysis of mtDNA haplogroup J in centenarians. Eur J Hum Genet. 2001;9(9):701–7.
95. Pyle A, Foltynie T, Tiangyou W, Lambert C, Keers SM, Allcock LM, et al. Mitochondrial DNA haplogroup cluster UKJT reduces the risk of PD. Ann Neurol. 2005;57(4):564–7.
96. Ross OA, McCormack R, Maxwell LD, Duguid RA, Quinn DJ, Barnett YA, et al. mt4216C variant in linkage with the mtDNA TJ cluster may confer a susceptibility to mitochondrial dysfunction resulting in an increased risk of Parkinson's disease in the Irish. Exp Gerontol. 2003;38(4):397–405.
97. Biffi A, Anderson CD, Nalls MA, Rahman R, Sonni A, Cortellini L, et al. Principal-component analysis for assessment of population stratification in mitochondrial medical genetics. Am J Hum Genet. 2010;86(6):904–17.
98. Hudson G, Nalls M, Evans JR, Breen DP, Winder-Rhodes S, Morrison KE, et al. Two-stage association study and meta-analysis of mitochondrial DNA variants in Parkinson disease. Neurology. 2013;80(22):2042–8.
99. Hudson G, Gomez-Duran A, Wilson IJ, Chinnery PF. Recent mitochondrial DNA mutations increase the risk of developing common late-onset human diseases. PLoS Genet. 2014;10(5):e1004369.
100. Ikebe S, Tanaka M, Ohno K, Sato W, Hattori K, Kondo T, et al. Increase of deleted mitochondrial DNA in the striatum in Parkinson's disease and senescence. Biochem Biophys Res Commun. 1990;170(3):1044–8.
101. Schapira AH, Holt IJ, Sweeney M, Harding AE, Jenner P, Marsden CD. Mitochondrial DNA analysis in Parkinson's disease. Mov Disord. 1990;5(4):294–7.
102. Lestienne P, Nelson I, Riederer P, Reichmann H, Jellinger K. Mitochondrial DNA in post-mortem brain from patients with Parkinson's disease. J Neurochem. 1991;56(5):1819.
103. Kraytsberg Y, Kudryavtseva E, McKee AC, Geula C, Kowall NW, Khrapko K. Mitochondrial DNA deletions are abundant and cause functional impairment in aged human substantia nigra neurons. Nat Genet. 2006;38(5):518–20.
104. Bender A, Krishnan KJ, Morris CM, Taylor GA, Reeve AK, Perry RH, et al. High levels of mitochondrial DNA deletions in substantia nigra neurons in aging and Parkinson disease. Nat Genet. 2006;38(5):515–7.
105. Kakiuchi C, Ishiwata M, Kametani M, Nelson C, Iwamoto K, Kato T. Quantitative analysis of mitochondrial DNA deletions in the brains of patients with bipolar disorder and schizophrenia. Int J Neuropsychopharmacol. 2005;8(4):515–22.
106. Corral-Debrinski M, Horton T, Lott MT, Shoffner JM, Beal MF, Wallace DC. Mitochondrial DNA deletions in human brain: regional variability and increase with advanced age. Nat Genet. 1992;2(4):324–9.
107. Nadasi E, Melegh B, Seress L, Kosztolányi G. Mitochondrial DNA deletions in newborn brain samples. Orv Hetil. 2004;145(25):1321–5.

108. Shoffner JM, Brown MD, Torroni A, Lott MT, Cabell MF, Mirra SS, et al. Mitochondrial DNA variants observed in Alzheimer disease and Parkinson disease patients. Genomics. 1993;17(1):171–84.
109. Parker Jr WD, Parks JK. Mitochondrial ND5 mutations in idiopathic Parkinson's disease. Biochem Biophys Res Commun. 2005;326(3):667–9.
110. Burns A, Iliffe S. Alzheimer's disease. BMJ. 2009;338:b158.
111. Anandatheerthavarada HK, Biswas G, Robin MA, Avadhani NG. Mitochondrial targeting and a novel transmembrane arrest of Alzheimer's amyloid precursor protein impairs mitochondrial function in neuronal cells. J Cell Biol. 2003;161(1):41–54.
112. Manczak M, Anekonda TS, Henson E, Park BS, Quinn J, Reddy PH. Mitochondria are a direct site of A beta accumulation in Alzheimer's disease neurons: implications for free radical generation and oxidative damage in disease progression. Hum Mol Genet. 2006;15(9):1437–49.
113. Melov S, Adlard PA, Morten K, Johnson F, Golden TR, Hinerfeld D, et al. Mitochondrial oxidative stress causes hyperphosphorylation of tau. PLoS One. 2007;2(6):e536.
114. Lakatos A, Derbeneva O, Younes D, Keator D, Bakken T, Lvova M, et al. Association between mitochondrial DNA variations and Alzheimer's disease in the ADNI cohort. Neurobiol Aging. 2010;31(8):1355–63.
115. Santoro A, Balbi V, Balducci E, Pirazzini C, Rosini F, Tavano F, et al. Evidence for subhaplogroup h5 of mitochondrial DNA as a risk factor for late onset alzheimer's disease. PLoS One. 2010;5(8):e12037.
116. Kruger J, Hinttala R, Majamaa K, Remes AM. Mitochondrial DNA haplogroups in early-onset alzheimer's disease and frontotemporal lobar degeneration. Mol Neurodegener. 2010;5:8.
117. Maruszak A, Canter JA, Styczynska M, Zekanowski C, Barcikowska M. Mitochondrial haplogroup H and Alzheimer's disease – is there a connection? Neurobiol Aging. 2009;30(11):1749–55.
118. van der Walt JM, Dementieva YA, Martin ER, Scott WK, Nicodemus KK, Kroner CC, et al. Analysis of European mitochondrial haplogroups with Alzheimer disease risk. Neurosci Lett. 2004;365(1):28–32.
119. Carrieri G, Bonafe M, De Luca M, Rose G, Varcasia O, Bruni A, et al. Mitochondrial DNA haplogroups and APOE4 allele are non-independent variables in sporadic Alzheimer's disease. Hum Genet. 2001;108(3):194–8.
120. Chagnon P, Gee M, Filion M, Robitaille Y, Belouchi M, Gauvreau D. Phylogenetic analysis of the mitochondrial genome indicates significant differences between patients with Alzheimer disease and controls in a French-Canadian founder population. Am J Med Genet. 1999;85(1):20–30.
121. Hudson G, Sims R, Harold D, Chapman J, Hollingworth P, Gerrish A, et al. No consistent evidence for association between mtDNA variants and Alzheimer disease. Neurology. 2012;78(14):1038–42.
122. Swerdlow RH, Parks JK, Cassarino DS, Maguire DJ, Maguire RS, Bennett Jr JP, et al. Cybrids in Alzheimer's disease: a cellular model of the disease? Neurology. 1997;49(4):918–25.
123. Tanaka N, Goto Y, Akanuma J, Kato M, Kinoshita T, Yamashita F, et al. Mitochondrial DNA variants in a Japanese population of patients with Alzheimer's disease. Mitochondrion. 2010;10(1):32–7.
124. Lustbader JW, Cirilli M, Lin C, Xu HW, Takuma K, Wang N, et al. ABAD directly links Abeta to mitochondrial toxicity in Alzheimer's disease. Science. 2004;304(5669):448–52.
125. Cortopassi GA, Shibata D, Soong NW, Arnheim N. A pattern of accumulation of a somatic deletion of mitochondrial-DNA in aging human tissues. Proc Natl Acad Sci U S A. 1992;89(16):7370–4.
126. Chang SW, Zhang DK, Chung HD, Zassenhaus HP. The frequency of point mutations in mitochondrial DNA is elevated in the Alzheimer's brain. Biochem Biophys Res Commun. 2000;273(1):203–8.

127. Coskun PE, Beal MF, Wallace DC. Alzheimer's brains harbor somatic mtDNA control-region mutations that suppress mitochondrial transcription and replication. Proc Natl Acad Sci U S A. 2004;101(29):10726–31.
128. Damiano M, Galvan L, Deglon N, Brouillet E. Mitochondria in Huntington's disease. Biochim Biophys Acta. 2010;1802(1):52–61.
129. Panov AV, Gutekunst CA, Leavitt BR, Hayden MR, Burke JR, Strittmatter WJ, et al. Early mitochondrial calcium defects in Huntington's disease are a direct effect of polyglutamines. Nat Neurosci. 2002;5(8):731–6.
130. Siddiqui A, Rivera-Sanchez S, Castro MD, Acevedo-Torres K, Rane A, Torres-Ramos CA, et al. Mitochondrial DNA damage is associated with reduced mitochondrial bioenergetics in Huntington's disease. Free Radic Biol Med. 2012;53(7):1478–88.
131. Banoei MM, Houshmand M, Panahi MS, Shariati P, Rostami M, Manshadi MD, et al. Huntington's disease and mitochondrial DNA deletions: event or regular mechanism for mutant huntingtin protein and CAG repeats expansion?! Cell Mol Neurobiol. 2007;27(7):867–75.
132. Chen CM, Wu YR, Cheng ML, Liu JL, Lee YM, Lee PW, et al. Increased oxidative damage and mitochondrial abnormalities in the peripheral blood of Huntington's disease patients. Biochem Biophys Res Commun. 2007;359(2):335–40.
133. Horton TM, Graham BH, Corral-Debrinski M, Shoffner JM, Kaufman AE, Beal MF, et al. Marked increase in mitochondrial DNA deletion levels in the cerebral cortex of Huntington's disease patients. Neurology. 1995;45(10):1879–83.
134. Oliveira JM, Chen S, Almeida S, Riley R, Gonçalves J, Oliveira CR, et al. Mitochondrial-dependent Ca2+ handling in Huntington's disease striatal cells: effect of histone deacetylase inhibitors. J Neurosci. 2006;26(43):11174–86.
135. Arning L, Haghikia A, Taherzadeh-Fard E, Saft C, Andrich J, Pula B, et al. Mitochondrial haplogroup H correlates with ATP levels and age at onset in Huntington disease. J Mol Med JMM. 2010;88(4):431–6.
136. Kiernan MC, Vucic S, Cheah BC, Turner MR, Eisen A, Hardiman O, et al. Amyotrophic lateral sclerosis. Lancet. 2011;377(9769):942–55.
137. Al-Chalabi A, Leigh PN. Recent advances in amyotrophic lateral sclerosis. Curr Opin Neurol. 2000;13(4):397–405.
138. Israelson A, Arbel N, Da Cruz S, Ilieva H, Yamanaka K, Shoshan-Barmatz V, et al. Misfolded mutant SOD1 directly inhibits VDAC1 conductance in a mouse model of inherited ALS. Neuron. 2010;67(4):575–87.
139. Warita H, Hayashi T, Murakami T, Manabe Y, Abe K. Oxidative damage to mitochondrial DNA in spinal motoneurons of transgenic ALS mice. Brain Res Mol Brain Res. 2001;89(1–2):147–52.
140. Dhaliwal GK, Grewal RP. Mitochondrial DNA deletion mutation levels are elevated in ALS brains. Neuroreport. 2000;11(11):2507–9.
141. Wiedemann FR, Manfredi G, Mawrin C, Beal MF, Schon EA. Mitochondrial DNA and respiratory chain function in spinal cords of ALS patients. J Neurochem. 2002;80(4):616–25.
142. Murakami T, Nagai M, Miyazaki K, Morimoto N, Ohta Y, Kurata T, et al. Early decrease of mitochondrial DNA repair enzymes in spinal motor neurons of presymptomatic transgenic mice carrying a mutant SOD1 gene. Brain Res. 2007;1150:182–9.
143. Marmolino D. Friedreich's ataxia: past, present and future. Brain Res Rev. 2011;67(1–2):311–30.
144. Karthikeyan G, Santos JH, Graziewicz MA, Copeland WC, Isaya G, Van Houten B, et al. Reduction in frataxin causes progressive accumulation of mitochondrial damage. Hum Mol Genet. 2003;12(24):3331–42.
145. Wilson RB, Roof DM. Respiratory deficiency due to loss of mitochondrial DNA in yeast lacking the frataxin homologue. Nat Genet. 1997;16(4):352–7.
146. Houshmand M, Panahi MSS, Nafisi S, Soltanzadeh A, Alkandari FM. Identification and sizing of GAA trinucleotide repeat expansion, investigation for D-loop variations and mitochondrial deletions in Iranian patients with Friedreich's ataxia. Mitochondrion. 2006;6(2):82–8.

147. Lo Giudice T, Lombardi F, Santorelli FM, Kawarai T, Orlacchio A. Hereditary spastic paraplegia: clinical-genetic characteristics and evolving molecular mechanisms. Exp Neurol. 2014;261:518–39.
148. Hansen JJ, Durr A, Cournu-Rebeix I, Georgopoulos C, Ang D, Nielsen MN, et al. Hereditary spastic paraplegia SPG13 is associated with a mutation in the gene encoding the mitochondrial chaperonin Hsp60. Am J Hum Genet. 2002;70(5):1328–32.
149. Almontashiri NA, Chen HH, Mailloux RJ, Tatsuta T, Teng AC, Mahmoud AB, et al. SPG7 variant escapes phosphorylation-regulated processing by AFG3L2, elevates mitochondrial ROS, and is associated with multiple clinical phenotypes. Cell Rep. 2014;7(3):834–47.
150. Verny C, Guegen N, Desquiret V, Chevrollier A, Prundean A, Dubas F, et al. Hereditary spastic paraplegia-like disorder due to a mitochondrial ATP6 gene point mutation. Mitochondrion. 2011;11(1):70–5.
151. Crosby AH, Patel H, Chioza BA, Proukakis C, Gurtz K, Patton MA, et al. Defective mitochondrial mRNA maturation is associated with spastic ataxia. Am J Hum Genet. 2010;87(5):655–60.
152. Sanchez-Ferrero E, Coto E, Corao AI, Díaz M, Gámez J, Esteban J, et al. Mitochondrial DNA polymorphisms/haplogroups in hereditary spastic paraplegia. J Neurol. 2012;259(2):246–50.
153. Berer K, Krishnamoorthy G. Microbial view of central nervous system autoimmunity. Febs Lett. 2014;588(22):4207–13.
154. Compston A, Coles A. Multiple sclerosis. Lancet. 2008;372(9648):1502–17.
155. Kutzelnigg A, Lucchinetti CF, Stadelmann C, Brück W, Rauschka H, Bergmann M, et al. Cortical demyelination and diffuse white matter injury in multiple sclerosis. Brain. 2005;128:2705–12.
156. Ingle GT, Stevenson VL, Miller DH, Thompson AJ. Primary progressive multiple sclerosis: a 5-year clinical and MR study. Brain. 2003;126(Pt 11):2528–36.
157. Signoretti S, Marmarou A, Tavazzi B, Lazzarino G, Beaumont A, Vagnozzi R. N-acetylaspartate reduction as a measure of injury severity and mitochondrial dysfunction following diffuse traumatic brain injury. J Neurotrauma. 2001;18(10):977–91.
158. Mahad D, Campbell G, Ziabreva I, Rosenstengel C, Siegfried Schroeder HW. Mitochondrial dysfunction as a cause of axonal degeneration in the progressive stage of multiple sclerosis. J Neurol Neurosurg Psychiatry. 2008;79(3):343–4.
159. Dutta R, McDonough J, Yin XG, Peterson J, Chang A, Torres T, et al. Mitochondrial dysfunction as a cause of axonal degeneration in multiple sclerosis patients. Ann Neurol. 2006;59(3):478–89.
160. Harding AE, Sweeney MG, Miller DH, Mumford CJ, Kellar-Wood H, Menard D, et al. Occurrence of a multiple sclerosis-like illness in women who have a Leber's hereditary optic neuropathy mitochondrial DNA mutation. Brain. 1992;115(Pt 4):979–89.
161. Ghezzi A, Baldini S, Zaffaroni M, Leoni G, Koudriavtseva T, Casini AR, et al. Devic's neuromyelitis optica and mitochondrial DNA mutation: a case report. Neurol Sci. 2004;25:S380–2.
162. Cock H, Mandler R, Ahmed W, Schapira AH. Neuromyelitis optica (Devic's syndrome): no association with the primary mitochondrial DNA mutations found in Leber hereditary optic neuropathy. J Neurol Neurosurg Psychiatry. 1997;62(1):85–7.
163. Kalman B, Mandler RN. Studies of mitochondrial DNA in Devic's disease revealed no pathogenic mutations, but polymorphisms also found in association with multiple sclerosis. Ann Neurol. 2002;51(5):661–2.
164. Otaegui D, Saenz A, Martinez-Zabaleta M, Villoslada P, Fernández-Manchola I, Alvarez de Arcaya A, et al. Mitochondrial haplogroups in Basque multiple sclerosis patients. Mult Scler. 2004;10(5):532–5.
165. Ban M, Elson J, Walton A, Turnbull D, Compston A, Chinnery P, et al. Investigation of the role of mitochondrial DNA in multiple sclerosis susceptibility. PLoS One. 2008;3(8):e2891.
166. Blokhin A, Vyshkina T, Komoly S, Kalman B. Lack of mitochondrial DNA deletions in lesions of multiple sclerosis. Neuromolecular Med. 2008;10(3):187–94.
167. Campbell GR, Reeve AK, Ziabreva I, Reynolds R, Turnbull DM, Mahad DJ. No excess of mitochondrial DNA deletions within muscle in progressive multiple sclerosis. Mult Scler J. 2013;19(14):1858–66.

Chapter 4
Mitochondrial Genes and Neurodegenerative Disease

Carlo Viscomi, Anna Ardissone, and Massimo Zeviani

Abstract Mitochondrial dysfunction is increasingly recognised as a cause of neurodegeneration in both primary mitochondrial diseases and common neurodegenerative diseases, including Parkinson's, Alzheimer's and Huntington's diseases and amyotrophic lateral sclerosis (ALS).

In this chapter, we will focus on the molecular basis of mitochondrial dysfunction and the mechanisms bridging it to neurodegeneration. In the first part, we will summarise some basic concepts of mitochondrial biology. We will then cover paediatric diseases, including different types of encephalopathies, Leigh disease, leukoencephalopathies and a variety of syndromes with peculiar features. Finally, we will cover diseases in adults, including MNGIE disease, Friedreich ataxia and the disorders associated with alterations in mitochondrial dynamics and quality control. These include axonal Charcot–Marie–Tooth neuropathy, hereditary spastic paraplegia, and autosomal dominant optic atrophy, as well as the inherited and idiopathic forms of common neurodegenerative diseases.

Keywords Mitochondria • Mitochondrial disease • Leigh disease • Leukoencephalopathy • Encephalocardiomyopathy • CoQ10 deficiency • Mitochondrial DNA depletion syndrome • Microphthalmia with linear skin defects • Neurodegeneration with iron accumulation in the brain • Friedreich ataxia • Mitochondrial neurogastrointestinal encephalomyopathy • Hereditary spastic paraplegia • Charcot–Marie–Tooth peripheral neuropathy • Dominant optic atrophy • Parkinson's disease • Alzheimer's disease • Amyotrophic lateral sclerosis

C. Viscomi, PhD • M. Zeviani, MD, PhD (✉)
Medical Research Council, Mitochondrial Biology Unit, Cambridge, UK
e-mail: mdz21@mrc-mbu.cam.ac.uk

A. Ardissone, MD
Unit of Child Neurology, IRCCS Foundation Neurological Institute "C. Besta", Milan, Italy

A.K. Reeve et al. (eds.), *Mitochondrial Dysfunction in Neurodegenerative Disorders*, DOI 10.1007/978-3-319-28637-2_4

4.1 Introduction

Mitochondria are essential organelles present in almost all eukaryotic organisms with key roles in a number of processes such as heat generation, calcium, iron and sulphur homeostasis, haem biosynthesis, cell quality control and apoptosis, etc. [1].

Since mitochondria are under the double genetic control of mitochondrial DNA (mtDNA) and nuclear DNA (nDNA), mutations in either genome can thus cause mitochondrial disease, by a number of different mechanisms. Defects of mtDNA are maternally inherited. Mutations can affect all mtDNA copies (homoplasmy) and in this case are often non-pathogenic variants or are present only in some copies of mtDNA (heteroplasmy). In this case, the percentage of mutated mtDNA is correlated to the severity of the disorder, and there is a threshold necessary for the expression of the biochemical defect and clinical phenotype. Mutations in nDNA are inherited as Mendelian traits, and therefore mitochondrial disorders may be transmitted as autosomal recessive (such as Leigh disease due to mutations in *SURF1*), dominant (for instance, progressive external ophthalmoplegia (PEO) due to mutations in *ANT1*) or X-linked (for instance, microphthalmia with linear skin lesions due to mutations in *COX7B*) traits. Tissue and organ function critically depends on adequate ATP production, especially in cells and tissues with high-energy demand, such as neurons and muscle fibres. This explains why primary disorders of mitochondrial bioenergetics usually cause neurodegeneration, heart abnormalities, and/or muscle weakness, leading to neuro(cardio)muscular disease in children and adults. Mitochondrial disorders are a group of highly heterogeneous conditions affecting humans at any age [2]. Individually rare, when taken as a whole, mitochondrial disorders are among the most frequent genetic diseases in humans, with an estimated prevalence of about 1 in 5,000 individuals in the European population, although healthy carriers are much more common [3].

In a large proportion of mitochondrial disease cases, it is still not possible to reach a molecular genetic diagnosis, so that more than 50 % of adult patients, and an even greater percentage of paediatric cases, remain genetically undefined. This gap between biochemical readout and genetic characterisation reflects the complexity and intricacy of the homeostatic pathways related to mitochondrial bioenergetics. In these cases, the diagnosis is thus based solely on biochemical and/or morphological findings, in skeletal muscle or, more rarely, in cultured fibroblasts, limiting the possibility of genetic counselling. Whilst the analysis of mtDNA is a relatively standardised procedure in most specialised centres, new technologies based on next-generation sequencing offer the realistic possibility of rapidly filling the gap between clinical/biochemical characterisation and genetic diagnosis in these diseases.

This chapter will focus on the repertoire of nuclear genes related to mitochondrial neurodegeneration in children and adults.

4.2 Diseases in Children

In mitochondrial disorders of infancy and childhood, clinical and genetic findings differ from those found in adults: in general, the phenotypes are much more severe, typically involving the brain, either in isolation or as part of a multisystem condition, and mutations in nDNA are more frequent than in adulthood.

The syndromes described in this chapter are related to mutations in nuclear genes, whose protein products are imported into the mitochondria. However, it should be noted that similar clinical phenotypes can also be observed in mtDNA mutations, as described in Chap. 3.

4.2.1 Leigh Disease

Leigh disease (LD) is the most frequent mitochondrial encephalopathy in infancy and childhood. LD is a neuropathological–neuroradiological entity characterised by focal bilateral lesions in one or more areas of the deep grey matter, including the basal ganglia, thalamus, brainstem, cerebellum and rostral spinal cord. The lesions are characterised by symmetrical areas of demyelination, gliosis, necrosis with spongiotic vacuolisation and capillary proliferation.

Neurological symptoms are related to the areas affected. Clinical signs at onset usually consist of hypotonia and psychomotor regression, followed by dystonia, ataxia, involuntary movements, seizures, spasticity, evidence of brainstem dysfunction as dysphagia and respiratory problems. Neurological signs are often accompanied by general symptoms, such as failure to thrive and recurrent vomiting. The peripheral nervous system is also frequently affected with axonal and demyelinating polyneuropathy. Lactic acidosis is a virtually constant finding.

MRI neuroimaging, an essential tool for diagnosis, typically shows high T2 signal in the involved areas, most frequently in the caudate nucleus and putamen, but also in the subthalamic nuclei, periaqueductal grey matter, tegmentum of the midbrain and pons, red nuclei and dentate nuclei (Fig. 4.1a, b).

In spite of the homogeneous clinical and neuroradiological presentation, at least 15 genes have been associated to LD. The most frequent biochemical findings are isolated defects of complex I, complex IV or complex V. Mutations have been identified in both nuclear and mitochondrial genes, encoding not only single respiratory chain (RC) subunits but also assembly factors of the RC complexes or enzymes catalysing the synthesis of haem moieties or cofactors (Table 4.1). Pyruvate dehydrogenase complex (PDHC) deficiency is also related to LD with a high number of mutations in *PDHA1*, an X-linked gene encoding for the alpha subunit of the first catalytic enzyme of the complex [4]. Sporadic de novo mutations and affected females due to skewed X inactivation are not uncommon. However, PDHC deficiency can also account for other mitochondrial encephalomyopathies, with

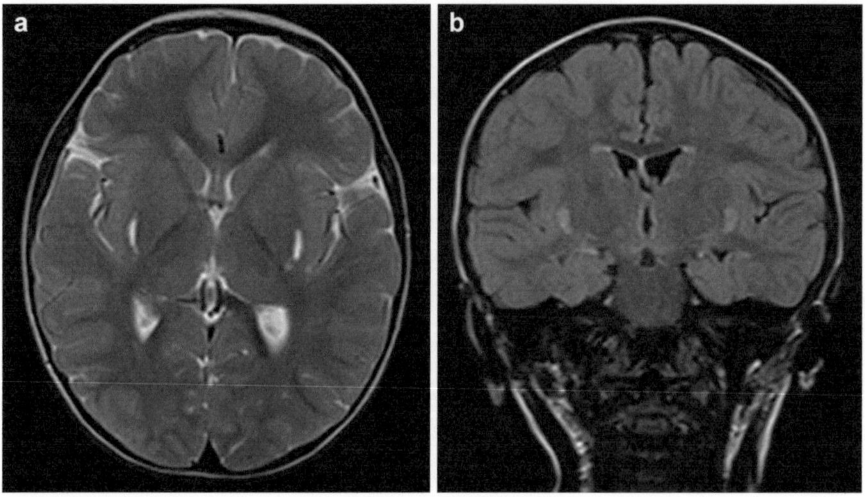

Fig. 4.1 Neuroradiological features of Leigh disease in a patient with complex IV deficiency due to a mutation in *SURF1*. (**a**) Axial T2-weighted and (**b**) coronal FLAIR. Note the bilateral hyperintensities in the putamen (posterolateral parts)

early-onset and neuroradiological findings different from those of typical LD (see below). *SURF1* is the most common nuclear gene involved in LD [5]. This gene encodes a 30 kDa protein involved in the formation of complex IV which explains the isolated, severe, generalised complex IV deficiency characterising the biochemical profile of this condition.

Although the precise mechanistic role of SURF1 is still poorly understood, it is thought to stabilise the S2–S3 assembly intermediate subcomplexes during COX assembly [6, 7]. The majority of the mutations (deletions, insertions, frameshift and nonsense) lead to a truncated protein, usually correlated to a severe phenotype, whereas patients carrying missense mutations may present with a milder phenotype characterised by later onset and longer survival [8–10].

SURF1 mutant patients often present some peculiar signs, such as open wide "staring" eyes with abnormal eye movements due to ocular ataxia, and hypertrichosis; the presence of these signs may orientate towards the correct diagnosis "at first glance".

Occasional patients may have normal MRI or different MRI patterns, such as severe cerebellar atrophy or diffuse white matter involvement [11, 12].

4.2.2 Ethylmalonic Encephalopathy

A particular form of Leigh-like disease is ethylmalonic encephalopathy (EE). EE is an example of mitochondrial disorder caused by genetically determined poisoning of the respiratory chain [13]. The *ETHE1* gene encodes the mitochondrial sulphur

Table 4.1 nDNA genes associated with Leigh disease

Gene	Inheritance	Protein function	Biochemical defects
NDUFS1, NDUFS3, NDUFS4, NDUFS7, NDUFS8, NDUFA2, NDUFA9, NDUFA10,NDUFA12, NDUFV1	AR	Subunits of complex I	Complex I
NDUFAF2,NDUFAF6, C20ORF7, FOXRED1	AR	Specific complex I assembly factors	Complex I
SDHA	AR	Flavoprotein subunit of complex II	Complex II
BCS1L	AR	Complex III assembly factor	Complex III
SURF1	AR	Complex IV assembly	Complex IV
COX10, COX15	AR	Synthesis of the haem prosthetic group for COX	Complex IV
TACO1	AR	Translation of subunit I of complex IV	Complex IV
ETHE1	AR	Mitochondrial sulphur dioxygenase	Complex IV
LRPPRC	AR	Chaperone for complex IV assembly	Complex IV (French-Canadian type of LD)
EFG1	AR	Mitochondrial translation factor	Multiple RC complexes
C12orf65	AR	Releasing peptides from mitochondrial ribosomes	Multiple RC complexes
PDHA1	X linked	α-subunit of the first catalytic enzyme of the PDHC	PDHC
COQ2	AR	Enzyme of biosynthesis of CoQ10	CoQ LD associated to nephrotic syndrome
AIF1	AR	Apoptotic-inducing factor	Multiple RC complexes

dioxygenase, an enzyme that takes part in the aerobic energetic exploitation of, and detoxification from, hydrogen sulphide, which is converted into sulphite and eventually fully oxidised to harmless sulphate. Mutations in *ETHE1* lead to accumulation of sulphide, a toxic compound that inhibits the activity of enzymes such as COX and short-chain acyl-CoA dehydrogenase, acts as a potent vasodilator and damages the colonic mucosa and the endothelia of small vessels. EE is characterised by an early onset with microangiopathy, chronic diarrhoea and neurological signs such as microcephaly, hypotonia, psychomotor delay, seizures and pyramidal signs. High excretion of ethylmalonic acid in urine and thiosulphate in plasma is a pathognomonic biomarker of the disease. Severe deficiency of complex IV is detected in the muscle biopsy [14].

4.2.3 Leukoencephalopathy

In about 10–20 % of infantile mitochondrial diseases, diffuse white matter involvement has been reported as a predominant neuroimaging feature, with little or no involvement of other structures [15].

The white matter damage is associated with early impairment of mitochondrial energy production that is crucial for myelination and maintenance of compact myelin.

Leukoencephalopathy may be related to different mitochondrial syndromes and molecular/biochemical phenotypes (Table 4.2): (i) late onset syndromes such as mitochondrial neurogastrointestinal encephalomyopathy (MNGIE); (ii) isolated or combined OXPHOS deficiencies; (iii) multiple mitochondrial dysfunction syndromes (MMDS), due to mutations in genes involved in iron–sulphur (Fe–S) cluster biosynthesis and associated to defects of complexes I, II and III and PDHC [16]; (iv) mitochondrial disorders due to mutations in mtDNA translation, particularly defects

Table 4.2 Mitochondrial leukoencephalopathies

	Molecular mutations	Biochemical defects
Late-onset syndromes	mtDNA point mutation (MELAS)	I-III-IV-V
	mtDNA deletion (KSS)	
	tRNAlys gene (MERRF)	
"Classical" OXPHOS diseases	*NDUFS1*, *NDUFV1* (and other subunit genes), *NUBPL*	I
	SDHAF1, *SDHB*, *SDHA*	II
	LYRM7	III
	COX6B1, *SURF1*, *APOPT1*	IV
	mtATP6	V
	PDHA1	PDH
	EF-Tu, *tRNA Glu*, *MPV17*	Multiple
MMDS	*NFU1*	I-II-III-PDH
	BOLA3	
	IBA57	
	ISCA2	
mtARS-related diseases	*DARS2*	I-III-IV-(V)
	EARS2	
	AARS2	
	MARS2	

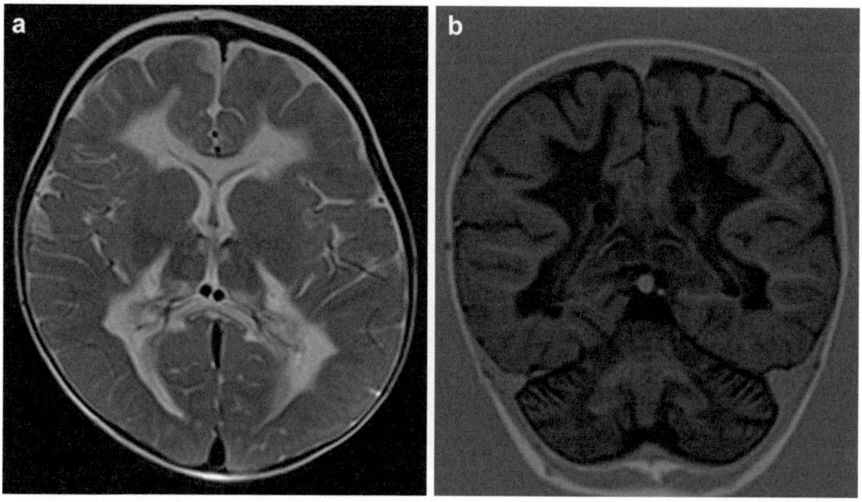

Fig. 4.2 Neuroradiological features of the leukoencephalopathy in a patient with complex II deficiency due to a mutation in *SDHAF1*. (**a**) Axial T2-weighted and (**b**) coronal T1-weighted images. Note the abnormalities of cerebral hemispheric white matter and cavitated appearance

of aminoacyl-tRNA synthetases, again associated with multiple defects in complex I-III-IV-V [17]; and (v) a still poorly defined miscellaneous group of conditions (e.g. mutations in *APOPT1* or *FBXL4*).

In MNGIE, white matter alterations are caused by an alteration of the blood–brain barrier and their clinical implications are unclear.

Early-onset OXPHOS deficiencies associated with leukoencephalopathy are most commonly characterised by isolated complex I or II defects or by MMDS. Two major clinical presentations are reported: (i) psychomotor delay since the first months of life, failure to thrive, growth impairment and a rapidly progressive course resulting in severe spastic quadriparesis and cognitive impairment and (ii) acute onset after a free period with focal motor signs, sometimes seizures, and slowly progressive course, motor impairment being usually more severe than cognitive impairment.

The MRI pattern (Fig. 4.2a, b) discloses diffuse involvement of supratentorial deep white matter, often associated with large cystic lesions within the affected white matter and progressive vacuolisation. In an increasing number of cases, involvement of brainstem or spinal cord white matter has been reported to form a specific pattern, e.g. leukoencephalopathy with brainstem and spinal cord involvement and lactate elevation (LBSL) or leukoencephalopathy with thalamus and brainstem involvement and high lactate (LTBL) [18, 19]. LBSL was first reported as a peculiar leukoencephalopathy and later recognised as a mitochondrial disease, due to mutations in *DARS2*, the gene encoding the mitochondrial aspartyl-tRNA synthetase [20, 21]. The MRI pattern shows signal abnormalities in the centrum semiovale, but sparing the U fibres, posterior arm of the internal capsule, splenium of the corpus callosum, the cerebellar peduncles, the intraparenchymal tract of the fifth cranial nerve, down to the pyramids, and the dorsal columns and corticospinal tracts of the rostral spinal cord, LBSL was initially described as a relatively mild disorder, characterised by juvenile onset of slowly progressive ataxia, spasticity and dorsal column dysfunction; an increasing number of infantile onset cases have later been reported, characterised by more rapid neurological deterioration and – in the most severe cases – early death [19]. LTBL is an early-onset leukoencephalopathy hallmarked by unique MRI features involving the centrum semiovale but consistently sparing the periventricular rim; the corpus callosum, basal ganglia, thalamus, midbrain, pons, medulla oblongata and cerebellar white matter are also consistently affected [18]. A milder phenotype was reported, with biphasic improvement of the MRI and clinical stabilisation. The responsible gene, identified by exome next-generation sequencing, is *EARS2* encoding the mitochondrial glutamyl-tRNA synthetase [18]. An autosomal recessive spastic ataxia frequently with leukoencephalopathy, thin corpus callosum and cerebellar atrophy has been associated with mutation in *MARS2* [22].

In patients with neonatal onset, particularly those affected by PDHC deficiency or *EF-Tu* gene mutation, leukoencephalopathy is frequently combined with brain structural malformations, e.g. polymicrogyria [23, 24].

Other inherited forms of mitochondrial leukodystrophy must be considered in the differential diagnosis of non-mitochondrial entities, such as vanishing white matter disease, Alexander's disease, Canavan's disease and megalencephalic leukoencephalopathy with subcortical cysts. Brain proton spectroscopy (H^+-MRSI) may be useful in these cases, as lactate concentration is usually high in brain lesions associated with mitochondrial disease [25]. However, a lactate peak may be found also in the active phase of other inherited leukodystrophies as well as in acute ischaemic or inflammatory lesions; likewise, its absence cannot rule out the diagnosis of a mitochondrial aetiology. A specific, albeit exceptionally rare, MRSI finding is a peak of succinate [26], hallmarking complex II deficiency, particularly *SDHAF1* mutations [27].

4.2.4 Encephalocardiomyopathy

Combined encephalocardiomyopathy is a severe, usually fatal, early-onset mitochondrial condition.

Subjects may present signs of the disease in utero, including intrauterine growth retardation, polyhydramnios and malformations. Patients are often critically ill at birth, with severe heart failure and lactic acidosis. Clinical findings include hypertrophic – rarely dilating – cardiomyopathy, severe hypotonia, failure to thrive and respiratory distress. Microcephaly, facial dysmorphism and liver failure are occasionally present. The clinical course may be fulminant in the neonatal period. Patients who survive usually develop neurological impairment with psychomotor delay, ataxia, pyramidal or extrapyramidal signs, cognitive stagnation and abnormal ocular movements, often associated with an MRI pattern of Leigh disease or leukoencephalopathy.

The most frequent biochemical abnormalities in encephalocardiomyopathies are isolated defects of complex I, complex IV or complex V [15]. Complex I deficiency was reported in patients with encephalocardiomyopathy, carrying mutations in genes encoding structural subunits (i.e. *NDUFS2*, *NDUFV2*, *NDUFA11*) or complex I assembly factors (*NDUFAF4*, *ACAD9*); defects of complex IV are often associated to mutations in *SCO2* or *COX15*, which are factors involved in the incorporation of copper or in the biosynthesis of the haem-a moiety, respectively, whereas defects of complex V are usually related to mutations in *TMEM70*, encoding a putative complex V assembly factor.

More recently other genotypes have been reported causing mitochondrial encephalomyopathies, with or no specific biochemical defects, for instance, *MTO1* and *GTPBP3* [28], encoding two partner proteins involved in the post-transcriptional maturation of mt-tRNAs, and mitochondrial ribosomal proteins *MRP22* and *MRPL44* [29].

4.2.5 CoQ10 Deficiency: Related Diseases

Coenzyme Q_{10} (CoQ_{10}), or ubiquinone, is a lipoidal quinone that shuttles electrons to complex III. Different syndromes associated with CoQ_{10} deficiency in the brain/muscle have been described [15], including encephalomyopathy with seizures, ataxia or mental retardation; multisystem infantile encephalopathy, cardiomyopathy and renal failure; ataxia and cerebellar atrophy; Leigh syndrome with growth retardation; and an isolated myopathy. Several cases have been reported with recurrent myoglobinuria and with ragged-red fibres/lipid storage in the muscle.

Mutations in the CoQ_{10} biosynthetic genes *COQ2*, *COQ4*, *COQ9*, *PDSS1* and *PDSS2* were reported in patients with severe infantile mitochondrial syndromes and tissue CoQ_{10} deficiency [30]. Mutations in *CABC1* (also known as *ADCK3*) usually cause spinocerebellar ataxia type 9. ADCK3 is the human ortholog of yeast CoQ8, whose specific function in CoQ10 biosynthesis is still unknown. Another form of ataxia with partial CoQ_{10} deficiency is due to mutations in *APTX* (aprataxin), causing ataxia-oculomotor apraxia syndrome, which suggest that CoQ10-associated ataxias are genetically heterogeneous and may be secondary [15].

4.2.6 Miscellaneous

The use of targeted sequencing of gene panels or exome sequencing is expanding the genetic characterisation of highly heterogeneous mitochondrial syndromes. Definite conclusions about genotype–phenotype correlation are difficult to draw in these cases, due to the exiguity of the cohort of patients so far reported in the literature. Here are a few examples.

Mutations in *AIFM1*, encoding a mitochondrial apoptosis-inducing factor (AIF), have been reported in early-onset, severe encephalopathy and axonal sensory–motor neuropathy [31], encephalomyopathy with moderate clinical severity, slow progressive course and cerebellar ataxia [32] and Charcot–Marie–Tooth X4 [33].

Mutations in *TTC19* – encoding a subunit of mitochondrial respiratory chain complex III – have been reported in few patients with heterogeneous phenotypes ranging from early-onset neurodegenerative disorders [34–37] to adult forms with psychiatric manifestations and cerebellar ataxia [38, 39].

An early-onset encephalomyopathy with progressive cerebral atrophy and involvement of the white matter, deep grey nuclei and brainstem structures has been associated with mutations in *FBXL4*, encoding a mitochondrial F-box domain-containing protein [40].

And a few individuals with isolated COX deficiency and slowly progressive, early-onset encephalomyopathy of variable severity had loss of function mutations in *APOPT1*, a gene encoding a protein of the inner mitochondrial compartment, the function of which is at present unknown [41].

4.2.7 Mitochondrial Disorders Due to Defects of Nuclear–Mitochondrial Intergenomic Signalling: mtDNA Depletion Syndromes (MDS)

Mitochondrial DNA depletion syndromes (MDS) result from a defect in nuclear-encoded factors involved in mtDNA maintenance. A spectrum of disorders of increasing severity is associated to instability of mtDNA. The most striking examples of this are the diseases associated with mutations in *POLG1*, ranging from relatively mild phenotypes associated to the presence of multiple deletions to very severe ones associated to profound mtDNA depletion.

Profound reduction of the mtDNA copy number is the molecular hallmark of MDS [42]. MDS differ from other respiratory chain disorders, as mtDNA depletion may manifest only in specific organs and so the manifestations vary from tissue-specific mtDNA depletion to widespread multisystemic disorders. MtDNA depletion causes a combined respiratory chain deficiency of the complex I, III, IV and V. Biochemical analysis of MRC enzyme activities may be normal in the muscle if the muscle is not the affected tissue. Nine genes have been so far implicated in MDS (Table 4.3).

4.2.7.1 Hepatocerebral Forms: MDS3, MDS4A, MDS6 and MDS7

Hepatocerebral MDS are probably the most common variant of MDS. Mutations have been found in four genes.

MDS4A, Mutation in *POLG1*

Mutations in *POLG1* are the most frequent cause of hepatocerebral MDS.

Table 4.3 Mitochondrial depletion syndromes

MDS forms	Phenotype	Gene
MDS1	MNGIE	*TYMP*
MDS2	Myopathic	*TK2*
MDS3	Hepatocerebral	*DGUOK*
MDS4A	Alpers	*POLG1*
MDS4B	MNGIE	*POLG1*
MDS5	Encephalomyopathic with methylmalonic aciduria	*SUCLA2*
MDS6	Hepatocerebral	*MPV17*
MDS7	Hepatocerebral	*C10orf2* (*TWINKLE*)
MDS8A	Encephalomyopathic type with renal tubulopathy	*RRM2B*
MDS8B	MNGIE	*RRM2B*
MDS9	Encephalomyopathic with methylmalonic aciduria	*SUCLG1*

POLG1 encodes pol gamma A, the catalytic subunit of polymerase gamma, which is the only known mtDNA polymerase and is essential for mtDNA replication [43]. More than a hundred mutations have been identified in *POLG1*, associated with a wide clinical spectrum raging from MDS4A [44], also known as Alpers-Huttenlocher syndrome (AHS), to juvenile-onset spinocerebellar ataxia and epilepsy (SCAE) syndrome, its variant sensory ataxia neuropathy, dysarthria, and ophthalmoparesis and dominant or recessive adult-onset progressive external ophthalmoplegia (PEO). AHS is the most severe phenotype among hepatocerebral MDS. It was described by Bernard Alpers in 1931, long before being recognised as a mitochondrial disease and associated with mutations in *POLG1* [44, 45]. AHS is hallmarked by diffuse progressive degeneration of the grey matter associated with a variable degree of liver involvement ranging from increased levels of hepatic enzymes in plasma to severe liver failure. The onset is in infancy or early childhood, although juvenile and adult cases have been reported [46, 47]. Principal clinical findings are psychomotor regression, refractory seizures and liver failure. Status epilepticus is common and poorly responsive to usual antiepileptic drugs; valproate should be avoided as it may precipitate hepatic failure [48]. In most patients with early onset, the course is rapidly progressive leading to death usually before 3 years of age [15, 49]. The MRI shows severe and progressive cortical and subcortical atrophy and often involvement of deep grey structures, particularly the thalami (Fig. 4.3 a, b). In the liver, mtDNA content can be decreased over 90 %.

MDS3, Mutation in *DGUOK*

Mutations in *DGUOK* are a frequent cause of infantile hepatocerebral MDS with over 80 reported patients and over 40 different mutations [50]. The mitochondrial deoxyguanosine kinase (dGK) mediates the phosphorylation of purine deoxyribonucleosides into the corresponding nucleotides. Most of the mutations found in *DGUOK* gene are nonsense changes, although missense mutations have also been

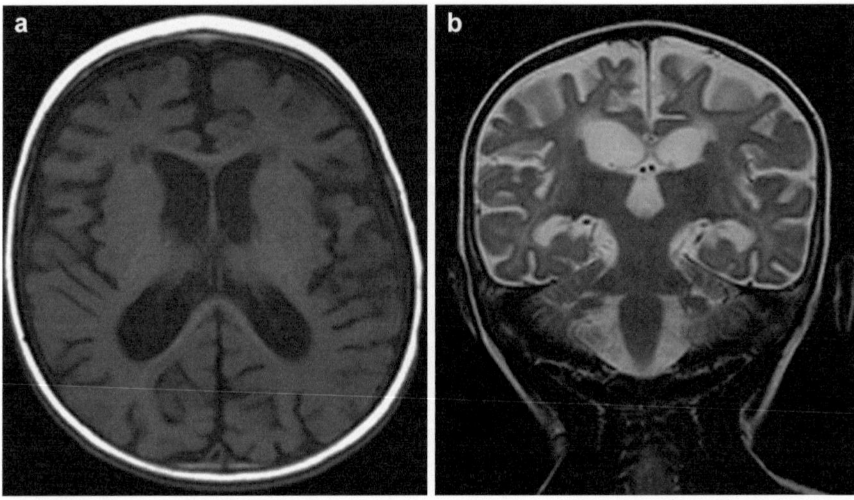

Fig. 4.3 Neuroradiological features of Alpers disease in a patient with complex PDH deficiency. (**a**) Axial T1-weighted and (**b**) coronal T2-weighted images. Note the extensive cortical atrophy

found. MDS3 manifests at birth or during the first months of life, and liver involvement seems to be the most prominent feature. The patients present at onset with hepatic cytolysis, cholestasis, jaundice and hypoglycaemia. Encephalopathy may be mild and neurological symptoms appear usually during the course of disease: the patients develop hypotonia, nystagmus and developmental delay. Null mutations tend to cause hepatocerebral disease, whilst missense mutations can cause pure hepatopathy. Patients with hepatocerebral forms usually suffer mortality due to liver failure around the age of 2 years; patients with isolated hepatopathy have better prognosis and survive longer. Liver transplantation is indicated for patients with isolated hepatopathy. mtDNA deletion in the liver is severe (<10 % residual mtDNA amount) [51].

MDS6, Mutation in *MPV17*

MPV17 mutations have been described in MDS6 and in Navajo neurohepatopathy which has a high prevalence in Navajo populations caused by the missense mutation W50Q with a founder effect. *MPV7* encodes for a protein that probably has multiple functions that are as yet still poorly understood; evidence obtained in yeast suggests its role in the structural preservation of the inner mitochondrial membrane and in the control of mtDNA maintenance and stability [52]. MDS6 may have different clinical presentations: an early phenotype with variable age of onset ranging from the first months of life till 5 years characterised by recurrent episodes of hypoglycaemia and severe progressive liver dysfunction and a later-onset form with moderate hepatopathy and progressive sensory-motor axonal neuropathy. All forms are also associated with variable degrees of demyelination both in central and peripheral nervous systems (see "leukoencephalopathy" previously in the text). The patients with early-onset forms usually go on to develop neurological symptoms as ataxia, hypotonia, dystonia and psychomotor regression. Liver transplant may be efficacious in the acute phase of liver dysfunction but will not prevent patients from developing neurological progressive syndrome. Marked mtDNA depletion in the liver, ranging from 5 % to 15 %, and multiple defects of mtDNA-related respiratory chain complexes are present in liver biopsy, whilst in the skeletal muscle, the amount of mtDNA may be only slightly reduced.

MDS7, Mutation in *C10ORF2*

C10ORF2 encodes for mtDNA helicase Twinkle protein that has a functional interaction with POLG and is necessary for mtDNA replication. Different phenotypes are related to mutations in *C10ORF2* including autosomal dominant PEO, infantile-onset spinocerebellar ataxia (IOSCA) and hepatocerebral phenotype (MDS7). Neonatal or infantile onset of MDS7 is very rare, and clinical manifestations are those of severe hepatocerebral forms reminiscent of Alpers disease [53]. Later in infancy and adolescent, patients develop ataxia, sensory neuropathy, athetosis, sensorineural deafness and epilepsy. MRI shows atrophy of the cerebellum, brainstem and spinal cord. Lactate is elevated in plasma and CSF. More recently a novel homozygous mutation in *C10ORF2* has been described associated with hepatocerebral forms and renal tubulopathy, suggesting an expanding phenotype in "hepato-cerebro-renal" [54].

4.2.7.2 Encephalomyopathic Forms: MDS5, MDS8A and MDS9

MDS5 and MDS9, Mutation in *SUCLA2* and *SUCLG1*

SUCLG1 and *SUCLA2* encode, respectively, the alpha- and beta-subunits of succinyl-CoA synthetase, a mitochondrial matrix enzyme that catalyses the reversible conversion of substrates succinyl-CoA and GDP/ADP to succinate and GTP/ATP in the tricarboxylic acid pathway. So far about 34 patients with *SUCLA2* mutations have been reported with a high prevalence in the Faroe Islands due to a founder effect [55]. At onset patients have psychomotor delay with irritability and hypotonia, and then as the disease progress, they develop sensorineural deafness, severe spastic dystonic quadriparesis, athetoid or choreiform movements, feeding difficulty, failure to thrive, growth retardation and respiratory insufficiency. MRI shows cerebral, cerebellar and medulla oblongata atrophy and bilateral basal ganglia involvement [56]. Almost all the patients have increased urinary excretion of methylmalonic acid that should be considered as a useful marker although recently two cases without methylmalonic aciduria have been reported [57, 58]. *SUCLG1* is rarer and presents with a similar, but usually worse phenotype and more severe mtDNA depletion in the muscle [59].

MDS8A, Mutation in *RRM2B*

RRM2B gene encodes the small subunit of p53-inducible ribonucleotide reductase, essential for the conversion of ribonucleoside diphosphates to deoxyribonucleoside diphosphates, which is crucial for mtDNA synthesis. Mutations in *RRM2B* have been reported in association to MNGIE-like phenotype or with a severe infantile encephalomyopathy with renal tubulopathy and mtDNA depletion or autosomal dominant progressive external ophthalmoplegia (PEO) with multiple mtDNA deletions. The infantile phenotype is characterised by early-onset hypotonia, failure to thrive, psychomotor delay, myopathy, central hypomyelination, sensorineural deafness and tubulopathy. The disease progresses leading to death in early months and lactic acidosis is a constant feature [60, 61].

4.2.8 Infantile Mitochondrial Neurological Diseases Not Related to MRC Dysfunction

4.2.8.1 Microphthalmia with Linear Skin Defects

The microphthalmia with linear skin defects (MLS) syndrome is an X-linked neurocutaneous disorder manifesting exclusively in females, suggesting embryonic lethality in hemizygous males. Affected individuals typically show uni- or bilateral microphthalmia and/or anophthalmia and linear skin lesions, usually associated with ocular anomalies, such as microcornea, coloboma, anterior chamber defects, optic nerve hypoplasia, etc. Developmental delay, abnormalities of the central nervous system, short stature and cardiac defects have also been observed.

Three mitochondrial genes have been found implicated in the pathogenesis of this disease [62]: *HCCS*, encoding the mitochondrial holocytochrome c-type synthase; *COX7B*, encoding a poorly characterised structural subunit of cytochrome c oxidase; and *NDUFB11*, encoding a poorly characterised supernumerary subunit of CI.

4.2.8.2 Neurodegeneration with Brain Iron Accumulation

Neurodegeneration with brain iron accumulation (NBIA) is a family of degenerative extrapyramidal monogenic disorders characterised by focal accumulation of iron in the brain, usually in the basal ganglia [63]. They are characterised by early or late onset, with the main symptoms associated to problems in movement, spasticity and cognitive impairment. NBIA is a general definition gathering a number of diseases due to mutations in genes, some of which are present in mitochondrial isoforms, including: pantothenate kinase-associated neurodegeneration, PKAN, due to mutations in *PANK2*; phospholipase 2, group VI-associated neurodegeneration (PLAN), due to mutations in *PLA2G6*; mitochondrial membrane protein-associated neurodegeneration (MPAN), due to mutations in *C19ORF12*, a gene of unknown function; fatty acid hydroxylase-associated neurodegeneration (FAHN), due to mutations in *FAH2*; and COASY protein-associated neurodegeneration (CoPAN), due to mutations in the CoA synthetase gene *COASY*.

A common feature of PKAN and PLAN is the fragmentation of the mitochondrial network with alteration of cristae morphology, probably linked to impaired synthesis (PanK2) or the remodelling (iPLA2β) of membrane lipids, suggesting a role for lipid metabolism alterations in the pathogenesis of the disease [64].

4.3 Diseases in Adults

4.3.1 Friedreich Ataxia

Friedreich ataxia (FRDA) is an autosomal recessive disorder caused by pathological GAA trinucleotide repeat expansions in the first intron of the FXN gene [65]. The encoded protein, frataxin, is a mitochondrial matrix, 21 kDa polypeptide involved in the formation of iron–sulphur clusters, which are critical components of the mitochondrial respiratory chain complexes and of many redox enzymes not only in mitochondria, but also in the cytoplasm and nucleus. *FXN* mutations impair OXPHOS and result in abnormal accumulation of intra-mitochondrial iron, which eventually reaches toxic levels. In addition, since frataxin has antioxidant properties, cellular defences against ROS are impaired in FRDA, further exacerbating neuronal damage. The onset of the disease is in the second decade of life, with progressive gait ataxia, loss of deep tendon reflexes, dysarthria, distal limb weakness, pes cavus, scoliosis and arrhythmias secondary to hypertrophic cardiomyopathy. In addition, optic nerve dysfunction is often present, although *FXN* mutations probably cause RGC loss via other disease mechanisms compared with *OPA1* and the primary mtDNA LHON mutations [66].

4.3.2 MNGIE

Mitochondrial neurogastrointestinal encephalomyopathy (MNGIE) is an autosomal recessive disease caused by loss-of-function mutations in *TYMP*, encoding thymidine phosphorylase (TP), a key enzyme in the degradation of pyrimidine nucleosides, thymidine (dThd) and deoxyuridine (dUrd) [67]. In MNGIE patients, TP deficiency leads to systemic accumulation of dThd and dUrd. This overload results in imbalanced mitochondrial deoxynucleotides, which impairs replication of mitochondrial DNA (mtDNA), and accumulation of point mutations and multiple deletions, resulting in mitochondrial dysfunction. The clinical phenotype is multisystemic, the skeletal and smooth muscle and peripheral nerves being exquisitely affected. As a result, polyneuropathy, myopathy with CPEO and severe impairment of intestinal peristalsis lead to progressive cachexia and neuromuscular impairment. As already mentioned, a functional leukoencephalopathy of uncertain significance is a virtually consistent MRI finding.

4.3.3 Diseases of Mitochondrial Dynamics and Quality Control

Mitochondria are tightly regulated by a complex quality control system, acting at either the organellar or protein level (see Chaps. 11 and 12). Mitochondria are highly dynamic organelles, undergoing continuous fusion and fission events [68] (see Chap. 7). In general, mitochondria undergoing fusion are protected from degradation, whereas fragmented organelles are prone to disposal [69]. Dynamin-related GTPases on the OM (mitofusins, MFN1 and MFN2) and IM (OPA1) control the fusion process, whereas fission is regulated by the cytosolic soluble dynamin-related protein 1 (DRP1). DRP1 interacts with docking adaptors (FIS1, MFF and MiD49/51), which form spiral filaments that constrict around mitochondria determining fission. This process is important not only for the distribution of mitochondria during cell division, but also for the elimination of dysfunctional mitochondria, as fragmented organelles are targeted for autophagic degradation, a process called mitophagy (see Chap. 11). Mitophagy is controlled by the PTEN-induced putative kinase 1 (PINK1) and by parkin [70]. PINK1 targets to mitochondria, but under normal conditions, it is rapidly degraded by a number of mitochondrial intermembrane and matrix peptidases [71], whereas parkin is a cytosolic component of E3 ubiquitin-protein ligase that ubiquitinates proteins targeted for degradation. When mitochondria depolarise, PINK1 is prevented from degradation and accumulates on the outer mitochondrial membrane, where it phosphorylates MFN2, which acts as a receptor for parkin. Parkin is then brought in close proximity to and phosphorylated by PINK1, initiating the ubiquitination of downstream targets (discussed in detail in Chap. 11).

A second line of defence operates within mitochondria and consists of chaperones and proteases, which promote folding of newly imported pre-proteins, protect mitochondrial proteins against stress and degrade irreversibly damaged polypeptides [72]. Once a mitochondrial precursor protein has been imported into the mitochondria, it is processed by the mitochondrial processing peptidase (MPP), which removes the mitochondrial signalling peptide that then gets degraded by the presequence peptidase (PreP). The mature protein is subsequently folded by the chaperones of the Hsp60 and Hsp70 families. Misfolded proteins are eventually degraded by the Lon protease.

In the IM, protein quality control is monitored by two AAA proteolytic complexes: the intermembrane space-ATPase Associated with various cellular Activities (i-AAA) protease and the matrix-ATPase Associated with various cellular Activities (m-AAA) protease. The i-AAA protease is a homo-oligomeric machine composed of a single protein subunit (YME1L). Conversely, the m-AAA protease is composed by either homo-oligomeric or hetero-oligomeric complexes composed of paraplegin (encoded by *SPG7*) and AFG3L2 [72].

Notably, the two quality control systems are somehow interconnected. In fact, the machinery controlling the dynamic behaviour of the mitochondrial network through balanced fusion and fission is controlled by the activity of the mitochondrial proteases, which, in turn, thus control the fission–fusion processes. An example is OPA1, a master regulator of mitochondrial fusion, whose processing is controlled by YME1L and by another IM metallopeptidase (OMA1) [73]. This regulation has a profound impact on cell survival. In fact, fusion is a pro-survival mechanism protecting against apoptosis and neurodegeneration, whilst fission occurs early during mitophagy and apoptosis.

4.3.3.1 Hereditary Spastic Paraplegia

Pathogenic mutations in the *SPG7* lead to an autosomal recessive form of hereditary spastic paraplegia (HSP), a progressive disorder characterised by weakness, spasticity (muscle rigidity) and loss of the vibratory sense of the lower limbs. HSP is caused by the selective retrograde degeneration of the longest motor and sensory axons of the central nervous system, the corticospinal tracts and the fasciculus gracilis.

Mutations in *AFG3L2* have been associated with an autosomal dominant form of spinocerebellar ataxia, SCA28 [74]. SCA28 patients are affected by a slowly progressive gait and limb incoordination of juvenile onset, dysarthria, hyperreflexia at lower limbs, nystagmus, ptosis and ophthalmoplegia. Accumulation of multiple mtDNA deletions has been described in patients with mutations in both *AFGL3L2* and *SPG7*, encoding paraplegin, the molecular partner of AFGL3L2 in the AAA+ complex [75, 76].

A homozygous mutation in *AFG3L2* in a highly conserved tyrosine at the beginning of the peptidase domain to a cysteine has been associated with an early-onset syndrome characterised by severe spastic paraplegia, ataxia, ptosis, oculomotor apraxia, dystonic movements and stimulus-induced myoclonus [77]. Strikingly, this phenotype combines manifestations of both HSP and SCA28 along with clinical features of other mitochondrial diseases (such as ptosis, ophthalmoplegia and myoclonic epilepsy). Defective mitochondrial translation has been documented in a *Afg3l2*$^{-/-}$ mouse model of the disease, suggesting a critical role of this process in the pathogenesis of the disease [78].

4.3.3.2 Charcot–Marie–Tooth Peripheral Neuropathy

Charcot–Marie–Tooth (CMT) disease is characterised by progressive degeneration of the peripheral nerves [79]. Patients develop distal muscle weakness and sensory loss, but depending on the actual causative gene, there can be quite marked variations in age of onset and rate of clinical progression. Mutations in MFN2 have been found in families with CMT2A, an autosomal dominant form of CMT [80]. The MFN2 gene consists of 19 exons and codes for a 757 amino acid, dynamin-related GTPase protein, which is anchored within the outer mitochondrial membrane. Although severe peripheral neuropathy with prominent proprioceptive loss is the main clinical sign, some patients show subacute visual failure and optic atrophy [79]. In other cases, a multisystem encephalomyopathy associated with accumulation of multiple deletions.

4.3.3.3 Dominant Optic Atrophy

Dominant optic atrophy (DOA) is the most common inherited optic nerve disorder. Although the onset may be in childhood, symptoms progress very slowly and poor vision becomes manifest in adulthood. Neuronal loss is typically limited to retinal ganglion cells (RGC), particularly those forming the papillo-macular bundle, leading to progressive optic nerve degeneration and visual failure. In most of the cases, DOA is due to heterozygous mutations in *OPA1*, encoding a dynamin-related GTPase localised in the inner mitochondrial membrane, although several other loci have been found but not characterised except *OPA3*, whose role and localisation are however still debated. OPA1 is by far the best characterised DOA gene [79]. It is expressed ubiquitously but particularly enriched within the RGC layer and photoreceptors. Accordingly, OPA1 mutations can affect other CNS populations, peripheral nerves and skeletal muscle (the so-called DOA+ phenotypes).

The majority of OPA1 mutations result in premature termination codons, leading to truncated mRNA species, which are rapidly degraded via nonsense-mediated mRNA decay. Haploinsufficiency, therefore, is thus a major disease mechanism in DOA. The spectrum of OPA1-linked phenotypes now encompasses a wide range of prominent neuromuscular features such as ataxia, myopathy, peripheral neuropathy and classical chronic progressive external ophthalmoplegia (CPEO), with unusually high frequency of late-onset Parkinsonism [81].

4.4 Secondary Mitochondrial Involvement in Neurodegeneration

Mitochondrial impairment has been proven to be involved in the pathogenesis of a number of classical neurodegenerative disorders including Parkinson's disease and Parkinsonisms. In other neurodegenerative conditions, such as Huntington's disease, Alzheimer's disease and amyotrophic lateral sclerosis, a role of mitochondrial dysfunction has been hypothesised, but more work is needed to test this mechanistically.

4.4.1 Parkinson's Disease

Parkinson's disease (PD) is one of the most common neurodegenerative diseases in the ageing population. It is characterised by the clinical triad of rigidity, bradykinesia and tremor, due to the loss of dopaminergic neurons in the substantia nigra with the presence of typical intracytoplasmic ubiquitin- and α-synuclein-positive inclusions, dubbed Lewy bodies. The link between mitochondrial dysfunction and PD stems from several observations, including those that neurotoxins which affect respiratory chain complex I induce specific death of dopaminergic neurons. Multiple mtDNA deletions accumulate in dopaminergic neurons, and a number of causative genes in familial forms of PD have been found in juvenile-onset familial Parkinson's disease [82]. These include PINK1, Parkin and DJ1, encoding a mitochondrial and cytosolic chaperone; and *LRRK2*, encoding a protein primarily localised to the cytoplasm and membranes, including the mitochondrial membrane, where it has multiple functions such as kinase, GTPase and scaffolding protein [83].

4.4.2 Huntington's Disease

HD is a devastating neurodegenerative disease resulting from an expansion of CAG triplet repeats in the *HTT* gene. This causes a corresponding expansion of the polyglutamine tract in the N-terminal region of the huntingtin protein. This polyQ expansion eventually leads to the accumulation of intracellular HTT aggregates [84]. Neurodegeneration occurs in the striatum during early pathogenesis, but broad areas of the brain are also affected during later stages of disease progression. Changes in mitochondrial dynamics have been found in HD patients and animal models of the disease, whose mitochondria are excessively fragmented and show decreased motility and respiration. This effect could be mediated by Drp1, a potential target of mutant HTT.

4.4.3 Alzheimer's Disease

Alzheimer's disease (AD) is the most common form of dementia, affecting over 24 million people worldwide [85]. The onset is usually between 60 and 70 years of age. AD is clinically characterised by memory loss and impairment of other cognitive including language, orientation, constructional abilities, abstract thinking, ability to solve problems, etc. The two main neuropathological hallmarks currently used to diagnose AD are the accumulation of intracellular neurofibrillary tangles, consisting mainly of hyperphosphorylated forms of the microtubule-associated protein tau, and the deposition of extracellular neuritic plaques mainly composed of β-amyloid (Aβ), a small peptide derived from processing of the amyloid-β precursor protein (APP; gene APP). The vast majority of cases of AD are sporadic, although in about 1 % of patients the disease is inherited (usually as an autosomal-dominant trait) as a result of mutations in one of three genes: APP, PSEN1 (presenilin-1; PS1) and PSEN2 (presenilin-2; PS2). PSEN1 and PSEN2 are components of the γ-secretase complex, which cleaves the C-terminal membrane domain of APP to produce Aβ peptides. Clinically, familial AD (FAD) is similar to the sporadic form (SAD), but has an earlier age of onset and is more aggressive [86].

The pathogenesis of the disease is highly debated, but two main hypotheses are nowadays the most accepted [85]. The first model is the "amyloid cascade" hypothesis, which proposes that the disease is due to the accumulation of $A\beta_{1-42}$ plaques, which are toxic to the cells. The resulting cellular stress triggers tau hyperphosphorylation via upregulation of calcium-sensitive kinases, leading to tangle formation.

The alternative is the mitochondrial cascade hypothesis, which stresses the role of age-dependent mitochondrial dysfunction and increased ROS production in Aβ deposition and AD pathogenesis. Mitochondrial dysfunction is among the earliest observed pathogenic alterations, detected well before the accumulation of neuritic plaques. A growing number of studies are emerging which report impaired mitochondrial functions including electron transfer, ATP synthesis, mitochondrial transcription, translation and protein synthesis, upregulation of voltage-dependent anion channel (VDAC) and increased production of reactive oxygen species (ROS) in AD patients and in AD transgenic mouse models. The Aβ accumulation in mitochondria from postmortem AD brains and cellular and transgenic mice models has been well documented, providing evidence that Aβ is physically localised within or on the surface of mitochondria, where it has been reported to interact with a number of factors, such as Drp1, Aβ-binding alcohol dehydrogenase (ABAD), cyclophilin D (CypD), cytochrome c oxidase, VDAC and hPreP. However, the connection between mitochondrial dysfunction and the progression of AD is not fully understood, especially whether mitochondrial dysfunction triggers Aβ accumulation, or it is the latter that cause mitochondrial impairment. In addition, it is still unclear how Aβ enters mitochondria [87]. Notably, it has been recently observed that the presenilins and the γ-secretase activity are predominantly located in the mitochondria-associated ER membranes (MAMs), a dynamic and highly specialised subdomain of the ER facing to and connected with the mitochondrial outer membrane, playing essential roles in lipid synthesis and transport between the two organelles, fatty acids, glucose and cholesterol metabolism, Ca^{2+} homeostasis and apoptosis [88].

4.4.4 Amyotrophic Lateral Sclerosis

Amyotrophic lateral sclerosis (ALS) is the most prevalent, adult-onset, motor neuron disease, characterised by the degeneration of upper and lower motor neurons, progressive muscle weakness and atrophy, leading to muscle paralysis.

Approximately, 90 % of patients develop ALS with unknown aetiology, classified as sporadic ALS (sALS), whereas the remaining 10 % are familial ALS cases (fALS) due to genetic defects [89]. An increasing number of genes are being linked to ALS. Superoxide dismutase 1 (SOD1) was the first gene to be discovered as responsible for fALS [90]. SOD1 mutations account for 20 % of fALS dominant cases, whereas they are rare in idiopathic cases. In the last decade, a number of mutations have been identified in several DNA-/RNA-binding proteins (e.g. TDP-43 and FUS). Other fALS genes, such as VCP and SQSTM1, are related to proteostatic pathways and cytoskeletal and axonal transport, such as PFN1 and DCTN1. Recently, the largest proportion of fALS cases (40 %) has been linked to intronic hexanucleotide repeat expansions in *C9ORF72*, a gene with still unknown function.

The proposed pathogenic mechanisms include oxidative stress, toxic activity of misfolded and aggregated proteins, endoplasmic reticulum stress, mitochondrial dysfunction and axonal disorganisation, including organelle transport defects [89].

4.5 Conclusion

The finding of numerous mitochondrial genes involved in mitochondrial neurodegeneration has highlighted the central role of mitochondria in neurodegenerative processes. The elucidation of the pathways in which these genes are involved will shed new light not only on the specific disease, but likely on the processes of idiopathic neurodegenerative diseases.

Glossary

Ataxia Lack of voluntary coordination of muscle movements that includes gait abnormality

Athetosis Slow, involuntary, convoluted, movements of the fingers, hands, toes and feet

Coloboma A hole in one of the structures of the eye (e.g. iris, retina, choroid or optic disc)

Demyelinating polyneuropathy Neurological disorder characterised by progressive weakness and impaired sensory function in the legs and arms due to the damage to the myelin sheath

Dysarthria Speech disorder due impaired coordination of movements in the muscles used for speech production

Dysphagia Difficulty in swallowing

Dystonia A state of abnormal muscle tone resulting in muscular spasm and abnormal posture

Hyperreflexia Abnormal reaction of the involuntary (autonomic) nervous system to stimulation

Hypertrichosis Abnormal amount of hair growing over the body

Hypotonia Low resistance to stretch of muscles

Leukoencephalopathy Disease affecting the brain white matter

Megalencephalic leukoencephalopathy Progressive condition that affects brain development and function, characterised by an enlarged brain

Microangiopathy Disease affecting small vessels

Myoclonus Brief, involuntary twitching of a muscle or a group of muscles

Myoglobinuria Presence of myoglobin in the urine
Nystagmus Repetitive, uncontrolled movements of the eyes
Ophthalmoplegia Slowly progressive paralysis of the extraocular muscles
Pes cavus High arch of the foot
Polyhydramnios Excessive accumulation of amniotic fluid
Pyramidal signs Symptoms related to the descending tract originated from pyramidal cells of motor cortex
Quadriparesis Weakness of the four limbs
Seizures Symptoms due to abnormal excessive or synchronous neuronal activity in the brain
Spasticity Abnormal muscle contraction characterised by a combination of paralysis, increased tendon reflex activity and hypertonia
Supratentorial The brain area located above the tentorium cerebelli, the membranous roof over the cerebellum
T1, T2, T2 FLAIR Neuroradiological terms that refer to the relaxation times that protons need to revert back to their resting states after the initial pulse at a certain radiofrequency

References

1. Wallace DC. A mitochondrial paradigm of metabolic and degenerative diseases, aging, and cancer: a dawn for evolutionary medicine. Ann Rev Genet. 2005;39:359–407.
2. Zeviani M, Di Donato S. Mitochondrial disorders. Brain. 2004;127(Pt 10):2153–72.
3. Elliott HR, Samuels DC, Eden JA, Relton CL, Chinnery PF. Pathogenic mitochondrial DNA mutations are common in the general population. Am J Hum Genet. 2008;83(2):254–60.
4. Matthews PM, Marchington DR, Squier M, Land J, Brown RM, Brown GK. Molecular genetic characterization of an X-linked form of Leigh's syndrome. Ann Neurol. 1993;33(6):652–5.
5. Finsterer J. Leigh and Leigh-like syndrome in children and adults. Pediatr Neurol. 2008;39(4):223–35.
6. Shoubridge EA. Cytochrome c oxidase deficiency. Am J Med Genet. 2001;106(1):46–52.
7. Zeviani M, Corona P, Nijtmans L, Tiranti V. Nuclear gene defects in mitochondrial disorders. Ital J Neurol Sci. 1999;20(6):401–8.
8. Lee IC, El-Hattab AW, Wang J, Li FY, Weng SW, Craigen WJ, et al. SURF1-associated Leigh syndrome: a case series and novel mutations. Hum Mutat. 2012;33(8):1192–200.
9. Piekutowska-Abramczuk D, Magner M, Popowska E, Pronicki M, Karczmarewicz E, Sykut-Cegielska J, et al. SURF1 missense mutations promote a mild Leigh phenotype. Clin Genet. 2009;76(2):195–204.
10. Tiranti V, Jaksch M, Hofmann S, Galimberti C, Hoertnagel K, Lulli L, et al. Loss-of-function mutations of SURF-1 are specifically associated with Leigh syndrome with cytochrome c oxidase deficiency. Ann Neurol. 1999;46(2):161–6.
11. Rahman S, Brown RM, Chong WK, Wilson CJ, Brown GK. A SURF1 gene mutation presenting as isolated leukodystrophy. Ann Neurol. 2001;49(6):797–800.
12. Salviati L, Freehauf C, Sacconi S, DiMauro S, Thoma J, Tsai AC. Novel SURF1 mutation in a child with subacute encephalopathy and without the radiological features of Leigh Syndrome. Am J Med Genet Part A. 2004;128A(2):195–8.
13. Tiranti V, Viscomi C, Hildebrandt T, Di Meo I, Mineri R, Tiveron C, et al. Loss of ETHE1, a mitochondrial dioxygenase, causes fatal sulfide toxicity in ethylmalonic encephalopathy. Nat Med. 2009;15(2):200–5.

14. Tiranti V, D'Adamo P, Briem E, Ferrari G, Mineri R, Lamantea E, et al. Ethylmalonic encephalopathy is caused by mutations in ETHE1, a gene encoding a mitochondrial matrix protein. Am J Hum Genet. 2004;74(2):239–52.
15. Uziel G, Ghezzi D, Zeviani M. Infantile mitochondrial encephalopathy. Semin Fetal Neonat Med. 2011;16(4):205–15.
16. Invernizzi F, Ardissone A, Lamantea E, Garavaglia B, Zeviani M, Farina L, et al. Cavitating leukoencephalopathy with multiple mitochondrial dysfunction syndrome and NFU1 mutations. Front Genet. 2014;5:412.
17. Diodato D, Ghezzi D, Tiranti V. The mitochondrial aminoacyl tRNA synthetases: genes and syndromes. Int J Cell Biol. 2014;2014:787956.
18. Steenweg ME, Ghezzi D, Haack T, Abbink TE, Martinelli D, van Berkel CG, et al. Leukoencephalopathy with thalamus and brainstem involvement and high lactate 'LTBL' caused by EARS2 mutations. Brain. 2012;135(Pt 5):1387–94.
19. van Berge L, Hamilton EM, Linnankivi T, Uziel G, Steenweg ME, Isohanni P, et al. Leukoencephalopathy with brainstem and spinal cord involvement and lactate elevation: clinical and genetic characterization and target for therapy. Brain. 2014;137(Pt 4):1019–29.
20. Scheper GC, van der Klok T, van Andel RJ, van Berkel CG, Sissler M, Smet J, et al. Mitochondrial aspartyl-tRNA synthetase deficiency causes leukoencephalopathy with brain stem and spinal cord involvement and lactate elevation. Nat Genet. 2007;39(4):534–9.
21. van der Knaap MS, van der Voorn P, Barkhof F, Van Coster R, Krageloh-Mann I, Feigenbaum A, et al. A new leukoencephalopathy with brainstem and spinal cord involvement and high lactate. Ann Neurol. 2003;53(2):252–8.
22. Bayat V, Thiffault I, Jaiswal M, Tetreault M, Donti T, Sasarman F, et al. Mutations in the mitochondrial methionyl-tRNA synthetase cause a neurodegenerative phenotype in flies and a recessive ataxia (ARSAL) in humans. PLoS Biol. 2012;10(3):e1001288.
23. Moroni I, Bugiani M, Bizzi A, Castelli G, Lamantea E, Uziel G. Cerebral white matter involvement in children with mitochondrial encephalopathies. Neuropediatrics. 2002;33(2):79–85.
24. Valente L, Tiranti V, Marsano RM, Malfatti E, Fernandez-Vizarra E, Donnini C, et al. Infantile encephalopathy and defective mitochondrial DNA translation in patients with mutations of mitochondrial elongation factors EFG1 and EFTu. Am J Hum Genet. 2007;80(1):44–58.
25. Sofou K, Steneryd K, Wiklund LM, Tulinius M, Darin N. MRI of the brain in childhood-onset mitochondrial disorders with central nervous system involvement. Mitochondrion. 2013;13(4):364–71.
26. Brockmann K, Bjornstad A, Dechent P, Korenke CG, Smeitink J, Trijbels JM, et al. Succinate in dystrophic white matter: a proton magnetic resonance spectroscopy finding characteristic for complex II deficiency. Ann Neurol. 2002;52(1):38–46.
27. Ghezzi D, Goffrini P, Uziel G, Horvath R, Klopstock T, Lochmuller H, et al. SDHAF1, encoding a LYR complex-II specific assembly factor, is mutated in SDH-defective infantile leukoencephalopathy. Nat Genet. 2009;41(6):654–6.
28. Kopajtich R, Nicholls TJ, Rorbach J, Metodiev MD, Freisinger P, Mandel H, et al. Mutations in GTPBP3 cause a mitochondrial translation defect associated with hypertrophic cardiomyopathy, lactic acidosis, and encephalopathy. Am J Hum Genet. 2014;95(6):708–20.
29. Baertling F, Haack TB, Rodenburg RJ, Schaper J, Seibt A, Strom TM, et al. MRPS22 mutation causes fatal neonatal lactic acidosis with brain and heart abnormalities. Neurogenetics. 2015;16(3):237–40.
30. Brea-Calvo G, Haack TB, Karall D, Ohtake A, Invernizzi F, Carrozzo R, et al. COQ4 mutations cause a broad spectrum of mitochondrial disorders associated with CoQ10 deficiency. Am J Hum Genet. 2015;96(2):309–17.
31. Ghezzi D, Sevrioukova I, Invernizzi F, Lamperti C, Mora M, D'Adamo P, et al. Severe X-linked mitochondrial encephalomyopathy associated with a mutation in apoptosis-inducing factor. Am J Hum Genet. 2010;86(4):639–49.
32. Ardissone A, Piscosquito G, Legati A, Langella T, Lamantea E, Garavaglia B, et al. A slowly progressive mitochondrial encephalomyopathy widens the spectrum of AIFM1 disorders. Neurology. 2015;84(21):2193–5.

33. Rinaldi C, Grunseich C, Sevrioukova IF, Schindler A, Horkayne-Szakaly I, Lamperti C, et al. Cowchock syndrome is associated with a mutation in apoptosis-inducing factor. Am J Hum Genet. 2012;91(6):1095–102.
34. Ardissone A, Granata T, Legati A, Diodato D, Melchionda L, Lamantea E, et al. Mitochondrial complex III deficiency caused by TTC19 defects: report of a novel mutation and review of literature. JIMD Rep. 2015;22:115–20.
35. Atwal PS. Mutations in the complex III assembly factor tetratricopeptide 19 gene TTC19 are a rare cause of Leigh syndrome. JIMD Rep. 2014;14:43–5.
36. Ghezzi D, Arzuffi P, Zordan M, Da Re C, Lamperti C, Benna C, et al. Mutations in TTC19 cause mitochondrial complex III deficiency and neurological impairment in humans and flies. Nat Genet. 2011;43(3):259–63.
37. Mordaunt DA, Jolley A, Balasubramaniam S, Thorburn DR, Mountford HS, Compton AG, et al. Phenotypic variation of TTC19-deficient mitochondrial complex III deficiency: a case report and literature review. Am J Med Genet Part A. 2015;167(6):1330–6.
38. Morino H, Miyamoto R, Ohnishi S, Maruyama H, Kawakami H. Exome sequencing reveals a novel TTC19 mutation in an autosomal recessive spinocerebellar ataxia patient. BMC Neurol. 2014;14:5.
39. Nogueira C, Barros J, Sa MJ, Azevedo L, Taipa R, Torraco A, et al. Novel TTC19 mutation in a family with severe psychiatric manifestations and complex III deficiency. Neurogenetics. 2013;14(2):153–60.
40. Gai X, Ghezzi D, Johnson MA, Biagosch CA, Shamseldin HE, Haack TB, et al. Mutations in FBXL4, encoding a mitochondrial protein, cause early-onset mitochondrial encephalomyopathy. Am J Hum Genet. 2013;93(3):482–95.
41. Melchionda L, Haack TB, Hardy S, Abbink TE, Fernandez-Vizarra E, Lamantea E, et al. Mutations in APOPT1, encoding a mitochondrial protein, cause cavitating leukoencephalopathy with cytochrome c oxidase deficiency. Am J Hum Genet. 2014;95(3):315–25.
42. Moraes CT, Shanske S, Tritschler HJ, Aprille JR, Andreetta F, Bonilla E, et al. mtDNA depletion with variable tissue expression: a novel genetic abnormality in mitochondrial diseases. Am J Hum Genet. 1991;48(3):492–501.
43. Korhonen JA, Pham XH, Pellegrini M, Falkenberg M. Reconstitution of a minimal mtDNA replisome in vitro. EMBO J. 2004;23(12):2423–9.
44. Naviaux RK, Nguyen KV. POLG mutations associated with Alpers' syndrome and mitochondrial DNA depletion. Ann Neurol. 2004;55(5):706–12.
45. Naviaux RK, Nyhan WL, Barshop BA, Poulton J, Markusic D, Karpinski NC, et al. Mitochondrial DNA polymerase gamma deficiency and mtDNA depletion in a child with Alpers' syndrome. Ann Neurol. 1999;45(1):54–8.
46. Uusimaa J, Hinttala R, Rantala H, Paivarinta M, Herva R, Roytta M, et al. Homozygous W748S mutation in the POLG1 gene in patients with juvenile-onset alpers syndrome and status epilepticus. Epilepsia. 2008;49(6):1038–45.
47. Wiltshire E, Davidzon G, DiMauro S, Akman HO, Sadleir L, Haas L, et al. Juvenile Alpers disease. Arch Neurol. 2008;65(1):121–4.
48. Horvath R, Hudson G, Ferrari G, Futterer N, Ahola S, Lamantea E, et al. Phenotypic spectrum associated with mutations of the mitochondrial polymerase gamma gene. Brain. 2006;129(Pt 7):1674–84.
49. Milone M, Massie R. Polymerase gamma 1 mutations: clinical correlations. Neurologist. 2010;16(2):84–91.
50. Poulton J, Hirano M, Spinazzola A, Arenas Hernandez M, Jardel C, Lombes A, et al. Collated mutations in mitochondrial DNA (mtDNA) depletion syndrome (excluding the mitochondrial gamma polymerase, POLG1). Biochim Biophys Acta. 2009;1792(12):1109–12.
51. Suomalainen A, Isohanni P. Mitochondrial DNA depletion syndromes – many genes, common mechanisms. Neuromuscul Disord. 2010;20(7):429–37.
52. Dallabona C, Marsano RM, Arzuffi P, Ghezzi D, Mancini P, Zeviani M, et al. Sym1, the yeast ortholog of the MPV17 human disease protein, is a stress-induced bioenergetic and morphogenetic mitochondrial modulator. Hum Mol Genet. 2010;19(6):1098–107.

53. Hakonen AH, Davidzon G, Salemi R, Bindoff LA, Van Goethem G, Dimauro S, et al. Abundance of the POLG disease mutations in Europe, Australia, New Zealand, and the United States explained by single ancient European founders. Eur J Hum Genet. 2007;15(7):779–83.
54. Prasad C, Melancon SB, Rupar CA, Prasad AN, Nunez LD, Rosenblatt DS, et al. Exome sequencing reveals a homozygous mutation in TWINKLE as the cause of multisystemic failure including renal tubulopathy in three siblings. Mol Genet Metab. 2013;108(3):190–4.
55. Ostergaard E, Hansen FJ, Sorensen N, Duno M, Vissing J, Larsen PL, et al. Mitochondrial encephalomyopathy with elevated methylmalonic acid is caused by SUCLA2 mutations. Brain. 2007;130(Pt 3):853–61.
56. Carrozzo R, Dionisi-Vici C, Steuerwald U, Lucioli S, Deodato F, Di Giandomenico S, et al. SUCLA2 mutations are associated with mild methylmalonic aciduria, Leigh-like encephalomyopathy, dystonia and deafness. Brain. 2007;130(Pt 3):862–74.
57. Jaberi E, Chitsazian F, Ali Shahidi G, Rohani M, Sina F, Safari I, et al. The novel mutation p.Asp251Asn in the beta-subunit of succinate-CoA ligase causes encephalomyopathy and elevated succinylcarnitine. J Hum Genet. 2013;58(8):526–30.
58. Lamperti C, Fang M, Invernizzi F, Liu X, Wang H, Zhang Q, et al. A novel homozygous mutation in SUCLA2 gene identified by exome sequencing. Mol Genet Metab. 2012;107(3):403–8.
59. Rouzier C, Le Guedard-Mereuze S, Fragaki K, Serre V, Miro J, Tuffery-Giraud S, et al. The severity of phenotype linked to SUCLG1 mutations could be correlated with residual amount of SUCLG1 protein. J Med Genet. 2010;47(10):670–6.
60. Acham-Roschitz B, Plecko B, Lindbichler F, Bittner R, Mache CJ, Sperl W, et al. A novel mutation of the RRM2B gene in an infant with early fatal encephalomyopathy, central hypomyelination, and tubulopathy. Mol Genet Metab. 2009;98(3):300–4.
61. Bornstein B, Area E, Flanigan KM, Ganesh J, Jayakar P, Swoboda KJ, et al. Mitochondrial DNA depletion syndrome due to mutations in the RRM2B gene. Neuromuscul Disord. 2008;18(6):453–9.
62. van Rahden VA, Fernandez-Vizarra E, Alawi M, Brand K, Fellmann F, Horn D, et al. Mutations in NDUFB11, encoding a complex I component of the mitochondrial respiratory chain, cause microphthalmia with linear skin defects syndrome. Am J Hum Genet. 2015;96(4):640–50.
63. Gregory A, Hayflick SJ. Genetics of neurodegeneration with brain iron accumulation. Curr Neurol Neurosci Rep. 2011;11(3):254–61.
64. Brunetti D, Dusi S, Morbin M, Uggetti A, Moda F, D'Amato I, et al. Pantothenate kinase-associated neurodegeneration: altered mitochondria membrane potential and defective respiration in Pank2 knock-out mouse model. Hum Mol Genet. 2012;21(24):5294–305.
65. Isaya G. Mitochondrial iron-sulfur cluster dysfunction in neurodegenerative disease. Front Pharmacol. 2014;5:29.
66. Yu-Wai-Man P, Griffiths PG, Chinnery PF. Mitochondrial optic neuropathies – disease mechanisms and therapeutic strategies. Prog Retin Eye Res. 2011;30(2):81–114.
67. Garone C, Tadesse S, Hirano M. Clinical and genetic spectrum of mitochondrial neurogastrointestinal encephalomyopathy. Brain. 2011;134(Pt 11):3326–32.
68. Mishra P, Chan DC. Mitochondrial dynamics and inheritance during cell division, development and disease. Nat Rev Mol Cell Biol. 2014;15(10):634–46.
69. Youle RJ, Narendra DP. Mechanisms of mitophagy. Nat Rev Mol Cell Biol. 2011;12(1):9–14.
70. Durcan TM, Fon EA. The three 'P's of mitophagy: PARKIN, PINK1, and post-translational modifications. Genes Dev. 2015;29(10):989–99.
71. Greene AW, Grenier K, Aguileta MA, Muise S, Farazifard R, Haque ME, et al. Mitochondrial processing peptidase regulates PINK1 processing, import and Parkin recruitment. EMBO Rep. 2012;13(4):378–85.
72. Rugarli EI, Langer T. Mitochondrial quality control: a matter of life and death for neurons. EMBO J. 2012;31(6):1336–49.
73. Anand R, Langer T, Baker MJ. Proteolytic control of mitochondrial function and morphogenesis. Biochim Biophys Acta. 2013;1833(1):195–204.

74. Di Bella D, Lazzaro F, Brusco A, Plumari M, Battaglia G, Pastore A, et al. Mutations in the mitochondrial protease gene AFG3L2 cause dominant hereditary ataxia SCA28. Nat Genet. 2010;42(4):313–21.
75. Gorman GS, Pfeffer G, Griffin H, Blakely EL, Kurzawa-Akanbi M, Gabriel J, et al. Clonal expansion of secondary mitochondrial DNA deletions associated with spinocerebellar ataxia type 28. JAMA Neurol. 2015;72(1):106–11.
76. Pfeffer G, Gorman GS, Griffin H, Kurzawa-Akanbi M, Blakely EL, Wilson I, et al. Mutations in the SPG7 gene cause chronic progressive external ophthalmoplegia through disordered mitochondrial DNA maintenance. Brain. 2014;137(Pt 5):1323–36.
77. Pierson TM, Adams D, Bonn F, Martinelli P, Cherukuri PF, Teer JK, et al. Whole-exome sequencing identifies homozygous AFG3L2 mutations in a spastic ataxia-neuropathy syndrome linked to mitochondrial m-AAA proteases. PLoS Genet. 2011;7(10):e1002325.
78. Almajan ER, Richter R, Paeger L, Martinelli P, Barth E, Decker T, et al. AFG3L2 supports mitochondrial protein synthesis and Purkinje cell survival. J Clin Invest. 2012;122(11):4048–58.
79. Burte F, Carelli V, Chinnery PF, Yu-Wai-Man P. Disturbed mitochondrial dynamics and neurodegenerative disorders. Nat Rev Neurol. 2015;11(1):11–24.
80. Zuchner S, Mersiyanova IV, Muglia M, Bissar-Tadmouri N, Rochelle J, Dadali EL, et al. Mutations in the mitochondrial GTPase mitofusin 2 cause Charcot-Marie-Tooth neuropathy type 2A. Nat Genet. 2004;36(5):449–51.
81. Carelli V, Musumeci O, Caporali L, Zanna C, La Morgia C, Del Dotto V, et al. Syndromic parkinsonism and dementia associated with OPA1 missense mutations. Ann Neurol. 2015;78(1):21–38.
82. Haelterman NA, Yoon WH, Sandoval H, Jaiswal M, Shulman JM, Bellen HJ. A mitocentric view of Parkinson's disease. Ann Rev Neurosci. 2014;37:137–59.
83. Pickrell AM, Youle RJ. The roles of PINK1, parkin, and mitochondrial fidelity in Parkinson's disease. Neuron. 2015;85(2):257–73.
84. Itoh K, Nakamura K, Iijima M, Sesaki H. Mitochondrial dynamics in neurodegeneration. Trend Cell Biol. 2013;23(2):64–71.
85. Pinho CM, Teixeira PF, Glaser E. Mitochondrial import and degradation of amyloid-beta peptide. Biochim Biophys Acta. 2014;1837(7):1069–74.
86. Jayadev S, Leverenz JB, Steinbart E, Stahl J, Klunk W, Yu CE, et al. Alzheimer's disease phenotypes and genotypes associated with mutations in presenilin 2. Brain. 2010;133(Pt 4):1143–54.
87. Schon EA, Area-Gomez E. Mitochondria-associated ER membranes in Alzheimer disease. Mol Cell Neurosci. 2013;55:26–36.
88. de Brito OM, Scorrano L. An intimate liaison: spatial organization of the endoplasmic reticulum-mitochondria relationship. EMBO J. 2010;29(16):2715–23.
89. Palomo GM, Manfredi G. Exploring new pathways of neurodegeneration in ALS: the role of mitochondria quality control. Brain Res. 1607;2015:36–46.
90. Rosen DR, Siddique T, Patterson D, Figlewicz DA, Sapp P, Hentati A, et al. Mutations in Cu/Zn superoxide dismutase gene are associated with familial amyotrophic lateral sclerosis. Nature. 1993;362(6415):59–62.

Chapter 5
Mitochondrial Signaling and Neurodegeneration

Martin Picard and Meagan J. McManus

Abstract The endosymbiosis of mitochondria and the resulting increase in energy supply thus conferred upon the eukaryotic cell enabled the evolution of multicellular organisms and complex organs, such as the brain. As a result, the brain and other organs with high energy demands, depend heavily upon mitochondrial metabolism for normal function, and as a consequence, defects in mitochondrial function lead to neurodegenerative disorders. However, the mechanisms linking mitochondrial defects to hallmarks of neurodegeneration, such as the protein aggregates are also extracellular epigenetic alterations, abnormal gene expression, systemic inflammation, and protein aggregates, remain unclear.

Emerging evidence demonstrates that mitochondria are not only power-generating organelles, but also engage in signaling at multiple levels. During the bioenergetic decline associated with aging, dysfunctional mitochondria generate signals of stress (SOS) that can trigger and/or amplify neurodegenerative processes. In this chapter, we describe emerging mechanisms for mitochondrial signaling at four different levels. Mitochondria communicate (1) with each other via fusion and specialized inter-mitochondrial junctions, (2) with cytoplasmic components via posttranslational modifications that shift signaling pathways and promote protein aggregation, (3) with the nucleus to regulate epigenetic modifications and gene expression, and (4) with the systemic environment where they alter neuroendocrine and inflammatory processes that impact neuronal function. The relevance of mitochondrial signaling to neurodegeneration is discussed.

Keywords Mitochondrial signaling • Mitochondrial signals of stress (SOS) • Epigenetics • Transcriptional regulation • Retrograde signaling • Mitochondrial damage-associated molecular patterns (mtDAMPs) • Inflammation • Protein aggregates • Microglia • Neuroendocrine regulation

---

M. Picard, PhD (✉)
College of Physicians and Surgeons, Division of Behavioral Medicine,
Departments of Psychiatry and Neurology, Columbia University Medical Center,
New York, NY, USA
e-mail: mp3484@columbia.edu

M.J. McManus, PhD (✉)
Center for Mitochondrial and Epigenomic Medicine (CMEM), Department of Pathology,
Children's Hospital of Philadelphia, Philadelphia, PA, USA
e-mail: meaganjm@gmail.com

A.K. Reeve et al. (eds.), *Mitochondrial Dysfunction in Neurodegenerative Disorders*, DOI 10.1007/978-3-319-28637-2_5

5.1 Introduction

Mitochondria evolved from an α-proteobacterium engulfed by the proto-eukaryotic cell over 1.5 billion years ago [1]. The endosymbiosis of this aerobic bacterium was a turning point in the evolution of complex life, with the larger amount of energy afforded by mitochondrial oxidative phosphorylation (OXPHOS), enabling the regulation of a complex genome comprised of >25,000 genes [2]. This biologically unprecedented marriage of OXPHOS with complex genetic material culminated in the development of multicellular organisms, tissues, and interdependent organ systems [3]. The brain, with its unparalleled structural and functional complexity, exemplifies the product of multicellular evolution enabled by mitochondria. Likely as a result of this evolutionary interdependence, the structure and function of the brain and of its constituent neurons are closely linked to energy metabolism in general, and to mitochondrial function in particular.

Unlike the traditional "powerhouse of the cell" concept would suggest, the role of mitochondria is not limited to energy production. As beneficial symbionts, mitochondria perform several other "non-energetic" (i.e., other than ATP synthesis) functions that impact cell function and fate. These crucial functions include, but are not limited to, calcium (Ca^{2+}) uptake and release that regulates energy production itself [4], intracellular signaling and vesicular endocytosis [5], and cell death [6]; reactive oxygen species (ROS) production to alter gene expression and protein function [7, 8]; macromolecule biosynthesis including heme, hormones, and purines/pyrimidines to enable cell growth and replication [9]; and the balance of specific cell death effectors [10].

Not unlike their bacterial ancestors that communicate and behave as colonies through regulated processes of quorum sensing [11–14], mitochondria participate in signaling at a number of levels. *Signaling* is defined as *the exchange of information between two compartments, involving transmission and/or reception of a chemical or molecular signal.* In this chapter, we consider mitochondrial signaling at four levels (Fig. 5.1).

First, we describe evidence of chemical and physical communication within the mitochondrial network. Second, we discuss how mitochondrial signals of stress (SOS) control the intracellular environment of the aging brain and alter protein function and folding in the cytoplasm. Third, mitochondrial signals that shift the epigenetic landscape in the nucleus and influence gene expression are discussed. Finally, we present emerging evidence that mitochondria exert systemic effects, in part via modulation of the immune and neuroendocrine networks relevant to the pathophysiology of neurodegenerative diseases. Overall, the emerging picture is one where mitochondria behave as signaling hubs, dynamically integrating endogenous and environmental factors to dictate neurological health by communicating with each other, within other cellular compartments, and systemically.

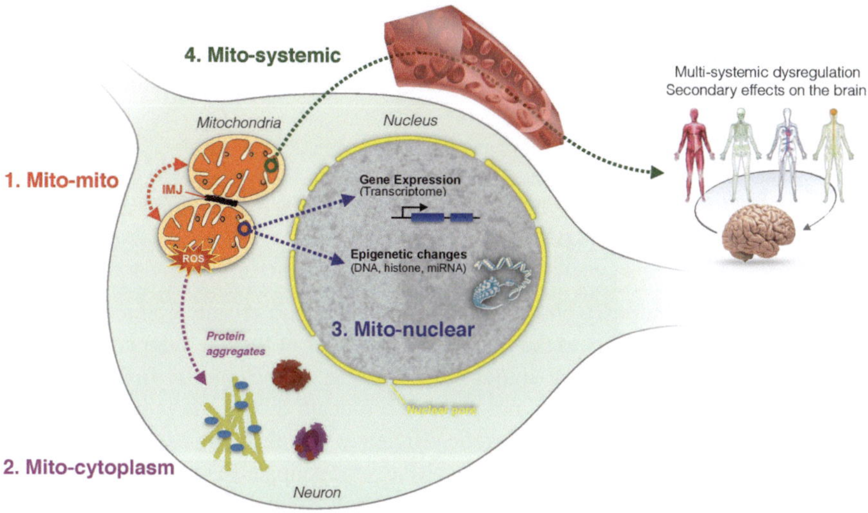

Fig. 5.1 Overview of mitochondrial signaling at four different levels. (1) Mitochondria exchange information with each other via inter-organelle fusion, the release of soluble mediators, and inter-mitochondrial junctions (IMJs) (see [38] for details). (2) Mitochondrial redox signaling shifts signal transduction pathways in the cytoplasm by modifying redox-sensitive amino acids and altering the activity of kinases, phosphatases, and proteases. (3) Mitochondrial metabolic intermediates and ROS travel to the nucleus where they impact epigenetic and transcriptional processes that alter gene expression profiles. (4) Mitochondria release signals that travel to the systemic circulation, influencing the behavior of surrounding cells and tissues, and the physiological function of neuroendocrine, cardiovascular, and nervous systems that indirectly feedback onto the brain to impact neuronal structure and function, and contribute to neurodegeneration

5.2 Mitochondria: Inter-mitochondrial Signaling

Within the cell cytoplasm, mitochondria interact with each other and engage in mitochondria-mitochondria signaling via a number of mechanisms. The field of research around inter-organellar signaling being relatively recent, some caveats must be pointed out as we discuss it in the context of neuronal dysfunction and the brain. Some of the mechanisms described in this section such as the inter-mitochondrial junctions (IMJs) have been reported in neurons, yet their functional relevance to neuronal pathophysiology is still a matter of speculation. In contrast, mitochondrial fusion as a means of exchanging molecular information has been extensively investigated in neurons and its involvement in neurodegeneration reviewed elsewhere (see Chap. 7). In different cell types, mitochondrial signaling involves the release of diffusible signals such as ROS, Ca^{2+}, and apoptotic inducers, which propagate in a wavelike fashion through the mitochondrial network [15–17]. These mechanisms have been identified and investigated mostly in certain cell types (e.g., cardiomyocytes), but may also occur to some extent in neurons. Thus, these mechanisms will not be discussed here.

This section focuses on two major mechanisms for inter-mitochondrial communication: (i) molecular exchange through mitochondrial fusion and (ii) communication through inter-mitochondrial junctions (IMJs) that connect adjacent mitochondria.

5.2.1 Mitochondria-Mitochondria Signaling Through Fusion and Molecular Exchanges

Mitochondria continuously change their size, shape, and cristae architecture [18], which allows them to adapt to the fluctuating demands of their cellular environment and react with the appropriate signal [19]. Whereas mitochondrial fusion appears important to preserve OXPHOS and prevent cell death during starvation [20], cristae remodeling may impact bioenergetic efficiency by enhancing respiratory supercomplex formation [21]. Mitochondrial cristae remodeling also regulates the release of mitochondrial SOS, including proapoptotic signals (reviewed in [18]). Furthermore, mitochondrial fusion and its reciprocal process of fission are necessary to maintain mitochondrial DNA (mtDNA) integrity and renew distal mitochondria for proper synaptic function [22, 23].

In addition to impacting intrinsic functional properties of mitochondria, the dynamic process of mitochondrial fusion is critical for the exchange of proteins and mtDNA among the mitochondrial network. Mitochondria contain >1,000 proteins that are normally distributed throughout the mitochondrial network within the cytoplasm. Ablation of the profusion factors mitofusin 1 and 2 and (Mfn1/Mfn2) leads to substantial protein heterogeneity among the mitochondrial network, with some mitochondria preferentially accumulating certain proteins and lacking others [24]. Protein heterogeneity is a distinguishing feature in neurons, with the proteome of synaptic and non-synaptic mitochondria differing by about 20 % [25]. However, the significance of selective fusion and fission in mitochondrial protein distribution for proper neuronal function, and the relevance to neurodegeneration, remains unknown.

Mitochondrial dynamics also contributes to mitochondrial axonal trafficking, the process by which mitochondria commute toward and away from synaptic terminals [26]. Because mitochondria constantly generate signals, the act of transporting mitochondria from one cellular compartment (e.g., cell body) to another (e.g., synaptic terminal) should be regarded as a form of intracellular signaling.

Importantly, mitochondrial fusion also results in the exchange of genetic material from neighboring mitochondria. In the context of mtDNA heteroplasmy, i.e., when some mitochondria contain mtDNA mutations and others have normal mtDNA, the mixing of mitochondrial contents through fusion enables functional complementation of gene products, thus maintaining mitochondrial homeostasis despite the presence of mitochondrial defects [27]. Mitochondrial functional complementation in heteroplasmic cells does not occur with "kiss-and-run" contact, but requires fusion via both Mfn2 (outer mitochondrial membrane fusion) and optic atrophy 1 (OPA1) (fusion of the inner membrane) and exchange of DNA nucleoids, although mtDNA transcription/translation appears dispensable [28]. Mitochondrial fusion also prevents severe heterogeneity in membrane potential among cells with mtDNA heteroplasmy [28]. Fusion-dependent mtDNA exchange can be regarded as a "stable" form of inter-mitochondrial signaling, where the normal mtDNA transmitted through fusion acts as the signal that promotes normal respiratory chain function among a mutant mtDNA bearing mitochondrion. Beyond permitting mtDNA exchange, mitochondrial fusion also appears essential to preserve mtDNA integrity, since mtDNA mutations accumulate in the absence of mitochondrial fusion [24].

As evidence that mitochondria-mitochondria signaling via membrane fusion is important in the context of neurodegeneration, ablation of mitofusins in the mouse promotes cell loss and cerebellar neurodegeneration [22]. In humans, a growing list of mutations in genes encoding the machinery necessary for mitochondrial fusion and cristae remodeling cause neurological disease affecting both the central and peripheral nervous systems [29].

5.2.2 Mitochondria-Mitochondria Signaling via Inter-mitochondrial Junctions

Even in the absence of fusion, electrochemical information can be passed from one mitochondrion to its neighbor(s) [30]. This form of signaling was originally described in cardiomyocytes [31] and more recently skeletal muscle [32], constituting the basis for the "cable" theory. Elongated "mitochondrial cables" extending up to >30 um have also been reported in dendrites of hippocampal neurons [33, 34]. This theory of cables stipulates that mitochondria behave as electrically coupled conduits that carry information about their electrical charge (membrane potential) throughout the mitochondrial network of a cell [35]. In various cell types such as cardiomyocytes, where mitochondrial membrane potential may spontaneously oscillate over time, clusters of physically juxtaposed mitochondria exhibit a substantial degree of coordination [36, 37], revealing the existence of inter-mitochondrial signaling without mitochondrial fusion.

Studies in the 1980s suggested the existence of physical structures connecting electrically coupled mitochondria in cardiomyocytes, termed *inter-mitochondrial junctions* (IMJs) [31]. Closer quantitative examination of IMJs by electron tomography, which provides substantial spatial resolution and three dimensionality, revealed the presence of structures characterized by enhanced electron density and physical tethering of juxtaposed mitochondria through a <10-nm gap [38]. These gap junction-like IMJs are evolutionary conserved from mollusk to mammal and inducible by the physical rapprochement of adjacent mitochondria through a synthetic linker molecular system [38]. More interestingly in the context of signaling is that cristae organization is altered at IMJs. When two mitochondria are joined by an IMJ, cristae from both mitochondria become coordinated in space, often exhibiting substantial deformation and bending to achieve near-perfect alignment [38].

Similar to synapses in the brain (which regulate the exchange of information from one cell to another), IMJs appear dynamically strengthened or weakened based on their activity. For example, presynaptic terminals of more metabolically demanding, tonic synapses contain more IMJs than phasic synapses that exhibit more temporally spaced firings in the brain [39]. Evidence from a different tissue, skeletal muscle, also indicates that low energy requirements during inactivity tend to reduce the number of IMJs [40], while exercise increases their numbers [41], indicating a certain level of IMJ plasticity. The transmitochondrial coordination of cristae represents the first physical demonstration of inter-mitochondrial signaling, but the functional significance for the brain, neurons, or mitochondria themselves remains unclear.

5.3 Mitochondria: Cellular Signaling

Mitochondria generate diffusible signals that are transmitted – or communicated – to both the cytoplasm and the cell nucleus. Many of these diffusible signals are propagated by redox-based posttranslational reactions that alter signal transduction pathways and protein processing in the cytoplasm, and activate transcription factors that translocate to the nucleus. In addition, mitochondrial signals, such as metabolic intermediates, are the epigenetic language that directs chromatin accessibility. This section discusses mitochondrial signaling to both cytoplasmic and nuclear compartments and the relevance of mitochondrial signals of stress (SOS) in neurodegenerative pathogenesis (Fig. 5.2).

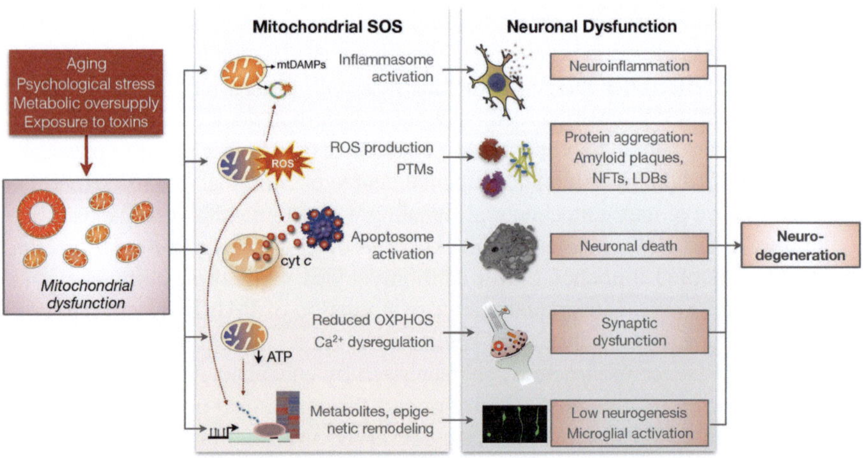

Fig. 5.2 Mitochondrial signals of stress: from mitochondrial dysfunction to neurodegenerative disease. Common stressors such as aging, psychological and metabolic stress, and exposure to toxins cause mitochondrial dysfunction and aberrant morphology, including donut-shaped mitochondria (O). Mitochondria respond by sending signals of stress (SOS), which induce defined molecular changes within the cytoplasm and nucleus that directly contribute to the established hallmarks of neurodegenerative disease, including protein aggregates and transcriptional dysregulation. *OXPHOS* oxidative phosphorylation, *Cyt c* cytochrome *c*, *PTMs* posttranslational modifications, *NFTs* neurofibrillary tangles, *LBs* Lewy bodies

5.3.1 Mitochondrial Bioenergetics and Redox Signaling

Mitochondrial redox chemistry is not only central to the production of ATP by OXPHOS, but also a critical means by which mitochondria communicate to the nucleus, regulate enzymatic activity, control cell death, and stimulate the immune system. During OXPHOS, up to 2% of the electrons leak to molecular oxygen and produce the superoxide radical ($O_2^{\cdot-}$) [8]. The mitochondrial matrix antioxidant, manganese superoxide dismutase (MnSOD), converts $O_2^{\cdot-}$ to hydrogen peroxide (H_2O_2), which may act as a signaling molecule in the cytosol/nucleus or generate more toxic ROS, such as the hydroxyl radical ($^{\cdot}OH$). The majority of mitochondrial ROS (mtROS) are produced by the OXPHOS complexes I and III, but other mitochondrial enzymes such as NADPH oxidase-4 (Nox4), monoamine oxidases (MAOs), and some TCA cycle enzymes are also important in neural redox regulation [7, 42].

While mtROS are often viewed as indiscriminate molecular villains, they are in fact mitochondrial signaling molecules that produce a wide range of physiological effects dependent on the cellular context in which they are produced. Low levels of mtROS can induce gene expression changes to preserve mitochondrial function during stress, such as those that extend the lifespan of *C. elegans* [43]. Physiologically appropriate fluctuations in mtROS modulate the redox status of the mitochondria and the cell via several key redox couples: NADPH/NADP$^+$, NADH/NAD$^+$, thioredoxins 1 and 2 (Trx 1 and Trx2) $(SH)_2$/SS, glutathione (GSH/GSSG), and cysteine/cysteine (CyS/CySS) [44]. These redox nodes are interconnected within and between cellular compartments. For instance, the mitochondrial and cytoplasmic NADH/NAD$^+$ redox states are linked via the activity of the malate/aspartate NAD(H) and α-glycerophosphate shuttles, which may be particularly important in neurons and less so in astrocytes [45]. Transient opening of the mitochondrial permeability transition pore (PTP) can also rapidly release NAD$^+$ and NADH into the cytoplasm [46], although the functional significance of this mechanism in the brain remains unclear.

The intramitochondrial redox state is intimately coupled to mitochondrial energetics by nicotinamide nucleotide transhydrogenase (NNT) [47]. NNT relies on the proton motive force generated during OXPHOS to regenerate NADPH from NADP$^+$, in order to power NADPH-dependent reductases that recycle the other redox couples in the mitochondrial matrix. Thus, by linking mitochondrial proton motive force to interconnected redox control nodes within and outside mitochondria, mitochondria may communicate the energetic state with the rest of the cell and organism. When mitochondrial bioenergetics is compromised, the redox nodes act as "barometers" that induce mitophagy or apoptosis, depending on the magnitude of the insult [48–51]. The importance of these redox nodes to cellular homeostasis is evidenced by the embryonic lethality that occurs when components are genetically deleted [52–55] and neuronal cell death that ensues when the system is overwhelmed by excess mtROS [56].

The downstream signaling pathways initiated by mtROS are determined by the reactivity, charge, and location of the particular species. In the aging brain, impairments in OXPHOS produce $O_2^{\cdot-}$ primarily within the mitochondrial matrix and inner membrane. In order for $O_2^{\cdot-}$ to react with mitochondrial components, the target must outcompete micromolar concentrations of MnSOD in the matrix [57]. One signaling molecule that meets these stringent requirements is nitric oxide ($NO^{\cdot}$). In contrast to $O_2^{\cdot-}$, $NO^{\cdot}$ possesses a relatively long half-life for a "reactive species," which allows it to readily diffuse into the mitochondria from cytoplasmic sources where it can induce neuroprotective pathways. In mitochondria, the close proximity of $NO^{\cdot}$ and $O_2^{\cdot-}$ generates the highly toxic species peroxynitrite ($ONOO^-$) by a spontaneous reaction that occurs seven times faster than that of $O_2^{\cdot-}$ and MnSOD [58]. This reaction occurs predominantly in neuroinflammatory conditions when $NO^{\cdot}$ is present at higher concentrations [59, 60]. In most contexts, $ONOO^-$ causes permanent protein damage, although it may protect the brain from ischemic episodes associated with stroke [61, 62].

The signaling potential of superoxide is increased in many neurodegenerative diseases due to nitration and permanent inactivation of MnSOD [63, 64]. $O_2^{\cdot-}$ is more selective than most mtROS, but readily reacts with iron-sulfur (Fe-S) centers of nonheme proteins, such as respiratory complexes I–III and TCA cycle enzymes. This reaction releases free ferrous iron (Fe^{2+}), which generates the highly reactive hydroxyl radical (·OH) via Fenton chemistry, thereby initiating a chain of lipid peroxidation, protein modifications, and DNA damage within or near the inner mitochondrial membrane (IMM) [57].

Cardiolipin, the major phospholipid of the IMM, is particularly susceptible to peroxidation by ROS due to its high degree of unsaturation. During mitochondrial stress, mtROS or cytochrome *c* (cyt *c*) may oxidize cardiolipin, which dissolves its association with cyt *c* [65, 66]. Oxidized cardiolipin then translocates to the outer membrane, recruits mitophagic or apoptotic machinery, and releases cyt *c* into the cytosol [67]. If the cytosolic redox balance is also shifted to a prooxidant state, the released cyt *c* will remain oxidized and able to initiate activation of the apoptosome [68, 69]. Therefore, mtROS initiate cell death pathways by direct modification of lipids and proteins, but the amplification of these signals necessary to effectuate intrinsic apoptosis requires modulation of redox nodes in multiple compartments. In addition, mtROS may also induce neuronal death via caspase-independent mechanisms involving IMM compromise and Ca^{2+} dysregulation [56].

Due to their transient nature, it is currently difficult to directly measure ROS *in vivo*, and therefore analysis of ROS levels is largely dependent on the molecular footprints left in their wake of destruction [42]. Recent efforts to develop mitochondria-targeted mass spectrometry probes are under development and could circumvent this technical limitation in preclinical studies [70]. Among all organs of the body, the brain is perhaps most vulnerable to oxidative damage due to its high utilization of oxygen, increased levels of polyunsaturated fatty acids and redox transition metals, and relatively low levels of antioxidants. Elevation of virtually every established marker of oxidative damage has been documented in brain tissue from the most prevalent neurodegenerative diseases, including Alzheimer's disease (AD), Parkinson's disease (PD), Lewy body disease (LBD), amyotrophic lateral sclerosis (ALS), Huntington's disease (HD), and Friedreich's ataxia (FA) [71, 72].

Interestingly, the evidence suggests that bioenergetic defects precede the onset of diagnostic symptoms [73]. Furthermore, neurological symptoms often emerge without or before the accumulation of proteinaceous lesions [74] that are generally considered the key determinants of neurodegenerative disease pathogenesis [72, 75, 76]. Therefore, the chronological disassociation between the emergence of diagnostic symptoms and proteinaceous lesions [77] suggests that the pathological processing of amyloid, tau, and alpha-synuclein could be a *consequence* of more proximal pathological stimuli, such as mitochondrial SOS produced by progressive mitochondrial dysfunction in aging [78].

5.3.2 Mitochondrial Distress Signals and Pathological Protein Processing

Several mechanisms link mitochondrial signaling with protein misfolding and aggregation. The causative nature of mitochondrial dysfunction in this association is supported by animal models and patients that develop aggregates and neurodegeneration after exposure to mitochondrial toxins [79, 80], as well as patient-derived transmitochondrial cybrids [81], in which the mtDNA from AD, PD, and ALS patients is combined with an otherwise healthy cell and induces neurodegenerative disease-associated proteinopathy and cell death [82]. The next three sections focus on recent insights linking signals from dysfunctional mitochondria to the most common protein lesions found in neurodegenerative disease patient brains: amyloid plaques, tau tangles, and α-synuclein Lewy bodies.

5.3.2.1 Amyloid Precursor Protein to Plaques

Since the discovery of amyloid plaques by Alois Alzheimer in the 1920s, the amyloid hypothesis has enjoyed the etiological spotlight. The main protein component of plaques is the amyloid-β (Aβ) peptide, a molecule of 40–42 amino acids with unclear function, derived from the proteolytic cleavage of the integral membrane amyloid precursor protein (APP). Interestingly, the key amyloidogenic proteases are influenced by fluctuations in mitochondria function and oxidative stress [83–88]. While mitochondrial SOS alter Aβ production and clearance in cellular and animal models, the pathophysiological relationship between mitochondrial distress and amyloid production remains ambiguous. As discussed below, since mitochondrial SOS occur early in AD and may act in concert to generate Aβ, which then localizes to mitochondria, one possibility is that the accumulation of "abnormal" Aβ represents a hormetic response to mitochondrial stress in neurons.

Indeed, APP cleavage products can act as antioxidants and Aβ monomers shift neuronal glucose metabolism away from mitochondria to glycolysis [89–91]. This metabolic shift would confer neuroprotection by decreasing mtROS produced by

OXPHOS while boosting endogenous antioxidant defenses by doubling NADPH levels and reducing the redox control nodes. In other studies, when the mtROS-responsive, Aβ-generating secretases are inhibited, neuronal death ensues, unless Aβ fragments are exogenously added [92]. Therefore, the induction of β- and γ-secretases by mitochondrial dysfunction may initially serve as a survival signal during oxidative stress by generating neuroprotective Aβ species. Aβ monomers preserve neuronal survival in classic models of mitochondria-dependent cell death that are mediated by mtROS [78, 80], but Aβ oligomers have the opposite effect [93–95], such that this pathophysiological role of Aβ depends on the particular species and the aggregation state [96, 97].

Interestingly, immunotherapeutic clearance of amyloid from transgenic mice and AD patients has adverse consequences [98–100]. Recent studies suggest that these effects involve the ability of amyloid to influence Ca^{2+} flux [101] associated with AD mitochondrial impairments, which points to a second potential role of amyloid in mediating mitochondrial SOS. Taken together, the available evidence suggests that dysfunctional mitochondria in the aging brain may initially signal to induce Aβ production via mtROS in order to prevent oxidative damage and Ca^{2+} dysregulation and preserve neuronal health.

5.3.2.2 Tau to Neurofibrillary Tangles

Mitochondrial redox signaling modulates aberrant processing of another key protein related to age-dependent dementia and AD: the microtubule-stabilizing protein *tau* [102]. Tau normally functions to promote microtubule assembly and stabilization for neurite trafficking. In neurodegeneration, tau disengages from the microtubule and becomes hyperphosphorylated into neurofibrillary tangles (NFTs). Similar to amyloid processing, mtROS may initiate the process in AD by inducing caspases 3 and 7 [88], which cleave tau into neurotoxic fragments [103]. Mitochondrial redox signaling activates the key kinases currently implicated in tau hyperphosphorylation, including the stress-activated kinases JNK and p38 [104], as well as glycogen synthase kinase-3β (GSK-3ß) [105, 106]. mtROS may also inhibit peptidyl-prolyl isomerase (Pin1), which is necessary for tau dephosphorylation and microtubule binding [107–109]. Oxidative impairment of Pin1 appears crucial in the clinical presentation of tau pathology, as it is increased in presymptomatic individuals and inversely correlates with NFT levels and neurodegeneration in the AD brain [110]. Therefore, early mitochondrial SOS may culminate in the pathological processing of tau, which then destabilizes microtubules and further impairs mitochondrial signaling by inhibiting mitochondrial dynamics and trafficking between the synapses and cell body [26, 111].

5.3.2.3 α-Synuclein to Lewy Bodies

Lewy bodies are composed primarily of aggregated *α-synuclein* (*α-syn*), along with ubiquitin and tau, and represent the key histological hallmark shared by PD and LBD. α-Syn is a ubiquitously expressed protein of unknown function that localizes to various cellular compartments, depending on the conformational changes induced by inherited mutations, or environmental conditions, such as increased oxidative stress [112–115].

In normal conditions, α-syn is found in the cytosol where it regulates the dopamine synthesis pathway [116] and within mitochondria-associated ER membranes (MAMs) where it may participate in a number of key signaling pathways, including the regulation of mitochondrial Ca^{2+} transients and mitochondrial dynamics [117]. PD-associated mitochondrial dysfunction induces α-syn [71, 118, 119], which may represent a retrograde signal designed to dampen the bioenergetic burden of synaptic transmission by blocking vesicle exocytosis and neurotransmitter release [120], to preserve neuronal Ca^{2+} homeostasis at MAMs [117], or to decrease oxidative stress associated with dopamine autoxidation [116] within the PD brain [119].

However, in the event that mitochondrial dysfunction persists, excess mtROS may cause α-syn nitration in dopaminergic neurons, thereby impairing its ability to bind membranes [121]. In this regard, nitration of α-syn in "sporadic" PD patients may produce the same effect as familial PD mutations in the α-syn gene that decrease its lipid-binding ability and redirect it from MAMs to the mitochondria [122]. Mitochondria-localized α-syn may impair complex I function directly or increase its sensitivity to other modifiers [123–126]. Complex I inhibition by α-syn could then serve a second signaling role by amplifying mtROS formation and changing α-syn physical conformation to promote aggregation and LB formation [127].

This section has thus far focused on the effect of mitochondrial signaling in the cytoplasmic compartment, where mitochondrial SOS lead to the accumulation and aggregation of proteins associated with neurodegenerative diseases. The following section discusses mitochondrial signaling to the nucleus, which affects gene expression.

5.3.3 *Metabolic Signaling and the Epigenetic Language*

Contrary to popular conception, mitochondria do not float randomly amid the cytoplasm. Their cellular distribution is highly regulated by energy needs and other intracellular cues [128]. While most attention is given to the synaptic mitochondria required to supply the estimated 4.7 billion ATP molecules per second of a single neuron's activity [129], mitochondria within the cell body are also critical for neuronal function via retrograde signaling. In the soma, mitochondria are positioned in the perinuclear region, in close proximity to the chromatin [130]. The chromatin is the DNA-protein structure that modulates gene accessibility by either "closing" portions of the genome to repress transcription or "opening" them to promote their expression. Mitochondrial regulation of gene expression is facilitated by two major factors. The first is "topological," consisting in the perinuclear positioning of mitochondria only a few hundred nanometers away from the nuclear pores through which molecular signals travel [130]. The second and more studied factor that enables mitochondria-nuclear signaling is the nature of biochemical signals that induce epigenetic modifications, most of which are produced by mitochondria. The latter aspect constitutes the focus of this section.

As part of their normal metabolism, mitochondria produce most of the biochemical substrates and cofactors required for epigenetic (i.e., "on top of" genes) modifications that remodel chromatin structure. These substrates and cofactors include, but are not limited to, acetyl-coenzyme A (Ac-CoA), succinyl-CoA, SAM these (S-adenosyl methionine), ATP, α-ketoglutarate, NAD^+ (nicotinamide adenine dinucleotide), and FAD (flavin adenine dinucleotide) [131, 132]. Posttranslational modifications require mitochondrial energy input and metabolism and consist of acetylation, succinylation, methylation, phosphorylation, and other labile alterations of the core chromatin component, the histones. The DNA itself is also modified via cytosine methylation, hydroxymethylation, and other modifications. This partially heritable collection of epigenetic modifications constitutes the relatively malleable "epigenetic landscape" that orchestrates gene expression, enabling a single genome to support multiple, drastically different cellular phenotypes (i.e., neurons vs. fibroblasts [133]). In neurodegenerative diseases, the epigenetic landscape of the brain is altered [134], including early changes in DNA methylation in Alzheimer's disease [135]. The link between mitochondrial intermediate metabolism and epigenetic modifications suggests a role for mitochondrial dysfunction and resulting signals in the transcriptional reprogramming that may underlie neurodegeneration.

In a study of human cells with varying levels of mitochondrial dysfunction, it was recently discovered that the majority (>67 %) of the human genome is under mitochondrial regulation [136]. Indeed, whereas low levels of mitochondrial dysfunction in the presence of 20–30 % of mtDNA mutation load induced epigenetic modifiers and repressed cell growth, higher levels (60–90 %) of the same mtDNA mutation caused opposing transcriptional changes, inducing anaerobic glycolytic metabolism and transcriptional programs associated with cell senescence [136]. A single nucleotide change in the mtDNA can thus exert a wide range of transcriptional changes in the cell nucleus.

In addition to metabolites, mtROS also travel to the nucleus where they promote an oxidized nuclear environment favorable to the expression of redox-responsive genes [137]. Beyond regulating gene expression under normal circumstances, mtROS signals may also accelerate telomere shortening and aging [138, 139]. MtROS further activate oxidative stress-sensitive, pro-inflammatory pathways such as nuclear factor-kappa B (Nf-kB) [140], involved in general inflammatory response to chronic stress in human monocytes and microglia [141].

Mitochondrial bioenergetic signals are also transduced to the nucleus via different transcription factors and co-activators including but not restricted to Nf-kB [142]. Upon activation, transcription factors located in the cytoplasm translocate to the nucleus through nuclear pores, where they interact with chromatin components and the DNA itself. Canonical mito-nuclear signaling mechanisms include the energy sensors AMP-activated protein kinase (AMPK) and peroxisome proliferator-activated receptor gamma co-activator 1 alpha (PGC-1α), nuclear factor erythroid 2-related factor (NRF2) and p53 induced by oxidative stress, and Ca^{2+}-sensitive calcineurin pathways [143, 144]. These and others have been shown to be altered in neurodegenerative conditions [145]. Therefore, multiple pathways exist whereby progressive mitochondrial dysfunction is communicated to the nucleus where it modifies gene expression via the epigenome.

Genetic and biochemical mitochondrial defects alter cellular bioenergetics and can lead to energy deficiency due to ATP depletion, which is generally believed to contribute to neurodegeneration. In addition, this section has outlined alternative pathways whereby dysfunctional mitochondria produce biochemical SOS, to which the cell and its plastic (epi)genome have evolved molecular sensitivity. These mitochondrial signals – oxidative stress and metabolic intermediates – travel to the nucleus to induce epigenetic modifications, many of which are post-translational modifications of regulatory proteins, that impact gene expression. From an evolutionary perspective, the coupling of genome remodeling epigenetic processes with mitochondrial signals in the brain was likely an important step for the evolution of species in an environment where energy supply and demand vary widely over time [146]. Multiple mechanisms thus exist to link mitochondrial metabolism to fundamental aspects of neuronal function.

5.3.4 Mitochondria as Synaptic Neuromodulators

Evidence for mitochondrial signaling at the synaptic level has recently expanded our conception of mitochondria not only as powerhouses but also as neuromodulators. In addition to mitochondrial ATP that influences synaptic function [147], mitochondrial Ca^{2+} uptake via the mitochondrial calcium uniporter (MCU) also regulates presynaptic neurotransmitter release [5]. On the other hand, abnormal mitochondrial morphology (i.e., donut shaped) in presynaptic terminals has been associated with abnormal synapse structure (smaller active zone, fewer docked neurotransmitter vesicles) and impaired short-term memory, indicating a link between mitochondrial morphology and higher cognitive function [148], although the mechanism by which this occurs remains unclear.

In mice, mitochondrial DNA heteroplasmy (i.e., a mixture of two mtDNA types differing by 91 nucleotides) impairs memory retention and alters animal behavior [149], demonstrating a link between mitochondrial genetics and cognitive function. Live cell imaging studies of synaptic neurotransmitter release further showed how the movement of mitochondria into a presynaptic terminal dynamically potentiates presynaptic vesicle release and reduces fatigability, whereas the absence of a mitochondrion in a presynaptic terminal leads to fatigability in response to consecutive stimulation [150]. Collectively, this positions mitochondria as neuromodulators within the brain [151, 152], illustrating a mechanism other than cell death and neurodegenerative changes by which mitochondrial signaling may affect cognitive function.

5.4 Mitochondria: Systemic Signaling

The previous section discussed how the state of mitochondria directs processes within the cell cytoplasm and nucleus and their role at the synapse. In addition, possibly as a result of the evolutionary intertwine between mitochondrial energetics and the development of multicellular organisms, organ systems such as the cardiovascular and endocrine axes are also functionally regulated by the bioenergetic state of mitochondria. In this scenario, mitochondrial signals act in a cell non-autonomous manner to influence systemic neuroendocrine processes linked to neuronal function.

Early evidence for mitochondrial signaling at the systemic level came from exercise physiology studies in patients with inherited mitochondrial myopathy. There is normally a tight coupling between whole body energy expenditure and cardiorespiratory responses during exercise, where a one-unit increase in oxygen consumption leads to a five-unit increase in cardiac output [153]. In contrast, patients with mitochondrial disease (m.3243A > G, single deletion) show exaggerated cardiac output and ventilatory responses that are disproportionate to energy demand [153], suggesting that dysfunctional mitochondria signal to the central nervous system, via an currently unknown mechanism, to exaggeratedly mobilize oxygen delivery systems (heart rate, and ventilation). A similar response was also observed at the level of skeletal muscle, where muscle fibers with respiratory chain-deficient mitochondria are selectively surrounded by a greater number of capillaries than adjacent fibers with normal mitochondria [154], indicating that defective mitochondria communicate their dysfunctional state via SOS to neighboring endothelial cells, stimulating the growth of capillaries (i.e., angiogenesis). Collectively, the evidence suggests that dysfunctional mitochondria engage in signaling not only with each other and the cell nucleus but also with surrounding cells and across the organism where they influence broad physiological functions.

The remaining section discusses recent lines of research that aim to understand the effect of mitochondrial function on neuroendocrine, metabolic, and inflammatory processes. Because neuroendocrine, metabolic, and inflammatory mediators impact neuronal health [155], mitochondrial systemic signaling represents an emerging pathway by which mitochondrial dysfunction may influence the etiology and progression of neurodegenerative diseases.

5.4.1 Mitochondrial Functions Regulate Neuroendocrine Systems

Two major neuroendocrine axes that release neuroactive hormones in the systemic circulation have been linked to mitochondrial function: the hypothalamic-pituitary-adrenal (HPA) axis, which leads to the secretion of *cortisol* (corticosterone in rodents), and the sympathetic-adrenal-medullary (SAM) axis, a division of the autonomic nervous system (ANS), which leads to the secretion of the catecholamines *noradrenaline* and *adrenaline*. Both cortisol and the catecholamines can influence brain structure and function by acting on the glucocorticoid receptor and adrenergic receptors on afferent sensory nerves, respectively [156, 157].

Recent evidence in animal models suggests that mtDNA variants can alter stress-reactive corticosterone (CORT) production in mice [158, 159] and the CORT-generating cells of the adrenal cortex are exquisitely sensitive to intramitochondrial oxidative stress [160]. Likewise, mutations or deletions of the redox-regulating enzyme NNT lead to hypocortisolemia in humans [160] and mice [161], respectively. In patients with mutations in the mitochondrial adenine nucleotide translocator (ANT), which impairs ATP/ADP exchange across the IMM, resting circulating catecholamine levels are also almost double that of healthy controls [162], indicating that both HPA and SAM axes' activities are modulated by mitochondrial dysfunction.

Mitochondrial respiratory chain dysfunction may also be associated with ANS dysfunction, involving increased sympathetic vagal nerve drive as evidenced by increased resting heart rate, and increased high frequency power RR interval variability (RRV) in patients with the m.3243A>G mutation [163]. Mitochondrial disorders are also associated with excessive epinephrine and norepinephrine secretion by the SAM axis during exercise [164].

A recent study in mice with either mtDNA point mutations affecting OXPHOS capacity or deletions of nuclear-encoded NNT or ANT1 genes demonstrated that in addition to gene expression in the hippocampus, the major neuroendocrine systems are under substantial mitochondrial regulation [161]. In response to psychological stress, mitochondrial defects may in some cases double corticosterone output while blunting catecholamine release [161]. Mitochondrial dysfunction may also promote stress-induced hypercortisolemia and hyperglycemia, which are associated with brain atrophy and cognitive decline [165]. These mitochondria-regulated neuroendocrine perturbations may cause neuronal insult directly by further damaging mitochondrial function and leading to the accumulation of mtDNA defects [165, 166] and indirectly by dysregulating the metabolic and immune systems [167], curtailing brain plasticity and adaptation to stress [156].

5.4.2 *Mitochondrial Immunogenic Signals*

Another common feature of late-onset neurodegenerative diseases that may be modulated by mitochondrial SOS is inflammation (Fig. 5.3). As we age, our metabolic and immune systems undergo a process of senescence characterized by a progressive decline in mitochondrial and immune function, which correlates with an increased frequency of infection and neurodegenerative disease [78, 168]. Inflammation in neurodegenerative diseases occurs without evidence of externally acquired pathogens, suggesting that endogenous factors may instead initiate inflammation (i.e., sterile inflammation) [169]. Interestingly, a growing body of literature indicates that mitochondrial SOS may trigger the immune systems and promote chronic low-grade inflammation implicated in neurodegenerative diseases.

The most common risk factors for neurodegenerative diseases include aging, metabolic and immune disorders, chemical toxin exposure, and psychological trauma [170]. These factors impact mitochondrial function by either promoting the accumulation of mtDNA mutations/deletions, decreased OXPHOS capacity, or excessive mtROS production, which can trigger the release of

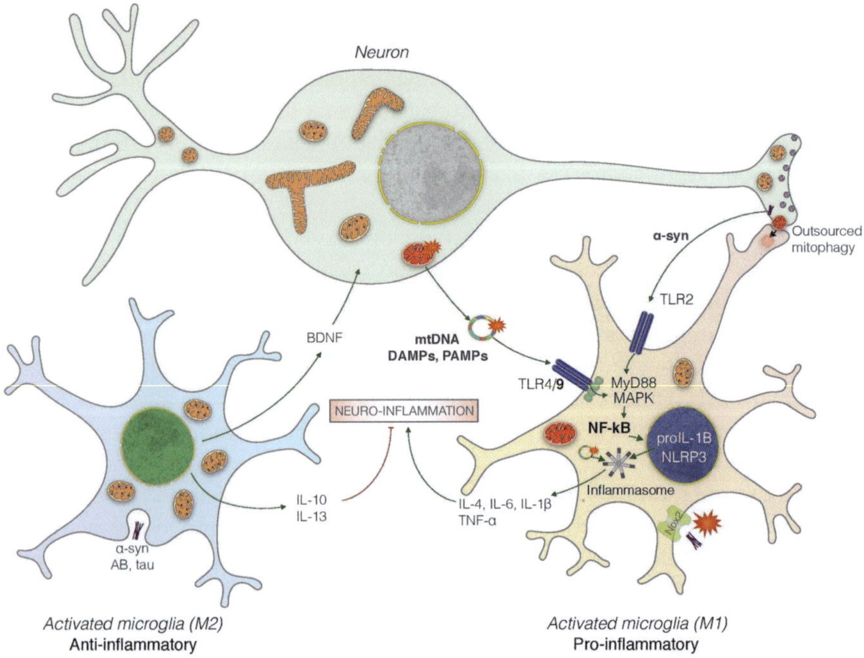

Fig. 5.3 Mitochondrial signals of stress and neuroinflammation. In neurodegenerative diseases, microglia become "primed" and contribute to inflammatory changes within the brain. Here, we propose that neuronal mitochondrial dysfunction initiates microglial activation by releasing immunogenic signals of stress (SOS). Mitochondrial SOS include reactive oxygen species (ROS) and damage-associated molecular patterns (DAMPs; mtDNA, RNA, and N-formyl peptides) that activate toll-like receptors 4 and 9 (TLR4/TLR9), inducing the innate immune response from microglia and systemic macrophages. In the aging brain, chronic stimulation by mitochondrial SOS shifts the response of "primed" microglia from primarily phagocytic and anti-inflammatory (M2) to pro-inflammatory (M1) and induces their proliferation. TLR4/TLR9 stimulation activates M1 microglia by MyD88 or MAPK, which sends NF-kB from the cytosol to the nucleus to induce proIL-1β and NLRP3 transcription. When activated by oxidized mtDNA, the NLRP3-inflammasome cleaves pro-caspase-1 and induces the release of cytokines (IL-4, IL-8, IL-1B, and TNF-α) that promote neuroinflammation, gliosis, and neuronal damage. Secondary insult from protein aggregates ( ) or glucocorticoids may exaggerate this response in sensitive brain regions, contributing to the mood disorders, cognitive decline, and cell death associated with neurodegenerative disease. Microglia may also contribute to elimination of dysfunctional mitochondria within the axon terminal, in a process whereby neurons outsource mitophagy [223]

mitochondria-derived damage-associated molecular patterns (mtDAMPs), such as mtDNA fragments and some mitochondrial proteins [169]. Progressive impairment of mitochondrial function and integrity and the associated release of immunogenic molecules may thus steadily stimulate the innate immune system and promote inflammation both systemically and within the brain [171]. Here, we discuss the interplay between mitochondrial SOS and the immune system, which can be divided into two main categories.

5.4.2.1 Mitochondrial SOS as Inflammatory Triggers

The circular mtDNA of bacterial origin and resultant N-formyl peptides are recognized as foreign molecules by the immune system [169]. N-formyl peptides and mtDNA itself constitute mtDAMPs that are released following mitochondrial stress, particularly oxidative stress [172]. The release of mtDAMPs engages the innate immune system through the intracellular DNA-sensing system cGAS [173] and toll-like receptors (TLRs) [174]. TLRs are present in dendritic cells, macrophages, and microglia in the CNS, as well as nonimmune cells such as neurons and epithelial cells, where they activate the inflammasome and pro-inflammatory gene expression [175]. In microglia and neuronal cultures, mtDAMPs cause dopaminergic cell death [176] and increased amyloidosis with AD-associated inflammation [177]. In animal models, mtDAMP inflammasome activation is associated with cognitive impairment [178] and neurodegenerative disease [176, 179, 180].

Other mitochondrial components have also been reported in human plasma and suggested to play signaling roles that may influence the brain (Table 5.1). They include cardiolipin, which may act as a mtDAMP when oxidized [181]; mitochondrial proteins encoded in the nuclear genome such as cyt *c* [182], heat shock protein 60 (Hsp60) [183], and prohibitins [184]; and mtDNA-encoded proteins derived from alternative open reading frames (altORFs) such as humanin [185] and MOTS-c [186] (see Table 5.1). Circulating Hsp60 levels correlate with neurodegenerative disease risk factors, such as psychological stress and circulating cholesterol levels [183], indicating a potential immunogenic link between psychosocial factors, mitochondrial stress, and neuronal demise [187]. Mitochondria-derived molecules are thus emerging as a source of signals that may chronically instigate systemic inflammation in neurodegenerative disease.

Just as mtDAMPs correlate with known mediators of neurodegeneration, new evidence suggests that established neurodegenerative disease therapeutics prevent mtDAMP release. The anti-inflammatory signal acetylcholine (ACh) prevents stress-induced release of mtDNA via binding a putative mitochondrial nicotinic

Table 5.1 Mitochondrial components reported in human blood

Molecule	Full name	Reference
Hsp60	Heat shock protein 60	[183]
Humanin	Humanin	[185]
MOTS-c	Mitochondrial open reading frame of the 12S RNA type c	[186]
PHB1/2	Prohibitin	[184]
mtDNA	Mitochondrial DNA	[171]
Cardiolipin	Cardiolipin	[181]
Cyt c	Cytochrome c	[182]

ACh receptor [172]. Therefore, it is plausible that the therapeutic benefit of nicotine and acetylcholinesterase (AChE) inhibitors for NDs [188] may not only be related to mitochondrial signaling in cell death [189], but also to mitochondria-related inflammatory signaling.

5.4.2.2 Mitochondria Regulate Intracellular Immunogenic Responses

Infected immune cells require energized mitochondria to recruit the mitochondrial antiviral signaling (MAVS) protein, which aggregates on the mitochondrial outer membrane and initiates antiviral signaling [190]. This signal stimulates the cellular antiviral response by activating nuclear translocation of NF-kB and interferon regulatory factors (IRFs) where they stimulate transcription of type I interferons and pro-inflammatory cytokine genes associated with "primed" microglia in aging and neurodegenerative diseases [191]. Ablating mitochondrial membrane potential inhibits this response [192], whereas mitochondrial ROS and mtDAMPs potentiate it [193, 194]. Interestingly, the mtDNA variants that govern mtROS and membrane potential also correlate with risk for neurological disorders and alter the metabolic-immune axis in a manner that mirrors the circulating cytokine profiles in patients with mood disorders and neurodegenerative disease [195–199].

Taken together, this relationship suggests that genetic or environmental insults to mitochondria may induce the release of SOS (mtDAMPs and mtROS) that drive neuroinflammation by activating the innate immune system and priming microglia (see Fig. 5.3). In turn, primed microglia in the aged brain inappropriately respond to immune challenges presented by mitochondrial dysfunction, as evident by their decreased phagocytic activity and sustained inflammatory signals in neurodegenerative disease [200]. The result of this faulty mitochondrial-immune network may amplify protein aggregates, neuronal death, and neurodegenerative disease progression.

5.4.3 Mitochondria as Mediators of Common Risk Factors in Neurodegenerative Disease

The risk of developing late-onset neurodegenerative disease is intimately tied to lifestyle factors such as physical activity [201] and psychosocial stress [202]. As discussed below, physical activity and chronic psychosocial stress induce biological patterns that correlate with downstream changes in biomarkers of increased [203] or decreased neurodegenerative disease risk [204, 205], respectively. Mitochondria may thus contribute to translate the lifestyle and environmental risk factors into the biological changes that cause neurodegenerative disease.

Physical inactivity, or sedentary behavior, promotes disease risk by producing a diabetes-like physiological state of metabolic oversupply that damages mitochondria [206]. Metabolic stress arises from excess energetic substrates (glucose, lipids), impairs mitochondrial dynamics, increases mtROS production, and leads to accumulation of mtDNA damage [206]. Diabetes exemplifies this state of metabolic dysregulation, where circulating glucose levels are continuously elevated. As a result, diabetes is associated with brain atrophy and cognitive decline in aging [165] and AD patients [207]. Hyperglycemia increases Aβ levels in the hippocampus [208]. On the other hand, exercise counteracts metabolic oversupply systemically [206] and within the brain by inducing mitochondrial biogenesis and neurogenesis and may stimulate clearance of Aβ [209–211]. The metabolic state, regulated by the combined effect of physical activity and diet, imparts substantial effects on neuronal health and the risk of neurodegenerative diseases [201].

Likewise, psychological stress in the social and work environment, which arises from real or perceived threat to the self, increases the risk of neurodegenerative disease. Psychosocial stress at work (i.e., low job control and high demand) is associated with increased risk for dementia and AD later in life [212]. Likewise, longitudinal studies demonstrated that chronic life stress during middle age is associated with brain atrophy and white matter lesions later in life [213] and increased risk for depression, dementia, and neurodegenerative diseases [202, 214], underscoring the negative impact of various stressors on neuronal health.

In seeking to resolve the biological connections responsible for these associations, it is relevant to note that hormones secreted during stress, including cortisol which becomes dysregulated during chronic stress and neurodegenerative disease, impact mitochondria both directly via the mitochondrial glucocorticoid receptor [166, 215] and indirectly by promoting metabolic stress systemically [167]. Excess stress-associated cortisol and metabolic stress can both impair neuronal mitochondria function and lead to neuroinflammation and dopaminergic death associated with PD [216]. On the other hand, physical activity/exercise confers protection against age-related hippocampal atrophy and memory decline [217] and reduces AD risk [218] and PD progression [219]. Interestingly, physical activity may also buffer against stress-associated cellular changes such as telomere shortening [220]. The exact mechanisms for the beneficial effects of physical activity and the damaging effect of chronic stress on brain structure and function remain unclear [156]. However, evidence that stress damages mitochondria, whereas physical activity promotes mitochondrial biogenesis [221], and that mitochondrial signaling is essential to exercise-induced neurogenesis (i.e., the formation of new neurons) in the adult brain [222], suggests that mitochondria constitute a hub where the action of stressors and resiliency factors intersect to influence brain structure and function.

5.5 Conclusion

The road from mitochondrial dysfunction to neurodegeneration takes many turns, and the trajectory from normal to abnormal brain function can be influenced by multiple mitochondrial signals. In addition to their central role in energy production, mitochondria perform a number of functions essential to neuronal activity and synaptic transmission. In this chapter, we have considered emerging facets of mitochondrial biology related to signal transduction from mitochondria to mitochondria, mitochondria to cytoplasm, mitochondria to nucleus, and mitochondria to the systemic circulation.

Mitochondrial dysfunction may result from primary inherited genetic defects or from acquired structural and functional changes with aging and chronic stressors (i.e., mitochondrial allostatic load) [167]. In turn, dysfunctional mitochondria produce abnormal signals of stress (i.e., SOS) that propagate to other cellular compartments and influence systemic regulatory processes.

This chapter integrated experimental, clinical, and epidemiological evidence that posits mitochondria as a proximal mediator of established pathogenic processes that define neurodegenerative disease. Why would mitochondrial signals influence such a broad number of physiological and molecular processes? A partial justification for this "mitocentric" perspective may in part rest in biological events that occurred as part of the evolution of complex life. To reiterate, mitochondria played a central and necessary role in the evolution of multicellular organ systems, of which the brain is arguably the most intricate product. The cellular machinery involved in neurological functioning, including synaptic transmission, and those of its basic cellular behaviors including replication, differentiation, and growth, as well as the underlying process of transcriptional regulation, have thus relied and become dependent upon mitochondrial biochemical outputs (ATP, mtROS, and intermediate metabolites).

As a result, multiple evolutionary-carved paths exist to translate signals of mitochondrial dysfunction into maladaptive cellular processes contributing to neurodegeneration. Future work will be needed to elucidate the molecular and physiological basis for mitochondrial signaling within and beyond the cell. A fuller understanding of the role of mitochondria in the process of neurodegeneration should provide new opportunities to design interventions to preserve optimal brain function throughout the lifespan.

References

1. Margulis L, Bermudes D. Symbiosis as a mechanism of evolution: status of cell symbiosis theory. Symbiosis. 1985;1:101–24.
2. Lane N, Martin W. The energetics of genome complexity. Nature. 2010;467(7318):929–34.
3. Wallace DC. Bioenergetics, the origins of complexity, and the ascent of man. Proc Natl Acad Sci U S A. 2010;107 Suppl 2:8947–53.
4. Glancy B, Balaban RS. Role of mitochondrial ca(2+) in the regulation of cellular energetics. Biochemistry. 2012;51(14):2959–73.
5. Marland JR, Hasel P, Bonnycastle K, Cousin MA. Mitochondrial calcium uptake modulates synaptic vesicle endocytosis in central nerve terminals. J Biol Chem. 2016;291(5):2080–6.
6. Rizzuto R, De Stefani D, Raffaello A, Mammucari C. Mitochondria as sensors and regulators of calcium signalling. Nat Rev Mol Cell Biol. 2012;13(9):566–78.
7. Lambert AJ, Brand MD. Reactive oxygen species production by mitochondria. Methods Mol Biol. 2009;554:165–81.

8. Balaban RS, Nemoto S, Finkel T. Mitochondria, oxidants, and aging. Cell. 2005;120(4):483–95.
9. Papadopoulos V, Miller WL. Role of mitochondria in steroidogenesis. Best Pract Res Clin Endocrinol Metab. 2012;26(6):771–90.
10. Kasahara A, Scorrano L. Mitochondria: from cell death executioners to regulators of cell differentiation. Trends Cell Biol. 2014;24(12):761–70.
11. Waters CM, Bassler BL. Quorum sensing: cell-to-cell communication in bacteria. Annu Rev Cell Dev Biol. 2005;21:319–46.
12. Nadell CD, Bucci V, Drescher K, Levin SA, Bassler BL, Xavier JB. Cutting through the complexity of cell collectives. Proc Biol Sci R Soc. 2013;280(1755):20122770.
13. Braschi E, McBride HM. Mitochondria and the culture of the Borg: understanding the integration of mitochondrial function within the reticulum, the cell, and the organism. Bioessays. 2010;32(11):958–66.
14. Picard M, Burelle Y. Mitochondria: starving to reach quorum?: insight into the physiological purpose of mitochondrial fusion. Bioessays. 2012;34(4):272–4.
15. Zhou L, Aon MA, Almas T, Cortassa S, Winslow RL, O'Rourke B. A reaction-diffusion model of ROS-induced ROS release in a mitochondrial network. PLoS Comput Biol. 2010;6(1):e1000657.
16. Pacher P, Hajnoczky G. Propagation of the apoptotic signal by mitochondrial waves. EMBO J. 2001;20(15):4107–21.
17. Zorov DB, Filburn CR, Klotz LO, Zweier JL, Sollott SJ. Reactive oxygen species (ROS)-induced ROS release: a new phenomenon accompanying induction of the mitochondrial permeability transition in cardiac myocytes. J Exp Med. 2000;192(7):1001–14.
18. Pernas L, Scorrano L. Mito-morphosis: mitochondrial fusion, fission, and cristae remodeling as key mediators of cellular function. Annu Rev Physiol. 2015;78:505–31.
19. Picard M, Shirihai OS, Gentil BJ, Burelle Y. Mitochondrial morphology transitions and functions: implications for retrograde signaling? Am J Physiol Regul Integr Comp Physiol. 2013;304(6):R393–406.
20. Gomes LC, Di Benedetto G, Scorrano L. During autophagy mitochondria elongate, are spared from degradation and sustain cell viability. Nat Cell Biol. 2011;13(5):589–98.
21. Cogliati S, Frezza C, Soriano ME, Varanita T, Quintana-Cabrera R, Corrado M, et al. Mitochondrial cristae shape determines respiratory chain supercomplexes assembly and respiratory efficiency. Cell. 2013;155(1):160–71.
22. Chen H, McCaffery JM, Chan DC. Mitochondrial fusion protects against neurodegeneration in the cerebellum. Cell. 2007;130(3):548–62.
23. Oettinghaus B, Schulz JM, Restelli LM, Licci M, Savoia C, Schmidt A, et al. Synaptic dysfunction, memory deficits and hippocampal atrophy due to ablation of mitochondrial fission in adult forebrain neurons. Cell Death Differ. 2016;23(1):18–28.
24. Chen H, Vermulst M, Wang YE, Chomyn A, Prolla TA, McCaffery JM, et al. Mitochondrial fusion is required for mtDNA stability in skeletal muscle and tolerance of mtDNA mutations. Cell. 2010;141(2):280–9.
25. Stauch KL, Purnell PR, Fox HS. Quantitative proteomics of synaptic and nonsynaptic mitochondria: insights for synaptic mitochondrial vulnerability. J Proteome Res. 2014;13(5):2620–36.
26. Sheng ZH. Mitochondrial trafficking and anchoring in neurons: new insight and implications. J Cell Biol. 2014;204(7):1087–98.
27. Ono T, Isobe K, Nakada K, Hayashi JI. Human cells are protected from mitochondrial dysfunction by complementation of DNA products in fused mitochondria. Nat Genet. 2001;28(3):272–5.
28. Yang L, Long Q, Liu J, Tang H, Li Y, Bao F, et al. Mitochondrial fusion provides an 'initial metabolic complementation' controlled by mtDNA. Cell Mol Life Sci. 2015;72(13):2585–98.
29. Archer SL. Mitochondrial dynamics—mitochondrial fission and fusion in human diseases. N Engl J Med. 2013;369(23):2236–51.

30. Santo-Domingo J, Giacomello M, Poburko D, Scorrano L, Demaurex N. OPA1 promotes pH flashes that spread between contiguous mitochondria without matrix protein exchange. EMBO J. 2013;32(13):1927–40.
31. Amchenkova AA, Bakeeva LE, Chentsov YS, Skulachev VP, Zorov DB. Coupling membranes as energy-transmitting cables. I. Filamentous mitochondria in fibroblasts and mitochondrial clusters in cardiomyocytes. J Cell Biol. 1988;107(2):481–95.
32. Glancy B, Hartnell LM, Malide D, Yu ZX, Combs CA, Connelly PS, et al. Mitochondrial reticulum for cellular energy distribution in muscle. Nature. 2015;523(7562):617–20.
33. Li Z, Okamoto K, Hayashi Y, Sheng M. The importance of dendritic mitochondria in the morphogenesis and plasticity of spines and synapses. Cell. 2004;119(6):873–87.
34. Popov V, Medvedev NI, Davies HA, Stewart MG. Mitochondria form a filamentous reticular network in hippocampal dendrites but are present as discrete bodies in axons: a three-dimensional ultrastructural study. J Comp Neurol. 2005;492(1):50–65.
35. Skulachev VP. Mitochondrial filaments and clusters as intracellular power-transmitting cables. Trends Biochem Sci. 2001;26(1):23–9.
36. Kurz FT, Aon MA, O'Rourke B, Armoundas AA. Spatio-temporal oscillations of individual mitochondria in cardiac myocytes reveal modulation of synchronized mitochondrial clusters. Proc Natl Acad Sci U S A. 2010;107(32):14315–20.
37. Kurz FT, Aon MA, O'Rourke B, Armoundas AA. Cardiac mitochondria exhibit dynamic functional clustering. Front Physiol. 2014;5:329.
38. Picard M, McManus MJ, Csordas G, Varnai P, Dorn 2nd GW, Williams D, et al. Trans-mitochondrial coordination of cristae at regulated membrane junctions. Nat Commun. 2015;6:6259.
39. Brodin L, Bakeeva L, Shupliakov O. Presynaptic mitochondria and the temporal pattern of neurotransmitter release. Philos Trans R Soc Lond Ser B Biol Sci. 1999;354(1381):365–72.
40. Picard M, Azuelos I, Jung B, Giordano C, Matecki S, Hussain S, et al. Mechanical ventilation triggers abnormal mitochondrial dynamics and morphology in the diaphragm. J Appl Physiol. 2015;118(9):1161–71.
41. Picard M, Gentil BJ, McManus MJ, White K, St Louis K, Gartside SE, et al. Acute exercise remodels mitochondrial membrane interactions in mouse skeletal muscle. J Appl Physiol. 2013;115(10):1562–71.
42. Andreyev AY, Kushnareva YE, Murphy AN, Starkov AA. Mitochondrial ROS metabolism: 10 years later. Biochem Biokhim. 2015;80(5):517–31.
43. Yee C, Yang W, Hekimi S. The intrinsic apoptosis pathway mediates the pro-longevity response to mitochondrial ROS in C. elegans. Cell. 2014;157(4):897–909.
44. Wallace DC, Fan W, Procaccio V. Mitochondrial energetics and therapeutics. Annu Rev Pathol. 2010;5:297–348.
45. McKenna MC, Waagepetersen HS, Schousboe A, Sonnewald U. Neuronal and astrocytic shuttle mechanisms for cytosolic-mitochondrial transfer of reducing equivalents: current evidence and pharmacological tools. Biochem Pharmacol. 2006;71(4):399–407.
46. Dumas JF, Argaud L, Cottet-Rousselle C, Vial G, Gonzalez C, Detaille D, et al. Effect of transient and permanent permeability transition pore opening on NAD(P)H localization in intact cells. J Biol Chem. 2009;284(22):15117–25.
47. Gameiro PA, Laviolette LA, Kelleher JK, Iliopoulos O, Stephanopoulos G. Cofactor balance by nicotinamide nucleotide transhydrogenase (NNT) coordinates reductive carboxylation and glucose catabolism in the tricarboxylic acid (TCA) cycle. J Biol Chem. 2013;288(18):12967–77.
48. Ma S, Zhang X, Zheng L, Li Z, Zhao X, Lai W, et al. Peroxiredoxin 6 is a crucial factor in the initial step of mitochondrial clearance and is upstream of the PINK1-Parkin pathway. Antioxid Redox Signal. 2015;24:486–501.
49. Meng F, Yao D, Shi Y, Kabakoff J, Wu W, Reicher J, et al. Oxidation of the cysteine-rich regions of parkin perturbs its E3 ligase activity and contributes to protein aggregation. Mol Neurodegener. 2011;6:34.

50. Sabens EA, Distler AM, Mieyal JJ. Levodopa deactivates enzymes that regulate thiol-disulfide homeostasis and promotes neuronal cell death: implications for therapy of Parkinson's disease. Biochemistry. 2010;49(12):2715–24.
51. Angeles DC, Gan BH, Onstead L, Zhao Y, Lim KL, Dachsel J, et al. Mutations in LRRK2 increase phosphorylation of peroxiredoxin 3 exacerbating oxidative stress-induced neuronal death. Hum Mutat. 2011;32(12):1390–7.
52. Shi ZZ, Osei-Frimpong J, Kala G, Kala SV, Barrios RJ, Habib GM, et al. Glutathione synthesis is essential for mouse development but not for cell growth in culture. Proc Natl Acad Sci U S A. 2000;97(10):5101–6.
53. Dalton TP, Dieter MZ, Yang Y, Shertzer HG, Nebert DW. Knockout of the mouse glutamate cysteine ligase catalytic subunit (Gclc) gene: embryonic lethal when homozygous, and proposed model for moderate glutathione deficiency when heterozygous. Biochem Biophys Res Commun. 2000;279(2):324–9.
54. Matsui M, Oshima M, Oshima H, Takaku K, Maruyama T, Yodoi J, et al. Early embryonic lethality caused by targeted disruption of the mouse thioredoxin gene. Dev Biol. 1996;178(1):179–85.
55. Jakupoglu C, Przemeck GK, Schneider M, Moreno SG, Mayr N, Hatzopoulos AK, et al. Cytoplasmic thioredoxin reductase is essential for embryogenesis but dispensable for cardiac development. Mol Cell Biol. 2005;25(5):1980–8.
56. McManus MJ, Murphy MP, Franklin JL. Mitochondria-derived reactive oxygen species mediate caspase-dependent and -independent neuronal deaths. Mol Cell Neurosci. 2014;63:13–23.
57. Murphy MP. How mitochondria produce reactive oxygen species. Biochem J. 2009;417(1):1–13.
58. Huie RE, Padmaja S. The reaction of no with superoxide. Free Radic Res Commun. 1993;18(4):195–9.
59. Khan M, Dhammu TS, Matsuda F, Annamalai B, Dhindsa TS, Singh I, et al. Targeting the nNOS/peroxynitrite/calpain system to confer neuroprotection and aid functional recovery in a mouse model of TBI. Brain Res. 2015;1630:159–70.
60. Di Filippo M, Chiasserini D, Tozzi A, Picconi B, Calabresi P. Mitochondria and the link between neuroinflammation and neurodegeneration. J Alzheimers Dis JAD. 2010;20 Suppl 2:S369–79.
61. Kalogeris T, Bao Y, Korthuis RJ. Mitochondrial reactive oxygen species: a double edged sword in ischemia/reperfusion vs preconditioning. Redox Biol. 2014;2:702–14.
62. Lacza Z, Snipes JA, Kis B, Szabó C, Grover G, Busija DW. Investigation of the subunit composition and the pharmacology of the mitochondrial ATP-dependent K+ channel in the brain. Brain Res. 2003;994(1):27–36.
63. Anantharaman M, Tangpong J, Keller JN, Murphy MP, Markesbery WR, Kiningham KK, et al. Beta-amyloid mediated nitration of manganese superoxide dismutase: implication for oxidative stress in a APPNLH/NLH X PS-1P264L/P264L double knock-in mouse model of Alzheimer's disease. Am J Pathol. 2006;168(5):1608–18.
64. Aoyama K, Matsubara K, Fujikawa Y, Nagahiro Y, Shimizu K, Umegae N, et al. Nitration of manganese superoxide dismutase in cerebrospinal fluids is a marker for peroxynitrite-mediated oxidative stress in neurodegenerative diseases. Ann Neurol. 2000;47(4):524–7.
65. Korytowski W, Basova LV, Pilat A, Kernstock RM, Girotti AW. Permeabilization of the mitochondrial outer membrane by Bax/truncated Bid (tBid) proteins as sensitized by cardiolipin hydroperoxide translocation: mechanistic implications for the intrinsic pathway of oxidative apoptosis. J Biol Chem. 2011;286(30):26334–43.
66. Orrenius S, Zhivotovsky B. Cardiolipin oxidation sets cytochrome c free. Nat Chem Biol. 2005;1(4):188–9.
67. Birk AV, Chao WM, Liu S, Soong Y, Szeto HH. Disruption of cytochrome c heme coordination is responsible for mitochondrial injury during ischemia. Biochim Biophys Acta. 2015;1847(10):1075–84.

68. Vaughn AE, Deshmukh M. Glucose metabolism inhibits apoptosis in neurons and cancer cells by redox inactivation of cytochrome c. Nat Cell Biol. 2008;10(12):1477–83.
69. Kirkland RA, Windelborn JA, Kasprzak JM, Franklin JL. A Bax-induced pro-oxidant state is critical for cytochrome c release during programmed neuronal death. J Neurosci Off J Soc Neurosci. 2002;22(15):6480–90.
70. Logan A, Shabalina IG, Prime TA, Rogatti S, Kalinovich AV, Hartley RC, et al. In vivo levels of mitochondrial hydrogen peroxide increase with age in mtDNA mutator mice. Aging Cell. 2014;13(4):765–8.
71. Pan-Montojo F, Schwarz M, Winkler C, Arnhold M, O'Sullivan GA, Pal A, et al. Environmental toxins trigger PD-like progression via increased alpha-synuclein release from enteric neurons in mice. Sci Rep. 2012;2:898.
72. Lin MT, Beal MF. Mitochondrial dysfunction and oxidative stress in neurodegenerative diseases. Nature. 2006;443(7113):787–95.
73. Murray J, Tsui WH, Li Y, McHugh P, Williams S, Cummings M, et al. FDG and amyloid PET in cognitively normal individuals at risk for late-onset alzheimer's disease. Adv Mol Imaging. 2014;4(2):15–26.
74. Mosconi L, Pupi A, De Leon MJ. Brain glucose hypometabolism and oxidative stress in preclinical Alzheimer's disease. Ann N Y Acad Sci. 2008;1147:180–95.
75. Majd S, Power JH, Grantham HJ. Neuronal response in Alzheimer's and Parkinson's disease: the effect of toxic proteins on intracellular pathways. BMC Neurosci. 2015;16:69.
76. Brettschneider J, Del Tredici K, Lee VM, Trojanowski JQ. Spreading of pathology in neurodegenerative diseases: a focus on human studies. Nat Rev Neurosci. 2015;16(2):109–20.
77. Su B, Wang X, Nunomura A, Moreira PI, Lee H, Perry G, et al. Oxidative stress signaling in Alzheimer's disease. Curr Alzheimer Res. 2008;5(6):525–32.
78. Wallace DC. Mitochondrial DNA, mutations in disease and aging. Environ Mol Mutagen. 2010;51(5):440–50.
79. Cannon JR, Greenamyre JT. Neurotoxic in vivo models of Parkinson's disease recent advances. Prog Brain Res. 2010;184:17–33.
80. Langston JW, Ballard P, Tetrud JW, Irwin I. Chronic parkinsonism in humans due to a product of meperidine-analog synthesis. Science. 1983;219(4587):979–80.
81. Swerdlow RH, Parks JK, Cassarino DS, Maguire DJ, Maguire RS, Bennett Jr JP, et al. Cybrids in Alzheimer's disease: a cellular model of the disease? Neurology. 1997;49(4):918–25.
82. Swerdlow RH. Role and treatment of mitochondrial DNA-related mitochondrial dysfunction in sporadic neurodegenerative diseases. Curr Pharm Des. 2011;17(31):3356–73.
83. Tan JL, Li QX, Ciccotosto GD, Crouch PJ, Culvenor JG, White AR, et al. Mild oxidative stress induces redistribution of BACE1 in non-apoptotic conditions and promotes the amyloidogenic processing of Alzheimer's disease amyloid precursor protein. PLoS One. 2013;8(4):e61246.
84. Tamagno E, Bardini P, Obbili A, Vitali A, Borghi R, Zaccheo D, et al. Oxidative stress increases expression and activity of BACE in NT2 neurons. Neurobiol Dis. 2002;10(3):279–88.
85. Tamagno E, Guglielmotto M, Aragno M, Borghi R, Autelli R, Giliberto L, et al. Oxidative stress activates a positive feedback between the gamma- and beta-secretase cleavages of the beta-amyloid precursor protein. J Neurochem. 2008;104(3):683–95.
86. Quiroz-Baez R, Rojas E, Arias C. Oxidative stress promotes JNK-dependent amyloidogenic processing of normally expressed human APP by differential modification of alpha-, beta- and gamma-secretase expression. Neurochem Int. 2009;55(7):662–70.
87. Bredesen DE, John V, Galvan V. Importance of the caspase cleavage site in amyloid-β protein precursor. J Alzheimers Dis JAD. 2010;22(1):57–63.
88. McManus MJ, Murphy MP, Franklin JL. The mitochondria-targeted antioxidant MitoQ prevents loss of spatial memory retention and early neuropathology in a transgenic mouse model of Alzheimer's disease. J Neurosci. 2011;31(44):15703–15.

89. Soucek T, Cumming R, Dargusch R, Maher P, Schubert D. The regulation of glucose metabolism by HIF-1 mediates a neuroprotective response to amyloid beta peptide. Neuron. 2003;39(1):43–56.
90. Zou K, Gong JS, Yanagisawa K, Michikawa M. A novel function of monomeric amyloid beta-protein serving as an antioxidant molecule against metal-induced oxidative damage. J Neurosci Off J Soc Neurosci. 2002;22(12):4833–41.
91. Kontush A, Berndt C, Weber W, Akopyan V, Arlt S, Schippling S, et al. Amyloid-beta is an antioxidant for lipoproteins in cerebrospinal fluid and plasma. Free Radic Biol Med. 2001;30(1):119–28.
92. Plant LD, Boyle JP, Smith IF, Peers C, Pearson HA. The production of amyloid beta peptide is a critical requirement for the viability of central neurons. J Neurosci Off J Soc Neurosci. 2003;23(13):5531–5.
93. Devi L, Ohno M. Mitochondrial dysfunction and accumulation of the beta-secretase-cleaved C-terminal fragment of APP in Alzheimer's disease transgenic mice. Neurobiol Dis. 2012;45(1):417–24.
94. Du H, Guo L, Yan S, Sosunov AA, McKhann GM, Yan SS. Early deficits in synaptic mitochondria in an Alzheimer's disease mouse model. Proc Natl Acad Sci U S A. 2010;107(43):18670–5.
95. Ronnback A, Pavlov PF, Mansory M, Gonze P, Marliere N, Winblad B, et al. Mitochondrial dysfunction in a transgenic mouse model expressing human amyloid precursor protein (APP) with the Arctic mutation. J Neurochem. 2015;136(3):497–502.
96. Kontush A, Atwood CS. Amyloid-beta: phylogenesis of a chameleon. Brain Res Brain Res Rev. 2004;46(1):118–20.
97. Giuffrida ML, Caraci F, De Bona P, Pappalardo G, Nicoletti F, Rizzarelli E, et al. The monomer state of beta-amyloid: where the Alzheimer's disease protein meets physiology. Rev Neurosci. 2010;21(2):83–93.
98. Salloway S, Sperling R, Fox NC, Blennow K, Klunk W, Raskind M, et al. Two phase 3 trials of bapineuzumab in mild-to-moderate Alzheimer's disease. N Engl J Med. 2014;370(4):322–33.
99. Doody RS, Thomas RG, Farlow M, Iwatsubo T, Vellas B, Joffe S, et al. Phase 3 trials of solanezumab for mild-to-moderate Alzheimer's disease. N Engl J Med. 2014;370(4):311–21.
100. Holmes C, Boche D, Wilkinson D, Yadegarfar G, Hopkins V, Bayer A, et al. Long-term effects of Abeta42 immunisation in Alzheimer's disease: follow-up of a randomised, placebo-controlled phase I trial. Lancet (London, England). 2008;372(9634):216–23.
101. Busche MA, Grienberger C, Keskin AD, Song B, Neumann U, Staufenbiel M, et al. Decreased amyloid-beta and increased neuronal hyperactivity by immunotherapy in Alzheimer's models. Nat Neurosci. 2015;18(12):1725–7.
102. Iqbal K, Liu F, Gong CX. Tau and neurodegenerative disease: the story so far. Nat Rev Neurol. 2015;12:15–27.
103. Cotman CW, Poon WW, Rissman RA, Blurton-Jones M. The role of caspase cleavage of tau in Alzheimer disease neuropathology. J Neuropathol Exp Neurol. 2005;64(2):104–12.
104. Gao F, Chen D, Hu Q, Wang G. Rotenone directly induces BV2 cell activation via the p38 MAPK pathway. PLoS One. 2013;8(8):e72046.
105. Chiara F, Gambalunga A, Sciacovelli M, Nicolli A, Ronconi L, Fregona D, et al. Chemotherapeutic induction of mitochondrial oxidative stress activates GSK-3alpha/beta and Bax, leading to permeability transition pore opening and tumor cell death. Cell Death Dis. 2012;3:e444.
106. Alavi Naini SM, Soussi-Yanicostas N. Tau hyperphosphorylation and oxidative stress, a critical vicious circle in neurodegenerative tauopathies? Oxidative Med Cell Longev. 2015;2015:151979.
107. Chen CH, Li W, Sultana R, You MH, Kondo A, Shahpasand K, et al. Pin1 cysteine-113 oxidation inhibits its catalytic activity and cellular function in Alzheimer's disease. Neurobiol Dis. 2015;76:13–23.
108. Innes BT, Sowole MA, Gyenis L, Dubinsky M, Konermann L, Litchfield DW, et al. Peroxide-mediated oxidation and inhibition of the peptidyl-prolyl isomerase Pin1. Biochim Biophys Acta. 2015;1852(5):905–12.

109. Butterfield DA, Abdul HM, Opii W, Newman SF, Joshi G, Ansari MA, et al. Pin1 in Alzheimer's disease. J Neurochem. 2006;98(6):1697–706.
110. Sultana R, Boyd-Kimball D, Poon HF, Cai J, Pierce WM, Klein JB, et al. Oxidative modification and down-regulation of Pin1 in Alzheimer's disease hippocampus: a redox proteomics analysis. Neurobiol Aging. 2006;27(7):918–25.
111. DuBoff B, Gotz J, Feany MB. Tau promotes neurodegeneration via DRP1 mislocalization in vivo. Neuron. 2012;75(4):618–32.
112. Hashimoto M, Hsu LJ, Xia Y, Takeda A, Sisk A, Sundsmo M, et al. Oxidative stress induces amyloid-like aggregate formation of NACP/alpha-synuclein in vitro. Neuroreport. 1999;10(4):717–21.
113. Perrin RJ, Woods WS, Clayton DF, George JM. Exposure to long chain polyunsaturated fatty acids triggers rapid multimerization of synucleins. J Biol Chem. 2001;276(45):41958–62.
114. Nam MK, Han JH, Jang JY, Yun SE, Kim GY, Kang S, et al. A novel link between the conformations, exposure of specific epitopes, and subcellular localization of alpha-synuclein. Biochim Biophys Acta. 2015;1850(12):2497–505.
115. Zhou W, Zhu M, Wilson MA, Petsko GA, Fink AL. The oxidation state of DJ-1 regulates its chaperone activity toward alpha-synuclein. J Mol Biol. 2006;356(4):1036–48.
116. Lou H, Montoya SE, Alerte TNM, Wang J, Wu J, Peng X, et al. Serine 129 phosphorylation reduces the ability of α-synuclein to regulate tyrosine hydroxylase and protein phosphatase 2A in vitro and in vivo. J Biol Chem. 2010;285(23):17648–61.
117. Cali T, Ottolini D, Negro A, Brini M. Alpha-synuclein controls mitochondrial calcium homeostasis by enhancing endoplasmic reticulum-mitochondria interactions. J Biol Chem. 2012;287(22):17914–29.
118. Cannon JR, Tapias V, Na HM, Honick AS, Drolet RE, Greenamyre JT. A highly reproducible rotenone model of Parkinson's disease. Neurobiol Dis. 2009;34(2):279–90.
119. McLean JR, Hallett PJ, Cooper O, Stanley M, Isacson O. Transcript expression levels of full-length alpha-synuclein and its three alternatively spliced variants in Parkinson's disease brain regions and in a transgenic mouse model of alpha-synuclein overexpression. Mol Cell Neurosci. 2012;49(2):230–9.
120. Lashuel HA, Overk CR, Oueslati A, Masliah E. The many faces of alpha-synuclein: from structure and toxicity to therapeutic target. Nat Rev Neurosci. 2013;14(1):38–48.
121. Hodara R, Norris EH, Giasson BI, Mishizen-Eberz AJ, Lynch DR, Lee VM, et al. Functional consequences of alpha-synuclein tyrosine nitration: diminished binding to lipid vesicles and increased fibril formation. J Biol Chem. 2004;279(46):47746–53.
122. Guardia-Laguarta C, Area-Gomez E, Rub C, Liu Y, Magrane J, Becker D, et al. Alpha-synuclein is localized to mitochondria-associated ER membranes. J Neurosci Off J Soc Neurosci. 2014;34(1):249–59.
123. Liu G, Zhang C, Yin J, Li X, Cheng F, Li Y, et al. Alpha-synuclein is differentially expressed in mitochondria from different rat brain regions and dose-dependently down-regulates complex I activity. Neurosci Lett. 2009;454(3):187–92.
124. Lee HJ, Shin SY, Choi C, Lee YH, Lee SJ. Formation and removal of alpha-synuclein aggregates in cells exposed to mitochondrial inhibitors. J Biol Chem. 2002;277(7):5411–7.
125. Devi L, Raghavendran V, Prabhu BM, Avadhani NG, Anandatheerthavarada HK. Mitochondrial import and accumulation of alpha-synuclein impair complex I in human dopaminergic neuronal cultures and Parkinson disease brain. J Biol Chem. 2008;283(14):9089–100.
126. Sherer TB, Kim JH, Betarbet R, Greenamyre JT. Subcutaneous rotenone exposure causes highly selective dopaminergic degeneration and alpha-synuclein aggregation. Exp Neurol. 2003;179(1):9–16.
127. Barrett PJ, Timothy Greenamyre J. Post-translational modification of alpha-synuclein in Parkinson's disease. Brain Res. 2015;1628(Pt B):247–53.
128. Lackner LL. Shaping the dynamic mitochondrial network. BMC Biol. 2014;12:35.
129. Zhu XH, Qiao H, Du F, Xiong Q, Liu X, Zhang X, et al. Quantitative imaging of energy expenditure in human brain. NeuroImage. 2012;60(4):2107–17.

130. Picard M. Mitochondrial synapses: intracellular communication and signal integration. Trends Neurosci. 2015;38(8):468–74.
131. Gut P, Verdin E. The nexus of chromatin regulation and intermediary metabolism. Nature. 2013;502(7472):489–98.
132. Salminen A, Kaarniranta K, Hiltunen M, Kauppinen A. Krebs cycle dysfunction shapes epigenetic landscape of chromatin: novel insights into mitochondrial regulation of aging process. Cell Signal. 2014;26(7):1598–603.
133. Bird A. Perceptions of epigenetics. Nature. 2007;447(7143):396–8.
134. Salminen A, Haapasalo A, Kauppinen A, Kaarniranta K, Soininen H, Hiltunen M. Impaired mitochondrial energy metabolism in Alzheimer's disease: impact on pathogenesis via disturbed epigenetic regulation of chromatin landscape. Prog Neurobiol. 2015;131:1–20.
135. De Jager PL, Srivastava G, Lunnon K, Burgess J, Schalkwyk LC, Yu L, et al. Alzheimer's disease: early alterations in brain DNA methylation at ANK1, BIN1, RHBDF2 and other loci. Nat Neurosci. 2014;17(9):1156–63.
136. Picard M, Zhang J, Hancock S, Derbeneva O, Golhar R, Golik P, et al. Progressive increase in mtDNA 3243A>G heteroplasmy causes abrupt transcriptional reprogramming. Proc Natl Acad Sci U S A. 2014;111(38):E4033–42.
137. Al-Mehdi AB, Pastukh VM, Swiger BM, Reed DJ, Patel MR, Bardwell GC, et al. Perinuclear mitochondrial clustering creates an oxidant-rich nuclear domain required for hypoxia-induced transcription. Sci Signal. 2012;5(231):ra47.
138. Passos JF, Saretzki G, Ahmed S, Nelson G, Richter T, Peters H, et al. Mitochondrial dysfunction accounts for the stochastic heterogeneity in telomere-dependent senescence. PLoS Biol. 2007;5(5):e110.
139. Velarde MC, Flynn JM, Day NU, Melov S, Campisi J. Mitochondrial oxidative stress caused by Sod2 deficiency promotes cellular senescence and aging phenotypes in the skin. Aging. 2012;4(1):3–12.
140. Hayashi Y, Yoshida M, Yamato M, Ide T, Wu Z, Ochi-Shindou M, et al. Reverse of age-dependent memory impairment and mitochondrial DNA damage in microglia by an overexpression of human mitochondrial transcription factor a in mice. J Neurosci Off J Soc Neurosci. 2008;28(34):8624–34.
141. Miller GE, Chen E, Sze J, Marin T, Arevalo JM, Doll R, et al. A functional genomic fingerprint of chronic stress in humans: blunted glucocorticoid and increased NF-kappaB signaling. Biol Psychiatry. 2008;64(4):266–72.
142. Shaughnessy DT, McAllister K, Worth L, Haugen AC, Meyer JN, Domann FE, et al. Mitochondria, energetics, epigenetics, and cellular responses to stress. Environ Health Perspect. 2014;122(12):1271–8.
143. Guha M, Avadhani NG. Mitochondrial retrograde signaling at the crossroads of tumor bioenergetics, genetics and epigenetics. Mitochondrion. 2013;13(6):577–91.
144. Arany Z. PGC-1 coactivators and skeletal muscle adaptations in health and disease. Curr Opin Genet Dev. 2008;18(5):426–34.
145. Cooper-Knock J, Kirby J, Ferraiuolo L, Heath PR, Rattray M, Shaw PJ. Gene expression profiling in human neurodegenerative disease. Nat Rev Neurol. 2012;8(9):518–30.
146. McEwen BS, Wingfield JC. The concept of allostasis in biology and biomedicine. Horm Behav. 2003;43(1):2–15.
147. Rangaraju V, Calloway N, Ryan TA. Activity-driven local ATP synthesis is required for synaptic function. Cell. 2014;156(4):825–35.
148. Hara Y, Yuk F, Puri R, Janssen WG, Rapp PR, Morrison JH. Presynaptic mitochondrial morphology in monkey prefrontal cortex correlates with working memory and is improved with estrogen treatment. Proc Natl Acad Sci U S A. 2014;111(1):486–91.
149. Sharpley MS, Marciniak C, Eckel-Mahan K, McManus M, Crimi M, Waymire K, et al. Heteroplasmy of mouse mtDNA is genetically unstable and results in altered behavior and cognition. Cell. 2012;151(2):333–43.
150. Sun T, Qiao H, Pan PY, Chen Y, Sheng ZH. Motile axonal mitochondria contribute to the variability of presynaptic strength. Cell Rep. 2013;4(3):413–9.

151. Vos M, Lauwers E, Verstreken P. Synaptic mitochondria in synaptic transmission and organization of vesicle pools in health and disease. Front Synaptic Neurosci. 2010;2:139.
152. Picard M, McEwen BS. Mitochondria impact brain function and cognition. Proc Natl Acad Sci U S A. 2014;111(1):7–8.
153. Taivassalo T, Jensen TD, Kennaway N, DiMauro S, Vissing J, Haller RG. The spectrum of exercise tolerance in mitochondrial myopathies: a study of 40 patients. Brain. 2003;126(Pt 2):413–23.
154. Taivassalo T, Ayyad K, Haller RG. Increased capillaries in mitochondrial myopathy: implications for the regulation of oxygen delivery. Brain. 2012;135(Pt 1):53–61.
155. McEwen BS. Physiology and neurobiology of stress and adaptation: central role of the brain. Physiol Rev. 2007;87(3):873–904.
156. McEwen BS, Bowles NP, Gray JD, Hill MN, Hunter RG, Karatsoreos IN, et al. Mechanisms of stress in the brain. Nat Neurosci. 2015;18(10):1353–63.
157. Tank AW, Lee WD. Peripheral and central effects of circulating catecholamines. Compr Physiol. 2015;5(1):1–15.
158. Gimsa U, Kanitz E, Otten W, Ibrahim SM. Behavior and stress reactivity in mouse strains with mitochondrial DNA variations. Ann N Y Acad Sci. 2009;1153:131–8.
159. Picard M, McManus MJ, Gray JD, Nasca C, Moffat C, Kopinski PK, et al. Mitochondrial functions modulate neuroendocrine, metabolic, inflammatory, and transcriptional responses to acute psychological stress. Proc Natl Acad Sci U S A. 2015;112(48):E6614–23.
160. Meimaridou E, Kowalczyk J, Guasti L, Hughes CR, Wagner F, Frommolt P, et al. Mutations in NNT encoding nicotinamide nucleotide transhydrogenase cause familial glucocorticoid deficiency. Nat Genet. 2012;44(7):740–2.
161. Picard M, McManus MJ, Gray JD, Nasca C, Moffat C, Kopinski PK, et al. Mitochondrial functions modulate neuroendocrine, metabolic, inflammatory, and transcriptional responses to acute psychological stress. Proc Natl Acad Sci. 2015;112(48):E6614–23.
162. Strauss KA, DuBiner L, Simon M, Zaragoza M, Sengupta PP, Li P, et al. Severity of cardiomyopathy associated with adenine nucleotide translocator-1 deficiency correlates with mtDNA haplogroup. Proc Natl Acad Sci U S A. 2013;110(9):3453–8.
163. Bates MG, Newman JH, Jakovljevic DG, Hollingsworth KG, Alston CL, Zalewski P, et al. Defining cardiac adaptations and safety of endurance training in patients with m.3243A>G-related mitochondrial disease. Int J Cardiol. 2013;168(4):3599–608.
164. Jeppesen TD, Orngreen MC, Van Hall G, Vissing J. Lactate metabolism during exercise in patients with mitochondrial myopathy. Neuromuscul Dis NMD. 2013;23(8):629–36.
165. van Elderen SG, de Roos A, de Craen AJ, Westendorp RG, Blauw GJ, Jukema JW, et al. Progression of brain atrophy and cognitive decline in diabetes mellitus: a 3-year follow-up. Neurology. 2010;75(11):997–1002.
166. Du J, Wang Y, Hunter R, Wei Y, Blumenthal R, Falke C, et al. Dynamic regulation of mitochondrial function by glucocorticoids. Proc Natl Acad Sci U S A. 2009;106(9):3543–8.
167. Picard M, Juster RP, McEwen BS. Mitochondrial allostatic load puts the 'gluc' back in glucocorticoids. Nat Rev Endocrinol. 2014;10(5):303–10.
168. Franceschi C, Campisi J. Chronic inflammation (inflammaging) and its potential contribution to age-associated diseases. J Gerontol Ser A Biol Med Sci. 2014;69 Suppl 1:S4–9.
169. Masters SL, Walsh PT. Release of the mitochondrial endosymbiont helps explain sterile inflammation. Proc Natl Acad Sci U S A. 2010;107(10):E32.
170. Brown RC, Lockwood AH, Sonawane BR. Neurodegenerative diseases: an overview of environmental risk factors. Environ Health Perspect. 2005;113(9):1250–6.
171. Pinti M, Cevenini E, Nasi M, De Biasi S, Salvioli S, Monti D, et al. Circulating mitochondrial DNA increases with age and is a familiar trait: Implications for "inflamm-aging". Eur J Immunol. 2014;44(5):1552–62.
172. Lu B, Kwan K, Levine YA, Olofsson PS, Yang H, Li J, et al. Alpha 7 nicotinic acetylcholine receptor signaling inhibits inflammasome activation by preventing mitochondrial DNA release. Mol Med. 2014;20:350–8.
173. West AP, Khoury-Hanold W, Staron M, Tal MC, Pineda CM, Lang SM, et al. Mitochondrial DNA stress primes the antiviral innate immune response. Nature. 2015;520(7548):553–7.

174. Zhang Q, Raoof M, Chen Y, Sumi Y, Sursal T, Junger W, et al. Circulating mitochondrial DAMPs cause inflammatory responses to injury. Nature. 2010;464(7285):104–7.
175. Shimada K, Crother TR, Karlin J, Dagvadorj J, Chiba N, Chen S, et al. Oxidized mitochondrial DNA activates the NLRP3 inflammasome during apoptosis. Immunity. 2012;36(3):401–14.
176. Gao X, Hu X, Qian L, Yang S, Zhang W, Zhang D, et al. Formyl-methionyl-leucyl-phenylalanine-induced dopaminergic neurotoxicity via microglial activation: a mediator between peripheral infection and neurodegeneration? Environ Health Perspect. 2008;116(5):593–8.
177. Wilkins HM, Carl SM, Weber SG, Ramanujan SA, Festoff BW, Linseman DA, et al. Mitochondrial lysates induce inflammation and Alzheimer's disease-relevant changes in microglial and neuronal cells. J Alzheimers Dis JAD. 2015;45(1):305–18.
178. Chen L, Na R, Boldt E, Ran Q. NLRP3 inflammasome activation by mitochondrial reactive oxygen species plays a key role in long-term cognitive impairment induced by paraquat exposure. Neurobiol Aging. 2015;36(9):2533–43.
179. Mathew A, Lindsley TA, Sheridan A, Bhoiwala DL, Hushmendy SF, Yager EJ, et al. Degraded mitochondrial DNA is a newly identified subtype of the damage associated molecular pattern (DAMP) family and possible trigger of neurodegeneration. J Alzheimers Dis JAD. 2012;30(3):617–27.
180. Heneka MT, Kummer MP, Stutz A, Delekate A, Schwartz S, Vieira-Saecker A, et al. NLRP3 is activated in Alzheimer's disease and contributes to pathology in APP/PS1 mice. Nature. 2013;493(7434):674–8.
181. Wan M, Hua X, Su J, Thiagarajan D, Frostegard AG, Haeggstrom JZ, et al. Oxidized but not native cardiolipin has pro-inflammatory effects, which are inhibited by Annexin A5. Atherosclerosis. 2014;235(2):592–8.
182. Pullerits R, Bokarewa M, Jonsson IM, Verdrengh M, Tarkowski A. Extracellular cytochrome c, a mitochondrial apoptosis-related protein, induces arthritis. Rheumatol (Oxford, England). 2005;44(1):32–9.
183. Shamaei-Tousi A, Steptoe A, O'Donnell K, Palmen J, Stephens JW, Hurel SJ, et al. Plasma heat shock protein 60 and cardiovascular disease risk: the role of psychosocial, genetic, and biological factors. Cell Stress Chaperones. 2007;12(4):384–92.
184. Mengwasser J, Piau A, Schlag P, Sleeman JP. Differential immunization identifies PHB1/PHB2 as blood-borne tumor antigens. Oncogene. 2004;23(44):7430–5.
185. Lee C, Yen K, Cohen P. Humanin: a harbinger of mitochondrial-derived peptides? Trends Endocrinol Metab TEM. 2013;24(5):222–8.
186. Lee C, Zeng J, Drew BG, Sallam T, Martin-Montalvo A, Wan J, et al. The mitochondrial-derived peptide MOTS-c promotes metabolic homeostasis and reduces obesity and insulin resistance. Cell Metab. 2015;21(3):443–54.
187. Lehnardt S, Schott E, Trimbuch T, Laubisch D, Krueger C, Wulczyn G, et al. A vicious cycle involving release of heat shock protein 60 from injured cells and activation of toll-like receptor 4 mediates neurodegeneration in the CNS. J Neurosci Off J Soc Neurosci. 2008;28(10):2320–31.
188. Perez XA. Preclinical evidence for a role of the nicotinic cholinergic system in Parkinson's disease. Neuropsychol Rev. 2015;25(4):371–83.
189. Akaike A, Takada-Takatori Y, Kume T, Izumi Y. Mechanisms of neuroprotective effects of nicotine and acetylcholinesterase inhibitors: role of alpha4 and alpha7 receptors in neuroprotection. J Mol Neurosci MN. 2010;40(1–2):211–6.
190. West AP, Shadel GS, Ghosh S. Mitochondria in innate immune responses. Nat Rev Immunol. 2011;11(6):389–402.
191. Cunningham C. Microglia and neurodegeneration: the role of systemic inflammation. Glia. 2013;61(1):71–90.
192. Koshiba T, Yasukawa K, Yanagi Y, Kawabata S. Mitochondrial membrane potential is required for MAVS-mediated antiviral signaling. Sci Signal. 2011;4(158):ra7.
193. West AP, Brodsky IE, Rahner C, Woo DK, Erdjument-Bromage H, Tempst P, et al. TLR signalling augments macrophage bactericidal activity through mitochondrial ROS. Nature. 2011;472(7344):476–80.

194. Frank MG, Weber MD, Fonken LK, Hershman SA, Watkins LR, Maier SF. The redox state of the alarmin HMGB1 is a pivotal factor in neuroinflammatory and microglial priming: a role for the NLRP3 inflammasome. Brain Behav Immun. 2015; doi: 10.1016/j.bbi.2015.10.009 (in press).
195. Gomez-Duran A, Pacheu-Grau D, Lopez-Gallardo E, Diez-Sanchez C, Montoya J, Lopez-Perez MJ, et al. Unmasking the causes of multifactorial disorders: OXPHOS differences between mitochondrial haplogroups. Hum Mol Genet. 2010;19(17):3343–53.
196. Karabatsiakis A, Bock C, Salinas-Manrique J, Kolassa S, Calzia E, Dietrich DE, et al. Mitochondrial respiration in peripheral blood mononuclear cells correlates with depressive subsymptoms and severity of major depression. Transl Psychiatr. 2014;4:e397.
197. Khusnutdinova E, Gilyazova I, Ruiz-Pesini E, Derbeneva O, Khusainova R, Khidiyatova I, et al. A mitochondrial etiology of neurodegenerative diseases: evidence from Parkinson's disease. Ann N Y Acad Sci. 2008;1147:1–20.
198. Coskun P, Wyrembak J, Schriner SE, Chen HW, Marciniack C, LaFerla F, et al. A mitochondrial etiology of Alzheimer and Parkinson disease. Biochim Biophys Acta. 2012;1820(5):553–64.
199. Kenney MC, Chwa M, Atilano SR, Falatoonzadeh P, Ramirez C, Malik D, et al. Inherited mitochondrial DNA variants can affect complement, inflammation and apoptosis pathways: insights into mitochondrial-nuclear interactions. Hum Mol Genet. 2014;23(13):3537–51.
200. Perry VH, Holmes C. Microglial priming in neurodegenerative disease. Nat Rev Neurol. 2014;10(4):217–24.
201. Mattson MP. Energy intake and exercise as determinants of brain health and vulnerability to injury and disease. Cell Metab. 2012;16(6):706–22.
202. Johansson L, Guo X, Waern M, Ostling S, Gustafson D, Bengtsson C, et al. Midlife psychological stress and risk of dementia: a 35-year longitudinal population study. Brain. 2010;133(Pt 8):2217–24.
203. Herrera AJ, Espinosa-Oliva AM, Carrillo-Jimenez A, Oliva-Martin MJ, Garcia-Revilla J, Garcia-Quintanilla A, et al. Relevance of chronic stress and the two faces of microglia in Parkinson's disease. Front Cell Neurosci. 2015;9:312.
204. Cheng A, Yang Y, Zhou Y, Maharana C, Lu D, Peng W, et al. Mitochondrial SIRT3 mediates adaptive responses of neurons to exercise and metabolic and excitatory challenges. Cell Metab. 2015;23(1):128–42.
205. Fiuza-Luces C, Garatachea N, Berger NA, Lucia A. Exercise is the real polypill. Physiol (Bethesda, Md). 2013;28(5):330–58.
206. Picard M, Turnbull DM. Linking the metabolic state and mitochondrial DNA in chronic disease, health, and aging. Diabetes. 2013;62(3):672–8.
207. Morris JK, Vidoni ED, Honea RA, Burns JM. Impaired glycemia increases disease progression in mild cognitive impairment. Neurobiol Aging. 2014;35(3):585–9.
208. Macauley SL, Stanley M, Caesar EE, Yamada SA, Raichle ME, Perez R, et al. Hyperglycemia modulates extracellular amyloid-beta concentrations and neuronal activity in vivo. J Clin Invest. 2015;125(6):2463–7.
209. Adlard PA, Perreau VM, Pop V, Cotman CW. Voluntary exercise decreases amyloid load in a transgenic model of Alzheimer's disease. J Neurosci Off J Soc Neurosci. 2005;25(17):4217–21.
210. Moore KM, Girens RE, Larson SK, Jones MR, Restivo JL, Holtzman DM, et al. A spectrum of exercise training reduces soluble Abeta in a dose-dependent manner in a mouse model of Alzheimer's disease. Neurobiol Dis. 2015;85:218–24.
211. Leal MC, Magnani N, Villordo S, Buslje CM, Evelson P, Castano EM, et al. Transcriptional regulation of insulin-degrading enzyme modulates mitochondrial amyloid beta (Abeta) peptide catabolism and functionality. J Biol Chem. 2013;288(18):12920–31.
212. Wang HX, Wahlberg M, Karp A, Winblad B, Fratiglioni L. Psychosocial stress at work is associated with increased dementia risk in late life. Alzheimers Dement. 2012;8(2):114–20.

213. Johansson L, Skoog I, Gustafson DR, Olesen PJ, Waern M, Bengtsson C, et al. Midlife psychological distress associated with late-life brain atrophy and white matter lesions: a 32-year population study of women. Psychosom Med. 2012;74(2):120–5.
214. Djamshidian A, Lees AJ. Can stress trigger Parkinson's disease? J Neurol Neurosurg Psychiatry. 2014;85(8):878–81.
215. Psarra AM, Sekeris CE. Glucocorticoids induce mitochondrial gene transcription in HepG2 cells: role of the mitochondrial glucocorticoid receptor. Biochim Biophys Acta. 2011;1813(10):1814–21.
216. de Pablos RM, Herrera AJ, Espinosa-Oliva AM, Sarmiento M, Munoz MF, Machado A, et al. Chronic stress enhances microglia activation and exacerbates death of nigral dopaminergic neurons under conditions of inflammation. J Neuroinflammation. 2014;11:34.
217. Erickson KI, Voss MW, Prakash RS, Basak C, Szabo A, Chaddock L, et al. Exercise training increases size of hippocampus and improves memory. Proc Natl Acad Sci U S A. 2011;108(7):3017–22.
218. Norton S, Matthews FE, Barnes DE, Yaffe K, Brayne C. Potential for primary prevention of Alzheimer's disease: an analysis of population-based data. Lancet Neurol. 2014;13(8):788–94.
219. Goldman JG, Weintraub D. Advances in the treatment of cognitive impairment in Parkinson's disease. Mov Disord. 2015;30(11):1471–89.
220. Puterman E, Lin J, Blackburn E, O'Donovan A, Adler N, Epel E. The power of exercise: buffering the effect of chronic stress on telomere length. PLoS One. 2010;5(5):e10837.
221. Steiner JL, Murphy EA, McClellan JL, Carmichael MD, Davis JM. Exercise training increases mitochondrial biogenesis in the brain. J Appl Physiol. 2011;111(4):1066–71.
222. Steib K, Schaffner I, Jagasia R, Ebert B, Lie DC. Mitochondria modify exercise-induced development of stem cell-derived neurons in the adult brain. J Neurosci Off J Soc Neurosci. 2014;34(19):6624–33.
223. Davis CH, Kim KY, Bushong EA, Mills EA, Boassa D, Shih T, et al. Transcellular degradation of axonal mitochondria. Proc Natl Acad Sci U S A. 2014;111(26):9633–8.

Part III
Functional Consequences of Mitochondrial Dysfunction

Chapter 6
Life on the Edge: Determinants of Selective Neuronal Vulnerability in Parkinson's Disease

James Surmeier, Enrico Zampese, Daniel Galtieri, and Paul T. Schumacker

Abstract One of the defining features of Parkinson's disease (PD) is its sparsely distributed pathology. This chapter explores the hypothesis that this selective vulnerability can be traced back to an unusual neuronal phenotype. This phenotype is best exemplified by dopaminergic neurons in the substantia nigra pars compacta (SNc), whose loss is responsible for the cardinal motor symptoms of PD. These neurons have extraordinarily long and branched axons, sustained autonomous spiking, and elevated levels of cytosolic Ca^{2+}, in addition to a chemically reactive neurotransmitter. This combination of features leads to sustained elevations in mitochondrial oxidative stress, possibly as a consequence of the reliance on feed-forward control of mitochondrial metabolism, and to increased susceptibility to alpha-synuclein (aSYN) aggregation. These derivative features could increase the impact of aging, genetic mutations, and environmental toxins linked to increased risk of PD, providing a unifying theory of PD pathogenesis. Although most of these traits are not amenable to therapeutic manipulation, Ca^{2+} loading is indeed, because it stems from opening of Cav1 (L-type) Ca^{2+} channels – channels that are antagonized by dihydropyridine drugs long used for the treatment of hypertension. Epidemiological studies have revealed that the use of dihydropyridines is associated with a reduced risk of developing PD. As a consequence, a large phase III clinical trial is underway in North America to determine if one of the dihydropyridines (isradipine) can slow PD progression.

Keywords Parkinson's disease • Calcium • Electrophysiology • Mitochondria Redox • Selective vulnerability • Dopamine • Substantia nigra • Pacemaking Oxidative stress • Bioenergetics • Two-photon microscopy

---

J. Surmeier, MS, PhD (✉) • E. Zampese, PhD • D. Galtieri, BS Psychology
P.T. Schumacker, PhD
Department of Physiology, Northwestern University, Feinberg School of Medicine, Chicago, IL, USA
e-mail: j-surmeier@northwestern.ed

A.K. Reeve et al. (eds.), *Mitochondrial Dysfunction in Neurodegenerative Disorders*, DOI 10.1007/978-3-319-28637-2_6

6.1 Introduction

PD is a disabling neurodegenerative disorder that is strongly associated with aging, increasing exponentially in incidence above the age of 65 [1]. Currently, there is no therapeutic strategy that is proven to slow or stop disease progression. The incidence of PD is expected to rise dramatically worldwide in the coming decades as life expectancy increases [2]. If this rising tide of PD is not countered with the development of a disease-modifying, neuroprotective therapy, it will pose an enormous burden on society in terms of financial cost and human suffering.

The best hope for developing a neuroprotective treatment lies in a better understanding of disease pathogenesis. In the last several decades, there has been a great deal of progress in identifying genetic risk factors for PD [3–6]. There is a reasonable belief that identifying the consequences of these mutations will lead to new "personalized" therapeutic strategies. However, the vast majority of PD cases are not associated with any known genetic mutations. Genome-wide association studies (GWASs) promise to put disease risk in these cases on a firmer genetic footing [5–7]. The working hypothesis of these studies is that the proteins linked to risk-increasing polymorphisms will begin to segregate into well-defined subcellular systems, providing insight into pathogenesis.

While they provide important clues about pathogenic mechanisms, genetic and environmental studies to date have not provided compelling clues about why PD manifests in a narrowly defined set of cell types. Indeed, all of the genes linked to familial forms of PD are expressed across diverse cell types, and expression level is not obviously correlated with the probability that a particular neuron will manifest pathology [8]. Moreover, environmental toxins and pesticides, which have been linked epidemiologically to PD [9, 10], are not known to be selectively taken up by at-risk neurons.

Which neurons are at risk in PD? The most widely accepted pathological marker in PD is Lewy pathology (LP) – Lewy body (LB) or Lewy neurite (LN) – an intracellular proteinaceous aggregate [11]. Nevertheless, LP is rare. First, it is found only in a narrowly circumscribed set of brain nuclei throughout much of the course of PD. At the time of the onset of cardinal motor symptoms, the affected nuclei are sparsely distributed primarily in the brainstem, mesencephalon, and diencephalon. Second, even within these at-risk nuclei, the percentage of neurons manifesting LP pathology is low [12]. Hence, LP pathology in the PD brain is a rare event at early stages in the disease process.

Although it is a defining feature of classical PD, the relationship between LP pathology and neuronal dysfunction and death is not established [13]. To be sure, it is possible to overexpress proteins found in LP, like alpha-synuclein, and induce neurodegeneration in animal models [14–16]. However, the aggregates created in these types of experiments recapitulate only some of the features of LBs and do not display the same anatomical distribution in the brain, unless this is enforced artificially (e.g., by selective overexpression of aSYN in an at-risk region, like the SNc). Injecting fragments of aSYN fibrils (generated either in the test tube or in the human brain) has been shown to produce neurodegeneration [17–19]. But again, the magnitude of the challenge and the time course of the neurodegenerative effects are dramatically different from those found in humans, keeping the question open as to whether under normal circumstances the degeneration in PD is driven by LB pathology. Other issues about the connection between LP and PD symptoms also arise, including the presence of LP in elderly patients lacking symptoms and the presence of LP in diseases lacking a PD phenotype, such as dementia with LB (DLB) [13, 20]. Definitive assessment of neuronal loss requires the use of stereological approaches and generic neuronal markers to distinguish neuronal death from loss of phenotypic markers, like the expression of tyrosine hydroxylase. In the best studied region – the substantia nigra – the degeneration of dopaminergic neurons begins *before* the appearance of substantial LP [13].

Nevertheless, if both neuronal degeneration and LP pathology are taken collectively as signs of PD, what determines the peculiar pattern seen in the PD brain? This is not resolved, but the determinants of this pattern surely are important clues about the principal pathogenic mechanisms in PD. The thesis of this review is that two cell phenotypic traits govern vulnerability. The first is sustained (autonomously generated) spiking that is supported by Ca^{2+}-dependent, feed-forward stimulation of mitochondrial oxidative phosphorylation to prevent bioenergetic crisis during periods of stress. The second is a long highly branched axon with thousands of transmitter release sites that creates a bioenergetic and proteostatic burden. The first trait might have evolved through evolutionary pressures related to the importance of SNc dopaminergic (DA) neuron function in coordinating muscular activity during locomotion associated with evading predators. To a large extent, this pressure no longer exists for humans, so it is conceivable they could be therapeutically modified to diminish the risk of PD or to slow its course. The second trait relates to the specific functions of these neurons requiring long and extensive axon fields – a characteristic that distinguishes these neurons from others and offers few opportunities for therapy.

6.2 The At-Risk Phenotype

The degeneration of SNc dopaminergic (DA) neurons is responsible for the cardinal motor symptoms of PD [21, 22]. So how do their phenotypic traits contribute to their selective vulnerability? Several interrelated hypotheses will be considered. In addition to the two traits mentioned above – their unusual physiological phenotype and their extraordinary axons – a third risk factor, their use of dopamine as a neurotransmitter, will be considered.

6.2.1 The Long and Winding Axonal Road

SNc DA neurons have massive axonal fields with hundreds of thousands of transmitter release sites [23–28]. A single SNc DA neuron can form 100,000 to 245,000 (or more) synapses in the striatum and have a total axonal length (with all its branches) of more than 40 cm [27, 29]. In humans, the axonal dimensions of SNc

DA neurons are even more impressive [29]. A human DA SNc neuron could, in principle, have 2.4 million synapses. Less vulnerable, ventral tegmental area (VTA) DA neurons appear to have less extensive branching [28, 30] and might "only" form 12,000–30,000 synapses [29]. Although smaller than SNc DA neurons, this is still considerably more than the synapses reported to be formed by most other neurons in the brain [31, 32].

Why might this massive axon be problematic? Modeling studies suggest that propagation of an action potential through a long unmyelinated axon is bioenergetically expensive [29, 33], potentially increasing the basal bioenergetic stress on the cell. Less is known about the bioenergetic cost arising from having so many DA release sites, all of which must maintain a pool of vesicles and the associated release and endocytosis machinery. That said, it would seem that the question is not whether the *aggregate* energy demand of an SNc DA neuron's axon is greater than that of, say, a cortical pyramidal neuron. After all, an elephant has much larger energy requirements than a mouse, but lives much longer. The question rather is whether it creates a need that renders the cell vulnerable to bioenergetic failure and death. Importantly, there is no direct evidence that the bioenergetic stress in the terminal field of an SNc DA neuron is higher than that of a cortical pyramidal neuron. There is no evidence that the number of mitochondria per release site or per micron of axon is lower in the terminal field of a SNc DA neuron. If it is not lower, then there is no reason to think that a priori the bioenergetic stress should be higher in SNc DA axons, regardless of their dimensions.

Setting aside the bioenergetic costs of spike propagation and transmitter release, the physical maintenance of such a complex architecture is in itself challenging. Newly synthesized proteins, lipids, and organelles must be transported from the soma to distal destination sites (anterograde transport), while damaged structures presumably need to be returned to the somatic compartment for degradation (retrograde transport) [34–39]. Endosomal trafficking could also be enormous for such a large axonal arbor [40]. All these transport processes are ATP dependent [41]. But perhaps of greater importance, all of this traffic is funneled through a single proximal axon. This is loosely analogous to funneling all the cars coming into and out of Chicago [with its eight million residents] on a two-lane road.

Such a bottleneck could render trafficking prone to disruption, an inference that is consistent with several lines of study [34, 38]. Moreover, reduction in the expression of motor proteins driving axonal trafficking has been reported in human SNc DA neuron axons in early stages of PD [42]. Misfolding or overexpression of aSYN – a PD-linked protein that is concentrated at transmitter release sites where it regulates vesicular cycling – could further strain SNc DA neurons [42]. The consequences of motor protein loss are recapitulated by overexpression of aSYN in rat models, which leads to early disruption in the axon and later in the cell body [42, 43]. In cell culture, axonal transport defects also are associated with expression of mutant aSYN [44] or treatment with toxins used to create PD models [45, 46]. Thus, accumulation of aSYN could either induce or reflect a dysfunction in the mechanisms involved in axon transport.

The trafficking problem envisioned for the axon of a DA neuron grows when mitochondria are added to the equation. Mitochondria are double-membrane organelles of endosymbiotic origin and have their own circular DNA (mtDNA). However, most of the genes necessary for mitochondrial assembly and function have been integrated in the nuclear DNA over the course of evolution; hence, they are assembled in the soma and trafficked to axonal sites where they provide a source of ATP in proximity to regions of bioenergetic need (for details, see Chaps. 1 and 8) [47–53]. It is not difficult to imagine that any defect in axonal transport of mitochondria could lead to localized axonal bioenergetic stress and defects in mitochondrial distribution have been described in several PD models [45, 54, 55].

Rather than existing as discrete autonomous organelles, mitochondria form a remarkable dynamic network that reflects continuous remodeling arising from the processes of fission and fusion (see Chap. 7). The process of fission is important in cells undergoing mitotic division, as it allows partitioning of organelles between the daughter cells. This is a regulated process as demonstrated in stem cells undergoing asymmetric cell division, where older mitochondria are segregated from newer organelles such that fewer old mitochondria are retained in the stemlike daughter cell. The importance of this for cell function is profound, as disruption of fission impairs this asymmetric segregation and leads to a loss of stem cell traits in the daughter cell [56]. Fission is also critical for the separation of damaged or affected mitochondria from normal organelles. This process allows cells to produce small dysfunctional mitochondrial bodies that can be degraded by mitophagy. Mitochondrial fusion promotes the redistribution of mtDNA and functional membrane complexes throughout the cellular mitochondrial pool. The balance between fission and fusion is dynamically regulated by expression of a core group of proteins, which are in turn regulated by posttranslational modifications [57, 58].

Impairing mitochondrial fission/fusion dynamics increases the vulnerability of SNc DA neurons. For example, deletion of mitofusin-2 (Mfn2), a protein controlling mitochondrial fusion and the association between mitochondria and endoplasmic reticulum (ER), decreases mitochondrial motility and the abundance of mitochondria in neuronal processes. More importantly, Mfn2 deletion causes retrograde degeneration of SNc DA neurons, but not VTA neurons [59, 60]. Deletion of Drp1, a protein critical to mitochondrial fission, has a similar effect [61]. Thus, deficits in mitochondrial dynamics could contribute to axonal degeneration of SNc DA neurons.

How does this translate to the human condition? Examination of human brains for PD-associated pathology reveals that loss of the terminal fields of SNc DA neurons precedes that of cell bodies [62–66]. This has led to the notion that the terminal fields are the most vulnerable part of SNc neurons and the site where degeneration begins in PD [19, 64, 67]. Does this mechanism generalize to other at-risk neurons? The data set is far from complete, but it is known that locus coeruleus, pedunculopontine nucleus, nucleus basalis of Meynert, the raphe nuclei, and the dorsal motor nucleus of the vagus all are populated with neurons with long branching dendrites [68]. It is reasonable to propose that defects in organelle trafficking could contribute to the loss of axonal function in these cells.

6.2.2 The Bioenergetic Burden of Neuronal Excitability

A defining feature of neurons, regardless of the length of their axons, is excitability. Neurons code information in complex patterns of action potentials or spikes. Spikes are transient regenerative events that rely upon the transmembrane electrochemical gradients for Na^+, K^+, and Ca^{2+} ions and are triggered when the membrane potential reaches a critical threshold. In most neurons, the spike is initiated by the transient opening of voltage-dependent membrane channels that are selectively permeable to Na^+ ions, allowing positively charged Na^+ ions to move from the extracellular space into the cytosol. This redistribution of charge pushes the transmembrane potential from relatively negative levels toward the equilibrium potential for sodium, typically near 0 mV. This depolarization causes voltage-dependent channels that are selectively permeable to K^+ ions to open, resulting in the movement of positively charged K^+ ions from the cytosol to the extracellular space, reestablishing the negative transmembrane potential. This sequence of events requires that the concentration of Na^+ ions is low in the cytosol, while the concentration of K^+ ions is high.

In most neurons, spikes are initiated in a specialized region at the junction between the cell body and the axon called the axon initial segment (AIS). The AIS functions as a site of integration for subthreshold depolarizing and hyperpolarizing inputs arriving at the cell body. The AIS is specialized in several respects, most notably in the concentration of voltage-dependent Na^+ channels, which are critical for initiating action potentials. Spikes initiated in the AIS propagate both down the axon to synaptic terminals (where they initiate transmitter release) and retrograde into the somatodendritic region, creating a temporally and spatially distributed propagation of the transmembrane electrochemical Na^+ and K^+ gradients.

Another cation that crosses the plasma membrane during spikes is Ca^{2+}. In most neurons, voltage-dependent Ca^{2+} channels are opened only by strong depolarization as occurs during a spike. With repolarization of the membrane, these channels close slowly, creating a period during which the driving force for influx of Ca^{2+} is large and the conductance remains high. This makes the total Ca^{2+} influx during a spike very sensitive to spike duration. Ca^{2+} entering during the spike triggers a number of events [69]. One is activation of calcium-dependent K^+ channels that help repolarize the membrane and impede the generation of another spike for some period of time (refractory period). This interaction takes place in a spatially restricted zone close to the plasma membrane. Neurons that need to spike at high frequencies typically restrict Ca^{2+} entry by keeping spikes very brief (<1 ms); Ca^{2+} entry during the spike is exquisitely dependent upon the duration of the spike itself, which varies considerably from one cell type to the next. To help manage the Ca^{2+} that enters the cell, neurons often express Ca^{2+}-buffering proteins [70, 71], like parvalbumin, in addition to the proteins that serve as Ca^{2+} signaling targets [69]. These buffering proteins may help to spatially restrict the increase in calcium to specific regions within the cell.

Exchangers and pumps are responsible for maintaining the electrochemical gradients for Na^+, K^+, Ca^{2+}, and Cl^-. These transmembrane proteins fall into two broad categories. The first are active pumps that rely upon adenosine triphosphate (ATP) to drive the movement of ions against electrochemical gradients. Pumps that fall into this category include the Na^+/K^+ ATPase, the plasma membrane Ca^{2+} ATPase (PMCA), and the smooth endoplasmic reticulum Ca^{2+} ATPase (SERCA). The other category utilizes the free energy stored in an existing electrochemical gradient to move ions. A good example of this type of protein is the Na^+/Ca^{2+} exchanger (NCX), which under physiological conditions uses the Na^+ gradient to move Ca^{2+} ions out of the cytosol. Another exchanger of this type is the Na^+/Ca^{2+}-K^+ exchanger (NCKX). While these exchangers do not consume ATP directly, they are dependent on the electrochemical gradients established and maintained by the active transport systems. Thus, this combination of pumps and exchangers contributes to the bioenergetic needs of neurons and enhances their reliance on ATP generation by mitochondria. Importantly, these energy requirements can render neurons sensitive to diseases or disorders that limit the ability of mitochondria to meet these demands.

Although the molecular events coupling ion movement to ATP hydrolysis are still not fully understood, the thermodynamics of ion movement are worth considering because they establish lower limits on the cost of pumping. For Na^+ and K^+ ions, the concentration differences maintained across the plasma membrane are similar, being 10- to 30-fold. In contrast, the concentration difference for the Ca^{2+} concentration is roughly 20,000-fold, being 2 mM in the extracellular space and around 100 nM in the cytosol. Because the free energy change needed to move an ion from one compartment to another depends upon the logarithm of the concentration ratio between the two compartments, Ca^{2+} should be roughly eight times more energetically expensive than a Na^+ ion (~4 times greater on a per charge basis). The Na^+/K^+ ATPase extrudes three Na^+ ions and takes up two K^+ per ATP molecule consumed. The PMCA pumps one Ca^{2+} molecule out for each ATP consumed. So overall, it takes about one ATP for each Ca^{2+} ion and one ATP for every three Na^+ ions. By this rough calculation, although Ca^{2+} is more expensive than Na^+, it is only about 1.5–2 times more expensive per charge than Na^+. Whether the bioenergetic cost of normal Ca^{2+} influx represents a significant fraction of the metabolic capacity of neurons remains to be seen.

Ca^{2+} that enters the cytosol is either rapidly pumped back out of the neuron or is sequestered in intracellular organelles. The most important of these for Ca^{2+} homeostasis is the endoplasmic reticulum (ER) [72]. The ER forms a continuous intracellular network, allowing it to regulate both local and global cellular Ca^{2+} signals. As in other neurons, the ER network in SNc DA neurons extends throughout the somatodendritic tree [73–76]. High-affinity ATP-dependent transporters move Ca^{2+} from the cytoplasm into the ER lumen. The absence of high-affinity, anchored intraluminal protein buffers of Ca^{2+} and the physical continuity of the lumen throughout the cell [73, 74] allows the ER to rapidly (~30 μm/s) redistribute Ca^{2+} between intracellular regions. Ca^{2+} sequestered in the ER is released at sites where it can be pumped back across the plasma membrane or where it can be used to modulate cellular activities, including mitochondrial function [77–81].

Mitochondria partner with the ER in Ca^{2+} homeostasis. Mitochondrial motility is controlled in such a way as to bring them close to sites of Ca^{2+} release [82–84].

Mitochondria also are commonly found physically coupled to the ER at sites known as the mitochondria-associated membrane (MAM) [85, 86]. These regions of juxtaposition create multifunctional signaling microdomains that coordinate ER-mitochondria communication and mitophagy initiation [85, 87–89]. For example, within these microdomains, Ca^{2+} released by the ER through ryanodine (RYR) and inositol trisphosphate (IP3) receptors moves down the steep potential gradient (~−200 mV) across the inner mitochondrial membrane into the matrix, through a pore called the mitochondrial Ca^{2+} uniporter [90]. Indeed, mitochondrial Ca^{2+} uptake would cease if the membrane potential was dissipated. The mitochondrial Ca^{2+} uptake channel is a multimeric complex consisting of a pore-forming subunit (MCU) and at least five other proteins that appear to regulate its sensitivity to Ca^{2+} and conductance under different conditions and at different local concentrations of Ca^{2+}. Growing evidence indicates that, while the mitochondrial membrane potential strongly promotes Ca^{2+} entry, the mitochondrial calcium uniporter is not an unregulated pathway for its uptake. The MCU has recently been cloned, opening an important door for detailed study of its role and regulation [91, 92]. In that regard, mice with targeted deletion of the MCU subunit exhibit a modest phenotype, which includes reduced performance of energy-intensive physical activity [93]. Importantly, MCU-lacking mitochondria still contain Ca^{2+}, indicating that there are other means of Ca^{2+} entering them.

Emerging evidence suggests that the mitochondrial calcium uniporter is localized to MAMs, allowing these channels to respond preferentially to Ca^{2+} released by the ER. Tissue-specific genetic deletion of mitofusin-2, which is required for mitochondrial fusion but also for mitochondria-ER MAMs, causes profound metabolic defects in the heart and muscle and has been linked to neurodegenerative disease [94, 95]. Tissue-specific deletion of mitofusin-1 produces the expected increase in mitochondrial fragmentation, but does not phenocopy the metabolic defects seen in mitofusin-2 knockouts [96]. These different phenotypes suggest that loss of mitochondrial fusion is not responsible for the neurodegeneration caused by mitofusin-2 mutations. One possibility is that mitofusin-2 somehow regulates the ability of ER to support Ca^{2+} flux through the mitochondrial uniporter.

6.2.3 Too Important to Rest: The At-Risk Pacemaking Phenotype

In most neurons, the challenge to ionic homeostasis created by spiking is episodic. Striatal spiny projection neurons, for example, reside very near the K^+ equilibrium potential some 40 mV away from spike threshold in the absence of synaptic potentials [97]. Excitatory synaptic input drives the membrane potential to spike threshold for hundreds of milliseconds at a time, but this activity is intermittent. However, there is a subset of neurons in the brain that spike frequently and autonomously, that is, they spike on their own in the absence of synaptic input. Typically, autonomously generated spiking is very regular (synaptic input decreases this regularity); as a consequence, these neurons are often referred to as autonomous pacemakers. Autonomous pacemakers can be divided into fast and slow spiking classes. Fast autonomous pacemakers, like globus pallidus or substantia nigra pars reticulata (SNr) GABAergic neurons, spike at rates above 20 Hz at physiological temperature [98, 99]; these neurons have very brief action potentials (<1 ms) and robustly express Ca^{2+}-binding proteins [100, 101]. Typically, autonomous spiking in these cells is driven by a combination of monovalent cation channels, like Nav1 Na^+, and hyperpolarization and nucleotide-activated cation (HCN) channels [102, 103]. Slow pacemakers spike at much lower frequencies (~2–5 Hz) and have broad spikes (>2 ms). Neurons of this class are found scattered in many regions of the brain, including the hypothalamus, striatum, mesencephalon, and medulla [104–108]. Although there has been much less work on pacemaking mechanisms in this class of neuron, it has been shown that their pacemaking typically depends upon Nav1, HCN channels, as well as cationic leak channels [109].

In SNc DA neurons, which are a type of slow, broad spike pacemaker, voltage-dependent L-type Ca^{2+} channels contribute to the inward current that promotes pacemaking [106, 110, 111]. These L-type channels have a distinctive pore-forming subunit (Cav1.3) that is coded for by the *CACNA1D* gene [112], which confers upon the channel the ability to open at relatively negative membrane potentials [113, 114]. These channels are relatively rare, and most L-type channels found in the brain have a Cav1.2 pore-forming subunit (coded for by the *CACNA1C* gene) [115]; this means that in most neurons, L-type channels open only during periods of strong depolarization, as occurs during a spike. This feature of Cav1.2 channels allows them to be used as "spike counters" in a wide range of neuronal processes, including homeostatic plasticity [116–118].

Although Cav1.3 channels contribute to pacemaking, they are not required as their antagonism does not change the rate and regularity of spiking [106]. Previous work arguing that Cav1 channels were necessary for pacemaking relied upon micromolar concentrations of dihydropyridine – concentrations that are well outside the range where specific antagonism of Cav1 channels is achieved and ones where Na^+ and K^+ channels are affected [106]. In this regard, it is important to recognize that dihydropyridines are not *blockers* of Cav1 Ca^{2+} channels; rather, they are negative allosteric modulators whose binding affinity increases roughly a thousandfold with membrane depolarization [119]. In the voltage range normally traversed by SNc DA neurons during pacemaking, the IC50 of the dihydropyridine drug isradipine is in the low nanomolar range. Guzman et al. [106] directly measured Ca^{2+} influx following dihydropyridine administration, demonstrating that Cav1 channels could be virtually eliminated without changing pacemaking rate or regularity. As pacemaking activity does not require Cav1.3 channel activity, it is reasonable to ask why SNc DA neurons would choose to assume the energetic costs related to the handling of Ca^{2+} entering by that pathway.

One way to conceptualize the pacemaking mechanism in these neurons is as redundant, coupled oscillators (Fig. 6.1). From this perspective, the primary oscillator is the "conventional" one driven by NALCN "leak" channels, Nav1 Na^+ chan-

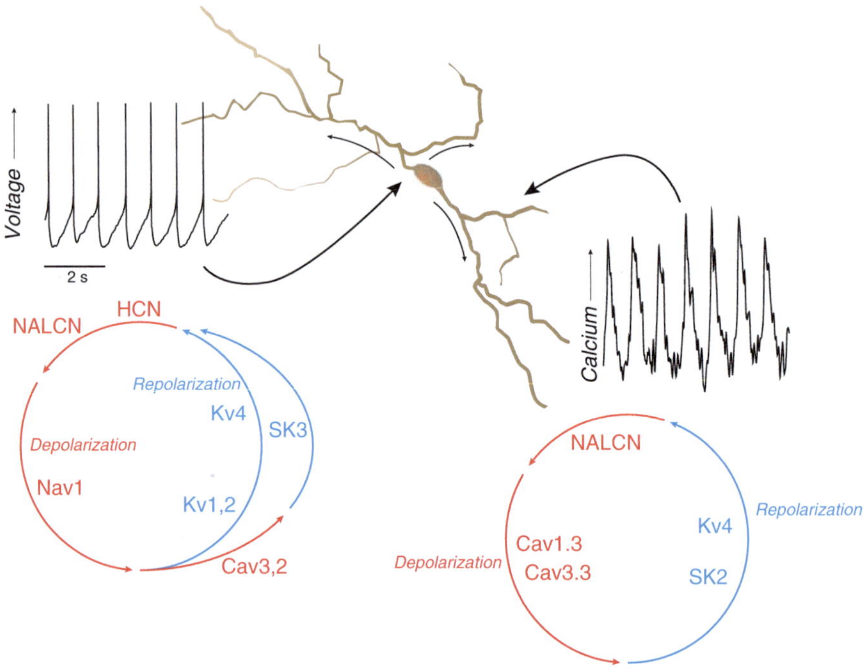

Fig. 6.1 Schematic of the dual pacemaking mechanisms in SNc dopaminergic neurons. In the somatic and axon initial segment region, monovalent cation channels predominate. In the dendrites and somatic regions, a Ca^{2+}-dependent mechanism is in place

nels, Cav3 Ca^{2+} channels, Kv1,2,4 K^+ channels, small conductance Ca^{2+}-activated K^+ (SK) channels, and HCN channels [106, 110, 120]. In addition, Ca^{2+} release from intracellular ER stores is likely to play a role in engagement of SK channels when Cav3 channels are opened by depolarization [78]. The deep, SK-mediated after-hyperpolarization serves to slow and regularize the pacemaking rate [121–123] and to de-inactivate Nav1 and Cav3 channels (both channels inactivate with depolarization), making them available for the next cycle of spiking. The channels contributing to this oscillator are localized in the soma, proximal dendrites, and AIS, as pacemaking remains fully functional in acutely isolated neurons in which dendrites have been truncated [110].

Kv4 K^+ channels also clearly slow the pacemaking rate of SNc DA neurons [124]. The Kv4.3 subunit is particularly prominent and is associated with KChIP3, a Kv4 channel-interacting protein that alters the gating properties of the channel and increases current amplitude [124]. KChIP3 promotes surface expression of Kv4 channels and also acts as a Ca^{2+} sensor, increasing Kv4 availability as Ca^{2+} rises. In other neurons, Cav3 channels are coupled to KChIP activity [125] raising the possibility that perisomatic Cav3 channels in SNc DA neurons promote the opening of both SK and Kv4 K^+ channels.

Complementing this somatic oscillator is a largely dendritic oscillator that relies upon Ca^{2+} entry through Cav1 and Cav3 channels [106, 110, 120, 126, 127]. Within the confined cytosolic space of the dendrite, the augmented elevation in cytosolic Ca^{2+} concentration leads to the opening of SK K^+ channels, which, along with Kv4 K^+ channels, repolarizes the membrane, deactivating Cav1 and Cav3 channels and setting the stage for the next cycle of the oscillation [106, 127]. The absence of strong intrinsic Ca^{2+} buffering is important for this mechanism to work [127], as it allows free Ca^{2+} concentration to reflect the spatiotemporal pattern of Ca^{2+} channel opening and intracellular release. In principle, this enables the surface area-to-volume ratio of a dendrite to determine its natural oscillation frequency [127]. As predicted by a relatively constant Cav1 channel density in the dendrites, the cytoplasmic swings in Ca^{2+} concentration during pacemaking steadily rise as dendrites taper [unpublished results].

In the presence of the Nav1 channel blocker tetrodotoxin (which disrupts the proximal Na^+ oscillator), the Ca^{2+} oscillator manifests itself as a slow oscillatory potential (SOP). The dominant frequency of the SOP is slightly slower than normal pacemaking, suggesting that in an intact cell the Nav1 oscillator "pulls" the Cav1 oscillator along. Antagonizing Cav1 channels with dihydropyridines completely eliminates the SOP, while antagonizing SK channels with apamin slows the SOP and increases its amplitude as the oscillation now depends upon Kv4 K^+ channels that are opened by depolarization.

In the somatic region, SK channels are primarily of the SK3 type [128, 129]. In the dendrites, SK2 channels predominate [130]. Although they have been reported to influence the regularity [not rate] of pacemaking, there are fundamental questions that are unresolved about how these channels are controlled and how they impact spiking in the face of normal synaptic input. Given the importance of Cav1 channels to the dendritic oscillator, a reasonable conjecture is that they are the major regulators of SK2 channels.

So why do these neurons express this rather peculiar dendritic oscillator? VTA DA neurons do not utilize this redundant oscillator and express high levels of calbindin, a Ca^{2+}-buffering protein [126, 131–135]. One possibility is that it confers a fail-safe robustness to pacemaking activity [108]. Pacemaking in SNc DA neurons is critical to maintaining "ambient" DA signaling in the basal ganglia where ongoing movement is controlled. This allows for bidirectional regulation of the basal ganglia circuitry. In that regard, synaptically driven burst spiking of DA neurons increases the excitability of action-promoting direct pathway neurons, whereas synaptically driven inhibition of pacemaking increases the excitability of action-suppressing indirect pathway neurons. Hence the failure of pacemaking would promote indirect pathway activity and suppress motor activity. In a prey-predator situation, it is critical that pacemaking is not stopped inappropriately. This would create a strong evolutionary pressure to develop ways to ensure this does not happen. The dendritic Ca^{2+} oscillator appears to accomplish this end by both providing a depolarizing force in the event the somatic oscillator falters and by intermittently hyperpolarizing the membrane enough to de-inactivate Nav1 and Cav3 channels. Moreover, this oscillator mechanism relies upon channels that are relatively resistant to depolarization. In contrast, the somatic/AIS oscillator depends upon Nav1 and Cav3 channels that inactivate with sustained depolarization. The Cav1.3 and SK2 channels in the dendritic oscillator do not manifest much voltage-dependent inactivation. It follows that the tolerance for instability in pacemaking should be much greater in neighboring VTA DA neurons. However, transient loss of activity in those cells might cause dysphoria but not an inability to keep moving when being pursued, so the consequences are less serious.

There is another possible explanation for why the dendritic Ca^{2+} oscillator is present in SNc DA neurons. In addition to its role in the regulation of membrane potential, Ca^{2+} has a wide range of signaling roles in all cells. Ca^{2+} entry is used by neurons to monitor spiking and to adjust biosynthetic processes to keep pace. For example, Ca^{2+} stimulates tyrosine hydroxylase and perhaps tryptophan decarboxylase to increase the synthesis of DA from tyrosine, allowing SNc DA neurons to up- or downregulate DA synthesis to match spike-driven DA release [136–139]. Another important role of Ca^{2+} is to stimulate mitochondrial oxidative phosphorylation [as described above]. This is a straightforward way of making sure that the supply of ATP meets the bioenergetics demands created by spiking (e.g., maintenance of ionic gradients, transmitter release, etc.).

In SNc DA neurons, the bioenergetic control system appears to operate in a *feed-forward* mode (Fig. 6.2). That is, Ca^{2+} entry drives oxidative phosphorylation in anticipation of need, rather than in response to a drop in ATP levels (feedback mode). This kind of feed-forward control is well established in cardiac cells [140, 141]. Understanding of how this regulates SNc DA neurons is emerging. Our current model proposes that Ca^{2+} influx through plasma membrane Cav1 channels during pacemaking triggers Ca^{2+}-induced Ca^{2+} release (CICR) from ER stores; this release is very prominent in SNc DA neurons for reasons that have yet to be fully

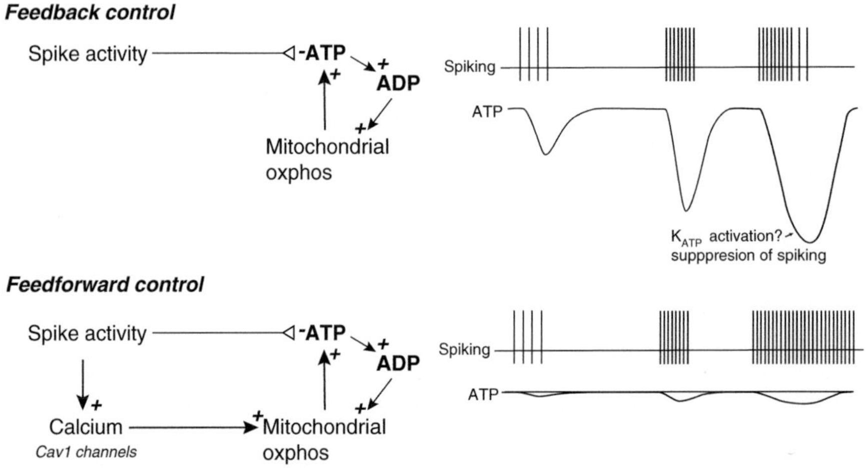

Fig. 6.2 Schematic representation of feedback (*top*) and feed-forward metabolic control of mitochondrial respiration. Hypothetical changes in cytosolic ATP levels are shown as well

understood. This repeated CICR results in Ca^{2+} being "fed" to the mitochondria, resulting in the sustained stimulation of oxidative phosphorylation [as described above]. Stimulation of the dehydrogenases of the TCA cycle leads to an increase in the pool of reducing equivalents available to supply the electron transport chain. This ensures that mitochondrial membrane potential, which drives ATP synthesis, will be maintained in the face of any sudden increase in bioenergetic demand. Such a situation could occur during a barrage of excitatory synaptic inputs associated with the appearance of a salient event in the environment. In the absence of that mechanism, the same barrage of activity could lead to a transient drop in cytosolic ATP levels [142]. SNc DA neurons are richly invested with K_{ATP} channels (Kir6, Sur) that would then open, hyperpolarize the membrane, and halt spiking [143–145]. Clearly this could be disastrous in a prey-predator situation.

By preventing ATP from falling, feed-forward bioenergetic control also sustains activity of the mammalian target of rapamycin complex, mTORC1, a critical regulator that couples cellular energy production to consumption. Activity of mTORC1 leads to enhanced protein translation and also to mitochondrial biogenesis by stimulating the activity of transcriptional regulators such as PPAR gamma coactivator-1alpha (PGC1a), a factor involved in the expression of nuclear-encoded mitochondrial proteins [146]. Loss of energy supply could lead to inhibition of mTORC1, thereby undermining its role mitochondrial biogenesis and thus ATP supply.

Although RYRs appear to be critical to CICR driven by pacemaking, it is highly likely that inositol trisphosphate receptors (IP3Rs), which are clustered with RYRs at MAMs [85], also participate in mitochondrial metabolic control. This regulation could be very important for the matching of mitochondrial metabolism to synaptically evoked activity; indeed, in other neurons, IP3R-mediated Ca^{2+} release and modulation of mitochondrial metabolism are necessary for neuronal survival [147]. In SNc DA neurons, activation of type 1 metabotropic glutamate receptors (mGluRs) (and other receptors in this signaling class) induces IP3R-dependent Ca^{2+} release [78, 148, 149]. However, this release is prominent only in neurons that have been silenced. In pacemaking neurons, IP3Rs are largely desensitized by Ca^{2+} [78]; the authors show that in this situation, burst stimulation is able to overcome the desensitization and induce Ca^{2+} release. This suggests that IP3Rs might only increase the gain of feed-forward mitochondrial control when SNc DA neurons are driven to spike in bursts by extrinsic input, but not during pacemaking. More work is needed on this point, particularly at the mitochondrial end of the signaling chain.

It is possible that axons use the same metabolic control mechanisms found in the somatodendritic membrane, increasing their vulnerability over time. Recent work has shown that Cav1 channels are present in the axonal terminal fields of SNc DA neurons and contribute to activity-evoked DA release, possibly by increasing mitochondrial ATP production [150]. In other axons, feed-forward, Ca^{2+}-mediated control of mitochondrial ATP production has been beautifully demonstrated using optical approaches [151]. Although important for local metabolic control, these mechanisms do have the potential to trigger pathology [152].

Do other at-risk neurons in PD exhibit similar physiological phenotypes? Although the data are sparse, the tentative answer appears to be yes. Neurons in the locus coeruleus (LC), dorsal motor nucleus of the vagus (DMV), and pedunculopontine nucleus (PPN) also exhibit calcium cycling and pacemaking activity [153–155]. Although not autonomous pacemakers, neurons in the red nucleus (RN) and nucleus basalis of Meynert (NBM) undergo continuous spiking during the waking state [156, 157]. Among these, LC and DMV have also been demonstrated to express Cav1.3 channels and to exhibit signs of elevated mitochondrial oxidative stress [105, 108]. Hence, there is reason to predict that these cells, like those in SNc, may sustain continuous mitochondrial stresses that, over time, lead to the development of disordered mitochondrial function. On the other side, VTA DA neurons, relatively spared in the disease, are also autonomous pacemakers but have been shown to have a minor engagement of CaV1.3 channels [126, 132, 135] and a higher abundance of cytosolic Ca^{2+} buffers [131, 134].

6.2.4 The Downside of Feed-Forward Metabolic Control

Although feed-forward control of mitochondrial metabolism should minimize the likelihood that cytosolic ATP levels would fall in stressful conditions, this process also has a downside. Mitochondrial electron flux and oxygen consumption are largely regulated by the electrochemical potential across the mitochondrial inner membrane. During states of increased metabolic demand (high ATP utilization), this potential tends to decrease as protons enter the matrix compartment in association with ATP synthase activity. The small decrease in potential "un-throttles" the electron transport chain, leading to a compensatory increase in electron flux and O_2 consumption. By contrast, when ATP demand and ATP synthase activity are low, the membrane potential increases, causing the electron flux and O_2 consumption to slow because the free energy derived from electron transfer through complexes I, III, and IV is not sufficient to translocate protons against the gradient. Ca^{2+}-mediated TCA stimulation results in sustained production of reducing equivalents for the ETC, even when complex V activity is low. This promotes hyperpolarization of the membrane, a condition known to increase the generation of reactive oxygen species [superoxide anion] as electrons stranded on iron-sulfur clusters or flavin groups are transferred to O_2 [158–160]. This effect is evident in SNc DA neurons, which exhibit enhanced mitochondrial oxidative stress when they are "idling" [pacemaking] in the absence of synaptic input. Mitochondrial oxidative stress is greater in dendrites than in the somatic region, paralleling the relative abundance of Cav1 channels and the swings in cytosolic Ca^{2+} concentration [161]. By contrast, VTA DA neurons and other neurons that are not at risk in PD do not exhibit this stress. Importantly, antagonizing Cav1 channels with dihydropyridines or blocking mitochondrial Ca^{2+} entry through the MCU dramatically diminishes this stress without changing pacemaking rate [132], indicating that it is a consequence of calcium uptake into the mitochondria.

Mitochondrial oxidative stress induced by superoxide production from the ETC spans a wide range, with some cell types exhibiting low basal levels and other types demonstrating higher levels. Multiple systems are expressed in mitochondria for scavenging ROS and for repairing the damage to proteins, lipids, and DNA. Hence, the existence of increased mitochondrial oxidative stress is not necessarily a prelude to imminent cell death. However, it is reasonable to predict that a chronic state of mitochondrial oxidative stress could be associated with a greater rate of organelle damage leading to an increased rate of mitochondrial turnover (mitophagy) that would require a corresponding increase in mitochondrial biogenesis. The relationship between the basal level of oxidative stress and mitochondrial turnover in SNc versus VTA DA neurons is not well understood. Nevertheless, sustained mitochondrial oxidative stress in SNc DA could explain the unusually high level of mtDNA deletions and declining respiratory function found in the SNc of PD patients and aged, non-symptomatic humans [162].

In addition to elevating the production of ROS, mitochondrial Ca^{2+} entry also increases the generation of nitric oxide (NO) and reactive nitrogen species (RNS). In both LC and SNc neurons, it appears that there is a mitochondrial form of nitric oxide synthase (NOS) that is stimulated by Ca^{2+} [108]. Mitochondrial NO could serve to protect mitochondria by acting to brake ETC activity and lower inner membrane potential [163–165], thereby limiting the driving force for Ca^{2+} entry and preventing opening of the permeability transition pore. However, nitrosylation of cytoplasmic proteins can promote proteostatic dysfunction, and aSYN nitrosylation is a hallmark of PD [166]. Although potentially important to pathogenesis, a mitochondrial NOS has yet to be genetically identified, raising the possibility that mitochondrial NO arises from another origin [167, 168].

Given the range of apparently deleterious consequences of using Ca^{2+} to support pacemaking and feed-forward control of mitochondrial respiration, why has it not been eliminated by evolution? It would seem that the benefits of this design for survival early in life during predation and reproduction outweigh the negative consequences late in life. Again, even transient loss of pacemaking during pursuit by a predator could compromise the ability to escape and survive. So, there clearly should be evolutionary pressure to make the system robust. The negative consequences of this design choice normally take six or seven decades to manifest themselves, well beyond the average lifespan of humans until very recently. Other examples of antagonistic pleiotropy have been described, where a genetic polymorphism is favored by selection because it promotes survival early in life, even though it compromises survival later in post-reproductive years [169]. Although the distinctive phenotype of SNc DA neurons is not the product of a single polymorphism but rather a complex genetic program [170], the principle of antagonist pleiotropy still seems apropos. The extent to which this model applies to other at-risk neurons remains to be seen. However, it is clear that many of them, like LC and DMV neurons, have critical roles to play in survival in the wild that could promote the selection of a similar phenotype.

6.2.5 Is Dopamine a Poisonous Transmitter?

Beyond its effects on mitochondria, sustained elevations in cytosolic Ca^{2+} concentration could have other damaging effects. One of these could be an elevation in DA synthesis. DA has long been thought to be a factor in the selective vulnerability of DA neurons [171]. DA is synthesized in the cytosol from tyrosine by reactions that include tyrosine hydroxylase (TH), which converts tyrosine into L-dihydroxyphenylalanine (L-DOPA), and aromatic amino acid decarboxylase (AADC), which converts L-DOPA to DA. DA is then sequestered into synaptic vesicles by the vesicular monoamine transporter 2 (VMAT2). TH, AADC, and VMAT2 are associated with vesicular membranes, suggesting that synthesis and uptake normally occur within a restricted space [172]. Once inside vesicles, the acidic pH stabilizes DA [173].

Why is this important? Several lines of evidence that suggest cytosolic DA can be toxic. For example, disrupting sequestration of DA by reducing VMAT2 expression or function increases the vulnerability of DA neurons [174–177]. DA can be oxidized to reactive quinones (DAQs) that are capable of modifying DNA, proteins, and lipids [178, 179]. DAQs can form DNA adducts or generate apurinic sites, disrupting gene expression and triggering DNA damage responses [180–182]. DAQs also target protein cysteine residues, causing misfolding and dysfunction [171]. Examples of proteins disrupted by DAQs include TH [183], parkin [184], aSYN [favoring protofibril formation] [185], DJ-1 [186], superoxide dismutase 2 (SOD2), and proteins of the mitochondrial electron transport chain (ETC) [187]. Translocation and degradation of proteins in lysosomes can also be impaired by DAQs [188]. The potential pathogenic role of DA and DAQs has been recently reviewed [189].

Although DAQs can be generated in a test tube or in a cell loaded with DA precursor, it is not clear whether this also occurs in vivo. Evidence supporting this possibility comes from the accumulation of the pigment neuromelanin (NM) in the SNc. DAQs are the precursors of the two portions of NM: the polymeric core known as pheomelanin and the polymeric surface or eumelanin [179, 189]. Cytosolic NM is found in double-membrane organelles, probably of autophagic origin [190]. Although neurons with high levels of NM, in particular those in the SNc, are the most susceptible to loss in PD [191, 192], the formation of NM and its incorporation into organelles are generally considered protective, since sequestration prevents DAQs from damaging cellular components [193]. Moreover, NM is able to chelate potentially toxic heavy metals [iron and lead] [194]. However, NM can inhibit proteasomal degradation, contributing to the accumulation of aSYN and thus to LB formation [189]. Furthermore, NM released into the extracellular space [upon loss of neuronal integrity] elevates the concentration of toxic metals and quinones, increasing the stress on neighboring healthy neurons and triggering neuroinflammation [179, 194].

DAs need not be oxidized to produce damage. Cytosolic DA is degraded by monoamine oxidases (MAOs) into 3,4-dihydroxyphenylacetaldehyde (DOPAL), which is toxic [195, 196]. However, at-risk neurons express high levels of aldehyde dehydrogenase 1 (ALDH1), which converts DOPAL into relatively innocuous

3,4-dihydroxyphenylacetic acid (DOPAC) [197, 198]; DOPAC is then metabolized into homovanillic acid (HVA) by catechol-O-methyltransferase (COMT) [199]. Genetic deletion of ALDH1 increases the vulnerability of SNc DA neurons [200] pointing to the importance of this route of metabolism. Although SNc neurons have long been assumed to express only MAO-A [189], recent work with human DA neurons derived from induced pluripotent stem cells suggests that both MAO-A and MAO-B are co-expressed [201]. This is of importance because of the clinical use of rasagiline, a selective MAO-B inhibitor, which has been used to treat PD patients [202–204]. Inhibition of MAO-B in DA neurons themselves could diminish the production of potentially damaging DOPAL. One piece in this puzzle that is not understood is whether the tethering of MAOs to mitochondria is of significance [205]. DA metabolism by MAOs might increase production of cytosolic hydrogen peroxide, in addition to DOPAL [206]. But the evidence that this happens in situ is incomplete at this point. An alternative hypothesis is that DA metabolism increases mitochondrial – but not cytosolic – oxidative stress [207, 208].

It can be argued that the evidence for DA as a pathogenic agent in PD is circumstantial, as no direct evidence in a physiological setting has linked DA production to the accumulation of cellular injury. In part, this could be because the focus of attention has been on dendrites and cell bodies, rather than on terminals, where there is a much greater density of vesicles. Given the possibility that degeneration of dopamine neurons begins in their terminal regions [66], this could prove to be an important omission.

What is the linkage between DA and Ca^{2+}? Ca^{2+} is known to stimulate tyrosine hydroxylase (TH), a key enzyme in the synthesis of DA from tyrosine [136–139]. Moreover, Mosharov et al. [209] found that Ca^{2+} entry through L-type channels stimulated the production of DA from L-DOPA, suggesting that enzymes downstream of TH are also stimulated by Ca^{2+}. This effect exacerbated the toxicity of combined L-DOPA treatment and aSYN overexpression, presumably by promoting the formation of DA adducts of aSYN that inhibited autophagy.

However, the role of DA in PD pathogenesis is far from established. Non-dopamine neurons manifest LP and are lost in PD, such as cholinergic neurons in the basal forebrain and PPN, while other dopamine neurons in hypothalamus, olfactory bulb, and ventral tegmental area (VTA) are at best affected to only a limited extent [13]. Furthermore, the use of L-DOPA to treat PD patients does not measurably increase the rate of disease progression [210–212].

6.3 Does the At-Risk Phenotype Amplify the Consequences of Genetic Mutations Associated with PD?

The list of genetic mutations linked to increased risk of developing PD has been growing [4, 6, 213]. Although they account for only about 10 % of PD cases overall [214], they provide important clues about pathogenic mechanisms. These mutations can be divided into recessive (loss of function) and dominant (gain of function) categories. The recessive mutations in *DJ-1*, *PINK1*, and *Parkin* are all associated with rare, early-onset forms of PD [213, 214]. All of these genes are widely expressed in the brain and, as a consequence, cannot explain the selective loss of SNc DA neurons on the basis of expression alone. Each of these genes has been tied to mitochondrial function. DJ-1 helps orchestrate oxidant defenses [215, 216]. PINK1 and Parkin play an important role in mitochondrial quality control and possibly biogenesis [217–222]. As a consequence, it is straightforward to hypothesize that their impact will be felt most acutely in neurons that have sustained mitochondrial stress, like SNc DA neurons. For each of the genes, there are demonstrated effects on SNc DA neurons that are consistent with this hypothesis [132, 219, 223].

The autosomal dominant mutations associated with PD are less well understood and appear to be more heterogeneous in their effects on neuronal function. Mutations and replication of aSYN are the best characterized of the dominant forms of PD [213]. aSYN is a major component of LP [224, 225]. As a consequence, our comments will be restricted to the possible role of aSYN in selective vulnerability.

As mentioned at the outset, the relationship between LP and neurodegeneration is uncertain. Hence, the relationship between aSYN aggregation and neurodegeneration is uncertain. Nevertheless, it is reasonable to assume that aSYN aggregation creates neuronal stress. So, is there reason to believe that the at-risk phenotype contributes to aSYN aggregation? A number of studies argue that elevated Ca^{2+} concentration promotes aggregation of aSYN [162, 169, 170, 226–229]. There are several ways this might happen. Ca^{2+} promotes the formation of fibrillar forms of aSYN, possibly by acting as a counterion for acidic residues in its C-terminal region [230]. Furthermore, cleavage of the C-terminal region of aSYN by the Ca^{2+}-activated protease calpain also promotes aSYN aggregation [231]. Inhibitors of calpain alleviate the toxicity of aSYN overexpression in SNc DA neurons [232]. Do cytosolic Ca^{2+} levels rise high enough in SNc DA neurons to activate calpain? Ratiometric imaging studies suggest that the Ca^{2+} concentration in SNc DA neurons rises into the neighborhood of 250 nM during normal activity [233]; unpublished work by our group has confirmed this result using alternative approaches and suggests that in more distal dendrites the Ca^{2+} concentration climbs even higher, particularly during synaptic stimulation. This is sufficient to partially activate calpain [71, 234–236], suggesting that there might be a sustained challenge to autophagy created by truncated aSYN in SNc DA neurons.

Aggregation of aSYN could lead to additional physiological adaptations. It is also possible that aSYN aggregation compromises Ca^{2+} homeostasis [71, 237–239]. In rats, aSYN overexpression leads to redox modification of Kv4 channels in SNc (but not VTA) DA neurons, resulting in an acceleration of pacemaking rate [240]. This is consistent with the observation that aSYN fibrils increase cytosolic oxidative stress in SNc dopaminergic neurons [161]. These physiological changes could create a "death spiral" for SNc DA neurons where growing oxidative stress begets more oxidative stress.

Studies using induced pluripotent stem cells (iPSCs) derived from a patient carrying the SNCA-A53T mutation reveal that the mutant cells exhibit signs of increased basal oxidative stress and are sensitized to the effects of mitochondrial toxins compared with isogenic wild-type cells [241]. The mechanism underlying this response was traced to the generation of nitrosative/oxidative stress that inhibited the activity of the transcription factor, myocyte enhancer factor 2c (MEF2C) through oxidation of a critical cysteine residue. MEF2C drives expression of peroxisome proliferator-activated receptor-gamma coactivator-1alpha (PGC1a) expression, a positive regulator of nuclear-encoded mitochondrial protein expression and mitochondrial biogenesis. The resulting suppression of PGC1a expression in turn led to a preferential increase in apoptotic cell death, possibly attributable to a decreased respiratory reserve relative to the wild-type cells. While the mechanism underlying the mutation-induced increase in oxidative stress still needs to be established, these results suggest that basal increases in oxidative stress may synergize with genetic mutations to augment the rate of SNc loss with aging and/or enhance vulnerability to environmental factors that would otherwise be well tolerated.

6.4 Toward a Human Neuroprotective Therapy

Although a number of clinical trials have been completed, no agent has been found to slow the progression of PD. Given the growing population afflicted with PD, there is increased urgency about this unmet medical need for both personal and societal reasons. Based upon the current state of knowledge about pathogenesis, what are the near-term prospects for such a treatment? One of the major impediments to the development of a disease-modifying therapy (DMT) is the lack of an animal model that recapitulates the pattern of PD pathology. Toxin models, like the 6-OHDA or MPTP models, recapitulate the consequences of SNc DA cell loss but do not accurately mimic pathogenic mechanisms in humans. Every compound that has failed in human clinical trials was found to ameliorate toxin-induced cell loss in rodents [242–244]. Viral models where a transgene of interest is selectively expressed in SNc hold promise [14–16, 245–251], as do some transgenic models, particularly those focusing on aSYN [252, 253]. However, the kinetics of pathology in these models is dramatically different from human PD; moreover, they fail to reproduce many of the other features of PD, including the distributed pattern of pathology.

In the absence of a compelling animal model in which potential therapeutics can be tested, what can be done beyond the identification of plausible mechanisms that are consistent with known features of the disease? Epidemiological data offer potential clues. There are a variety of factors that are associated with reduced PD incidence: smoking tobacco, coffee consumption, and the use of statins, anti-inflammatories, and dihydropyridines [9, 226, 227, 254]. Of these, dihydropyridines (DHPs) are of particular interest because of the connection between Ca^{2+} entry through Cav1 channels in SNc DA neurons and other factors thought to be important in PD pathogenesis. DHPs have a long record of safe use in humans treated for hypertension and have good brain bioavailability [228]. DHPs are negative allosteric modulators [not blockers] of Cav1 channel opening, with generally higher affinity for Cav1.2 channels than Cav1.3 channels. One exception to this is isradipine, which has nearly equal affinity for Cav1.2 and Cav1.3 channels [255–258]. The nearly 30 % reduction in risk seen in the epidemiological studies of isradipine is therefore surprising, given the modest antagonism of Cav1.3 channels [the principal source of Ca^{2+} entry into SNc DA neurons] that must have been achieved. In addition to being associated with a reduced probability of PD when taken prior to diagnosis, DHP use also is associated with slowing of progression after diagnosis [229]. This effect is certainly consistent with the putative mechanism involving a chronic suppression of mitochondrial Ca^{2+} entry, with a subsequent lessening of mitochondrial oxidative stress.

But even strongly suggestive epidemiological studies are not a substitute for a prospective controlled clinical trial. A phase II clinical trial with early-stage PD patients found that isradipine was well tolerated at doses up to 10 mg/day [259]. In this dose range, isradipine has relatively minor side effects (e.g., ankle edema) [260]. The question is whether it will prove neuroprotective at these doses. Systemic administration of isradipine to rodents at doses that achieve serum concentrations near those achieved in humans protects against MPTP and 6-OHDA toxicity [112, 261]. Using a modulated receptor model, Ilijic et al. [261] estimated that roughly 40–60 % of the Cav1.3 channels were antagonized at the half-maximal dose for protection from intrastriatal 6-OHDA injection. But as noted above, success in toxin models does not guarantee success in humans. Nevertheless, given the evidence that Ca^{2+} entry through Cav1 channels could create both mitochondrial and proteostatic stress in SNc DA neurons, the epidemiological evidence, and the modest side effect profile, the National Institutes of Health in 2014 funded a 5-year, 56-center, phase III clinical trial in North America to determine whether isradipine will slow the progression of early-stage PD. Unfortunately, this study, like others that have been conducted to date, has two major shortcomings. One is that patients enrolled in the study will be symptomatic because there are no validated biomarkers for near-term PD onset. At this point, much of the damage is done, at least to SNc DA neurons [262]. Second, the study will have to rely upon changes in the UPDRS score of patients to determine progression and ultimately efficacy. Not only does this introduce experimenter noise into the trial (increasing the magnitude of the clinical effect required for significance), it also lengthens the observation period required to determine progression.

Acknowledgments This work was supported by NIH Grant NS 047085 and awards from the JPB and IDP Foundations (DJS) and HL35440 and HL122062 (PTS).

References

1. de Lau LM, Giesbergen PC, de Rijk MC, Hofman A, Koudstaal PJ, Breteler MM. Incidence of Parkinsonism and Parkinson disease in a general population: the Rotterdam study. Neurology. 2004;63(7):1240–4.
2. Dorsey ER, Constantinescu R, Thompson JP, Biglan KM, Holloway RG, Kieburtz K, et al. Projected number of people with Parkinson disease in the most populous nations, 2005 through 2030. Neurology. 2007;68(5):384–6.

3. Lin MK, Farrer MJ. Genetics and genomics of Parkinson's disease. Genome Med. 2014;6(6):48.
4. Bras JM, Singleton A. Genetic susceptibility in Parkinson's disease. Biochim Biophys Acta. 2009;1792(7):597–603.
5. Cookson MR, Bandmann O. Parkinson's disease: insights from pathways. Hum Mol Genet. 2010;19(R1):R21–7.
6. Mullin S, Schapira A. The genetics of Parkinson's disease. Br Med Bull. 2015;114(1):39–52. Epub 2015/05/23.
7. Papapetropoulos S, Shehadeh L, McCorquodale D. Optimizing human post-mortem brain tissue gene expression profiling in Parkinson's disease and other neurodegenerative disorders: from target "fishing" to translational breakthroughs. J Neurosci Res. 2007;85(14):3013–24.
8. Lewis PA, Cookson MR. Gene expression in the Parkinson's disease brain. Brain Res Bull. 2012;88(4):302–12.
9. Tanner CM. Advances in environmental epidemiology. Mov Disord. 2010;25 Suppl 1:S58–62. Epub 2010/02/27.
10. Betarbet R, Canet-Aviles RM, Sherer TB, Mastroberardino PG, McLendon C, Kim JH, et al. Intersecting pathways to neurodegeneration in Parkinson's disease: effects of the pesticide rotenone on DJ-1, alpha-synuclein, and the ubiquitin-proteasome system. Neurobiol Dis. 2006;22(2):404–20.
11. Berg D, Postuma RB, Bloem B, Chan P, Dubois B, Gasser T, et al. Time to redefine PD? Introductory statement of the MDS task force on the definition of Parkinson's disease. Mov Disord. 2014;29(4):454–62.
12. Halliday G, Lees A, Stern M. Milestones in Parkinson's disease – clinical and pathologic features. Mov Disord. 2011;26(6):1015–21. Epub 2011/06/01.
13. Halliday G, McCann H, Shepherd C. Evaluation of the Braak hypothesis: how far can it explain the pathogenesis of Parkinson's disease? Expert Rev Neurother. 2012;12(6):673–86.
14. Kirik D, Bjorklund A. Modeling CNS neurodegeneration by overexpression of disease-causing proteins using viral vectors. Trend Neurosci. 2003;26(7):386–92.
15. Kirik D, Rosenblad C, Burger C, Lundberg C, Johansen TE, Muzyczka N, et al. Parkinson-like neurodegeneration induced by targeted overexpression of alpha-synuclein in the nigrostriatal system. J Neurosci. 2002;22(7):2780–91.
16. Kirik D, Annett LE, Burger C, Muzyczka N, Mandel RJ, Bjorklund A. Nigrostriatal alpha-synucleinopathy induced by viral vector-mediated overexpression of human alpha-synuclein: a new primate model of Parkinson's disease. Proc Natl Acad Sci U S A. 2003;100(5):2884–9. Epub 2003/02/26.
17. Luk KC, Kehm V, Carroll J, Zhang B, O'Brien P, Trojanowski JQ, et al. Pathological alpha-synuclein transmission initiates Parkinson-like neurodegeneration in nontransgenic mice. Science. 2012;338(6109):949–53.
18. Recasens A, Dehay B, Bove J, Carballo-Carbajal I, Dovero S, Perez-Villalba A, et al. Lewy body extracts from Parkinson disease brains trigger alpha-synuclein pathology and neurodegeneration in mice and monkeys. Ann Neurol. 2014;75(3):351–62.
19. Volpicelli-Daley LA, Luk KC, Patel TP, Tanik SA, Riddle DM, Stieber A, et al. Exogenous alpha-synuclein fibrils induce Lewy body pathology leading to synaptic dysfunction and neuron death. Neuron. 2011;72(1):57–71. Epub 2011/10/11.
20. Burke RE, Dauer WT, Vonsattel JP. A critical evaluation of the Braak staging scheme for Parkinson's disease. Ann Neurol. 2008;64(5):485–91.
21. Hornykiewicz O. Dopamine (3-hydroxytyramine) and brain function. Pharmacol Rev. 1966;18(2):925–64.
22. Riederer P, Wuketich S. Time course of nigrostriatal degeneration in Parkinson's disease. A detailed study of influential factors in human brain amine analysis. J Neural Transm. 1976;38(3–4):277–301.
23. Groves PM, Linder JC, Young SJ. 5-hydroxydopamine-labeled dopaminergic axons: three-dimensional reconstructions of axons, synapses and postsynaptic targets in rat neostriatum. Neuroscience. 1994;58(3):593–604.

24. Gauthier J, Parent M, Levesque M, Parent A. The axonal arborization of single nigrostriatal neurons in rats. Brain Res. 1999;834(1–2):228–32.
25. Prensa L, Parent A. The nigrostriatal pathway in the rat: a single-axon study of the relationship between dorsal and ventral tier nigral neurons and the striosome/matrix striatal compartments. J Neurosci. 2001;21(18):7247–60.
26. Arbuthnott GW, Wickens J. Space, time and dopamine. Trend Neurosci. 2007;30(2):62–9. Epub 2006/12/19.
27. Matsuda W, Furuta T, Nakamura KC, Hioki H, Fujiyama F, Arai R, et al. Single nigrostriatal dopaminergic neurons form widely spread and highly dense axonal arborizations in the neostriatum. J Neurosci. 2009;29(2):444–53. Epub 2009/01/16.
28. Fallon JH. Collateralization of monoamine neurons: mesotelencephalic dopamine projections to caudate, septum, and frontal cortex. J Neurosci. 1981;1(12):1361–8.
29. Bolam JP, Pissadaki EK. Living on the edge with too many mouths to feed: why dopamine neurons die. Mov Disord. 2012;27(12):1478–83. Epub 2012/09/26.
30. Loughlin SE, Fallon JH. Substantia nigra and ventral tegmental area projections to cortex: topography and collateralization. Neuroscience. 1984;11(2):425–35.
31. Kawaguchi Y, Wilson CJ, Emson PC. Projection subtypes of rat neostriatal matrix cells revealed by intracellular injection of biocytin. J Neurosci. 1990;10(10):3421–8.
32. Wu Y, Richard S, Parent A. The organization of the striatal output system: a single-cell juxtacellular labeling study in the rat. Neurosci Res. 2000;38(1):49–62.
33. Pissadaki EK, Bolam JP. The energy cost of action potential propagation in dopamine neurons: clues to susceptibility in Parkinson's disease. Front Comput Neurosci. 2013;7:13. Epub 2013/03/22.
34. De Vos KJ, Grierson AJ, Ackerley S, Miller CC. Role of axonal transport in neurodegenerative diseases. Ann Rev Neurosci. 2008;31:151–73. Epub 2008/06/19.
35. Salinas S, Bilsland LG, Schiavo G. Molecular landmarks along the axonal route: axonal transport in health and disease. Curr Opin Cell Biol. 2008;20(4):445–53. Epub 2008/05/23.
36. Morfini GA, Burns M, Binder LI, Kanaan NM, LaPointe N, Bosco DA, et al. Axonal transport defects in neurodegenerative diseases. J Neurosci. 2009;29(41):12776–86. Epub 2009/10/16.
37. Staff NP, Benarroch EE, Klein CJ. Neuronal intracellular transport and neurodegenerative disease. Neurology. 2011;76(11):1015–20.
38. Millecamps S, Julien JP. Axonal transport deficits and neurodegenerative diseases. Nat Rev Neurosci. 2013;14(3):161–76. Epub 2013/01/31.
39. Ashrafi G, Schlehe JS, LaVoie MJ, Schwarz TL. Mitophagy of damaged mitochondria occurs locally in distal neuronal axons and requires PINK1 and Parkin. J Cell Biol. 2014;206(5):655–70. Epub 2014/08/27.
40. Schmieg N, Menendez G, Schiavo G, Terenzio M. Signalling endosomes in axonal transport: travel updates on the molecular highway. Semin Cell Dev Biol. 2014;27:32–43. Epub 2013/11/01.
41. Gennerich A, Vale RD. Walking the walk: how kinesin and dynein coordinate their steps. Curr Opin Cell Biol. 2009;21(1):59–67. Epub 2009/01/31.
42. Chu Y, Morfini GA, Langhamer LB, He Y, Brady ST, Kordower JH. Alterations in axonal transport motor proteins in sporadic and experimental Parkinson's disease. Brain. 2012;135(Pt 7):2058–73. Epub 2012/06/22.
43. Chung CY, Koprich JB, Siddiqi H, Isacson O. Dynamic changes in presynaptic and axonal transport proteins combined with striatal neuroinflammation precede dopaminergic neuronal loss in a rat model of AAV alpha-synucleinopathy. J Neurosci. 2009;29(11):3365–73. Epub 2009/03/20.
44. Saha AR, Hill J, Utton MA, Asuni AA, Ackerley S, Grierson AJ, et al. Parkinson's disease alpha-synuclein mutations exhibit defective axonal transport in cultured neurons. J Cell Sci. 2004;117(Pt 7):1017–24. Epub 2004/03/05.
45. Kim-Han JS, Antenor-Dorsey JA, O'Malley KL. The Parkinsonian mimetic, MPP+, specifically impairs mitochondrial transport in dopamine axons. J Neurosci. 2011;31(19):7212–21. Epub 2011/05/13.

46. Lu X, Kim-Han JS, Harmon S, Sakiyama-Elbert SE, O'Malley KL. The Parkinsonian mimetic, 6-OHDA, impairs axonal transport in dopaminergic axons. Mol Neurodegener. 2014;9:17. Epub 2014/06/03.
47. Mironov SL, Symonchuk N. ER vesicles and mitochondria move and communicate at synapses. J Cell Sci. 2006;119(Pt 23):4926–34. Epub 2006/11/16.
48. Saotome M, Safiulina D, Szabadkai G, Das S, Fransson A, Aspenstrom P, et al. Bidirectional Ca2+ -dependent control of mitochondrial dynamics by the Miro GTPase. Proc Natl Acad Sci U S A. 2008;105(52):20728–33. Epub 2008/12/23.
49. Macaskill AF, Rinholm JE, Twelvetrees AE, Arancibia-Carcamo IL, Muir J, Fransson A, et al. Miro1 is a calcium sensor for glutamate receptor-dependent localization of mitochondria at synapses. Neuron. 2009;61(4):541–55. Epub 2009/03/03.
50. Wang X, Schwarz TL. The mechanism of Ca2+ -dependent regulation of kinesin-mediated mitochondrial motility. Cell. 2009;136(1):163–74. Epub 2009/01/13.
51. Saxton WM, Hollenbeck PJ. The axonal transport of mitochondria. J Cell Sci. 2012;125(Pt 9):2095–104. Epub 2012/05/24.
52. Lovas JR, Wang X. The meaning of mitochondrial movement to a neuron's life. Biochim Biophys Acta. 2013;1833(1):184–94. Epub 2012/05/03.
53. Sheng ZH. Mitochondrial trafficking and anchoring in neurons: new insight and implications. J Cell Biol. 2014;204(7):1087–98. Epub 2014/04/02.
54. Wang X, Winter D, Ashrafi G, Schlehe J, Wong YL, Selkoe D, et al. PINK1 and Parkin target Miro for phosphorylation and degradation to arrest mitochondrial motility. Cell. 2011;147(4):893–906. Epub 2011/11/15.
55. Liu S, Sawada T, Lee S, Yu W, Silverio G, Alapatt P, et al. Parkinson's disease-associated kinase PINK1 regulates Miro protein level and axonal transport of mitochondria. PLoS Genet. 2012;8(3), e1002537. Epub 2012/03/08.
56. Katajisto P, Dohla J, Chaffer CL, Pentinmikko N, Marjanovic N, Iqbal S, et al. Stem cells. Asymmetric apportioning of aged mitochondria between daughter cells is required for stemness. Science. 2015;348(6232):340–3.
57. Friedman JR, Nunnari J. Mitochondrial form and function. Nature. 2014;505(7483):335–43. Epub 2014/01/17.
58. Campello S, Scorrano L. Mitochondrial shape changes: orchestrating cell pathophysiology. EMBO Rep. 2010;11(9):678–84. Epub 2010/08/21.
59. Lee S, Sterky FH, Mourier A, Terzioglu M, Cullheim S, Olson L, et al. Mitofusin 2 is necessary for striatal axonal projections of midbrain dopamine neurons. Hum Mol Genet. 2012;21(22):4827–35. Epub 2012/08/24.
60. Pham AH, Meng S, Chu QN, Chan DC. Loss of Mfn2 results in progressive, retrograde degeneration of dopaminergic neurons in the nigrostriatal circuit. Hum Mol Genet. 2012;21(22):4817–26. Epub 2012/08/04.
61. Berthet A, Margolis EB, Zhang J, Hsieh I, Zhang J, Hnasko TS, et al. Loss of mitochondrial fission depletes axonal mitochondria in midbrain dopamine neurons. J Neurosci. 2014;34(43):14304–17. Epub 2014/10/24.
62. Kish SJ, Shannak K, Hornykiewicz O. Uneven pattern of dopamine loss in the striatum of patients with idiopathic Parkinson's disease. Pathophysiologic and clinical implications. N Engl J Med. 1988;318(14):876–80.
63. Scherman D, Desnos C, Darchen F, Pollak P, Javoy-Agid F, Agid YA. Striatal dopamine deficiency in Parkinson's disease: role of aging. Ann Neurol. 1989;26(4):551–7.
64. Garcia-Reitbock P, Anichtchik O, Bellucci A, Iovino M, Ballini C, Fineberg E, et al. SNARE protein redistribution and synaptic failure in a transgenic mouse model of Parkinson's disease. Brain. 2010;133(Pt 7):2032–44. Epub 2010/06/11.
65. Burke RE, O'Malley K. Axon degeneration in Parkinson's disease. Exp Neurol. 2013;246:72–83. Epub 2012/01/31.
66. Kordower JH, Olanow CW, Dodiya HB, Chu Y, Beach TG, Adler CH, et al. Disease duration and the integrity of the nigrostriatal system in Parkinson's disease. Brain. 2013;136(Pt 8):2419–31. Epub 2013/07/26.

67. Li Y, Liu W, Oo TF, Wang L, Tang Y, Jackson-Lewis V, et al. Mutant LRRK2(R1441G) BAC transgenic mice recapitulate cardinal features of Parkinson's disease. Nat Neurosci. 2009;12(7):826–8. Epub 2009/06/09.
68. Sulzer D, Surmeier DJ. Neuronal vulnerability, pathogenesis, and Parkinson's disease. Mov Disord. 2013;28(1):41–50. Epub 2012/07/14.
69. Clapham DE. Calcium signaling. Cell. 2007;131(6):1047–58.
70. Augustine GJ, Santamaria F, Tanaka K. Local calcium signaling in neurons. Neuron. 2003;40(2):331–46.
71. Mattson MP. Calcium and neurodegeneration. Aging Cell. 2007;6(3):337–50.
72. Berridge MJ. The endoplasmic reticulum: a multifunctional signaling organelle. Cell Calcium. 2002;32(5–6):235–49.
73. Mogami H, Nakano K, Tepikin AV, Petersen OH. Ca2+ flow via tunnels in polarized cells: recharging of apical Ca2+ stores by focal Ca2+ entry through basal membrane patch. Cell. 1997;88(1):49–55.
74. Park MK, Petersen OH, Tepikin AV. The endoplasmic reticulum as one continuous Ca(2+) pool: visualization of rapid Ca(2+) movements and equilibration. EMBO J. 2000;19(21):5729–39.
75. Choi YM, Kim SH, Chung S, Uhm DY, Park MK. Regional interaction of endoplasmic reticulum Ca2+ signals between soma and dendrites through rapid luminal Ca2+ diffusion. J Neurosci. 2006;26(47):12127–36. Epub 2006/11/24.
76. Schwyn RC, Fox CA. The primate substantia nigra: a Golgi and electron microscopic study. J Hirnforsch. 1974;15(1):95–126.
77. Bardo S, Cavazzini MG, Emptage N. The role of the endoplasmic reticulum Ca2+ store in the plasticity of central neurons. Trend Pharmacol Sci. 2006;27(2):78–84.
78. Cui G, Bernier BE, Harnett MT, Morikawa H. Differential regulation of action potential- and metabotropic glutamate receptor-induced Ca2+ signals by inositol 1,4,5-trisphosphate in dopaminergic neurons. J Neurosci. 2007;27(17):4776–85. Epub 2007/04/27.
79. Rose CR, Konnerth A. Stores not just for storage. Intracellular calcium release and synaptic plasticity. Neuron. 2001;31(4):519–22.
80. Verkhratsky A. Physiology and pathophysiology of the calcium store in the endoplasmic reticulum of neurons. Physiol Rev. 2005;85(1):201–79.
81. Park MK, Choi YM, Kang YK, Petersen OH. The endoplasmic reticulum as an integrator of multiple dendritic events. Neuroscientist. 2008;14(1):68–77. Epub 2007/10/04.
82. Chen H, Chan DC. Critical dependence of neurons on mitochondrial dynamics. Curr Opin Cell Biol. 2006;18(4):453–9. Epub 2006/06/20.
83. Schwarz TL. Mitochondrial trafficking in neurons. Cold Spring Harb Perspect Biol. 2013;5(6):1–16. Epub 2013/06/05.
84. Boldogh IR, Pon LA. Mitochondria on the move. Trend Cell Biol. 2007;17(10):502–10.
85. Hayashi T, Rizzuto R, Hajnoczky G, Su TP. MAM: more than just a housekeeper. Trend Cell Biol. 2009;19(2):81–8. Epub 2009/01/16.
86. Pizzo P, Pozzan T. Mitochondria-endoplasmic reticulum choreography: structure and signaling dynamics. Trend Cell Biol. 2007;17(10):511–7. Epub 2007/09/14.
87. Csordas G, Renken C, Varnai P, Walter L, Weaver D, Buttle KF, et al. Structural and functional features and significance of the physical linkage between ER and mitochondria. J Cell Biol. 2006;174(7):915–21.
88. de Brito OM, Scorrano L. An intimate liaison: spatial organization of the endoplasmic reticulum-mitochondria relationship. EMBO J. 2010;29(16):2715–23. Epub 2010/08/19.
89. Rizzuto R, Pozzan T. Microdomains of intracellular Ca2+: molecular determinants and functional consequences. Physiol Rev. 2006;86(1):369–408.
90. Kirichok Y, Krapivinsky G, Clapham DE. The mitochondrial calcium uniporter is a highly selective ion channel. Nature. 2004;427(6972):360–4.
91. De Stefani D, Raffaello A, Teardo E, Szabo I, Rizzuto R. A forty-kilodalton protein of the inner membrane is the mitochondrial calcium uniporter. Nature. 2011;476(7360):336–40. Epub 2011/06/21.

92. Baughman JM, Perocchi F, Girgis HS, Plovanich M, Belcher-Timme CA, Sancak Y, et al. Integrative genomics identifies MCU as an essential component of the mitochondrial calcium uniporter. Nature. 2011;476(7360):341–5. Epub 2011/06/21.
93. Pan X, Liu J, Nguyen T, Liu C, Sun J, Teng Y, et al. The physiological role of mitochondrial calcium revealed by mice lacking the mitochondrial calcium uniporter. Nat Cell Biol. 2013;15(12):1464–72.
94. Mourier A, Motori E, Brandt T, Lagouge M, Atanassov I, Galinier A, et al. Mitofusin 2 is required to maintain mitochondrial coenzyme Q levels. J Cell Biol. 2015;208(4): 429–42.
95. Sebastian D, Hernandez-Alvarez MI, Segales J, Sorianello E, Munoz JP, Sala D, et al. Mitofusin 2 (Mfn2) links mitochondrial and endoplasmic reticulum function with insulin signaling and is essential for normal glucose homeostasis. Proc Natl Acad Sci U S A. 2012;109(14):5523–8.
96. Papanicolaou KN, Ngoh GA, Dabkowski ER, O'Connell KA, Ribeiro Jr RF, Stanley WC, et al. Cardiomyocyte deletion of mitofusin-1 leads to mitochondrial fragmentation and improves tolerance to ROS-induced mitochondrial dysfunction and cell death. Am J Physiol Heart Circ Physiol. 2012;302(1):H167–79.
97. Gerfen CR, Surmeier DJ. Modulation of striatal projection systems by dopamine. Ann Rev Neurosci. 2011;34:441–66.
98. Atherton JF, Bevan MD. Ionic mechanisms underlying autonomous action potential generation in the somata and dendrites of GABAergic substantia nigra pars reticulata neurons in vitro. J Neurosci. 2005;25(36):8272–81.
99. Chan CS, Glajch KE, Gertler TS, Guzman JN, Mercer JN, Lewis AS, et al. HCN channelopathy in external globus pallidus neurons in models of Parkinson's disease. Nat Neurosci. 2011;14(1):85–92.
100. Gross A, Sims RE, Swinny JD, Sieghart W, Bolam JP, Stanford IM. Differential localization of GABA(A) receptor subunits in relation to rat striatopallidal and pallidopallidal synapses. Eur J Neurosci. 2011;33(5):868–78.
101. Lee CR, Tepper JM. Morphological and physiological properties of parvalbumin- and calretinin-containing gamma-aminobutyric acidergic neurons in the substantia nigra. J Comp Neurol. 2007;500(5):958–72.
102. Chan CS, Shigemoto R, Mercer JN, Surmeier DJ. HCN2 and HCN1 channels govern the regularity of autonomous pacemaking and synaptic resetting in globus pallidus neurons. J Neurosci. 2004;24(44):9921–32.
103. Mercer JN, Chan CS, Tkatch T, Held J, Surmeier DJ. Nav1.6 sodium channels are critical to pacemaking and fast spiking in globus pallidus neurons. J Neurosci. 2007;27(49):13552–66.
104. Ding JB, Guzman JN, Peterson JD, Goldberg JA, Surmeier DJ. Thalamic gating of corticostriatal signaling by cholinergic interneurons. Neuron. 2010;67(2):294–307.
105. Goldberg JA, Guzman JN, Estep CM, Ilijic E, Kondapalli J, Sanchez-Padilla J, et al. Calcium entry induces mitochondrial oxidant stress in vagal neurons at risk in Parkinson's disease. Nat Neurosci. 2012;15(10):1414–21. Epub 2012/09/04.
106. Guzman JN, Sanchez-Padilla J, Chan CS, Surmeier DJ. Robust pacemaking in substantia nigra dopaminergic neurons. J Neurosci. 2009;29(35):11011–9. Epub 2009/09/04.
107. Paladini CA, Beckstead MJ, Weinshenker D. Electrophysiological properties of catecholaminergic neurons in the norepinephrine-deficient mouse. Neuroscience. 2007;144(3):1067–74.
108. Sanchez-Padilla J, Guzman JN, Ilijic E, Kondapalli J, Galtieri DJ, Yang B, et al. Mitochondrial oxidant stress in locus coeruleus is regulated by activity and nitric oxide synthase. Nat Neurosci. 2014;17(6):832–40. Epub 2014/05/13.
109. Bean BP. The action potential in mammalian central neurons. Nat Rev Neurosci. 2007;8(6):451–65.
110. Puopolo M, Raviola E, Bean BP. Roles of subthreshold calcium current and sodium current in spontaneous firing of mouse midbrain dopamine neurons. J Neurosci. 2007;27(3):645–56. Epub 2007/01/20.

111. Putzier I, Kullmann PH, Horn JP, Levitan ES. Cav1.3 channel voltage dependence, not Ca2+ selectivity, drives pacemaker activity and amplifies bursts in nigral dopamine neurons. J Neurosci. 2009;29(49):15414–9. Epub 2009/12/17.
112. Chan CS, Guzman JN, Ilijic E, Mercer JN, Rick C, Tkatch T, et al. 'Rejuvenation' protects neurons in mouse models of Parkinson's disease. Nature. 2007;447(7148):1081–6. Epub 2007/06/15.
113. Xu W, Lipscombe D. Neuronal Ca(V)1.3alpha(1) L-type channels activate at relatively hyperpolarized membrane potentials and are incompletely inhibited by dihydropyridines. J Neurosci. 2001;21(16):5944–51.
114. Surmeier DJ, Guzman JN, Sanchez-Padilla J. Calcium, cellular aging, and selective neuronal vulnerability in Parkinson's disease. Cell Calcium. 2010;47(2):175–82. Epub 2010/01/08.
115. Sinnegger-Brauns MJ, Huber IG, Koschak A, Wild C, Obermair GJ, Einzinger U, et al. Expression and 1,4-dihydropyridine-binding properties of brain L-type calcium channel isoforms. Mol Pharmacol. 2009;75(2):407–14.
116. Marder E, Abbott LF, Turrigiano GG, Liu Z, Golowasch J. Memory from the dynamics of intrinsic membrane currents. Proc Natl Acad Sci U S A. 1996;93(24):13481–6.
117. Turrigiano G, Abbott LF, Marder E. Activity-dependent changes in the intrinsic properties of cultured neurons. Science. 1994;264(5161):974–7.
118. Turrigiano GG. The self-tuning neuron: synaptic scaling of excitatory synapses. Cell. 2008;135(3):422–35. Epub 2008/11/06.
119. Bean BP. Nitrendipine block of cardiac calcium channels: high-affinity binding to the inactivated state. Proc Natl Acad Sci U S A. 1984;81:6388–92.
120. Neuhoff H, Neu A, Liss B, Roeper J. I(h) channels contribute to the different functional properties of identified dopaminergic subpopulations in the midbrain. J Neurosci. 2002;22(4):1290–302.
121. Shepard PD, Bunney BS. Repetitive firing properties of putative dopamine-containing neurons in vitro: regulation by an apamin-sensitive Ca(2+)-activated K+ conductance. Exp Brain Res. 1991;86(1):141–50.
122. Nedergaard S, Flatman JA, Engberg I. Nifedipine- and omega-conotoxin-sensitive Ca2+ conductances in guinea-pig substantia nigra pars compacta neurones. J Physiol. 1993;466:727–47.
123. Ping HX, Shepard PD. Apamin-sensitive Ca(2+)-activated K+ channels regulate pacemaker activity in nigral dopamine neurons. Neuroreport. 1996;7(3):809–14.
124. Liss B, Franz O, Sewing S, Bruns R, Neuhoff H, Roeper J. Tuning pacemaker frequency of individual dopaminergic neurons by Kv4.3L and KChip3.1 transcription. EMBO J. 2001;20(20):5715–24.
125. Anderson D, Mehaffey WH, Iftinca M, Rehak R, Engbers JD, Hameed S, et al. Regulation of neuronal activity by Cav3-Kv4 channel signaling complexes. Nat Neurosci. 2010;13(3):333–7. Epub 2010/02/16.
126. Khaliq ZM, Bean BP. Pacemaking in dopaminergic ventral tegmental area neurons: depolarizing drive from background and voltage-dependent sodium conductances. J Neurosci. 2010;30(21):7401–13. Epub 2010/05/28.
127. Wilson CJ, Callaway JC. Coupled oscillator model of the dopaminergic neuron of the substantia nigra. J Neurophysiol. 2000;83(5):3084–100.
128. Sarpal D, Koenig JI, Adelman JP, Brady D, Prendeville LC, Shepard PD. Regional distribution of SK3 mRNA-containing neurons in the adult and adolescent rat ventral midbrain and their relationship to dopamine-containing cells. Synapse. 2004;53(2):104–13.
129. Wolfart J, Neuhoff H, Franz O, Roeper J. Differential expression of the small-conductance, calcium-activated potassium channel SK3 is critical for pacemaker control in dopaminergic midbrain neurons. J Neurosci. 2001;21(10):3443–56.
130. Deignan J, Lujan R, Bond C, Riegel A, Watanabe M, Williams JT, et al. SK2 and SK3 expression differentially affect firing frequency and precision in dopamine neurons. Neuroscience. 2012;217:67–76.

131. Gaspar P, Ben Jelloun N, Febvret A. Sparing of the dopaminergic neurons containing calbindin-D28k and of the dopaminergic mesocortical projections in weaver mutant mice. Neuroscience. 1994;61(2):293–305.
132. Guzman JN, Sanchez-Padilla J, Wokosin D, Kondapalli J, Ilijic E, Schumacker PT, et al. Oxidant stress evoked by pacemaking in dopaminergic neurons is attenuated by DJ-1. Nature. 2010;468(7324):696–700. Epub 2010/11/12.
133. Liang C-L, Sinton CM, German DC. Midbrain dopaminergic neurons in the mouse: co-localization with Calbindin-D28K and calretinin. Neuroscience. 1996;75(2):523–33.
134. Liang CL, Sinton C, Sonsalla PK, German DC. Midbrain dopaminergic neurons in the mouse that contain calbindin-D28k exhibit reduced vulnerability to MPTP-induced neurodegeneration. Neurodegeneration. 1996;5(4):313–8.
135. Liu Y, Harding M, Pittman A, Dore J, Striessnig J, Rajadhyaksha A, et al. Cav1.2 and Cav1.3 L-type calcium channels regulate dopaminergic firing activity in the mouse ventral tegmental area. J Neurophysiol. 2014;112(5):1119–30.
136. Morgenroth VH, Boadle-Biber M, Roth RH. Tyrosine hydroxylase: activation by nerve stimulation. Proc Natl Acad Sci U S A. 1974;71(11):4283–7.
137. Menezes A, Zeman R, Sabban E. Involvement of intracellular or extracellular calcium in activation of tyrosine hydroxylase gene expression in PC12 cells. J Neurochem. 1996;67(6):2316–24.
138. Aumann TD, Egan K, Lim J, Boon WC, Bye CR, Chua HK, et al. Neuronal activity regulates expression of tyrosine hydroxylase in adult mouse substantia nigra pars compacta neurons. J Neurochem. 2011;116(4):646–58. Epub 2010/12/21.
139. Aumann T, Horne M. Activity-dependent regulation of the dopamine phenotype in substantia nigra neurons. J Neurochem. 2012;121(4):497–515. Epub 2012/02/24.
140. Balaban RS. Domestication of the cardiac mitochondrion for energy conversion. J Mol Cell Cardiol. 2009;46(6):832–41.
141. Liu T, O'Rourke B. Regulation of mitochondrial Ca2+ and its effects on energetics and redox balance in normal and failing heart. J Bioenerg Biomembr. 2009;41(2):127–32.
142. Wellstead P, Cloutier M. Modelling and simulation of brain energy metabolism: energy and Parkinson's disease. In: Wellstead P, Cloutier M, editors. Systems biology of Parkinson's disease. New York: Springer; 2012. p. 19–38.
143. Liss B, Bruns R, Roeper J. Alternative sulfonylurea receptor expression defines metabolic sensitivity of K-ATP channels in dopaminergic midbrain neurons. EMBO J. 1999;18(4):833–46.
144. Liss B, Haeckel O, Wildmann J, Miki T, Seino S, Roeper J. K-ATP channels promote the differential degeneration of dopaminergic midbrain neurons. Nat Neurosci. 2005;8(12):1742–51. Epub 2005/11/22.
145. Schiemann J, Schlaudraff F, Klose V, Bingmer M, Seino S, Magill PJ, et al. K-ATP channels in dopamine substantia nigra neurons control bursting and novelty-induced exploration. Nat Neurosci. 2012;15(9):1272–80. Epub 2012/08/21.
146. Morita M, Gravel SP, Chenard V, Sikstrom K, Zheng L, Alain T, et al. mTORC1 controls mitochondrial activity and biogenesis through 4E-BP-dependent translational regulation. Cell Metab. 2013;18(5):698–711.
147. Cardenas C, Foskett JK. Mitochondrial Ca(2+) signals in autophagy. Cell Calcium. 2012;52(1):44–51. Epub 2012/03/31.
148. Morikawa H, Khodakhah K, Williams JT. Two intracellular pathways mediate metabotropic glutamate receptor-induced Ca2+ mobilization in dopamine neurons. J Neurosci. 2003;23(1):149–57.
149. Morikawa H, Imani F, Khodakhah K, Williams JT. Inositol 1,4,5-triphosphate-evoked responses in midbrain dopamine neurons. J Neurosci. 2000;20(20):RC103.
150. Brimblecombe KR, Gracie CJ, Platt NJ, Cragg SJ. Gating of dopamine transmission by calcium and axonal N-, Q-, T- and L-type voltage-gated calcium channels differs between striatal domains. J Physiol. 2015;593(4):929–46. Epub 2014/12/24.

151. Rangaraju V, Calloway N, Ryan TA. Activity-driven local ATP synthesis is required for synaptic function. Cell. 2014;156(4):825–35. Epub 2014/02/18.
152. Villegas R, Martinez NW, Lillo J, Pihan P, Hernandez D, Twiss JL, et al. Calcium release from intra-axonal endoplasmic reticulum leads to axon degeneration through mitochondrial dysfunction. J Neurosci. 2014;34(21):7179–89. Epub 2014/05/23.
153. Williams JT, North RA, Shefner SA, Nishi S, Egan TM. Membrane properties of rat locus coeruleus neurones. Neuroscience. 1984;13(1):137–56.
154. Mo ZL, Katafuchi T, Muratani H, Hori T. Effects of vasopressin and angiotensin II on neurones in the rat dorsal motor nucleus of the vagus, in vitro. J Physiol. 1992;458:561–77.
155. Takakusaki K, Kitai ST. Ionic mechanisms involved in the spontaneous firing of tegmental pedunculopontine nucleus neurons of the rat. Neuroscience. 1997;78(3):771–94.
156. McGinty DJ, Harper RM. Dorsal raphe neurons: depression of firing during sleep in cats. Brain Res. 1976;101(3):569–75.
157. Hedrick T, Waters J. Physiological properties of cholinergic and non-cholinergic magnocellular neurons in acute slices from adult mouse nucleus basalis. PLoS ONE. 2010;5(6), e11046. Epub 2010/06/16.
158. Korshunov SS, Skulachev VP, Starkov AA. High protonic potential actuates a mechanism of production of reactive oxygen species in mitochondria. FEBS Lett. 1997;416(1):15–8.
159. Müller M, Cheung KH, Foskett JK. Enhanced ROS generation mediated by Alzheimer's disease presenilin regulation of InsP3R Ca2+ signaling. Antioxid Redox Signal. 2011;14(7):1225–35.
160. Sabharwal SS, Schumacker PT. Mitochondrial ROS in cancer: initiators, amplifiers or an Achilles' heel? Nat Rev Cancer. 2014;14(11):709–21. Epub 2014/10/25.
161. Dryanovski DI, Guzman JN, Xie Z, Galteri DJ, Volpicelli-Daley LA, Lee VM, et al. Calcium entry and alpha-synuclein inclusions elevate dendritic mitochondrial oxidant stress in dopaminergic neurons. J Neurosci. 2013;33(24):10154–64. Epub 2013/06/14.
162. Reeve A, Simcox E, Turnbull D. Ageing and Parkinson's disease: why is advancing age the biggest risk factor? Ageing Res Rev. 2014;14:19–30. Epub 2014/02/08.
163. Brown GC. Nitric oxide regulates mitochondrial respiration and cell functions by inhibiting cytochrome oxidase. FEBS Lett. 1995;369(2–3):136–9.
164. Brown GC. Regulation of mitochondrial respiration by nitric oxide inhibition of cytochrome c oxidase. Biochim Biophys Acta. 2001;1504(1):46–57.
165. Clementi E, Brown GC, Feelisch M, Moncada S. Persistent inhibition of cell respiration by nitric oxide: crucial role of S-nitrosylation of mitochondrial complex I and protective action of glutathione. Proc Natl Acad Sci U S A. 1998;95(13):7631–6.
166. Giasson BI, Duda JE, Murray IV, Chen Q, Souza JM, Hurtig HI, et al. Oxidative damage linked to neurodegeneration by selective alpha-synuclein nitration in synucleinopathy lesions. Science. 2000;290(5493):985–9.
167. Ghafourifar P, Richter C. Nitric oxide synthase activity in mitochondria. FEBS Lett. 1997;418(3):291–6.
168. Lacza Z, Snipes JA, Zhang J, Horváth EM, Figueroa JP, Szabó C, et al. Mitochondrial nitric oxide synthase is not eNOS, nNOS or iNOS. Free Radic Biol Med. 2003;35(10):1217–28.
169. Kirkwood TB, Austad SN. Why do we age? Nature. 2000;408(6809):233–8.
170. Brichta L, Shin W, Jackson-Lewis V, Blesa J, Yap EL, Walker Z, et al. Identification of neurodegenerative factors using translatome-regulatory network analysis. Nat Neurosci. 2015;18(9):1325–33. Epub 2015/07/28.
171. Bisaglia M, Filograna R, Beltramini M, Bubacco L. Are dopamine derivatives implicated in the pathogenesis of Parkinson's disease? Ageing Res Rev. 2014;13:107–14. Epub 2014/01/07.
172. Cartier EA, Parra LA, Baust TB, Quiroz M, Salazar G, Faundez V, et al. A biochemical and functional protein complex involving dopamine synthesis and transport into synaptic vesicles. J Biol Chem. 2010;285(3):1957–66. Epub 2009/11/12.
173. Alter SP, Lenzi GM, Bernstein AI, Miller GW. Vesicular integrity in Parkinson's disease. Curr Neurol Neurosci Rep. 2013;13(7):362. Epub 2013/05/22.

174. Takahashi N, Miner LL, Sora I, Ujike H, Revay RS, Kostic V, et al. VMAT2 knockout mice: heterozygotes display reduced amphetamine-conditioned reward, enhanced amphetamine locomotion, and enhanced MPTP toxicity. Proc Natl Acad Sci U S A. 1997;94(18):9938–43.
175. Larsen KE, Fon EA, Hastings TG, Edwards RH, Sulzer D. Methamphetamine-induced degeneration of dopaminergic neurons involves autophagy and upregulation of dopamine synthesis. J Neurosci. 2002;22(20):8951–60.
176. Caudle WM, Richardson JR, Wang MZ, Taylor TN, Guillot TS, McCormack AL, et al. Reduced vesicular storage of dopamine causes progressive nigrostriatal neurodegeneration. J Neurosci. 2007;27(30):8138–48. Epub 2007/07/27.
177. Guillot TS, Shepherd KR, Richardson JR, Wang MZ, Li Y, Emson PC, et al. Reduced vesicular storage of dopamine exacerbates methamphetamine-induced neurodegeneration and astrogliosis. J Neurochem. 2008;106(5):2205–17. Epub 2008/07/23.
178. Asanuma M, Miyazaki I, Diaz-Corrales FJ, Ogawa N. Quinone formation as dopaminergic neuron-specific oxidative stress in the pathogenesis of sporadic Parkinson's disease and neurotoxin-induced parkinsonism. Acta Med Okayama. 2004;58(5):221–33.
179. Zucca FA, Basso E, Cupaioli FA, Ferrari E, Sulzer D, Casella L, et al. Neuromelanin of the human substantia nigra: an update. Neurotox Res. 2014;25(1):13–23. Epub 2013/10/25.
180. Lévay G, Ye Q, Bodell WJ. Formation of DNA adducts and oxidative base damage by copper mediated oxidation of dopamine and 6-hydroxydopamine. Exp Neurol. 1997;146(2):570–4.
181. Cavalieri EL, Li KM, Balu N, Saeed M, Devanesan P, Higginbotham S, et al. Catechol orthoquinones: the electrophilic compounds that form depurinating DNA adducts and could initiate cancer and other diseases. Carcinogenesis. 2002;23(6):1071–7.
182. Zahid M, Saeed M, Yang L, Beseler C, Rogan E, Cavalieri EL. Formation of dopamine quinone-DNA adducts and their potential role in the etiology of Parkinson's disease. IUBMB Life. 2011;63(12):1087–93. Epub 2011/11/03.
183. Xu Y, Stokes AH, Roskoski RJ, Vrana KE. Dopamine, in the presence of tyrosinase, covalently modifies and inactivates tyrosine hydroxylase. J Neurosci Res. 1998;54(5):691–7.
184. LaVoie MJ, Ostaszewski BL, Weihofen A, Schlossmacher MG, Selkoe DJ. Dopamine covalently modifies and functionally inactivates parkin. Nat Med. 2005;11(11):1214–21. Epub 2005/10/18.
185. Conway KA, Rochet JC, Bieganski RM, Lansbury PTJ. Kinetic stabilization of the alpha-synuclein protofibril by a dopamine-alpha-synuclein adduct. Science. 2001;294(5545):1346–9. Epub 1346.
186. Van Laar VS, Mishizen AJ, Cascio M, Hastings TG. Proteomic identification of dopamine-conjugated proteins from isolated rat brain mitochondria and SH-SY5Y cells. Neurobiol Dis. 2009;34(3):487–500. Epub 2009/04/01.
187. Van Laar VS, Dukes AA, Cascio M, Hastings TG. Proteomic analysis of rat brain mitochondria following exposure to dopamine quinone: implications for Parkinson disease. Neurobiol Dis. 2008;29(3):477–89. Epub 2008/01/30.
188. Martinez-Vicente M, Talloczy Z, Kaushik S, Massey AC, Mazzulli J, Mosharov EV, et al. Dopamine-modified α-synuclein blocks chaperone-mediated autophagy. J Clin Invest. 2008;118(2):777–88.
189. Segura-Aguilar J, Paris I, Munoz P, Ferrari E, Zecca L, Zucca FA. Protective and toxic roles of dopamine in Parkinson's disease. J Neurochem. 2014;129(6):898–915. Epub 2014/02/20.
190. Sulzer D, Mosharov E, Talloczy Z, Zucca FA, Simon JD, Zecca L. Neuronal pigmented autophagic vacuoles: lipofuscin, neuromelanin, and ceroid as macroautophagic responses during aging and disease. J Neurochem. 2008;106(1):24–36. Epub 2008/04/04.
191. Mann DM, Yates PO. Possible role of neuromelanin in the pathogenesis of Parkinson's disease. Mech Ageing Dev. 1983;21(2):193–203.
192. Hirsch E, Graybiel AM, Agid YA. Melanized dopaminergic neurons are differentially susceptible to degeneration in Parkinson's disease. Nature. 1988;334(6180):345–8.

193. Sulzer D, Bogulavsky J, Larsen KE, Behr G, Karatekin E, Kleinman MH, et al. Neuromelanin biosynthesis is driven by excess cytosolic catecholamines not accumulated by synaptic vesicles. Proc Natl Acad Sci U S A. 2000;97(22):11869–74.
194. Karlsson O, Lindquist NG. Melanin affinity and its possible role in neurodegeneration. J Neural Transm. 2013;120(12):1623–30. Epub 2013/07/04.
195. Burke W. Neurotoxicity of MAO metabolites of catecholamine neurotransmitters: role in neurodegenerative diseases. Neurotoxicology. 2004;25(1–2):101–15.
196. Burke WJ, Li SW, Williams EA, Nonneman R, Zahm DS. 3,4-Dihydroxyphenylacetaldehyde is the toxic dopamine metabolite in vivo: implications for Parkinson's disease pathogenesis. Brain Res. 2003;989(2):205–13.
197. Galter D, Buervenich S, Carmine A, Anvret M, Olson L. ALDH1 mRNA: presence in human dopamine neurons and decreases in substantia nigra in Parkinson's disease and in the ventral tegmental area in schizophrenia. Neurobiol Dis. 2003;14(3):637–47.
198. Westerlund M, Galter D, Carmine A, Olson L. Tissue- and species-specific expression patterns of class I, III, and IV Adh and Aldh 1 mRNAs in rodent embryos. Cell Tissue Res. 2005;322(2):227–36. Epub 2005/07/28.
199. Marchitti SA, Deitrich RA, Vasiliou V. Neurotoxicity and metabolism of the catecholamine-derived 3,4-dihydroxyphenylacetaldehyde and 3,4-dihydroxyphenylglycolaldehyde: the role of aldehyde dehydrogenase. Pharmacol Rev. 2007;59(2):125–50. Epub 2007/03/24.
200. Liu G, Yu J, Ding J, Xie C, Sun L, Rudenko I, et al. Aldehyde dehydrogenase 1 defines and protects a nigrostriatal dopaminergic neuron subpopulation. J Clin Invest. 2014;124(7):3032–46.
201. Woodard CM, Campos BA, Kuo SH, Nirenberg MJ, Nestor MW, Zimmer M, et al. iPSC-derived dopamine neurons reveal differences between monozygotic twins discordant for Parkinson's disease. Cell Rep. 2014;9(4):1173–82. Epub 2014/12/03.
202. Olanow CW, Hauser RA, Jankovic J, Langston W, Lang A, Poewe W, et al. A randomized, double-blind, placebo-controlled, delayed start study to assess rasagiline as a disease modifying therapy in Parkinson's disease (the ADAGIO study): rationale, design, and baseline characteristics. Mov Disord. 2008;23(15):2194–201. Epub 2008/10/22.
203. Olanow CW, Rascol O, Hauser R, Feigin PD, Jankovic J, Lang A, et al. A double-blind, delayed-start trial of rasagiline in Parkinson's disease. N Engl J Med. 2009;361(13):1268–78.
204. Jankovic J, Berkovich E, Eyal E, Tolosa E. Symptomatic efficacy of rasagiline monotherapy in early Parkinson's disease: post-hoc analyses from the ADAGIO trial. Parkinsonism Relat Disord. 2014;20(6):640–3. Epub 2014/03/19.
205. Binda C, Mattevi A, Edmondson DE. Structural properties of human monoamine oxidases A and B. Int Rev Neurobiol. 2011;100:1–11. Epub 2011/10/06.
206. Hauptmann N, Grimsby J, Shih JC, Cadenas E. The metabolism of tyramine by monoamine oxidase A/B causes oxidative damage to mitochondrial DNA. Arch Biochem Biophys. 1996;335(2):295–304.
207. Cohen G, Farooqui R, Kesler N. Parkinson disease: a new link between monoamine oxidase and mitochondrial electron flow. Proc Natl Acad Sci U S A. 1997;94(10):4890–4.
208. Cohen G, Kesler N. Monoamine oxidase and mitochondrial respiration. J Neurochem. 1999;73(6):2310–5.
209. Mosharov EV, Larsen KE, Kanter E, Phillips KA, Wilson K, Schmitz Y, et al. Interplay between cytosolic dopamine, calcium, and alpha-synuclein causes selective death of substantia nigra neurons. Neuron. 2009;62(2):218–29. Epub 2009/05/05.
210. Fahn S. Does levodopa slow or hasten the rate of progression of Parkinson's disease? J Neurol. 2005;252 Suppl 4:IV37–42. Epub 2005/10/14.
211. Parkkinen L, O'Sullivan SS, Kuoppamäki M, Collins C, Kallis C, Holton JL, et al. Does levodopa accelerate the pathologic process in Parkinson disease brain? Neurology. 2011;77(15):1420–6.
212. Olanow CW. Levodopa: effect on cell death and the natural history of Parkinson's disease. Mov Disord. 2015;30(1):37–44. Epub 2014/12/17.

213. Bonifati V. Genetics of Parkinson's disease – state of the art, 2013. Parkinsonism Relat Disord. 2014;20:S23–8.
214. Thomas B, Beal MF. Parkinson's disease. Hum Mol Genet. 2007;16(2):R183–94. Epub 2007/10/04.
215. Giaime E, Yamaguchi H, Gautier CA, Kitada T, Shen J. Loss of DJ-1 does not affect mitochondrial respiration but increases ROS production and mitochondrial permeability transition pore opening. PLoS ONE. 2012;7(7), e40501. Epub 2012/07/14.
216. Krebiehl G, Ruckerbauer S, Burbulla LF, Kieper N, Maurer B, Waak J, et al. Reduced basal autophagy and impaired mitochondrial dynamics due to loss of Parkinson's disease-associated protein DJ-1. PLoS ONE. 2010;5(2), e9367. Epub 2010/02/27.
217. McLelland GL, Soubannier V, Chen CX, McBride HM, Fon EA. Parkin and PINK1 function in a vesicular trafficking pathway regulating mitochondrial quality control. EMBO J. 2014;33(4):282–95. Epub 2014/01/22.
218. Narendra DP, Jin SM, Tanaka A, Suen DF, Gautier CA, Shen J, et al. PINK1 is selectively stabilized on impaired mitochondria to activate Parkin. PLoS Biol. 2010;8(1), e1000298. Epub 2010/02/04.
219. Pickrell AM, Huang CH, Kennedy SR, Ordureau A, Sideris DP, Hoekstra JG, et al. Endogenous Parkin preserves dopaminergic substantia nigral neurons following mitochondrial DNA mutagenic stress. Neuron. 2015;87(2):371–81. Epub 2015/07/17.
220. Pickrell AM, Youle RJ. The roles of PINK1, Parkin, and mitochondrial fidelity in Parkinson's disease. Neuron. 2015;85(2):257–73. Epub 2015/01/23.
221. Shin JH, Ko HS, Kang H, Lee Y, Lee YI, Pletinkova O, et al. PARIS (ZNF746) repression of PGC-1alpha contributes to neurodegeneration in Parkinson's disease. Cell. 2011;144(5):689–702. Epub 2011/03/08.
222. Vives-Bauza C, Zhou C, Huang Y, Cui M, de Vries RL, Kim J, et al. PINK1-dependent recruitment of Parkin to mitochondria in mitophagy. Proc Natl Acad Sci U S A. 2010;107(1):378–83. Epub 2009/12/08.
223. Bishop MW, Chakraborty S, Matthews GA, Dougalis A, Wood NW, Festenstein R, et al. Hyperexcitable substantia nigra dopamine neurons in PINK1- and HtrA2/Omi-deficient mice. J Neurophysiol. 2010;104(6):3009–20. Epub 2010/10/12.
224. Spillantini MG, Crowther RA, Jakes R, Hasegawa M, Goedert M. alpha-Synuclein in filamentous inclusions of Lewy bodies from Parkinson's disease and dementia with lewy bodies. Proc Natl Acad Sci U S A. 1998;95(11):6469–73.
225. Spillantini MG, Schmidt ML, Lee VM, Trojanowski JQ, Jakes R, Goedert M. Alpha-synuclein in Lewy bodies. Nature. 1997;388(6645):839–40.
226. Pasternak B, Svanstrom H, Nielsen NM, Fugger L, Melbye M, Hviid A. Use of calcium channel blockers and Parkinson's disease. Am J Epidemiol. 2012;175(7):627–35. Epub 2012/03/06.
227. Becker C, Jick SS, Meier CR. Use of antihypertensives and the risk of Parkinson disease. Neurology. 2008;70(16 Pt2):1438–44.
228. Uchida S, Yamada S, Nagai K, Deguchi Y, Kimura R. Brain pharmacokinetics and in vivo receptor binding of 1,4-dihydropyridine calcium channel antagonists. Life Sci. 1997;61(21):2083–90.
229. Marras C, Gruneir A, Rochon P, Wang X, Anderson G, Brotchie J, et al. Dihydropyridine calcium channel blockers and the progression of Parkinsonism. Ann Neurol. 2012;71(3):362–9. Epub 2012/03/28.
230. Nielsen MS, Vorum H, Lindersson E, Jensen PH. Ca2+ binding to alpha-synuclein regulates ligand binding and oligomerization. J Biol Chem. 2001;276(25):22680–4. Epub 2001/04/20.
231. Dufty BM, Warner LR, Hou ST, Jiang SX, Gomez-Isla T, Leenhouts KM, et al. Calpain-cleavage of alpha-synuclein: connecting proteolytic processing to disease-linked aggregation. Am J Pathol. 2007;170(5):1725–38. Epub 2007/04/26.
232. Diepenbroek M, Casadei N, Esmer H, Saido TC, Takano J, Kahle PJ, et al. Overexpression of the calpain-specific inhibitor calpastatin reduces human alpha-Synuclein processing, aggregation and synaptic impairment in [A30P]alphaSyn transgenic mice. Hum Mol Genet. 2014;23(15):3975–89. Epub 2014/03/13.

233. Foehring RC, Zhang XF, Lee JC, Callaway JC. Endogenous calcium buffering capacity of substantia nigral dopamine neurons. J Neurophysiol. 2009;102(4):2326–33. Epub 2009/08/14.
234. Chan SL, Mattson MP. Caspase and calpain substrates: roles in synaptic plasticity and cell death. J Neurosci Res. 1999;58(1):167–90.
235. Goll DE, Thompson VF, Li H, Wei W, Cong J. The calpain system. Physiol Rev. 2003;83(3):731–801.
236. Goll DE, Thompson VF, Taylor RG, Zalewska T. Is calpain activity regulated by membranes and autolysis or by calcium and calpastatin? Bioessays. 1992;14(8):549–56.
237. Belal C, Ameli NJ, El Kommos A, Bezalel S, Al'Khafaji AM, Mughal MR, et al. The homocysteine-inducible endoplasmic reticulum (ER) stress protein Herp counteracts mutant alpha-synuclein-induced ER stress via the homeostatic regulation of ER-resident calcium release channel proteins. Hum Mol Genet. 2012;21(5):963–77. Epub 2011/11/03.
238. Hettiarachchi NT, Parker A, Dallas ML, Pennington K, Hung CC, Pearson HA, et al. alpha-Synuclein modulation of Ca2+ signaling in human neuroblastoma (SH-SY5Y) cells. J Neurochem. 2009;111(5):1192–201. Epub 2009/10/29.
239. Vekrellis K, Xilouri M, Emmanouilidou E, Rideout HJ, Stefanis L. Pathological roles of alpha-synuclein in neurological disorders. Lancet Neurol. 2011;10(11):1015–25.
240. Subramaniam M, Althof D, Gispert S, Schwenk J, Auburger G, Kulik A, et al. Mutant alpha-synuclein enhances firing frequencies in dopamine substantia nigra neurons by oxidative impairment of A-type potassium channels. J Neurosci. 2014;34(41):13586–99. Epub 2014/10/10.
241. Ryan SD, Dolatabadi N, Chan SF, Zhang X, Akhtar MW, Parker J, et al. Isogenic human iPSC Parkinson's model shows nitrosative stress-induced dysfunction in MEF2-PGC1alpha transcription. Cell. 2013;155(6):1351–64. Epub 2013/12/03.
242. Anderson DW, Bradbury KA, Schneider JS. Neuroprotection in Parkinson models varies with toxin administration protocol. Eur J Neurosci. 2006;24(11):3174–82. Epub 2006/12/13.
243. Cleren C, Yang L, Lorenzo B, Calingasan NY, Schomer A, Sireci A, et al. Therapeutic effects of coenzyme Q10 (CoQ10) and reduced CoQ10 in the MPTP model of Parkinsonism. J Neurochem. 2008;104(6):1613–21. Epub 2007/11/02.
244. Schintu N, Frau L, Ibba M, Caboni P, Garau A, Carboni E, et al. PPAR-gamma-mediated neuroprotection in a chronic mouse model of Parkinson's disease. Eur J Neurosci. 2009;29(5):954–63. Epub 2009/02/28.
245. Dawson TM, Ko HS, Dawson VL. Genetic animal models of Parkinson's disease. Neuron. 2010;66(5):646–61. Epub 2010/06/16.
246. Decressac M, Mattsson B, Lundblad M, Weikop P, Bjorklund A. Progressive neurodegenerative and behavioural changes induced by AAV-mediated overexpression of alpha-synuclein in midbrain dopamine neurons. Neurobiol Dis. 2012;45(3):939–53. Epub 2011/12/21.
247. Dong Z, Ferger B, Paterna JC, Vogel D, Furler S, Osinde M, et al. Dopamine-dependent neurodegeneration in rats induced by viral vector-mediated overexpression of the parkin target protein, CDCrel-1. Proc Natl Acad Sci U S A. 2003;100(21):12438–43. Epub 2003/10/08.
248. Dusonchet J, Bensadoun JC, Schneider BL, Aebischer P. Targeted overexpression of the parkin substrate Pael-R in the nigrostriatal system of adult rats to model Parkinson's disease. Neurobiol Dis. 2009;35(1):32–41. Epub 2009/04/08.
249. Dusonchet J, Kochubey O, Stafa K, Young Jr SM, Zufferey R, Moore DJ, et al. A rat model of progressive nigral neurodegeneration induced by the Parkinson's disease-associated G2019S mutation in LRRK2. J Neurosci. 2011;31(3):907–12. Epub 2011/01/21.
250. Low K, Aebischer P. Use of viral vectors to create animal models for Parkinson's disease. Neurobiol Dis. 2012;48(2):189–201. Epub 2012/01/10.
251. Yamada M, Iwatsubo T, Mizuno Y, Mochizuki H. Overexpression of alpha-synuclein in rat substantia nigra results in loss of dopaminergic neurons, phosphorylation of alpha-synuclein and activation of caspase-9: resemblance to pathogenetic changes in Parkinson's disease. J Neurochem. 2004;91(2):451–61. Epub 2004/09/28.
252. Johnson SJ, Wade-Martins R. A BACwards glance at neurodegeneration: molecular insights into disease from LRRK2, SNCA and MAPT BAC-transgenic mice. Biochem Soc Trans. 2011;39(4):862–7. Epub 2011/07/27.

253. Taylor TN, Potgieter D, Anwar S, Senior SL, Janezic S, Threlfell S, et al. Region-specific deficits in dopamine, but not norepinephrine, signaling in a novel A30P alpha-synuclein BAC transgenic mouse. Neurobiol Dis. 2014;62:193–207. Epub 2013/10/15.
254. Ritz B, Rhodes SL, Qian L, Schernhammer E, Olsen JH, Friis S. L-type calcium channel blockers and Parkinson disease in Denmark. Ann Neurol. 2010;67(5):600–6. Epub 2010/05/04.
255. Kupsch A, Sautter J, Schwarz J, Riederer P, Gerlach M, Oertel WH. 1-Methyl-4-phenyl-1,2,3,6-tetrahydropyridine-induced neurotoxicity in non-human primates is antagonized by pretreatment with nimodipine at the nigral, but not at the striatal level. Brain Res. 1996;741(1–2):185–96.
256. Eisenberg MJ, Brox A, Bestawros AN. Calcium channel blockers: an update. Am J Med. 2004;116(1):35–43.
257. Mannhold R, Rekker RF, Sonntag C, ter Laak AM, Dross K, Polymeropoulos EE. Comparative evaluation of the predictive power of calculation procedures for molecular lipophilicity. J Pharm Sci. 1995;84(12):1410–9.
258. Striessnig J, Koschak A, Sinnegger-Brauns MJ, Hetzenauer A, Nguyen NK, Busquet P, et al. Role of voltage-gated L-type Ca2+ channel isoforms for brain function. Biochem Soc Trans. 2006;34(Pt 5):903–9.
259. Simuni T, Borushko E, Avram MJ, Miskevics S, Martel A, Zadikoff C, et al. Tolerability of isradipine in early Parkinson's disease: a pilot dose escalation study. Mov Disord. 2010;25(16):2863–6.
260. Fitton A, Benfield P. Isradipine. A review of its pharmacodynamic and pharmacokinetic properties, and therapeutic use in cardiovascular disease. Drugs. 1990;40(1):31–74.
261. Ilijic E, Guzman JN, Surmeier DJ. The L-type channel antagonist isradipine is neuroprotective in a mouse model of Parkinson's disease. Neurobiol Dis. 2011;43(2):364–71. Epub 2011/04/26.
262. Fahn S. Description of Parkinson's disease as a clinical syndrome. Ann N Y Acad Sci. 2003;991:1–14.

Chapter 7
Mitochondrial Dynamics and Neurodegeneration

Arnaud Mourier

Abstract In vivo, mitochondria display a high degree of connectivity and mobility. Within the cell, mitochondrial fusion and fission machineries tightly control the dynamics and distribution of the mitochondrial network. Due to their key energetic role, the localization of mitochondria at intracellular sites of high-energy demand is crucial to maintain cell energy metabolism. Neurons are metabolically active cells with high-energy demands at locations distant from the cell body (see Chaps. 8 and 9). Consequently, they are particularly dependent on mitochondrial distribution and function. Accordingly, new evidence identifies defective mitochondrial dynamics as a central pathological event underpinning a number of early and late-onset neurodegenerative disorders. Mutations in genes encoding proteins playing central roles in mitochondrial dynamics and functions have been identified in patients with peripheral neuropathies such as Charcot-Marie-Tooth (CMT) and dominant inherited optic atrophy. Moreover, defects of mitochondrial dynamics have recently been associated with common neurodegenerative diseases such as Parkinson's, Alzheimer's, and Huntington's diseases. Understanding the regulation of mitochondrial dynamics in neurons may open new avenues for the development of therapies in neurodegenerative diseases.

Keywords Neurodegenerative diseases • Mitochondria • Mitochondrial dynamics Bioenergetics

7.1 Introduction

Mitochondria are organelles originally observed in spermatocytes more than a century ago [1]. The name *mitochondrion* is based on their microscopic appearance and is derived from the Greek language for the words thread (*mitos*) and granules (*chondros*). Mitochondria are double membrane bound: the inner membrane delimits the

A. Mourier, PhD
Mitochondrial Biology, Max Planck Institute for Biology of Ageing, Cologne, Germany
e-mail: arnaud.mourier@age.mpg.de

A.K. Reeve et al. (eds.), *Mitochondrial Dysfunction in Neurodegenerative Disorders*, DOI 10.1007/978-3-319-28637-2_7

matrix and intermembrane space (IMS), whereas the outer membrane separates the IMS from the cytosol. The inner mitochondrial membrane has the highest density of protein in the cell and can be dissociated in two domains: the boundary region, which constitutes flattened membranes that are in close proximity to the outer membrane, and the cristae membranes, which are lamellar invaginations with highly curved edges. The cristae invaginations house the oxidative phosphorylation system (OXPHOS) [2–4]. This system is composed of two functional entities, i.e., the respiratory/electron transport chain (RC) and the phosphorylation system, which includes the ATP synthase and membrane carriers, such as the ATP/ADP carrier (ANT) and the phosphate carrier (PiC). The RC is historically defined as consisting of mobile electron carriers: coenzyme Q and cytochrome *c* and four complexes, denoted complex I–IV, which perform substrate oxidation to drive proton extrusion from the mitochondrial matrix to the IMS. The proton electrochemical potential difference across the inner membrane (ΔP) is then used by the ATP synthase to drive ATP synthesis thus coupling proton transport to ATP production [5] (see Chap. 1 for details).

Interestingly, complexes I, III, and IV of the RC and ATP synthase are under dual genetic control. The mitochondrial genome (mtDNA) only encodes 13 proteins that are all components of the OXPHOS system, and nuclear genes encode the remaining mitochondrial proteins. The nuclear genome encodes the mitochondrial proteome required for the maintenance and expression of mtDNA [6], protein synthesis [7], import and degradation [8, 9], iron-sulfur cluster synthesis [10], citric acid and urea cycles, fatty acid oxidation, and additional metabolic pathways.

Mitochondria form a dynamic network inside the cell, and specialized transport machineries ensure their mobility and proper subcellular localizations. Due to their key energetic role, mitochondria are often positioned at intracellular sites of high-energy demand. In the muscle, mitochondria are embedded between myofibrils that consume ATP during contraction. Likewise, in neurons mitochondria are transported to and accumulate in synapses to provide the energy required to maintain and regulate neurotransmission. Thus, proper control of mitochondrial subcellular localization and network morphology is an absolute requisite to maintain energy homeostasis and cell functions. This chapter will review recent findings on the energetic relevance of mitochondrial dynamics and its implication in neurodegenerative disorders.

7.2 Mitochondrial Dynamics

"Mitochondrial dynamics" describes the continuous changes in the position, size, and shape of mitochondria within cells. In eukaryotic cells, mitochondria are arranged in a wide variety of shapes, ranging from long interconnected tubules to individual spheres [11–13]. Mitochondrial morphology is highly plastic and dynamic, and mitochondria can travel long distances on the cytoskeletal track. Depending on the cell type, mitochondrial mobility can be mediated through either

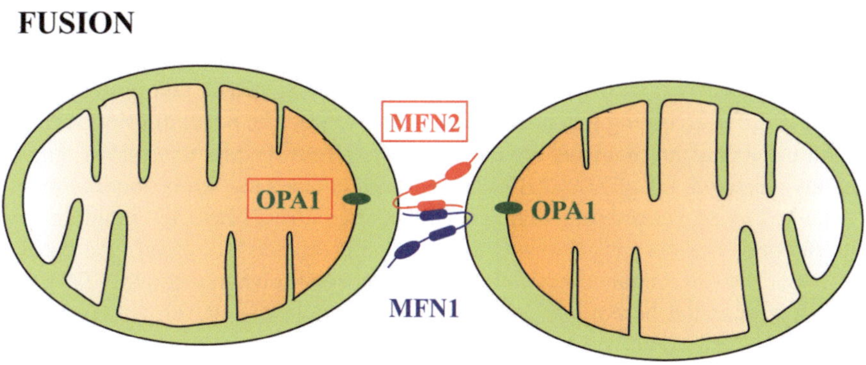

Fig. 7.1 The mammalian machineries involved in mitochondrial fusion. Heterotypic interaction between the mitochondrial outer membrane proteins MFN1 and MFN2 is depicted. OPA1 is located in the intermembrane space and associated with the inner mitochondrial membrane. A *red square* highlights protein implicated in diseases

the microtubules or the actin cytoskeleton [14, 15]. In neurons, mitochondrial transport occupies a key role in the delivery and renewal of axonal and synaptic mitochondria [16]. Interestingly, while the cytoskeleton is important to ensure proper mitochondrial intracellular distribution and transport, it is not required to maintain mitochondrial network morphology [17–20]. Mitochondrial network morphology is in fact the result of balanced fusion and fission controlled by protein members of the dynamin-related protein (DRP) family. Recent microscopic, structural, and biochemical studies have led to the characterization of the core machinery of mitochondrial fusion and fission.

7.2.1 Mitochondrial Fusion

The identification of fuzzy onion (FZO) as an essential protein mediating mitochondrial fusion events occurring during spermatogenesis in *drosophila* drove rapid advances in our knowledge of the mitochondrial dynamic machinery [21]. The characterization of this evolutionary conserved GTPase protein in budding yeast (FZO1) showed that this outer mitochondrial membrane protein functions in mitochondrial fusion [22, 23]. Genetic screens, performed in yeast, led to the identification of the protein machineries involved in mitochondrial fusion and fission [24]. This approach revealed that the core of the yeast mitochondrial fusion machinery is composed of three proteins. FZO1 and MGM1 (Mitochondrial Genome Maintenance 1) [25–27], respectively, control fusion of the outer and inner mitochondrial membrane, whereas UGO1 (UGO is Japanese for fusion) is proposed to be a two membrane-spanning protein mediating the interaction between FZO1 and MGM1 [28–30]. In contrast to UGO1, both FZO1 and MGM1 present mammalian homologues known as mitofusin 1 and mitofusin 2 (MFN1 and MFN2) [31] and OPA1 (optic atrophy 1) [32, 33], respectively (Fig. 7.1). MFN1 and MFN2 show 81 % similarity to each other and are 52 % similar to *drosophila* FZO [19]. Both MFNs are ubiquitously expressed in mammals, although their mRNA and protein levels strongly differ according to the tissues [17, 31, 34, 35]. The posttranscriptional and posttranslational mechanisms regulating MFNs tissue-specific expression remain largely unknown. Like FZO1, MFNs are anchored to the outer mitochondrial membrane by two transmembrane segments and contain one GTPase and multiple predicted coiled-coil-forming domains which are exposed and face toward the cytosol [17, 36]. The C terminal coiled-coil domain has been shown to mediate MFN1 and MFN2 antiparallel homotypic and heterotypic complexes [19, 37]. Even though both MFN1 and MFN2 are expressed in mouse embryonic fibroblasts, each protein is essential to maintain mitochondrial network morphology [19]. However, despite their high similarity, MFN1 and MFN2 exhibit different GTPase and membrane-tethering capacities [38]. These functional disparities could explain why the mitochondrial network morphology aberration in *Mfn2* knockout MEFs can be efficiently rescued by MFN1 overexpression, whereas, overexpression of MFN2 only mildly rescues the diffracted mitochondrial network of *Mfn1* knockout MEFs [19]. Interestingly, activity or degradation of FZO1/MFNs proteins is highly regulated by phosphorylation and ubiquitination at distinct lysine residues [39–42].

In mammals, OPA1 is the main actor controlling the fusion of the mitochondrial inner membrane. MGM1/OPA1 presents multiple isoforms that are located in the IMS or associated with the inner membrane [25–27, 43]. The proteolytic processing of OPA1 generates so-called non-cleaved or long OPA1 (L-OPA1) and cleaved or short OPA1 (S-OPA1) isoforms. The regulated cleavage of OPA1 isoforms by OMA1 (metallopeptidase that exerts activities overlapping with the m-AAA protease) [44, 45] or YME1L (mammalian orthologue of the yeast *Yme1*) [46–48] results in the loss of the transmembrane domain of the protein and controls the role of OPA1 in mitochondrial fusion. Furthermore, the alternative splicing of OPA1 pre-mRNA introduces additional complexity and yields a total of eight isoforms presenting one or two processing sites and expressed in a tissue-specific manner [49]. However, in contrast with the proteolytic processing, which is common to both orthologues OPA1 and MGM1, the alternative splicing is uniquely involved in the generation of mammalian OPA1 isoforms.

7.2.2 Mitochondrial Fission

Yeast genetic screens for an extragenic suppressor of fusion mutants led to the identification of key regulators of mitochondrial fission. The key component of the fission machinery DNM1 (dynamin-1) is a cytosolic protein which can be recruited into punctuate structures on the outer mitochondrial membrane [50, 51]. According to the most recent model, this DNM1p recruitment to the outer mitochondrial membrane is mediated by FIS1 (mitochondrial fission protein 1) [52] and causes membrane constriction through its interaction with adaptor proteins MDV1 (mitochondrial

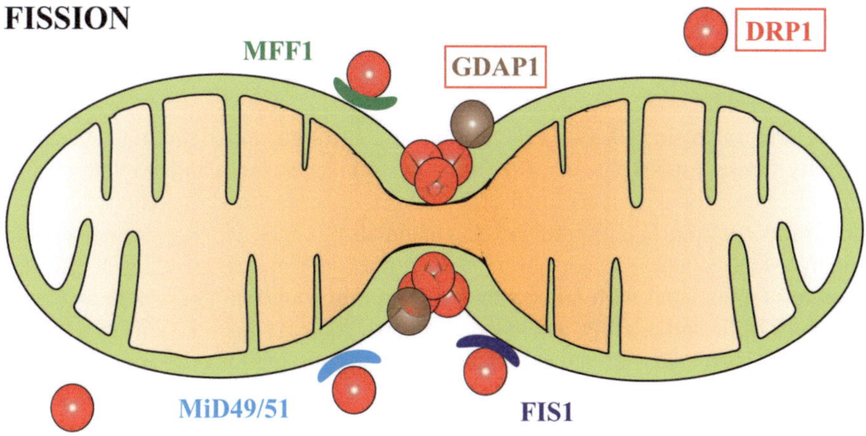

Fig. 7.2 Mammalian mitochondrion under fission. The scheme illustrates DRP1 recruitment to the outer membrane constriction sites by adaptor proteins: FIS1, MFF1, MiD49/51. GDAP1 is located at the mitochondrial outer membrane to mediate mitochondrial fission. A *red square* highlights protein implicated in diseases

division 1) [53–58] and CAF4 (CCR4-associated factor 4) [59, 60]. After its recruitment, DNM1 forms extended spirals [61], which undergo conformational change upon GTP hydrolysis leading to the constriction and division of mitochondria [62].

Extension of these studies to mammalian cells led to the identification of dynamin-related protein 1 (DRP1), also called DLP1 (dynamin-like protein 1) in humans, and FIS1 as components of the mammalian fission machinery [63–65] (Fig. 7.2). However, no orthologues of MDV1 and CAF4 were found in mammals; instead, a growing body of evidence indicates that the mitochondrial fission factor, MFF [66], and the mitochondrial elongation factors, MID49 and MID51 (mitochondrial dynamic proteins of 49 and 51 kDa) [67–71], are further components of the fission machinery. In addition, GDAP1 (ganglioside-induced differentiation-associated protein 1) is a tail-anchored protein of the mitochondrial outer membrane that regulates mitochondrial fission [72, 73]. Whereas GDAP1 recessive mutations are associated with decreased mitochondrial fission activity, dominant mutations result in impairment of mitochondrial fusion [74]. Phylogenetic and structural analyses suggest that GDAP1 belongs to a subfamily of glutathione-S-transferases (GSTs). However, no functional GST activity associated to this protein has been found so far [75]. Unlike other proteins involved in mitochondrial fusion and fission containing GTPase and dynamin domains, GDAP1 sequence analysis does not suggest any involvement in mitochondrial dynamics.

Mitochondrial fission is highly regulated and controlled. For instance, DRP1 undergoes several posttranslational modifications such as phosphorylation [76–79], S-nitrosylation [80, 81], ubiquitination [82–84], and sumoylation [85–87]. These modifications control the activity and subcellular localization of DRP1 [19]. Furthermore, in contrast to mitochondrial fusion, the core of the mitochondrial fission machinery plays a similar role in peroxisomes [65, 88, 89].

7.3 What Are the Physiological and Bioenergetic Roles of Mitochondrial Dynamics?

Mitochondrial fusion plays a critical role in controlling mitochondrial OXPHOS activity through maintenance of the mitochondrial genome. The complete loss of mitochondrial fusion is associated with a loss of the mitochondrial genome in yeast and with a partial loss of mtDNA in mammals [22, 23, 90–92]. Moreover, the mtDNA maintenance defect observed in mitochondrial fusion deficient skeletal muscle is associated with an accumulation of mitochondrial point mutations and deletions [92]. Remarkably, where emerging data indicate that the majority of mitochondrial fission sites are located in close proximity to mtDNA molecules [93, 94] and endoplasmic reticulum contact sites [95, 96], the loss of mitochondrial fission has no deleterious effect on mtDNA levels [51, 97, 98]. Understanding the role of mitochondrial dynamics in the maintenance and protection of mtDNA continues to attract great scientific interest. Intriguingly, partial impairment of mitochondrial fusion caused by the loss of MFN2 in different mammalian tissues or cultured cells affects mitochondrial bioenergetics without drastically affecting OXPHOS subunits or mtDNA levels [99–101].

Despite their high similarity and their common role in mitochondrial fusion through their physical interaction, MFN1 and MFN2 seem to functionally differ. Ubiquitous knockout of the *Mfn1* or *Mfn2* genes results in embryo lethality in mid-gestation [19], due to placental dysfunction [19, 102]. Remarkably, by using a conditional knockout allele in conjunction with cre-recombinase expression only in the embryo, *Mfn1* or *Mfn2* knockout mice are born alive [102]. Mice ubiquitously lacking MFN1 are apparently healthy, whereas loss of MFN2 causes mouse lethality in the early postnatal period and triggers cerebellar atrophy causing severe defects in movement and balance [102]. Despite its well-established role in mitochondrial fusion, a growing body of evidence suggests that MFN2 has additional functions, such as tethering mitochondria with the endoplasmic reticulum (ER) [103], lipid droplets [104], and with the Miro/kinesin system [105, 106]. However, the role of MFN2 in mediating ER-mitochondrial tethering has been recently revised [107, 108]. The different functions of MFN2 are supported by in vivo studies in mice, showing that MFN2 is required for normal glucose homeostasis [99], steroidogenesis [104, 109], is essential for cerebellum development and function [102], and for the axonal projections of dopaminergic neurons [110, 111]. Interestingly, the loss of MFN2 in Purkinje or dopaminergic neurons is associated with a loss of complex IV activity [102, 110]. In contrast, loss of MFN2 in the liver, skeletal muscle, and heart is associated with mild OXPHOS dysfunctions, which are not due to the loss of OXPHOS protein complex levels or activities [99, 112]. Similarly, loss of function of MFN2 in mouse and human fibroblasts cultivated with glucose and high serum shows no gross bioenergetic defects [19, 108, 113]. Interestingly, respiratory chain deficiency observed in MEFs cultivated under low-serum conditions and in mouse heart lacking MFN2 is caused by a deficiency of the terpenoid synthesis pathway affecting the levels of coenzyme Q [101]. This interesting observation demonstrates that bioenergetic defects associated with loss of mitochondrial fusion are not only related to mitochondrial genetic impairments but could originate from anabolic pathway dysfunctions.

Mitochondrial bioenergetics and mitochondrial fusion are heavily interdependent since mitochondrial fusion requires the membrane potential generated across the inner mitochondrial membrane by the respiratory chain [18, 114–116]. Moreover, bioenergetic defects impair mitochondrial dynamics [117] through the proteolytic processing of OPA1 [118]. The strong connection between mitochondrial fusion and bioenergetics could explain the mitochondrial network remodeling observed when cells are metabolizing non-fermentable [119, 120] or high-energy demanding carbon sources [121, 122]. In addition, downregulation of DRP1 activity under starvation has been shown to increase mitochondrial fusion to prevent the degradation of mitochondria through autophagy [123, 124]. In contrast, DRP1 upregulation [77, 85] and MFN2 downregulation [39] under apoptotic conditions promote mitochondrial fragmentation and the following release of proapoptotic factors from mitochondria mediated by BAK [BCL2-antagonist/killer] and BAX [BCL2-associated protein X] [125].

7.4 Mitochondrial Dynamics and Neurodegeneration

Mitochondrial dynamics hold a central role in maintaining proper mitochondrial distribution and function. Therefore, it is not surprising that alterations in mitochondrial fusion and fission significantly impair neuronal functions. The importance of mitochondrial fusion and fission proteins in the biology of neurons is underscored by the occurrence of several neurodegenerative diseases that result from mutation of these genes. For example, mutations in *Mfn2* are associated with Charcot-Marie-Tooth disease type 2A, and *Opa*1 mutations cause autosomal dominant atrophy characterized by progressive degeneration of the optic nerve.

7.4.1 Charcot-Marie-Tooth Neuropathy

Charcot-Marie-Tooth [CMT] diseases represent a group of clinically and genetically heterogeneous inherited neuropathies affecting motor and/or sensory neurons. With a prevalence between 1 and 8 cases per 10,000 people, CMT is the most prevalent inherited neuropathy [126–128]. CMT type 2A accounts for 22 % of autosomal dominant neuropathies. Typical clinical symptoms of CMT2A are progressive distal limb muscle weakness and/or atrophy, stepping gait, distal sensory loss, and mobility impairment, which can lead to wheelchair dependency. In addition, optic atrophy can be associated with CMT2A presenting an unusually severe phenotype with an early age of onset [129]. However, while affecting a highly specific set of neurons, i.e., neurons with the longest axons [peripheral sensory and motor neurons], the clinical and electrophysiological phenotypes of CMT2A are very diverse. A major step in the comprehension of the disease was made when several studies identified pathogenic mutations in the *Mfn2* gene [129–131], and to date about 60 different *Mfn*2 mutations have been identified in CMT2A patients [132]. Remarkably, homozygous expression of mutated $MFN2^{T105M}$ in mouse motor neurons recapitulates the key clinical signs of CMT2A. As previously observed in CMT2A neurons [133, 134], affected motor neurons exhibit improper mitochondrial distribution [135]. However, since MFN2 is important for the maintenance of mitochondrial bioenergetics and mitochondrial transport in different types of neuron [102, 110], the molecular role of MFN2 in axonal integrity maintenance remains to be elucidated.

MFN2 is not the only protein involved in mitochondrial dynamics to be associated with the Charcot-Marie-Tooth disease. Unlike other Charcot-Marie-Tooth disease-linked genes, the various GDAP1-associated mutations are associated with demyelinating [136], axonal, or mixed forms of CMT disease with recessive or dominant modes of inheritance, showing a wide range of severity and onset of disease [137–139]. CMT diseases caused by autosomal recessive GDAP1 mutations are generally severe and lead to aggressive disorders appearing during early childhood. The rapid progression of these disorders leads to functional disability. However, the phenotypic presentations of patients carrying GDAP1 mutations are heterogeneous. In most cases patients can present with vocal cord paresis, diaphragmatic paralysis, and facial weakness. In contrast, dominant inherited GDAP1 mutations are less severe and associated with a later onset. Interestingly, GDAP1 knockout mice present with an age-related hypomyelinating peripheral neuropathy. Furthermore, loss of GDAP1 impairs mitochondrial morphology and axonal transport in peripheral neurons [140, 141].

7.4.2 Autosomal Dominant Optic Atrophy

Autosomal dominant optic atrophy (ADOA) is a hereditary optic neuropathy characterized by a bilateral degeneration of optic nerves causing symmetrical visual loss, typically starting during the first decade of life. The disease primarily affects the retinal ganglion cells (RGCs) and their axons which form the optic nerve [142]. ADOA is mainly linked to *OPA1* mutations [32, 49] and is a relatively common form of inherited optic neuropathy; its prevalence is around 3 cases in 100,000 people [143, 144]. In 2007, two independent groups generated mice models expressing a different mutated version of *OPA1*. These studies showed that OPA1 was required during early mouse embryonic development. Interestingly, heterozygous *OPA1* mutants are viable, but exhibit an age-dependent loss of RGCs that eventually progresses to a severe degeneration of ganglion cells and nerve fiber layer [145, 146]. These works demonstrate that the phenotype was not caused by a specific OPA1 proteolytic processing defect, but was associated with a concerted decrease of both L-OPA1 and S-OPA1 levels. In addition, loss of OPA1 was associated with abnormal mitochondrial cristae morphology, mitochondrial swelling, and a reduced number of optic nerve axons. These mouse models showed that mitochondrial defects caused by the loss of OPA1 affect high-energy glutamatergic synapses which lead to dendritic degeneration of ganglion cells [147].

7.4.3 Neuropathy Linked to DLP1

Human diseases linked to DRP1/DLP1 mutations are extremely rare. In 2007, a de novo mutation in one DLP1 allele was identified in a neonate patient presenting microcephaly, abnormal brain development, optic atrophy, and elevated plasma concentration of lactic acid as well as very long-chain fatty acids causing their death at 37 days of age [148]. Despite the drastic elongation of the mitochondrial and peroxisomal networks, OXPHOS activities were not affected in skin fibroblasts and skeletal muscle biopsies. This mutation was later reported to prevent higher-order assembly of DLP1, thus precluding organelle fission [149]. Interestingly, independent mouse knockouts showed that DRP1 was required for embryonic and brain development. Neuron-specific *Drp1* knockout presents aberrant mitochondrial morphology and distribution preventing proper dendritic and axonal development [98, 150]. Loss of DRP1 in dopaminergic neurons affects mitochondrial mobility and depletes axonal mitochondria. These mitochondrial defects cause degeneration of synaptic terminals and cell loss [151].

The pathological spectrum associated with disturbed mitochondrial fission has recently expanded to include Parkinson's, Huntington's, and Alzheimer's diseases. Pathogenic mutations involved in Parkinson's and Huntington's diseases have been associated with an increased recruitment of DRP1 to mitochondria resulting in excessive mitochondrial fragmentation [152, 153]. Furthermore, DRP1 has also been shown to interact with a Parkinson disease-related protein, LRRK2 (leucine-rich repeat kinase 2) [154]. This interaction enhances mitochondrial translocation of DRP1 to promote mitochondrial fragmentation. In the case of Alzheimer's disease, mitochondrial fragmentation is progressively increased during the progression of the disease both in patients and in transgenic mouse models [155–157]. However, the molecular mechanism involved is still unclear. Interestingly, it has been reported that DRP1 posttranslational modifications such as S-nitrosylation and phosphorylation hold a crucial role in Alzheimer's disease [80, 158].

7.5 Conclusion

Overall, it is clear that perturbations in mitochondrial dynamics and activity are directly or indirectly involved in neurodegenerative diseases. However, although in vitro studies have provided crucial information regarding mutations, structure, and molecular response, almost nothing is known about the in vivo regulation of mitochondrial dynamics in healthy adult axons or synapses. Furthermore, it is becoming increasingly clear that multiple factors can affect mitochondrial dynamics, and the strong interdependence between mitochondrial activity, dynamics, and transport introduces additional complexity. A future challenge will be to unravel the molecular nature of these interactions. To this end, availability of established disease models will allow rapid progress in our understanding of the physiological regulation of mitochondrial dynamics and will open promising new avenues to test both in vitro and in vivo efficacy of potential new agents to find treatment options for these progressive neurodegenerative diseases.

References

1. Benda C. Ueber die Spermatogenese der Vertebraten und höherer Evertebraten, II. Theil: Die Histiogenese der Spermien. Arch Anat Physiol. 1989;73:393–8.
2. Gilkerson RW, Selker JML, Capaldi RA. The cristae membrane of mitochondria is the principal site of oxidative phosphorylation. FEBS Lett. 2003;546(2–3):355–8.
3. Appelhans T, Richter CP, Wilkens V, Hess ST, Piehler J, Busch KB. Nanoscale organization of mitochondrial microcompartments revealed by combining tracking and localization microscopy. Nano Lett. 2012;12(2):610–6.
4. Busch KB, Deckers-Hebestreit G, Hanke GT, Mulkidjanian AY. Dynamics of bioenergetic microcompartments. Biol Chem. 2013;394(2):163–88.
5. Mitchell P. Coupling of phosphorylation to electron and hydrogen transfer by a chemi-osmotic type of mechanism. Nature. 1961;191:144–8.
6. Falkenberg M, Larsson N-G, Gustafsson CM. DNA replication and transcription in mammalian mitochondria. Annu Rev Biochem. 2007;76:679–99.
7. Rorbach J, Soleimanpour-Lichaei R, Lightowlers RN, Chrzanowska-Lightowlers ZMA. How do mammalian mitochondria synthesize proteins? Biochem Soc Trans. 2007;35(Pt 5):1290–1.
8. Bolender N, Sickmann A, Wagner R, Meisinger C, Pfanner N. Multiple pathways for sorting mitochondrial precursor proteins. EMBO Rep. 2008;9(1):42–9.
9. Tatsuta T, Langer T. Quality control of mitochondria: protection against neurodegeneration and ageing. EMBO J. 2008;27(2):306–14.
10. Lill R, Mühlenhoff U. Maturation of iron-sulfur proteins in eukaryotes: mechanisms, connected processes, and diseases. Annu Rev Biochem. 2008;77:669–700.
11. Koning AJ, Lum PY, Williams JM, Wright R. DiOC6 staining reveals organelle structure and dynamics in living yeast cells. Cell Motil Cytoskeleton. 1993;25(2):111–28.
12. Bereiter-Hahn J, Vöth M. Dynamics of mitochondria in living cells: shape changes, dislocations, fusion, and fission of mitochondria. Microsc Res Tech. 1994;27(3):198–219.
13. Chen H, Chan DC. Mitochondrial dynamics in mammals. Curr Top Dev Biol. 2004;59:119–44.
14. Morris RL, Hollenbeck PJ. Axonal transport of mitochondria along microtubules and F-actin in living vertebrate neurons. J Cell Biol. 1995;131(5):1315–26.
15. Grafstein B, Forman DS. Intracellular transport in neurons. Physiol Rev. 1980;60(4):1167–283.
16. Hollenbeck PJ, Saxton WM. The axonal transport of mitochondria. J Cell Sci. 2005;118(Pt 23):5411–9.
17. Rojo M, Legros F, Chateau D, Lombès A. Membrane topology and mitochondrial targeting of mitofusins, ubiquitous mammalian homologs of the transmembrane GTPase Fzo. J Cell Sci. 2002;115(Pt 8):1663–74.
18. Legros F, Lombès A, Frachon P, Rojo M. Mitochondrial fusion in human cells is efficient, requires the inner membrane potential, and is mediated by mitofusins. Mol Biol Cell. 2002;13(12):4343–54.

19. Chen H, Detmer SA, Ewald AJ, Griffin EE, Fraser SE, Chan DC. Mitofusins Mfn1 and Mfn2 coordinately regulate mitochondrial fusion and are essential for embryonic development. J Cell Biol. 2003;160(2):189–200.
20. Mattenberger Y, James DI, Martinou JC. Fusion of mitochondria in mammalian cells is dependent on the mitochondrial inner membrane potential and independent of microtubules or actin. FEBS Lett. 2003;538(1–3):53–9.
21. Hales KG, Fuller MT. Developmentally regulated mitochondrial fusion mediated by a conserved, novel, predicted GTPase. Cell. 1997;90(1):121–9.
22. Hermann GJ, Thatcher JW, Mills JP, Hales KG, Fuller MT, Nunnari J, et al. Mitochondrial fusion in yeast requires the transmembrane GTPase Fzo1p. J Cell Biol. 1998;143(2):359–73.
23. Rapaport D, Brunner M, Neupert W, Westermann B. Fzo1p is a mitochondrial outer membrane protein essential for the biogenesis of functional mitochondria in Saccharomyces cerevisiae. J Biol Chem. 1998;273(32):20150–5.
24. Okamoto K, Shaw JM. Mitochondrial morphology and dynamics in yeast and multicellular eukaryotes. Annu Rev Genet. 2005;39:503–36.
25. Wong ED, Wagner JA, Gorsich SW, McCaffery JM, Shaw JM, Nunnari J. The dynamin-related GTPase, Mgm1p, is an intermembrane space protein required for maintenance of fusion competent mitochondria. J Cell Biol. 2000;151(2):341–52.
26. Herlan M, Vogel F, Bornhövd C, Neupert W, Reichert AS. Processing of Mgm1 by the rhomboid-type protease Pcp1 is required for maintenance of mitochondrial morphology and of mitochondrial DNA. J Biol Chem. 2003;278(30):27781–8.
27. Griparic L, van der Wel NN, Orozco IJ, Peters PJ, van der Bliek AM. Loss of the intermembrane space protein Mgm1/OPA1 induces swelling and localized constrictions along the lengths of mitochondria. J Biol Chem. 2004;279(18):18792–8.
28. Sesaki H, Jensen RE. UGO1 encodes an outer membrane protein required for mitochondrial fusion. J Cell Biol. 2001;152(6):1123–34.
29. Sesaki H, Jensen RE. Ugo1p links the Fzo1p and Mgm1p GTPases for mitochondrial fusion. J Biol Chem. 2004;279(27):28298–303.
30. Coonrod EM, Karren MA, Shaw JM. Ugo1p is a multipass transmembrane protein with a single carrier domain required for mitochondrial fusion. Traffic. 2007;8(5):500–11.
31. Eura Y, Ishihara N, Yokota S, Mihara K. Two mitofusin proteins, mammalian homologues of FZO, with distinct functions are both required for mitochondrial fusion. J Biochem. 2003;134(3):333–44.
32. Delettre C, Lenaers G, Griffoin JM, Gigarel N, Lorenzo C, Belenguer P, et al. Nuclear gene OPA1, encoding a mitochondrial dynamin-related protein, is mutated in dominant optic atrophy. Nat Genet. 2000;26(2):207–10.
33. Alexander C, Votruba M, Pesch UE, Thiselton DL, Mayer S, Moore A, et al. OPA1, encoding a dynamin-related GTPase, is mutated in autosomal dominant optic atrophy linked to chromosome 3q28. Nat Genet. 2000;26(2):211–5.
34. Santel A, Frank S, Gaume B, Herrler M, Youle RJ, Fuller MT. Mitofusin-1 protein is a generally expressed mediator of mitochondrial fusion in mammalian cells. J Cell Sci. 2003;116(Pt 13):2763–74.
35. Kawalec M, Zabłocka B, Kabzińska D, Neska J, Beręsewicz M. Mitofusin 2 expression dominates over mitofusin 1 exclusively in mouse dorsal root ganglia – a possible explanation for peripheral nervous system involvement in Charcot-Marie-Tooth 2A. Folia Neuropathol. 2014;52(4):436–42.
36. Fritz S, Rapaport D, Klanner E, Neupert W, Westermann B. Connection of the mitochondrial outer and inner membranes by Fzo1 is critical for organellar fusion. J Cell Biol. 2001;152(4):683–92.
37. Koshiba T, Detmer SA, Kaiser JT, Chen H, McCaffery JM, Chan DC. Structural basis of mitochondrial tethering by mitofusin complexes. Science. 2004;305(5685):858–62.
38. Ishihara N, Eura Y, Mihara K. Mitofusin 1 and 2 play distinct roles in mitochondrial fusion reactions via GTPase activity. J Cell Sci. 2004;117(Pt 26):6535–46.

39. Leboucher GP, Tsai YC, Yang M, Shaw KC, Zhou M, Veenstra TD, et al. Stress-induced phosphorylation and proteasomal degradation of mitofusin 2 facilitates mitochondrial fragmentation and apoptosis. Mol Cell. 2012;47(4):547–57.
40. Anton F, Dittmar G, Langer T, Escobar-Henriques M. Two deubiquitylases Act on mitofusin and regulate mitochondrial fusion along independent pathways. Mol Cell. 2013;49(3):487–98.
41. Cohen MMJ, Leboucher GP, Livnat-Levanon N, Glickman MH, Weissman AM. Ubiquitin-proteasome-dependent degradation of a mitofusin, a critical regulator of mitochondrial fusion. Mol Biol Cell. 2008;19(6):2457–64.
42. Park Y-Y, Cho H. Mitofusin 1 is degraded at G(2)/M phase through ubiquitylation by MARCH5. Cell Div. 2012;7(7:25):1–6.
43. Olichon A, Emorine LJ, Descoins E, Pelloquin L, Brichese L, Gas N, et al. The human dynamin-related protein OPA1 is anchored to the mitochondrial inner membrane facing the inter-membrane space. FEBS Lett. 2002;523(1–3):171–6.
44. Ehses S, Raschke I, Mancuso G, Bernacchia A, Geimer S, Tondera D, et al. Regulation of OPA1 processing and mitochondrial fusion by m-AAA protease isoenzymes and OMA1. J Cell Biol. 2009;187(7):1023–36.
45. Head B, Griparic L, Amiri M, Gandre-Babbe S, van der Bliek AM. Inducible proteolytic inactivation of OPA1 mediated by the OMA1 protease in mammalian cells. J Cell Biol. 2009;187(7):959–66.
46. Griparic L, Kanazawa T, van der Bliek AM. Regulation of the mitochondrial dynamin-like protein Opa1 by proteolytic cleavage. J Cell Biol. 2007;178(5):757–64.
47. Song Z, Chen H, Fiket M, Alexander C, Chan DC. OPA1 processing controls mitochondrial fusion and is regulated by mRNA splicing, membrane potential, and Yme1L. J Cell Biol. 2007;178(5):749–55.
48. Anand R, Wai T, Baker MJ, Kladt N, Schauss AC, Rugarli E, et al. The i-AAA protease YME1L and OMA1 cleave OPA1 to balance mitochondrial fusion and fission. J Cell Biol. 2014;204(6):919–29.
49. Delettre C, Griffoin JM, Kaplan J, Dollfus H, Lorenz B, Faivre L, et al. Mutation spectrum and splicing variants in the OPA1 gene. Hum Genet. 2001;109(6):584–91.
50. Otsuga D, Keegan BR, Brisch E, Thatcher JW, Hermann GJ, Bleazard W, et al. The dynamin-related GTPase, Dnm1p, controls mitochondrial morphology in yeast. J Cell Biol. 1998;143(2):333–49.
51. Bleazard W, McCaffery JM, King EJ, Bale S, Mozdy A, Tieu Q, et al. The dynamin-related GTPase Dnm1 regulates mitochondrial fission in yeast. Nat Cell Biol. 1999;1(5):298–304.
52. Mozdy AD, McCaffery JM, Shaw JM. Dnm1p GTPase-mediated mitochondrial fission is a multi-step process requiring the novel integral membrane component Fis1p. J Cell Biol. 2000;151(2):367–80.
53. Tieu Q, Nunnari J. Mdv1p is a WD repeat protein that interacts with the dynamin-related GTPase, Dnm1p, to trigger mitochondrial division. J Cell Biol. 2000;151(2):353–66.
54. Cerveny KL, McCaffery JM, Jensen RE. Division of mitochondria requires a novel DMN1-interacting protein, Net2p. Mol Biol Cell. 2001;12(2):309–21.
55. Tieu Q, Okreglak V, Naylor K, Nunnari J. The WD repeat protein, Mdv1p, functions as a molecular adaptor by interacting with Dnm1p and Fis1p during mitochondrial fission. J Cell Biol. 2002;158(3):445–52.
56. Cerveny KL, Jensen RE. The WD-repeats of Net2p interact with Dnm1p and Fis1p to regulate division of mitochondria. Mol Biol Cell. 2003;14(10):4126–39.
57. Karren MA, Coonrod EM, Anderson TK, Shaw JM. The role of Fis1p-Mdv1p interactions in mitochondrial fission complex assembly. J Cell Biol. 2005;171(2):291–301.
58. Naylor K, Ingerman E, Okreglak V, Marino M, Hinshaw JE, Nunnari J. Mdv1 interacts with assembled dnm1 to promote mitochondrial division. J Biol Chem. 2006;281(4):2177–83.
59. Schauss AC, Bewersdorf J, Jakobs S. Fis1p and Caf4p, but not Mdv1p, determine the polar localization of Dnm1p clusters on the mitochondrial surface. J Cell Sci. 2006;119(Pt 15):3098–106.

60. Guo Q, Koirala S, Perkins EM, McCaffery JM, Shaw JM. The mitochondrial fission adaptors Caf4 and Mdv1 are not functionally equivalent. PLoS One. 2012;7(12):e53523.
61. Ingerman E, Perkins EM, Marino M, Mears JA, McCaffery JM, Hinshaw JE, et al. Dnm1 forms spirals that are structurally tailored to fit mitochondria. J Cell Biol. 2005;170(7):1021–7.
62. Mears JA, Lackner LL, Fang S, Ingerman E, Nunnari J, Hinshaw JE. Conformational changes in Dnm1 support a contractile mechanism for mitochondrial fission. Nat Struct Mol Biol. 2011;18(1):20–6.
63. James DI, Parone PA, Mattenberger Y, Martinou JC. hFis1, a novel component of the mammalian mitochondrial fission machinery. J Biol Chem. 2003;278(38):36373–9.
64. Smirnova E, Griparic L, Shurland DL, van der Bliek AM. Dynamin-related protein Drp1 is required for mitochondrial division in mammalian cells. Mol Biol Cell. 2001;12(8):2245–56.
65. Yoon Y, Krueger EW, Oswald BJ, McNiven MA. The mitochondrial protein hFis1 regulates mitochondrial fission in mammalian cells through an interaction with the dynamin-like protein DLP1. Mol Cell Biol. 2003;23(15):5409–20.
66. Gandre-Babbe S, van der Bliek AM. The novel tail-anchored membrane protein Mff controls mitochondrial and peroxisomal fission in mammalian cells. Mol Biol Cell. 2008;19(6):2402–12.
67. Palmer CS, Osellame LD, Laine D, Koutsopoulos OS, Frazier AE, Ryan MT. MiD49 and MiD51, new components of the mitochondrial fission machinery. EMBO Rep. 2011;12(6):565–73.
68. Zhao J, Liu T, Jin S, Wang X, Qu M, Uhlén P, et al. Human MIEF1 recruits Drp1 to mitochondrial outer membranes and promotes mitochondrial fusion rather than fission. EMBO J. 2011;30(14):2762–78.
69. Losón OC, Song Z, Chen H, Chan DC. Fis1, Mff, MiD49, and MiD51 mediate Drp1 recruitment in mitochondrial fission. Mol Biol Cell. 2013;24(5):659–67.
70. Losón OC, Liu R, Rome ME, Meng S, Kaiser JT, Shan S-O, et al. The mitochondrial fission receptor MiD51 requires ADP as a cofactor. Structure. 2014;22(3):367–77.
71. Losón OC, Meng S, Ngo H, Liu R, Kaiser JT, Chan DC. Crystal structure and functional analysis of MiD49, a receptor for the mitochondrial fission protein Drp1. Protein Sci. 2015;24(3):386–94.
72. Niemann A, Ruegg M, La Padula V, Schenone A, Suter U. Ganglioside-induced differentiation associated protein 1 is a regulator of the mitochondrial network: new implications for Charcot-Marie-Tooth disease. J Cell Biol. 2005;170(7):1067–78.
73. Pedrola L, Espert A, Wu X, Claramunt R, Shy ME, Palau F. GDAP1, the protein causing Charcot-Marie-Tooth disease type 4A, is expressed in neurons and is associated with mitochondria. Hum Mol Genet. 2005;14(8):1087–94.
74. Niemann A, Wagner KM, Ruegg M, Suter U. GDAP1 mutations differ in their effects on mitochondrial dynamics and apoptosis depending on the mode of inheritance. Neurobiol Dis. 2009;36(3):509–20.
75. Shield AJ, Murray TP, Board PG. Functional characterisation of ganglioside-induced differentiation-associated protein 1 as a glutathione transferase. Biochem Biophys Res Commun. 2006;347(4):859–66.
76. Taguchi N, Ishihara N, Jofuku A, Oka T, Mihara K. Mitotic phosphorylation of dynamin-related GTPase Drp1 participates in mitochondrial fission. J Biol Chem. 2007;282(15):11521–9.
77. Cribbs JT, Strack S. Reversible phosphorylation of Drp1 by cyclic AMP-dependent protein kinase and calcineurin regulates mitochondrial fission and cell death. EMBO Rep. 2007;8(10):939–44.
78. Chang C-R, Blackstone C. Drp1 phosphorylation and mitochondrial regulation. EMBO Rep. 2007;8(12):1088–9, author reply 1089–90.
79. Cereghetti GM, Stangherlin A, Martins de Brito O, Chang CR, Blackstone C, Bernardi P, et al. Dephosphorylation by calcineurin regulates translocation of Drp1 to mitochondria. Proc Natl Acad Sci U S A. 2008;105(41):15803–8.

80. Cho D-H, Nakamura T, Fang J, Cieplak P, Godzik A, Gu Z, et al. S-nitrosylation of Drp1 mediates beta-amyloid-related mitochondrial fission and neuronal injury. Science. 2009;324(5923):102–5.
81. Nakamura T, Cieplak P, Cho D-H, Godzik A, Lipton SA. S-nitrosylation of Drp1 links excessive mitochondrial fission to neuronal injury in neurodegeneration. Mitochondrion. 2010;10(5):573–8.
82. Nakamura N, Kimura Y, Tokuda M, Honda S, Hirose S. MARCH-V is a novel mitofusin 2- and Drp1-binding protein able to change mitochondrial morphology. EMBO Rep. 2006;7(10):1019–22.
83. Karbowski M, Neutzner A, Youle RJ. The mitochondrial E3 ubiquitin ligase MARCH5 is required for Drp1 dependent mitochondrial division. J Cell Biol. 2007;178(1):71–84.
84. Horn SR, Thomenius MJ, Johnson ES, Freel CD, Wu JQ, Coloff JL, et al. Regulation of mitochondrial morphology by APC/CCdh1-mediated control of Drp1 stability. Mol Biol Cell. 2011;22(8):1207–16.
85. Wasiak S, Zunino R, McBride HM. Bax/Bak promote sumoylation of DRP1 and its stable association with mitochondria during apoptotic cell death. J Cell Biol. 2007;177(3):439–50.
86. Braschi E, Zunino R, McBride HM. MAPL is a new mitochondrial SUMO E3 ligase that regulates mitochondrial fission. EMBO Rep. 2009;10(7):748–54.
87. Figueroa-Romero C, Iñiguez-Lluhí JA, Stadler J, Chang C-R, Arnoult D, Keller PJ, et al. SUMOylation of the mitochondrial fission protein Drp1 occurs at multiple nonconsensus sites within the B domain and is linked to its activity cycle. FASEB J. 2009;23(11):3917–27.
88. Koch A, Thiemann M, Grabenbauer M, Yoon Y, McNiven MA, Schrader M. Dynamin-like protein 1 is involved in peroxisomal fission. J Biol Chem. 2003;278(10):8597–605.
89. Li X, Gould SJ. The dynamin-like GTPase DLP1 is essential for peroxisome division and is recruited to peroxisomes in part by PEX11. J Biol Chem. 2003;278(19):17012–20.
90. Jones BA, Fangman WL. Mitochondrial DNA maintenance in yeast requires a protein containing a region related to the GTP-binding domain of dynamin. Genes Dev. 1992; 6(3):380–9.
91. Shepard KA, Yaffe MP. The yeast dynamin-like protein, Mgm1p, functions on the mitochondrial outer membrane to mediate mitochondrial inheritance. J Cell Biol. 1999;144(4):711–20.
92. Chen H, Vermulst M, Wang YE, Chomyn A, Prolla TA, McCaffery JM, et al. Mitochondrial fusion is required for mtDNA stability in skeletal muscle and tolerance of mtDNA mutations. Cell. 2010;141(2):280–9.
93. Murley A, Lackner LL, Osman C, West M, Voeltz GK, Walter P, et al. ER-associated mitochondrial division links the distribution of mitochondria and mitochondrial DNA in yeast. Elife. 2013;2:e00422.
94. Ban-Ishihara R, Ishihara T, Sasaki N, Mihara K, Ishihara N. Dynamics of nucleoid structure regulated by mitochondrial fission contributes to cristae reformation and release of cytochrome c. Proc Natl Acad Sci U S A. 2013;110(29):11863–8.
95. Meeusen S, Nunnari J. Evidence for a two membrane-spanning autonomous mitochondrial DNA replisome. J Cell Biol. 2003;163(3):503–10.
96. Kornmann B, Currie E, Collins SR, Schuldiner M, Nunnari J, Weissman JS, et al. An ER-mitochondria tethering complex revealed by a synthetic biology screen. Science. 2009;325(5939):477–81.
97. Sesaki H, Jensen RE. Division versus fusion: Dnm1p and Fzo1p antagonistically regulate mitochondrial shape. J Cell Biol. 1999;147(4):699–706.
98. Ishihara N, Nomura M, Jofuku A, Kato H, Suzuki SO, Masuda K, et al. Mitochondrial fission factor Drp1 is essential for embryonic development and synapse formation in mice. Nat Cell Biol. 2009;11(8):958–66.
99. Sebastián D, Hernández-Alvarez MI, Segalés J, Sorianello E, Muñoz JP, Sala D, et al. Mitofusin 2 (Mfn2) links mitochondrial and endoplasmic reticulum function with insulin signaling and is essential for normal glucose homeostasis. Proc Natl Acad Sci U S A. 2012;109(14):5523–8.

100. Loiseau D, Chevrollier A, Verny C, Guillet V, Gueguen N, Pou de Crescenzo M-A, et al. Mitochondrial coupling defect in Charcot-Marie-Tooth type 2A disease. Ann Neurol. 2007;61(4):315–23.
101. Mourier A, Motori E, Brandt T, Lagouge M, Atanassov I, Galinier A, et al. Mitofusin 2 is required to maintain mitochondrial coenzyme Q levels. J Cell Biol. 2015;208(4):429–42.
102. Chen H, McCaffery JM, Chan DC. Mitochondrial fusion protects against neurodegeneration in the cerebellum. Cell. 2007;130(3):548–62.
103. de Brito OM, Scorrano L. Mitofusin 2 tethers endoplasmic reticulum to mitochondria. Nature. 2008;456(7222):605–10.
104. Wasilewski M, Semenzato M, Rafelski SM, Robbins J, Bakardjiev AI, Scorrano L. Optic atrophy 1-dependent mitochondrial remodeling controls steroidogenesis in trophoblasts. Curr Biol. 2012;22(13):1228–34.
105. Misko A, Jiang S, Wegorzewska I, Milbrandt J, Baloh RH. Mitofusin 2 is necessary for transport of axonal mitochondria and interacts with the Miro/Milton complex. J Neurosci. 2010;30(12):4232–40.
106. Misko AL, Sasaki Y, Tuck E, Milbrandt J, Baloh RH. Mitofusin2 mutations disrupt axonal mitochondrial positioning and promote axon degeneration. J Neurosci. 2012;32(12):4145–55.
107. Cosson P, Marchetti A, Ravazzola M, Orci L. Mitofusin-2 independent juxtaposition of endoplasmic reticulum and mitochondria: an ultrastructural study. PLoS One. 2012;7(9):e46293.
108. Filadi R, Greotti E, Turacchio G, Luini A, Pozzan T, Pizzo P. Mitofusin 2 ablation increases endoplasmic reticulum-mitochondria coupling. Proc Natl Acad Sci USA. 2015;112(17):2174–81.
109. Duarte A, Poderoso C, Cooke M, Soria G, Cornejo Maciel F, Gottifredi V, et al. Mitochondrial fusion is essential for steroid biosynthesis. PLoS One. 2012;7(9):e45829.
110. Lee S, Sterky FH, Mourier A, Terzioglu M, Cullheim S, Olson L, et al. Mitofusin 2 is necessary for striatal axonal projections of midbrain dopamine neurons. Hum Mol Genet. 2012;21(22):4827–35.
111. Pham AH, Meng S, Chu QN, Chan DC. Loss of Mfn2 results in progressive, retrograde degeneration of dopaminergic neurons in the nigrostriatal circuit. Hum Mol Genet. 2012;21(22):4817–26.
112. Papanicolaou KN, Khairallah RJ, Ngoh GA, Chikando A, Luptak I, O'Shea KM, et al. Mitofusin-2 maintains mitochondrial structure and contributes to stress-induced permeability transition in cardiac myocytes. Mol Cell Biol. 2011;31(6):1309–28.
113. Amiott EA, Lott P, Soto J, Kang PB, McCaffery JM, Dimauro S, et al. Mitochondrial fusion and function in Charcot-Marie-Tooth type 2A patient fibroblasts with mitofusin 2 mutations. Exp Neurol. 2008;211(1):115–27.
114. Meeusen S, McCaffery JM, Nunnari J. Mitochondrial fusion intermediates revealed in vitro. Science. 2004;305(5691):1747–52.
115. Malka F, Guillery O, Cifuentes-Diaz C, Guillou E, Belenguer P, Lombès A, et al. Separate fusion of outer and inner mitochondrial membranes. EMBO Rep. 2005;6(9):853–9.
116. Sauvanet C, Duvezin-Caubet S, di Rago J-P, Rojo M. Energetic requirements and bioenergetic modulation of mitochondrial morphology and dynamics. Semin Cell Dev Biol. 2010;21(6):558–65.
117. Benard G, Bellance N, James D, Parrone P, Fernandez H, Letellier T, et al. Mitochondrial bioenergetics and structural network organization. J Cell Sci. 2007;120(Pt 5):838–48.
118. Duvezin-Caubet S, Jagasia R, Wagener J, Hofmann S, Trifunovic A, Hansson A, et al. Proteolytic processing of OPA1 links mitochondrial dysfunction to alterations in mitochondrial morphology. J Biol Chem. 2006;281(49):37972–9.
119. Egner A, Jakobs S, Hell SW. Fast 100-nm resolution three-dimensional microscope reveals structural plasticity of mitochondria in live yeast. Proc Natl Acad Sci U S A. 2002;99(6):3370–5.
120. Jimenez L, Laporte D, Duvezin-Caubet S, Courtout F, Sagot I. Mitochondrial ATP synthases cluster as discrete domains that reorganize with the cellular demand for oxidative phosphorylation. J Cell Sci. 2014;127(Pt 4):719–26.

121. Rossignol R, Gilkerson R, Aggeler R, Yamagata K, Remington SJ, Capaldi RA. Energy substrate modulates mitochondrial structure and oxidative capacity in cancer cells. Cancer Res. 2004;64(3):985–93.
122. Mishra P, Carelli V, Manfredi G, Chan DC. Proteolytic cleavage of Opa1 stimulates mitochondrial inner membrane fusion and couples fusion to oxidative phosphorylation. Cell Metab. 2014;19(4):630–41.
123. Gomes LC, Di Benedetto G, Scorrano L. During autophagy mitochondria elongate, are spared from degradation and sustain cell viability. Nat Cell Biol. 2011;13(5):589–98.
124. Rambold AS, Kostelecky B, Elia N, Lippincott-Schwartz J. Tubular network formation protects mitochondria from autophagosomal degradation during nutrient starvation. Proc Natl Acad Sci U S A. 2011;108(25):10190–5.
125. Cassidy-Stone A, Chipuk JE, Ingerman E, Song C, Yoo C, Kuwana T, et al. Chemical inhibition of the mitochondrial division dynamin reveals its role in Bax/Bak-dependent mitochondrial outer membrane permeabilization. Dev Cell. 2008;14(2):193–204.
126. Skre H. Genetic and clinical aspects of Charcot-Marie-Tooth disease. Clin Genet. 1974;6(2):98–118.
127. Emery AE. Population frequencies of inherited neuromuscular diseases – a world survey. Neuromuscul Disord. 1991;1(1):19–29.
128. Hoyer H, Braathen GJ, Busk OL, Holla OL, Svendsen M, Hilmarsen HT, et al. Genetic diagnosis of Charcot-Marie-Tooth disease in a population by next-generation sequencing. Biomed Res Int. 2014;2014:210401.
129. Chung KW, Kim SB, Park KD, Choi KG, Lee JH, Eun HW, et al. Early onset severe and late-onset mild Charcot-Marie-Tooth disease with mitofusin 2 (MFN2) mutations. Brain. 2006;129(Pt 8):2103–18.
130. Züchner SS, Mersiyanova IVI, Muglia MM, Bissar-Tadmouri NN, Rochelle JJ, Dadali ELE, et al. Mutations in the mitochondrial GTPase mitofusin 2 cause Charcot-Marie-Tooth neuropathy type 2A. Nat Genet. 2004;36(5):449–51.
131. Kijima K, Numakura C, Izumino H, Umetsu K, Nezu A, Shiiki T, et al. Mitochondrial GTPase mitofusin 2 mutation in Charcot-Marie-Tooth neuropathy type 2A. Hum Genet. 2005;116(1–2):23–7.
132. Cartoni R, Martinou JC. Role of mitofusin 2 mutations in the physiopathology of Charcot-Marie-Tooth disease type 2A. Exp Neurol. 2009;218(2):268–73.
133. Vallat J-M, Ouvrier RA, Pollard JD, Magdelaine C, Zhu D, Nicholson GA, et al. Histopathological findings in hereditary motor and sensory neuropathy of axonal type with onset in early childhood associated with mitofusin 2 mutations. J Neuropathol Exp Neurol. 2008;67(11):1097–102.
134. Verhoeven K, Claeys KG, Züchner S, Schröder JM, Weis J, Ceuterick C, et al. MFN2 mutation distribution and genotype/phenotype correlation in Charcot-Marie-Tooth type 2. Brain. 2006;129(Pt 8):2093–102.
135. Detmer SA, Vande Velde C, Cleveland DW, Chan DC. Hind limb gait defects due to motor axon loss and reduced distal muscles in a transgenic mouse model of Charcot-Marie-Tooth type 2A. Hum Mol Genet. 2008;17(3):367–75.
136. Cuesta A, Pedrola L, Sevilla T, García-Planells J, Chumillas MJ, Mayordomo F, et al. The gene encoding ganglioside-induced differentiation-associated protein 1 is mutated in axonal Charcot-Marie-Tooth type 4A disease. Nat Genet. 2002;30(1):22–5.
137. Baxter RV, Ben Othmane K, Rochelle JM, Stajich JE, Hulette C, Dew-Knight S, et al. Ganglioside-induced differentiation-associated protein-1 is mutant in Charcot-Marie-Tooth disease type 4A/8q21. Nat Genet. 2002;30(1):21–2.
138. Claramunt R, Pedrola L, Sevilla T, López de Munain A, Berciano J, Cuesta A, et al. Genetics of Charcot-Marie-Tooth disease type 4A: mutations, inheritance, phenotypic variability, and founder effect. J Med Genet. 2005;42(4):358–65.
139. Senderek J, Bergmann C, Ramaekers VT, Nelis E, Bernert G, Makowski A, et al. Mutations in the ganglioside-induced differentiation-associated protein-1 (GDAP1) gene in intermediate type autosomal recessive Charcot-Marie-Tooth neuropathy. Brain. 2003;126(Pt 3):642–9.

140. Niemann A, Huber N, Wagner KM, Somandin C, Horn M, Lebrun-Julien F, et al. The Gdap1 knockout mouse mechanistically links redox control to Charcot-Marie-Tooth disease. Brain. 2014;137(Pt 3):668–82.
141. Barneo-Muñoz M, Juárez P, Civera-Tregón A, Yndriago L, Pla-Martin D, Zenker J, et al. Lack of GDAP1 induces neuronal calcium and mitochondrial defects in a knockout mouse model of Charcot-Marie-Tooth neuropathy. PLoS Genet. 2015;11(4):e1005115.
142. Lenaers G, Hamel C, Delettre C, Amati-Bonneau P, Procaccio V, Bonneau D, et al. Dominant optic atrophy. Orphanet J Rare Dis. 2012;7:46.
143. Thiselton DL, Alexander C, Taanman J-W, Brooks S, Rosenberg T, Eiberg H, et al. A comprehensive survey of mutations in the OPA1 gene in patients with autosomal dominant optic atrophy. Invest Ophthalmol Vis Sci. 2002;43(6):1715–24.
144. Yu-Wai-Man P, Sitarz KS, Samuels DC, Griffiths PG, Reeve AK, Bindoff LA, et al. OPA1 mutations cause cytochrome c oxidase deficiency due to loss of wild-type mtDNA molecules. Hum Mol Genet. 2010;19(15):3043–52.
145. Davies VJ, Hollins AJ, Piechota MJ, Yip W, Davies JR, White KE, et al. Opa1 deficiency in a mouse model of autosomal dominant optic atrophy impairs mitochondrial morphology, optic nerve structure and visual function. Hum Mol Genet. 2007;16(11):1307–18.
146. Alavi MV, Bette S, Schimpf S, Schuettauf F, Schraermeyer U, Wehrl HF, et al. A splice site mutation in the murine Opa1 gene features pathology of autosomal dominant optic atrophy. Brain. 2007;130(Pt 4):1029–42.
147. Williams PA, Piechota M, von Ruhland C, Taylor E, Morgan JE, Votruba M. Opa1 is essential for retinal ganglion cell synaptic architecture and connectivity. Brain. 2012;135(Pt 2):493–505.
148. Waterham HR, Koster J, van Roermund CWT, Mooyer PAW, Wanders RJA, Leonard JV. A lethal defect of mitochondrial and peroxisomal fission. N Engl J Med. 2007;356(17):1736–41.
149. Chang C-R, Manlandro CM, Arnoult D, Stadler J, Posey AE, Hill RB, et al. A lethal de novo mutation in the middle domain of the dynamin-related GTPase Drp1 impairs higher order assembly and mitochondrial division. J Biol Chem. 2010;285(42):32494–503.
150. Wakabayashi J, Zhang Z, Wakabayashi N, Tamura Y, Fukaya M, Kensler TW, et al. The dynamin-related GTPase Drp1 is required for embryonic and brain development in mice. J Cell Biol. 2009;186(6):805–16.
151. Berthet A, Margolis EB, Zhang J, Hsieh I, Zhang J, Hnasko TS, et al. Loss of mitochondrial fission depletes axonal mitochondria in midbrain dopamine neurons. J Neurosci. 2014;34(43):14304–17.
152. Shirendeb U, Reddy AP, Manczak M, Calkins MJ, Mao P, Tagle DA, et al. Abnormal mitochondrial dynamics, mitochondrial loss and mutant huntingtin oligomers in Huntington's disease: implications for selective neuronal damage. Hum Mol Genet. 2011;20(7):1438–55.
153. Costa V, Giacomello M, Hudec R, Lopreiato R, Ermak G, Lim D, et al. Mitochondrial fission and cristae disruption increase the response of cell models of Huntington's disease to apoptotic stimuli. EMBO Mol Med. 2010;2(12):490–503.
154. Wang X, Yan MH, Fujioka H, Liu J, Wilson-Delfosse A, Chen SG, et al. LRRK2 regulates mitochondrial dynamics and function through direct interaction with DLP1. Hum Mol Genet. 2012;21(9):1931–44.
155. Su B, Wang X, Bonda D, Perry G, Smith M, Zhu X. Abnormal mitochondrial dynamics – a novel therapeutic target for Alzheimer's disease? Mol Neurobiol. 2010;41(2–3):87–96.
156. Wang X, Su B, Lee H-G, Li X, Perry G, Smith MA, et al. Impaired balance of mitochondrial fission and fusion in Alzheimer's disease. J Neurosci. 2009;29(28):9090–103.
157. Calkins MJ, Manczak M, Mao P, Shirendeb U, Reddy PH. Impaired mitochondrial biogenesis, defective axonal transport of mitochondria, abnormal mitochondrial dynamics and synaptic degeneration in a mouse model of Alzheimer's disease. Hum Mol Genet. 2011;20(23):4515–29.
158. Wang S, Song J, Tan M, Albers KM, Jia J. Mitochondrial fission proteins in peripheral blood lymphocytes are potential biomarkers for Alzheimer's disease. Eur J Neurol. 2012;19(7):1015–22.

Chapter 8
Mitochondrial Dysfunction and Transport in Demyelinating Disease with Inflammation

Marija Sajic

Abstract Demyelinating diseases with inflammation affect both the central and peripheral nervous systems. The histopathological hallmark of these devastating, and sometimes life-threatening, diseases is segmental demyelination of long axonal tracts and varying degrees of inflammation, with important long-term consequences for the function and survival of axons and neurons. Mitochondria are essential for the function of all cells, but maybe more so for neurons, in both physiological and pathological conditions. Healthy neurons utilise a sensitive, responsive and dynamic mitochondrial population for their varied basal metabolism. It is not surprising, then, that mitochondria also play a fundamental role in adaptive mechanisms which enable neurons to survive pathological events such as loss of myelin sheath and exposure to inflammatory mediators. Indeed, it is believed that the survival of axons in such an environment, and perhaps recovery to full functionality, critically depends on the adaptive capacity of its mitochondrial population. This chapter will present and interpret currently available, published knowledge relating to the role of neuronal mitochondria in neuroprotection and neurodegeneration during demyelinating inflammatory diseases of the central and peripheral nervous systems.

Keywords Multiple sclerosis • Guillain-Barré syndrome • Mitochondria Mitochondrial dynamics • Demyelination • Inflammation

8.1 Multiple Sclerosis

Multiple sclerosis (MS) is the most common demyelinating inflammatory disease of the central nervous system (CNS) and the leading cause of non-traumatic disability in young adults in the developed world [1, 2]. As a distinct neurological

M. Sajic, MD, PhD
Department of Neuroinflammation,
University College London Institute of Neurology, London, UK
e-mail: m.sajic@ucl.ac.uk

A.K. Reeve et al. (eds.), *Mitochondrial Dysfunction in Neurodegenerative Disorders*, DOI 10.1007/978-3-319-28637-2_8

entity, MS was based on clinical criteria established by the French physician Jean-Martin Charcot more than 150 years ago [3]. The disease typically follows a relapsing-remitting course, whereby periods of loss of neurological function called relapses or attacks are followed by periods of complete or incomplete recovery (remissions). Over time, complete remissions become rare and neurological dysfunction gradually accumulates, leading to permanent, and often severe, disability.

Over 100 years ago, Otto Marburg described the pathology of what we now recognise as a partially remyelinated MS 'plaque'. Since then, a large number of descriptive histopathological studies of post-mortem tissue have provided a detailed picture of MS lesions. These and, in more recent years, MRI-based imaging studies in patients together revealed unique, and to date unexplained, characteristics of MS – frequently appearing and disappearing inflammatory-demyelinating lesions that are well demarcated from the normal-appearing surrounding tissue and scattered throughout the CNS in a seemingly random fashion. Widespread neurodegeneration is a more recently recognised major feature of MS and often suspected as the likely cause of permanent disability.

The ubiquitous presence of infiltrating immune cells in the examined lesions has led to the often stated but as yet unproven hypothesis that multiple sclerosis is an autoimmune disease. Great efforts have been dedicated to identifying a common, specific autoimmune target antigen in MS, with particular focus on components of the myelin sheath. However, to date, all attempts to identify an autoantigen have failed [4].

The main pathological characteristics of MS lesions are (1) presence of scattered inflammatory foci in the brain and spinal cord parenchyma, mainly consisting of macrophages and lymphocytes and often centred on a blood vessel, (2) segmental demyelination, (3) neuronal and axonal degeneration and (4) focal gliosis [5]. Although a major feature, inflammation is not pathognomonic of MS. Other neurological diseases with progressive loss of neurons, such as Alzheimer's [6] and Parkinson's disease [7], also show a degree of inflammation [8, 9]. Although a varying degree of white matter loss has been reported in both of these neurodegenerative conditions [10–12], the feature that distinguishes MS from all other CNS diseases is the type of demyelination. Namely, MS is the only CNS disease characterised by the so-called segmental (i.e. spreading from one node to the next one) demyelination. Indeed, recent histopathological evidence from exceptionally early MS lesions suggests that primary demyelination is the earliest event in the lesion formation [13–15]. The loss of the myelin sheath along a section of an axon has considerable consequences for its function, most obviously for impulse conduction. This early event sets off a series of adaptive mechanisms aimed at preserving the axon structure, restoring its primary function in impulse conduction and restoring myelin. Mitochondria play a central role in a number of these adaptive, survival mechanisms. This chapter will review currently available, published knowledge on the role of mitochondria in axonal and neuronal protection against degeneration during inflammatory-demyelinating disease.

8.1.1 The Role of Myelin in Axonal Energy Homeostasis

The main role of myelin is generally believed to be in enabling saltatory impulse conduction in myelinated axons. This is separate from any metabolic support for the axon which may be provided by the myelinating glial cell. As in all cells, the neuron maintains ionic disequilibrium across its plasma membrane via the activity of ATP utilising pumps. The maintained ionic concentration gradient is measurable as a resting potential. A feature more specific to neurons however is the presence in the membrane of voltage-dependent Na^+ and K^+ gates. These gates open transiently in response to a critical level of depolarisation, particularly the Na^+ gates, allowing ions to flow along the concentration gradient and producing a transient reversal of potential – the action potential – before closing and allowing restoration of the resting potential. The opening of the channels, and hence the action potential, is an all-or-nothing phenomenon. The depolarisation and accompanying ionic flow last long enough to open adjacent gates. However, gates once opened and closed require a finite time to achieve the initial configuration and cannot be immediately reopened; consequently, the action potential progresses along the axon in one direction, giving impulse conduction. The immediate free energy required for the action potential is derived from the local change in entropy. The resting potential relies on the continuous activity of the Na^+/K^+ ATPase, which is distributed evenly along the axolemma (axonal membrane) in both myelinated and nonmyelinated axons. This would establish a requirement for an even distribution of mitochondria.

In contrast to unmyelinated axons, the distribution of sodium channels in myelinated axons (Fig. 8.1) is restricted to narrow gaps between the segments of the myelin sheath called nodes of Ranvier (~1 μm) where there may be 1,000–2,000 channels per μm^2. Neighbouring nodes, for example, in human sural nerve, a sensory nerve which descends down the posterior lateral side of the calf and lateral side of foot dorsum, may be separated by up to 1.4 mm of continuous myelin, called internodal segments [16]. Such clustering of channels at regular, but distant, intervals means that the flux of sodium and potassium 'jumps' from node to node skipping long internodal stretches, thus vastly increasing the speed of conduction. Furthermore, myelin acts as an electrical insulator, reducing the overall capacitance of the axolemma and thereby maximising the effect of potential change brought about by the inward sodium current at the node [17–19]. In other words, in the absence of myelin, the depolarising current quickly dissipates as it flows along the axons rather than staying 'focused' to the narrow nodal gap, resulting in insufficient current available to depolarise the next node and saltatory conduction fails. Having such an effect on preventing the loss of current, it follows that myelin reduces the net amount of ions required for successful membrane depolarisation across the axolemma. In this way, myelin also minimises axonal energy consumption, as far fewer sodium and potassium ions are required to be actively transported by Na^+/K^+ATPase in order to re-establish resting potential than in unmyelinated axons.

As mitochondrial oxidative phosphorylation is the main source of ATP in neurons, both neuronal cell bodies, dendrites and axons are highly dependent on correct

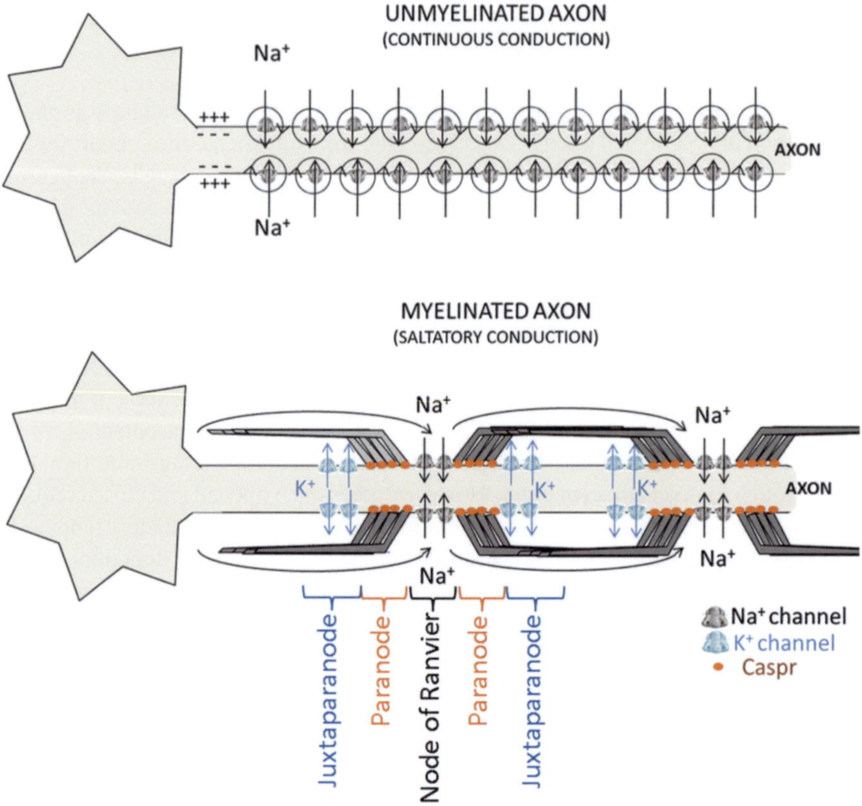

Fig. 8.1 Impulse conduction in unmyelinated and myelinated axons. In *unmyelinated axons* (*top panel*), sodium channels are distributed along the whole surface of axonal membrane. During impulse conduction, sodium ions enter intra-axonal space from extracellular compartment along its entire length, sequentially changing membrane potential and thus creating small, continuous depolarising currents which spread slowly from one end of the axon to another. In *myelinated axons* (*bottom panel*), channel distribution is highly organised within specific subregions. Sodium channels are restricted to narrow gaps between the segments of the myelin sheath, called nodes of Ranvier. Myelin is attached to axolemma at paranodal regions by a number of anchoring proteins such as Caspr. Voltage-gated potassium channels and sodium/potassium pump are located in the juxtaparanodal regions. During impulse conduction, ion fluxes are limited to these narrow regions, thus greatly increasing speed of conduction

functioning of these organelles for reliable and constant ATP supply. Furthermore, neurons are morphologically the most complex cells in the body, many having thousands of dendrites and axons of considerable length; thus, their function is critically dependent not only on the correct function but also on the precise and correct *location* of mitochondria. For example, neurons exhibit great variations in impulse activity over time, thus the relative energy demands of different neuronal subcompartments are often rapidly changing. Although mitochondrial DNA synthesis has been reported to occur within some vertebrate peripheral axons [20], given that the majority of mitochondrial proteins are encoded in the nuclear genome [21], the bulk of mitochondrial biogenesis is thought to occur in the vicinity of the cell body. Therefore, any increase in metabolic demand at a distance from the soma would require increased mitochondrial transport via the long-distance, axonal transport system. This complex transport machinery provides optimal mitochondrial supply to all neuronal areas simultaneously, including sites remote from the cell body. Indeed, axonal impulse activity and mitochondrial axonal transport rate are positively correlated in both myelinated and unmyelinated peripheral nerve fibres [22]. Clearly, delivering mitochondria to areas of high metabolic demand (the right place) at the right time is essential. One such area appears to be the unmyelinated or demyelinated segment. This difference in axonal energy requirements depending on the presence of myelin is shown clearly by the example of optic nerve fibres crossing *lamina cribrosa*, a mesh-like structure at the posterior part of the sclera. Namely, the axons of the optic nerve are in many mammals unmyelinated on the intraocular side, but the same axons are myelinated on the extraocular side of *lamina cribrosa*. Interestingly, both the total amount and the activity of mitochondrial complex IV (cytochrome *c* oxidase) have been found to be significantly greater in the unmyelinated part of these axons than in their myelinated segments [23, 24]. As complex IV is the major consumer of cellular oxygen, this increase in complex IV activity suggests that the demand for the oxygen, and by inference the ATP, is higher in the unmyelinated than in the myelinated segment of the axon. Furthermore, given that the diameter of the optic nerve axons on either side of the *lamina cribrosa* is the same [25], the increase in the level of the complex IV subunit II [23] suggests a greater density of mitochondria in the unmyelinated segment of optic nerve axons. A similar high relative density of mitochondria in unmyelinated, compared with myelinated, axons has been observed in peripheral nerve axons (Fig. 8.2). This inverse correlation between myelination and mitochondrial density is entirely consistent with the energy-saving effect of the myelin sheath on axonal metabolic demand arising directly from impulse conduction, i.e. if conduction increases axonal transport, then this is an additional demand for energy consequential to conduction.

8.1.2 Energy Metabolism of Demyelinated Axon

Having in mind the above, it is not surprising that the loss of myelin sheath (i.e. demyelination during multiple sclerosis) renders axons vulnerable to a range of insults, including relative energy insufficiency. One of the earliest, and the most obvious, consequences of myelin loss is the impairment of impulse conduction propagation across the demyelinated segment. The effect of demyelination on impulse conduction has been studied in a range of focal, experimentally induced demyelinating lesions, typically using Schwann cell and oligodendrocyte toxins which inhibit protein synthesis such as diphtheria toxin and ethidium bromide or agents that cause destruction of myelin without affecting the myelin-forming cells,

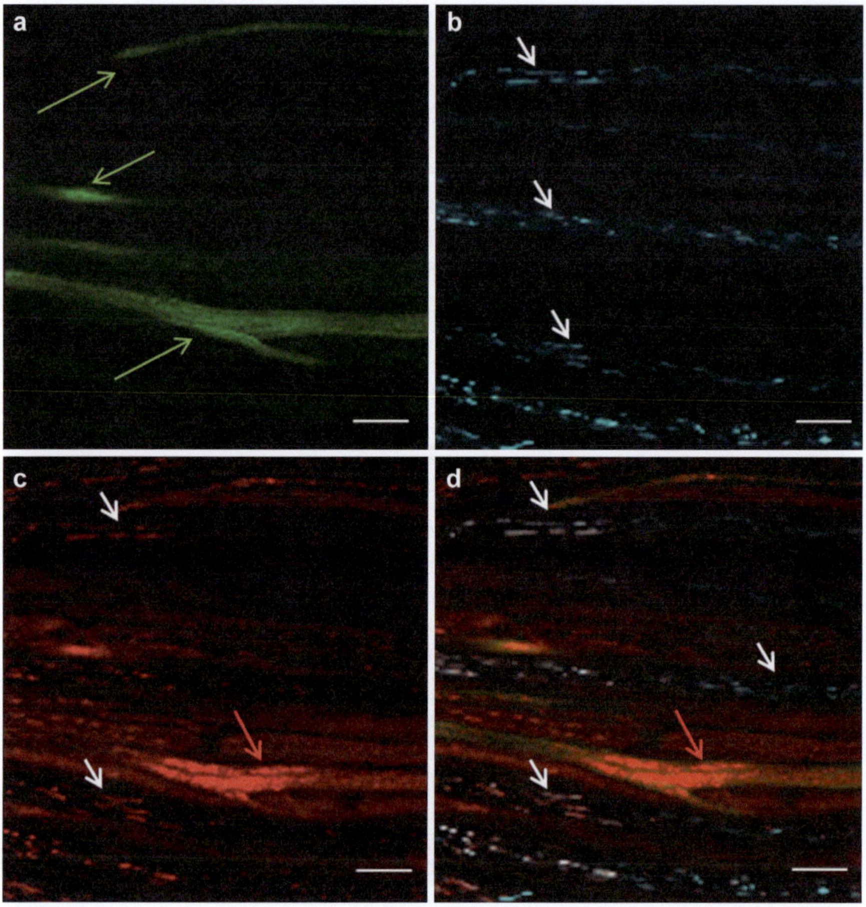

Fig. 8.2 In vivo confocal image of mitochondria in myelinated and unmyelinated fibres in Thy1-CFP-S positive (mito-S) mouse peripheral nerve. Exposed saphenous nerve axons are labelled in situ with Alexa Fluor 488 conjugated Griffonia isolectin IB4 (*green*) and tetramethylrhodamine methyl ester (TMRM, *red*). Isolectin IB4 is a glycoprotein isolated from the seeds of the African legume, *Griffonia simplicifolia*, known to specifically bind to a subgroup of unmyelinated fibres, hence named IB4$^+$ C fibres. TMRM is a potentiometric dye which preferentially accumulates within polarised (healthy) mitochondria. This combination of unmyelinated fibre labelling, functional mitochondrial labelling and constitutively expressed mitochondrial CFP enables visualisation of polarised mitochondria (**c**, **d**) within unmyelinated fibres (**a**, **d**). In CFP-S mice, the cyan fluorescent protein is constitutively expressed only in mitochondria within 40–60 % of myelinated axons (not expressed in unmyelinated fibres), thus enabling distinction between mitochondria in myelinated fibres and mitochondria in Schwann (and other) cells (Note that the density of mitochondria within unmyelinated fibres (*green arrows* in **a**) is noticeably greater (*red arrows* in **c** and in merged image in **d**) than in larger diameter, myelinated fibres (*white arrows*). Scale bar = 10 μm)

such as lysolecithin [26]. In such lesions, demyelination begins at the paranodal regions and spreads in both directions along the juxtaparanode and internode, as myelin loses its compact structure and detaches from axons [18]. This structural change affects the electrical properties of the axolemma in such a way that the overall membrane capacitance increases [17]. Owing to this increase in capacitance of a demyelinating axon, greater inward current is necessary in order to depolarise the next node, which means that more sodium moves into the axon and it takes a longer time to reach the excitation threshold than in myelinated axons. It has been shown in experimentally induced demyelinated lesions that internodal conduction may be 30 times slower than in normal, myelinated internodes [17]. Furthermore, repetitive impulse activity along such demyelinated segments leads to a gradual build-up of intracellular sodium, eventually overwhelming the sodium pump, and subsequently completely blocking the conduction [17]. Nonetheless, as the loss of myelin progresses and axons become completely denuded, impulse conduction is established again, albeit with different characteristics. Instead of being saltatory conduction, over the 2 or 3 weeks of persistent demyelination, the conduction re-establishes as slow and continuous, reminiscent of conduction in unmyelinated axons [27]. Furthermore, the demyelinated axons may exhibit spontaneous, ectopic impulse activity [28].

These elegant studies of axonal conduction along demyelinated segments, performed over 30 years ago, predicted that profound changes in distribution of ion channels along the axonal membrane must underlie the re-establishment, and the change in the nature, of impulse conduction. Indeed, it was later reported in animal models of multiple sclerosis [29, 30] and in MS lesions [31, 32] that the pattern of distribution of voltage-gated sodium channels changes dramatically following demyelination. Specifically, a subgroup of sodium channels called persistent current $Na_V1.6$ channels, and other Na channels such as Nav1.2, become diffusely distributed along the demyelinated axolemma, instead of highly concentrated in a narrow area of the node [31]. Similarly, it has been shown that the distribution of potassium channels changes from their strict paranodal location in myelinated axons to diffuse localisation along demyelinated or dysmyelinated internodes [33–36]. As in healthy unmyelinated fibres, the widespread distribution of these changes allow for greater amounts of ion movements across the axolemma during impulse conduction, thereby necessitating greater activity of Na^+/K^+ ATPase, which accordingly increases ATP demand. It has been shown in models of ischemia [37, 38] and neurotrauma [39] that when such increased energy demand cannot be met due to insufficient ATP, the ensuing energy imbalance leads to accumulation of sodium ions in the axoplasm. Although harder to prove, high impulse conduction load is presumed to have a similar effect on intra-axonal sodium concentration, as the repetitive firing may lead to fast depletion of ATP. The high intra-axonal concentration of sodium, in turn, activates another pump called the sodium/calcium exchanger (Na^+/Ca^{2+} exchanger) to work in reverse [39], importing calcium ions from the extracellular space in exchange for intracellular sodium, whereas it normally exports calcium from the axoplasm. These changes lead to pathologically high intracellular concentrations of calcium within demyelinated axons [38, 39] and activation of dangerous Ca-mediated cascades which, in experimental models of inflammatory-demyelination, result in axonal degeneration [30, 32].

From all the above, it is clear that two critical factors that determine the future of demyelinated axons are (1) the availability of ATP for the increased demand in the demyelinated segment and (2) the capacity of that segment to buffer excess calcium load. Both of these functions are performed by mitochondria; hence, unsurprisingly, mitochondria play a major role in determining the long-term outcome of demyelination. Importantly, however, following demyelination, the increased demand for mitochondria arises within an area that may be a long way away from the cell body. Therefore, other aspects of axonal function, such as those regulating mitochondrial transport from cell body along the axon, are essential for delivering mitochondria to the vulnerable site and thus at least equally important for long-term axonal survival.

8.1.3 The Role of Mitochondria in Survival of Demyelinated Axon

How long demyelinated axons survive in vivo is unclear. Answering this question is particularly difficult due to extremely limited availability of human tissue from demyelinated lesions. Even so, it has been apparent from the earliest descriptions of MS lesions, such as those described by Otto Marburg in 1906 [40], as well as from the more recent ones [41, 42] that some demyelinated axons can persist within the lesions long enough to become remyelinated. Autopsy material from patients who died early in the course of MS revealed crucial information about natural evolution of demyelinated axons in human brains during early stages of the disease. These studies showed that demyelinated axons that had newly arisen within previously unaffected white matter often become remyelinated, forming so-called shadow plaques [43, 44]. Furthermore, in one study of 98 lesions, approximately 15 % of old shadow plaques showed evidence that axons within this area had been recently demyelinated for the second time [42]. Similar focal recurrence of new and chronic lesions has been reported in longitudinal MRI examinations [45–47].

Experimental models of demyelination and dysmyelination further elucidated the role of myelin in axonal survival. A number of significant differences between these, typically genetically modified, rodent models of myelin loss and multiple sclerosis, a disease that is yet to be faithfully reproduced in experimental animals, preclude us from drawing parallels between human and rodent demyelinated axons. Nonetheless, studies in which one myelin component has been replaced by another provided an important insight into the role of myelin as trophic support for axons. For example, a study in which the CNS myelin protein called proteolipid protein has been replaced by a peripheral nervous system (PNS) myelin protein called P0 in the CNS of mice showed that despite appropriate myelination and apparently intact compact myelin, these mice show increased rate of axonal degeneration and subsequent neurological deficit in comparison with wild-type mice [48]. This and other studies [49, 50] have emphasised the role of myelin-forming cells in axonal health and survival independently of the role of myelin itself.

Given that myelin-forming cells provide trophic support [51] and the role of myelin in energy saving, successful remyelination of demyelinated axons, enables better and more long-lasting functionality and overall protection to axons than would be the case for the chronically demyelinated axons, as experimentally confirmed in mice [52]. Indeed, axonal degeneration represents a major feature of multiple sclerosis, as reported originally by Charcot [3]. Furthermore, axonal loss increases progressively over the duration of the disease [53, 54] and has been identified as a major determinant in the irreversible loss of neurological function [55]. Changes in mitochondrial activity and in organisation of the mitochondrial network help to promote axonal survival and thus play an important role in determining the overall disease progression rate.

8.1.3.1 Changes in Mitochondrial Distribution Associated with Demyelination

As mitochondria are the main source of ATP in neurons, the increased demand for ATP brought about by myelin loss and subsequent changes in the nature of impulse conduction and the expression pattern of ion channels are closely accompanied by changes in the pattern and distribution of axonal mitochondria. More specifically, axonal mitochondria aggregate within demyelinated regions. For example, in experimentally induced segmental demyelination of cat optic nerve, the number of mitochondria per unit area, as assessed by the number of mitochondrial profiles on the cross-section of electron micrograph, was found to be up to 2.5 times greater in demyelinated than in control optic nerve axons [56]. Interestingly, the greatest mitochondrial density within these demyelinated segments coincided with periods of impulse conduction re-establishment (i.e. the transition from conduction block to slow, continuous conduction) [57] when the energy demand is predicted to be the highest (as long as the frequency of conduction is not greatly reduced) and then decreased in parallel with advancing remyelination and thus the decrease in axonal energy demand. Furthermore, the peak density of mitochondria in the demyelinated segments of the same optic nerve axons was nearly identical to the density of mitochondria in their unmyelinated, intraocular, counterparts which were not affected by this experimental demyelination [57]. Such closely comparable mitochondrial densities suggest that the levels of energy demand in demyelinated and unmyelinated axons of similar calibres are similar.

In vitro, the effect of demyelination on mitochondrial distribution within axons has been examined in organotypic cerebellar slice cultures and myelinating dorsal root ganglion (DRG) cultures [58]. Although being examined at postnatal day 8 or 9, axons in organotypic cerebellar slice cultures do not possess all the characteristics of adult myelinated axons. Nonetheless, like in the adult cat axons described above, their exposure to demyelinating agent lysolecithin resulted in significant increase in mitochondrial content in comparison with axons from control cultures, as assessed by electron microscopy [59]. A similar effect has been reported in cultures of rat embryonic peripheral nerve axons. Namely, mixed cultures of DRG and Schwann cells, isolated from embryonic day 16, can be maintained in culture for up to 9 weeks during which time Schwann cells can be induced to myelinate the growing DRG axons by addition of ascorbic acid to the culture medium [60]. When lysolecithin is added to these cultures, axons become demyelinated. Electron microscopy assessment has shown that following exposure to lysolecithin, the size of stationary mitochondrial sites within demyelinated axons significantly increases compared with myelinated axons [58]. Also, larger mitochondria become significantly more abundant in demyelinated than in myelinated axons. However, in this model system unlike in adult mice, despite this increase in size, the number of stationary mitochondrial sites per unit of axonal length remained identical in myelinated and demyelinated axons, suggesting that some smaller, newly aggregated, mitochondria may have fused in order to form large mitochondria. This difference between axons demyelinated in culture and demyelinated in situ likely reflects the fact that axonal growth is continuing in vitro and other differences in the axon maturity and type of myelination between these model systems. In another words, developing axons and adult axons may employ different mechanisms to increase mitochondrial mass within demyelinated region.

A significant increase in axonal mitochondrial content has also been observed in the corpus callosum of mice fed cuprizone [59], a toxin which causes demyelination by selectively killing oligodendrocytes [61], in comparison with control fed mice. In this elegant study, authors used a serial block scanning electron microscope to precisely reconstruct individual mitochondria within both demyelinated and myelinated axons and quantify mitochondrial number, length and volume. Using this technique, the study found that the number, length and volume of individual mitochondria were significantly greater in demyelinated than in myelinated axons. Overall mitochondrial volume per axonal volume of demyelinated axons was found to be double that of myelinated axons [59]. Another experimental model of demyelination also showed the increase in mitochondrial content in demyelinated axons. Targeted demyelination induced by stereotactic injection of ethidium bromide (EB) to the caudal cerebellar peduncle was also associated with increased axonal mitochondrial content, reaching the peak levels during demyelination and early remyelination, but staying high even in fully remyelinated axons [62]. The difference between remyelinated and myelinated axons was attributed to a significant increase in mitochondrial number rather than their size [62]. It is worth noting that the glial trophic support to the axons is removed in the above models, thus the increased myelin content may represent a response to the combination of increased ATP and trophic demand. In vivo, real-time confocal imaging data from our laboratory showed that in demyelination of the peripheral nerve induced with lysolecithin, the increase in mitochondrial density in the demyelinated segment faithfully coincides with the period of re-establishment of continuous conduction and thus the greatest energy demand, in such axons.

Finally, the change in mitochondrial density has also been reported in axons in multiple sclerosis lesions. Immunohistochemical labelling for a mitochondrial marker called endoplasmic reticulum-associated binding protein (ERAB) showed a significant increase in the level of this protein in axons within active demyelinating lesions in comparison with normal-appearing white matter [63]. Similarly, a significant increase in mitochondrial density judged by the level of voltage-gated anion channel (VDAC), a protein specific for the outer membrane of all mitochondria [64], has been found in axons within chronically demyelinated lesions compared with control and normal-appearing white matter [65]. The increase in density of mitochondria in some human lesions seems to persist beyond the duration of demyelination, remaining significantly greater in remyelinated than in normally myelinated axons [62]. Although the stage and the length of human lesions are difficult to assign and immunohistochemical studies are less powerful in determining the precise number and volume of mitochondria per axon area than serial electron microscopy, these studies nonetheless suggest that an increase in mitochondrial mass is an important mechanism for matching increased energy and metabolic demand in human, as well as in experimental, axonal demyelination.

8.1.4 Mechanisms Underlying Mitochondrial Redistribution Along Demyelinated Axons

Mitochondria are remarkably dynamic organelles. In excised, intercostal nerve axons, about 15 % of all axonal mitochondria appear mobile [66]. In electrically active PNS axons conducting within physiological range, the number, the size and the velocity of mobile mitochondria increase proportionally to increase in conduction frequency. Remarkably, this increased mobilisation of mitochondria within conducting axons preferentially affects anterogradely moving mitochondria, thus resulting in accumulation of mitochondria in peripheral nerve terminals [22]. Thus, expectedly, the mechanism underlying the redistribution of mitochondria to energetically demanding regions of axons following demyelination is closely coupled with axonal transport of these organelles.

8.1.4.1 Axonal Transport of Mitochondria

Mitochondrial axonal transport is complex and highly regulated process which involves specialised molecular machinery (Fig. 8.3). The two major components of this machinery are (1) microtubules, which serve as tracks, and (2) molecular motors, which carry mitochondria along the tracks.

Microtubules are long, cylindrical structures spanning the length of axon, formed by polymerisation of α- and β-tubulin heterodimers. The heterodimers are assembled in such a way that α-tubulin is always exposed at one end of the microtubule

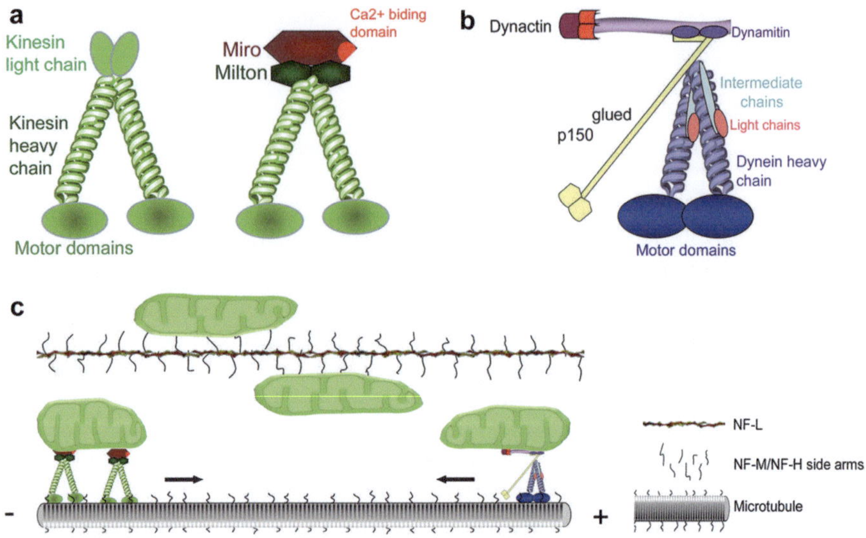

Fig. 8.3 Mitochondrial axonal transport machinery. (**a**) The canonical structure of kinesin 1, the main anterograde motor. Kinesin 1 consists of motor domains, heavy and light chains. The motor domains contain ATPase-hydrolysing moiety and bind to microtubules (shown in part **c**). Adaptor proteins Milton and Miro1, which also contain Ca^{2+}-binding domain, bind kinesin to mitochondria (shown in part **c**) and play a part in regulation of mitochondrial arrest at sites of high Ca^{2+} concentration. Using energy from ATP hydrolysis, kinesin makes step-like movements along microtubules towards its positive end (**c**), carrying mitochondria along the microtubule track. (**b**) Dynein is a multi-protein complex which acts as the major retrograde motor. Dynein also uses ATP for movement along the microtubules. For detailed structure of dynein see reference [76]. (**c**) Mitochondrial trafficking. Mitochondria are transported along microtubules by molecular motors kinesin 1 and dynein. Kinesin 1 moves towards the 'plus end' of the microtubule, which is typically furthest from the cell body. In contrast, dynein predominantly moves towards the 'minus end' of the microtubule, typically nearest the cell body. Stationary mitochondria appear to be bound to cytoskeletal proteins neurofilament medium (*NF-M*) and neurofilament heavy (*NF-H*) chain side arm bound to neurofilament light (*NF-L*) chain

and β-tubulin is exposed on the opposite end, thus giving intrinsic polarity to microtubules. By convention, the cell body-anchored end is called the 'minus end', and the growing end (with exposed α-tubulin) is called the 'plus end'. Such orientation whereby the 'plus end' points towards the axon terminal is nearly uniform in all axons [67] and plays an important part in axonal transport as molecular motors move preferentially in one direction. The major molecular motors for axonal transport are kinesin 1 and dynein [68]. These motors comprise of a number of specialised domains assembled into large, multiunit complexes (see Fig. 8.3). Two of the domains associate with microtubules and are thus called microtubule-binding domains. They are linked by a hinge region, resembling two feet (microtubule-binding domains), legs (light chains) and the hip (hinge region). The microtubule-binding domains exert ATP-hydrolysing activity [69]. Using the energy released by ATP hydrolysis, the 'legs' alternate in attaching and detaching from microtubules in a stepwise motion, thus pulling along the cargo [70, 71]. One ATP molecule is hydrolysed for every 8 nm step of kinesin [72]. The kinesin family of molecular motors predominantly moves towards the 'plus end' of microtubules, thus enabling anterograde (from cell body towards the terminal) transport of cargoes [74, 75].

Other domains of molecular motors, often called adaptor domains, have functions such as binding cargoes and sensing changes in the local environment. In the kinesin family, major adaptor proteins are cargo-binding Milton and Miro1. Other adaptor proteins connecting kinesin and mitochondria include syntabulin, fasciculation and elongation protein zeta 1 (FEZ1) and RAN binding protein 2 (RANBP2). Miro1 contains a Ca^{2+}-sensing and Ca^{2+}-binding domain which plays a critical part in mitochondrial arrest at sites of high Ca^{2+} concentration. Namely, the linkage between Miro1 and mitochondria is disrupted by Ca^{2+} binding to Miro1, e.g. at sites of high (micromolar) intra-axonal concentrations of Ca^{2+} [73]. In this way, the Ca^{2+}-dependent binding of Miro1 to mitochondria regulates mitochondrial retention at sites of high Ca^{2+} concentrations.

Dynein is a particularly complex multi-protein motor (for detailed structure, see review by Hirokawa et al. [76]), which also uses ATP for movement along the microtubules. The dynein family, and a related large complex called dynactin, can move in both directions, but favour movement towards the 'minus end' of microtubules, enabling retrograde (towards the cell body) transport [77–79].

A number of studies have (reviewed in [34]) proposed that the purpose of mobilisation of axonal mitochondria by means of axonal transport is a physiological response which enables delivery of sufficient number of mitochondria to sites of high-energy demand or high Ca^{2+} concentration, such as peripheral nerve terminals, active growth cones and synapses. Data from our lab has shown that mitochondria are preferentially mobilised from the areas proximal to the segment depleted of ATP (i.e. from the side of the cell body or through the anterograde axonal transport). Also, it has been shown in cultured rat cortical and cerebellar neurons that mitochondrial motility affects other aspects of mitochondrial dynamics, such as fission and fusion [80]. For example, it has been shown that the probability that a mobile mitochondrion will get engaged in fusion with a stationary mitochondrion increases proportionally to the velocity of the moving partner, whereby the higher the speed at impact, the higher the chance that the colliding mitochondria will fuse [80].

Studies of mitochondrial transport in myelinating DRG cultures showed that the velocity of mitochondrial transported along the axons increases by almost 50 % following demyelination, returning to the same level as prior to demyelination once axons become remyelinated. On the other hand, data from our lab shows that, in contrast to myelinating DRG cultures, axonal mitochondrial transport within demyelinated peripheral nerve axons in vivo becomes significantly reduced just prior to demyelination. It is unclear what accounts for this difference, as the same agent, lysolecithin, is used in both systems. It seems likely that in cultured DRGs, which are unlikely to be conducting, and which are still growing while being examined, mitochondria need to be transported towards the growing cone [81] to populate the growing axons, whereas in adult axons imaged in vivo mitochondria had already been positioned along axons prior to demyelination [22].

Once recruited from other parts of the axon and transported towards the demyelinated segments, a specific mechanism for mitochondrial anchoring is activated to ensure that mitochondria remain within the demyelinated area. The protein which tethers mitochondria to microtubules and thus immobilises them is called syntaphilin [82]. Immunoreactivity for this protein has been shown to significantly increase in experimentally demyelinated *corpus callosum* in mice, when compared with controls [59]. Indeed, deletion of syntaphilin prevented mitochondrial accumulation within demyelinated axons in mice [59]. Furthermore, demyelinated axons in these syntaphilin-deficient mice show significantly higher susceptibility to degeneration than those in wild-type controls [59]. Also, the study showed that the degeneration of demyelinated axons in syntaphilin-deficient mice could be minimised when mice were treated with sodium channel blocker flecainide, thus supporting the hypothesis that impulse load-induced sodium/calcium cascade initiates axonal degeneration in such axons. In human axons obtained from post-mortem brains from patients with multiple sclerosis, increased immunohistological labelling for syntaphilin associated with mitochondrial labelling has also been reported [65]. Therefore, focal mitochondrial immobilisation mediated by the anchoring protein syntaphilin appears to represent a common and important adaptation mechanism for providing energy, and/or a local calcium buffer to areas where ion fluxes of sodium and Ca^{2+} into denuded axon segments may become augmented.

8.1.4.2 Changes in Mitochondrial Function Associated with Demyelination

Apart from the mitochondrial mass, changes in mitochondrial function have also been reported within demyelinated axons. In experimental demyelination, induced in mice fed a cuprizone diet, mitochondrial complex IV activity was significantly increased in comparison with the control diet-fed mice [83]. However, in this study, the increase in complex IV activity was found only in the early stages of the treatment, i.e. prior to immunohistochemically detectible demyelination, the level of this complex activity being similar between the groups during the overt demyelination [83]. In another study, using ethidium bromide injection to induce focal demyelination, the activity of mitochondrial complex IV has been found to be significantly increased in remyelinated axons in comparison with myelinated axons [62]. However, the study does not report the complex IV activity in demyelinated axons.

In human axons, an increase in activity of complex IV has been reported within MS lesions negative for proteolipid protein (PLP), one of the major components of myelin, using immunohistochemistry, histochemistry and biochemical assays [84]. Also, the increase in activity of complex IV was found in a subset of axons within inactive lesions [65]. Importantly, the axons showing increased complex IV activity appeared structurally intact and had no pathological accumulation of amyloid precursor protein or dephosphorylated neurofilament. It is possible, therefore, that the greater mitochondrial activity in these axons reflects the successful adaptation to conditions in such lesions, in the form of high local ATP production to enable survival. However, only direct measurements of produced ATP, a methodology that is currently not available for human tissue, could directly address this hypothesis. Furthermore, studies that have examined lesions at different stages have described the opposite effect, i.e. the decrease of activity of mitochondrial complexes, as discussed below.

8.1.5 Effects of Inflammation on Mitochondrial Function and Transport in Inflammatory-Demyelinating Lesions

Inflammation is a major characteristic of MS. Importantly, a haematogenous inflammatory reaction develops within the CNS rapidly following initiation of lesions [85, 86]. The relative abundance of inflammatory cells has led to one of the most widely used classifications of MS lesions into active or inactive lesions. Typically, active MS lesions are characterised by the abundant presence of infiltrating inflammatory cells, mainly phagocytic macrophages filled with the initial degradation products of myelin, interspersed between segmentally demyelinated, relatively spared axons [87]. On the other hand, inactive (or chronic) lesions are typically characterised by fewer inflammatory cells scattered between demyelinated, thinly myelinated and damaged axons and intense glial scar [87]. Importantly, the extent of inflammation strongly and positively correlates with the magnitude of axonal degeneration [88, 89]. The mechanism underlying this high rate of axonal degeneration is partly attributed to inflammatory damage to mitochondria. Inflammatory cells produce a wide range of mediators [90–94], of which reactive nitrogen [95] and oxygen [96] species have been shown to be particularly toxic to mitochondria [97–102]. Indeed, while the increase in the activity of complex IV has been found in a subset of chronic, inactive white matter lesions [65, 84], at the active edge of these lesions where the inflammation is prominent, the complex IV activity has been found to be significantly decreased [65]. This pattern of complex IV activity is consistent with findings that reactive nitrogen and oxygen species produced by activated microglia and macrophages are potent inhibitors of complexes I and IV [103]. In the motor cortex of patients with MS, activity of complexes I, III and V (ATP synthase) has been found to be significantly decreased in comparison with non-neurological controls, even in the areas that had been classified as 'not lesioned' by conventional ultra- and immunohistochemical methods [48]. Interestingly, the activity of complex IV in patient tissue was comparable to that of controls, raising the possibility that the increase in activity of complex IV is a compensatory mechanism to maintain mitochondrial membrane potential. Nonetheless, the decrease in activity of complex V in this study suggests that neurons in the cortex of MS patients were deprived of ATP even before they become demyelinated. The decrease in the activity of the complexes in this study may be explained by the low expression of nuclear genes encoding for mitochondrial respiratory chain proteins [48]. However, it is unclear what caused the low expression pattern of these genes. Another study that reported a decrease in activity of complex I in chronic active lesions suggested that the cause may be the oxidative damage to mitochondrial DNA [104]. Experimental studies in acute, lipopolysaccharide (LPS)-induced model of systemic inflammation (sepsis) in rats showed that, compared with controls, systemic inflammation reduces activity of all mitochondrial complexes and oxygen consumption and significantly lowers production of ATP in mitochondria isolated from the liver [105] or heart muscle [106] and that this effect is prevented with antioxidant therapy. Also, injection of LPS in mouse brain has been shown to increase lipid peroxidation and decrease glutathione content at the site of injection, causing loss of mitochondrial membrane potential and mitochondrial integrity [107].

Experimental studies have further elucidated mechanisms by which mitochondria participate in axonal damage within active MS lesions. For example, in vivo confocal imaging of spinal cord axons of mice with experimental autoimmune encephalomyelitis (EAE), a widely used model of MS, demonstrated changes in the shape of intra-axonal mitochondria consistent with mitochondrial fragmentation [108]. Similar morphological changes in axonal mitochondria could be reproduced by external addition of reactive oxygen species, implicating this mediator in mitochondrial damage in EAE [108]. Functional status of such fragmented mitochondria is difficult to assess, although the fragmentation occurred preferentially within the axonal ovoids, the formations often preceding axonal transection. Studies from our lab suggest that fragmentation of axonal mitochondria occurs at the onset of neurological deficit in EAE and that it is associated with loss of mitochondrial membrane potential. Further in vivo confocal imaging studies showed that axonal transport of mitochondria and other cargoes is significantly reduced in EAE in comparison with axons from control mice, leading to loss of mitochondria in distal segments of affected axons [109]. Similarly to mitochondrial fragmentation, this axonal transport deficit preceded structural deficits and could be rescued by redox scavengers [109].

Inducible nitric oxide synthase that synthesises nitric oxide (NO) production in macrophages, another potent mitochondrial toxin, is often found in inflammatory foci in MS [95, 110]. Apart from possibly inhibiting mitochondrial respiration and thus ATP depletion, NO has another detrimental effect. Specifically, NO [111], as well as ROS [112], has been shown to play a role in the induction of the so-called mitochondrial permeability transition pore (mPTP), an inner mitochondrial membrane channel permeable for small molecules and solutes [113]. Abrupt opening of mPTP leads to loss of mitochondrial membrane potential, uncoupling of oxidative phosphorylation and rapid mitochondrial swelling. Perhaps most dangerously for the axon, the opening of mPTP releases intra-mitochondrial Ca^{2+} into the axoplasm, raising its concentration to pathologically high levels. Bearing in mind the capacity of mitochondria to store large amount of Ca^{2+} [114], the opening of mPTP is expected to be particularly detrimental within demyelinated segments of axons which already exhibit high mitochondrial density. Furthermore, mitochondria in such segments are likely to contain high Ca^{2+} concentration, given the altered ionic balance of demyelinated axon, including the described reversal of Na^+/Ca^{2+} exchanger.

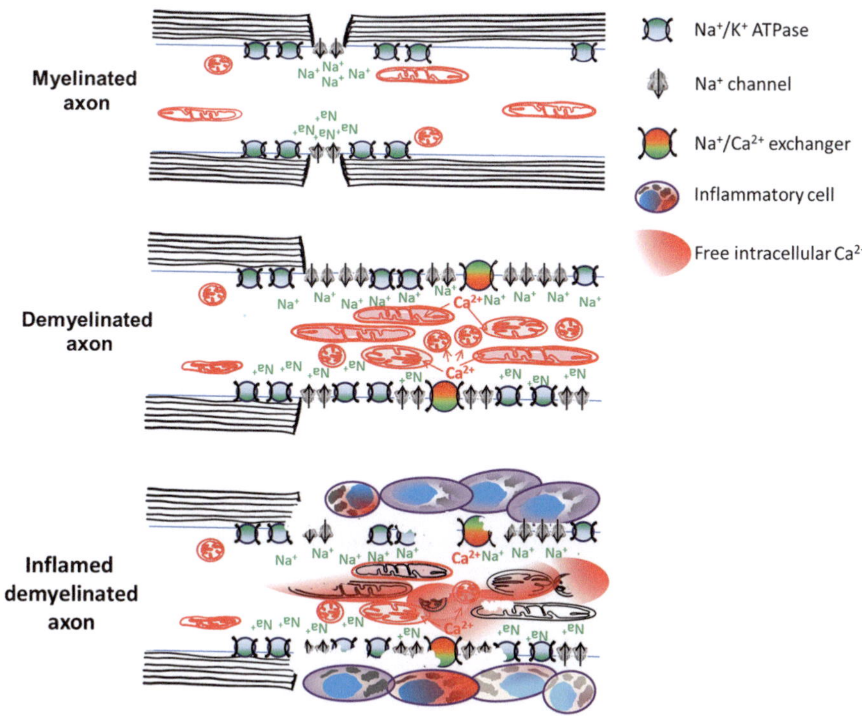

Fig. 8.4 Schematic representation of main axonal and mitochondrial changes following demyelination. *Top panel*: *In myelinated axon*, sodium channels are concentrated within narrow nodes of Ranvier. Sodium/potassium ATPase pumps sodium ions out of the axons following action potential propagation, thus re-establishing resting membrane potential. Mitochondria are distributed along the internode to provide ATP for the action of sodium pump. *Middle panel*: *Following demyelination*, sodium channels become distributed along the whole denuded surface of axolemma, thus facilitating continuous conduction of action potentials. The increased axoplasmic concentration of sodium ions brought about by continuous conduction necessitates greater activity of sodium pump. In addition, Na^+/Ca^{2+} exchanger operates in reverse, i.e. exports Na^+ in exchange for Ca^{2+}. Increased ATP demand and Ca^{2+} concentration induce increase in the number and size of mitochondria within the demyelinated segment, thus enabling its function and survival. *Bottom panel*: *Inflammation* disturbs the balance between the energy demand and supply established within demyelinated axons. Inflammatory mediators inhibit mitochondrial respiration, thus depleting ATP, and open mitochondrial permeability transition pore, releasing small solutes from mitochondrial matrix, including Ca^{2+}, which triggers a number of pathological processes leading to rapid axonal degeneration

Taken together, the strong association between inflammation and impairment of mitochondrial function demonstrated in studies above suggests that inflammatory environment may be the deciding factor with regard to mitochondrial involvement in axonal survival vs. axonal degeneration (summarised in Fig. 8.4). More specifically, in the absence of inflammation, intrinsic capacity of axonal mitochondrial network to match the increase of the local demand for energy production and Ca^{2+} handling appears sufficient to enable long-term axonal survival in adverse conditions such as demyelination. However, the presence of focal inflammation may disturb this balance by damaging mitochondria. This effect of inflammation on mitochondria may be particularly harmful in areas of axons in which mitochondrial mass had previously increased in order to buffer excess Ca^{2+}.

8.2 Guillain-Barré Syndrome and Inflammatory Neuropathies

Guillain-Barré syndrome (GBS) is a life-threatening, postinfectious, immune-mediated disease of the peripheral nervous system characterised by ascending muscle weakness. It is believed that the disease arises when the subject's immune response generates pathogen-specific antibodies that are cross-reactive with components of the peripheral nerve myelin or axolemma, resulting in myelin and/or axonal damage. Most patients recover; however, about a half retain residual neurological deficits [115]; and a proportion develop a chronic form of the disease known as chronic inflammatory-demyelinating polyradiculoneuropathy (CIDP) [116]. Although the triggers of GBS and CIDP have not been identified, and may be different, both of these diseases are characterised by inflammatory infiltration of the peripheral nerves, predominantly by lymphocytes and activated macrophages [117–121]. It has been shown that these cells produce the usual pro-inflammatory mediators, such as nitric oxide, ROS, proteases and pro-inflammatory cytokines [122]. However, little is known about the effect(s) of inflammation on mitochondria within the affected nerves. Also, it is not known whether mitochondrial dysfunction, either pre-existing (perhaps hidden by compensatory mechanisms active in noninflamed nerves) or secondary (caused by the inflammation), contributes to the pathogenesis of immune-mediated peripheral neuropathies. As mentioned above, some of these pro-inflammatory factors interfere with mitochondrial function and dynamics. For example, nitric oxide, produced by inducible nitric oxide synthase (iNOS) in activated macrophages, has been shown to reversibly inhibit mitochondrial respiration [97, 98]. It is likely that such NO-induced damage to mitochondria occurs in inflamed peripheral nerve axons, as the presence of nitrotyrosine, a marker of nitric oxide-mediated peroxynitrite formation, has been identified in nerve biopsies from patients with CIDP [122]. In neuron and Schwann cell cocultures, neurons seem intrinsically more susceptible than Schwann cells to detrimental effects of pathological concentrations of NO [122]. Furthermore, nitric oxide has been shown to lead to excessive mitochondrial fission in cultured cortical neurons via S-nitrosylation of dynamin-related protein 1 [123, 124]. It is possible that NO produced by iNOS-positive inflammatory cells may result in excessive mitochondrial fission in peripheral nerve axons during the course of inflammatory neuropathies. However, whether and how inflammation alters mitochondrial function and transport in axons in vivo, it is entirely unknown. It could be speculated that, as in the CNS [109] and other cells [125], certain inflammatory signals may impair mitochondrial transport in peripheral nerve axons. For example, in cultured fibroblasts, tumour necrosis factor alpha (TNFα) has been shown to inhibit mitochondrial transport by function of hyperphosphorylation of motor protein kinesin 1 [125]. Disruption of microtubule networks, as shown to occur during experimental autoimmune encephalomyelitis in mice [126], may further contribute to impairment of mitochondrial transport in peripheral nerve axons.

8.3 Conclusion

Impulse conduction, myelination status and effects of inflammatory insult to the axons are closely interconnected and interdependent. Highly dynamic axonal mitochondria represent a major adaptive mechanism for maintaining functionality and structural integrity of axons under pathological conditions. To this effect, mitochondria act as a network rather than as individual organelles. Utilising upregulation in mitochondrial complex activity, changes in fusion/fission balance and redistribution of mitochondrial mass, mitochondrial network can be effectively reorganised to achieve optimal Ca^{2+} storage and supply of ATP to vulnerable regions of long axonal tracts. Such reorganisation and functional adaptation enables both short- and long-term protection against neurodegeneration. These adaptive modifications may, however, become exhausted under sustained or aggressive inflammatory attack. Furthermore, toxic effects of inflammatory mediators on Ca^{2+}-laden axonal mitochondria may result in destructive rise in intra-axonal Ca^{2+} levels and subsequent rapid axonal degeneration. Further evidence is hence needed to fully establish the role of mitochondria in the pathogenesis and/or progression of the inflammatory diseases of central and peripheral nervous systems.

References

1. Pugliatti M, Rosati G, Carton H, Riise T, Drulovic J, Vécsei L, et al. The epidemiology of multiple sclerosis in Europe. Eur J Neurol. 2006;7:700–22.
2. Tullman MJ. Overview of the epidemiology, diagnosis, and disease progression associated with multiple sclerosis. Am J Manag Care [Internet]. 2013;19(2 Suppl):S15–20.
3. Charcot JM. Histologie de la sclerose en plaques. Gaz Hop. 1868;41:554–66.
4. Trapp BD. Pathogenesis of multiple sclerosis: the eyes only see what the mind is prepared to comprehend. Ann Neurol. 2004;55(4):455–7.
5. Compston A, McDonald I, Noseworthy J, Lassmann H, Miller D, Smith K, Wekerle H, Confavreux C, editors. McAlpine's multiple sclerosis. 4th ed. Philadelphia: Elsevier; 2006.
6. Heppner FL, Ransohoff RM, Becher B. Immune attack: the role of inflammation in Alzheimer disease. Nat Rev Neurosci. 2015;16(6):358–72.
7. Hirsch EC, Hunot S. Neuroinflammation in Parkinson's disease: a target for neuroprotection? Lancet Neurol. 2009;8(4):382–97.
8. Amor S, Puentes F, Baker D, van der Valk P. Inflammation in neurodegenerative diseases. Immunology. 2010;129(2):154–69.

9. Glass CK, Saijo K, Winner B, Marchetto MC, Gage FH. Mechanisms underlying inflammation in neurodegeneration. Cell. 2010;140(6):918–34.
10. Bohnen NI, Albin RL. White matter lesions in Parkinson disease. Nat Rev Neurol. 2011;7(4):229–36.
11. Agosta F, Dalla Libera D, Spinelli EG, Finardi A, Canu E, Bergami A, et al. Myeloid microvesicles in cerebrospinal fluid are associated with myelin damage and neuronal loss in mild cognitive impairment and Alzheimer disease. Ann Neurol. 2014;76(6):813–25.
12. Kutzelnigg A, Lucchinetti CF, Stadelmann C, Brück W, Rauschka H, Bergmann M, et al. Cortical demyelination and diffuse white matter injury in multiple sclerosis. Brain. 2005;128(Pt 11):2705–12.
13. Gay FW, Drye TJ, Dick GWA, Esiri MM. The application of multifactorial cluster analysis in the staging of plaques in early multiple sclerosis: identification and characterization of the primary demyelinating lesion. Brain. 1997;120(8):1461–83.
14. Barnett MH, Prineas JW. Relapsing and remitting multiple sclerosis: pathology of the newly forming lesion. Ann Neurol. 2004;55(4):458–68.
15. Henderson APD, Barnett MH, Parratt JDE, Prineas JW. Multiple sclerosis: distribution of inflammatory cells in newly forming lesions. Ann Neurol. 2009;66(6):739–53.
16. Jacobs JM. On internodal length. J Anat. 1988;157:153–62.
17. Rasminsky M, Sears TA. Internodal conduction in undissected demyelinated nerve fibres. J Physiol. 1972;227(2):323–50.
18. Bostock H, Sears TA. The internodal axon membrane: electrical excitability and continuous conduction in segmental demyelination. J Physiol. 1978;280:273–301.
19. Smith KJ, McDonald WI. The pathophysiology of multiple sclerosis: the mechanisms underlying the production of symptoms and the natural history of the disease. Philos Trans R Soc Lond B Biol Sci. 1999;354(1390):1649–73.
20. Amiri M, Hollenbeck PJ. Mitochondrial biogenesis in the axons of vertebrate peripheral neurons. Dev Neurobiol [Internet]. 2008;68(11):1348–61.
21. Cotter D, Guda P, Fahy E, Subramaniam S. MitoProteome: mitochondrial protein sequence database and annotation system. Nucleic Acids Res. 2004;32(Database issue):D463–7.
22. Sajic M, Mastrolia V, Lee CY, Trigo D, Sadeghian M, Mosley AJ, et al. Impulse conduction increases mitochondrial transport in adult mammalian peripheral nerves in vivo. PLoS Biol. 2013;11(12):e1001754.
23. Bristow EA. The distribution of mitochondrial activity in relation to optic nerve structure. Arch Ophthalmol. 2002;120(6):791.
24. Barron MJ, Griffiths P, Turnbull DM, Bates D, Nichols P. The distributions of mitochondria and sodium channels reflect the specific energy requirements and conduction properties of the human optic nerve head. Br J Ophthalmol. 2004;88(2):286–90.
25. Minckler DS, McLean IW, Tso MO. Distribution of axonal and glial elements in the rhesus optic nerve head studied by electron microscopy. Am J Ophthalmol. 1976;82(2):179–87.
26. Hall SM, Gregson NA. The in vivo and ultrastructural effects of injection of lysophosphatidylcholine into myelinated peripheral nerve fibres of the adult mouse. J Cell Sci. 1971;9(3):769–89.
27. Smith KJ, Bostock H, Hall SM. Saltatory conduction precedes remyelination in axons demyelinated with lysophosphatidyl choline. J Neurol Sci. 1982;54(1):13–31.
28. Smith KJ, McDonald WI. Spontaneous and evoked electrical discharges from a central demyelinating lesion. J Neurol Sci. 1982;55(1):39–47.
29. Noebels JL, Marcom PK, Jalilian-Tehrani MH. Sodium channel density in hypomyelinated brain increased by myelin basic protein gene deletion. Nature. 1991;352(6334):431–4.
30. Craner MJ, Lo AC, Black JA, Waxman SG. Abnormal sodium channel distribution in optic nerve axons in a model of inflammatory demyelination. Brain. 2003;126(Pt 7):1552–61.
31. Moll C, Mourre C, Lazdunski M, Ulrich J. Increase of sodium channels in demyelinated lesions of multiple sclerosis. Brain Res. 1991;556(2):311–6.

32. Craner MJ, Newcombe J, Black JA, Hartle C, Cuzner ML, Waxman SG. Molecular changes in neurons in multiple sclerosis: altered axonal expression of Nav1.2 and Nav1.6 sodium channels and Na+/Ca2+ exchanger. Proc Natl Acad Sci U S A. 2004;101(21):8168–73.
33. Dupree JL, Girault JA, Popko B. Axo-glial interactions regulate the localization of axonal paranodal proteins. J Cell Biol. 1999;147(6):1145–52.
34. Mathis C, Denisenko-Nehrbass N, Girault J-A, Borrelli E. Essential role of oligodendrocytes in the formation and maintenance of central nervous system nodal regions. Development. 2001;128(23):4881–90.
35. Boyle MET, Berglund EO, Murai KK, Weber L, Peles E, Ranscht B. Contactin orchestrates assembly of the septate-like junctions at the paranode in myelinated peripheral nerve. Neuron. 2001;30(2):385–97.
36. Poliak S, Gollan L, Salomon D, Berglund EO, Ohara R, Ranscht B, et al. Localization of Caspr2 in myelinated nerves depends on axon-glia interactions and the generation of barriers along the axon. J Neurosci. 2001;21(19):7568–75.
37. Stys PK, Waxman SG, Ransom BR. Na(+)-Ca2+ exchanger mediates Ca2+ influx during anoxia in mammalian central nervous system white matter. Ann Neurol. 1991;30(3):375–80.
38. Stys P, Waxman S, Ransom B. Ionic mechanisms of anoxic injury in mammalian CNS white matter: role of Na+ channels and Na(+)-Ca2+ exchanger. J Neurosci. 1992;12(2):430–9.
39. Li S, Jiang Q, Stys PK. Important role of reverse Na+-Ca2+ exchange in spinal cord white matter injury at physiological temperature. J Neurophysiol. 2000;84(2):1116–9.
40. Die Sogenannte OM. Akute multiple sklerose. Jahrb Psychiatr. 1906;27:211–312.
41. Prineas JW, Connell F. Remyelination in multiple sclerosis. Ann Neurol. 1979;5(1):22–31.
42. Prineas JW, Barnard RO, Revesz T, Kwon EE, Sharer L, Cho E-S. Multiple sclerosis. Brain. 1993;116(3):681–93.
43. Lassmann H. Comparative neuropathology of chronic experimental allergic encephalomyelitis and multiple sclerosis. Schriftenr Neurol. 1983;25(1):135.
44. Moore GR, Neumann PE, Suzuki K, Lijtmaer HN, Traugott U, Raine CS. Balo's concentric sclerosis: new observations on lesion development. Ann Neurol. 1985;17(6):604–11.
45. Willoughby EW, Grochowski E, Li DK, Oger J, Kastrukoff LF, Paty DW. Serial magnetic resonance scanning in multiple sclerosis: a second prospective study in relapsing patients. Ann Neurol. 1989;25(1):43–9.
46. Koopmans RA, Li DK, Oger JJ, Kastrukoff LF, Jardine C, Costley L, et al. Chronic progressive multiple sclerosis: serial magnetic resonance brain imaging over six months. Ann Neurol. 1989;26(2):248–56.
47. Thompson AJ, Kermode AG, Wicks D, MacManus DG, Kendall BE, Kingsley DP, et al. Major differences in the dynamics of primary and secondary progressive multiple sclerosis. Ann Neurol. 1991;29(1):53–62.
48. Dutta R, McDonough J, Yin X, Peterson J, Chang A, Torres T, et al. Mitochondrial dysfunction as a cause of axonal degeneration in multiple sclerosis patients. Ann Neurol. 2006;59(3):478–89.
49. Lappe-Siefke C, Goebbels S, Gravel M, Nicksch E, Lee J, Braun PE, et al. Disruption of Cnp1 uncouples oligodendroglial functions in axonal support and myelination. Nat Genet. 2003;33(3):366–74.
50. Li C, Tropak MB, Gerlai R, Clapoff S, Abramow-Newerly W, Trapp B, et al. Myelination in the absence of myelin-associated glycoprotein. Nature. 1994;369(6483):747–50.
51. Fünfschilling U, Supplie LM, Mahad D, Boretius S, Saab AS, Edgar J, et al. Glycolytic oligodendrocytes maintain myelin and long-term axonal integrity. Nature. 2012;485(7399):517–21.
52. Irvine KA, Blakemore WF. Remyelination protects axons from demyelination-associated axon degeneration. Brain. 2008;131(Pt 6):1464–77.
53. Lovas G, Szilagyi N, Majtenyi K, Palkovits M, Komoly S. Axonal changes in chronic demyelinated cervical spinal cord plaques. Brain. 2000;123(2):308–17.

54. Bjartmar C, Kidd G, Mörk S, Rudick R, Trapp BD. Neurological disability correlates with spinal cord axonal loss and reduced N-acetyl aspartate in chronic multiple sclerosis patients. Ann Neurol. 2000;48(6):893–901.
55. De Stefano N, Matthews PM, Fu L, Narayanan S, Stanley J, Francis GS, et al. Axonal damage correlates with disability in patients with relapsing-remitting multiple sclerosis. Results of a longitudinal magnetic resonance spectroscopy study. Brain. 1998;121(Pt 8):1469–77.
56. Mutsaers SE, Carroll WM. Focal accumulation of intra-axonal mitochondria in demyelination of the cat optic nerve. Acta Neuropathol. 1998;96(2):139–43.
57. Carroll WM, Jennings AR, Mastaglia FL. Galactocerebroside antiserum causes demyelination of cat optic nerve. Brain Res. 1985;330(2):378–81.
58. Kiryu-Seo S, Ohno N, Kidd GJ, Komuro H, Trapp BD. Demyelination increases axonal stationary mitochondrial size and the speed of axonal mitochondrial transport. J Neurosci. 2010;30(19):6658–66.
59. Ohno N, Chiang H, Mahad DJ, Kidd GJ, Liu L, Ransohoff RM, et al. Mitochondrial immobilization mediated by syntaphilin facilitates survival of demyelinated axons. Proc Natl Acad Sci U S A. 2014;111(27):9953–8.
60. Fex Svenningsen A, Shan W-S, Colman DR, Pedraza L. Rapid method for culturing embryonic neuron-glial cell cocultures. J Neurosci Res. 2003;72(5):565–73.
61. Matsushima GK, Morell P. The neurotoxicant, cuprizone, as a model to study demyelination and remyelination in the central nervous system. Brain Pathol. 2001;11(1):107–16.
62. Zambonin JL, Zhao C, Ohno N, Campbell GR, Engeham S, Ziabreva I, et al. Increased mitochondrial content in remyelinated axons: implications for multiple sclerosis. Brain. 2011;134(Pt 7):1901–13.
63. Witte ME, Mahad DJ, Lassmann H, van Horssen J. Mitochondrial dysfunction contributes to neurodegeneration in multiple sclerosis. Trends Mol Med. 2014;20(3):179–87.
64. Colombini M. Voltage gating in the mitochondrial channel, VDAC. J Membr Biol. 1989;111(2):103–11.
65. Mahad DJ, Ziabreva I, Campbell G, Lax N, White K, Hanson PS, et al. Mitochondrial changes within axons in multiple sclerosis. Brain. 2009;132(Pt 5):1161–74.
66. Misgeld T, Kerschensteiner M, Bareyre FM, Burgess RW, Lichtman JW. Imaging axonal transport of mitochondria in vivo. Nat Methods. 2007;4(7):559–61.
67. Dombeck DA, Kasischke KA, Vishwasrao HD, Ingelsson M, Hyman BT, Webb WW. Uniform polarity microtubule assemblies imaged in native brain tissue by second-harmonic generation microscopy. Proc Natl Acad Sci U S A. 2003;100(12):7081–6.
68. Pilling AD, Horiuchi D, Lively CM, Saxton WM. Kinesin-1 and dynein are the primary motors for fast transport of mitochondria in drosophila motor axons. Mol Biol Cell. 2006;17(2):2057–68.
69. Hua W, Young EC, Fleming ML, Gelles J. Coupling of kinesin steps to ATP hydrolysis. Nature. 1997;388(6640):390–3. [Internet]. Available from: file:///Users/mbzcf/Documents/Papers/1997/Hua/Nature 1997 Hua.pdf\npapers://37ea81e1-bf1a-4112-927e-b768b470c325/Paper/p778.
70. Pfister KK, Shah PR, Hummerich H, Russ A, Cotton J, Annuar AA. Genetic analysis of the cytoplasmic dynein subunit families. PLoS Genet. 2006;2(1):e1.
71. Hirokawa N, Takemura R. Molecular motors and mechanisms of directional transport in neurons. Nat Rev Neurosci. 2005;6(3):201–14.
72. Schnitzer MJ, Block SM. Kinesin hydrolyses one ATP per 8-nm step. Nature. 1997;388(6640):386–90.
73. MacAskill AF, Rinholm JE, Twelvetrees AE, Arancibia-Carcamo IL, Muir J, Fransson A, et al. Miro1 is a calcium sensor for glutamate receptor-dependent localization of mitochondria at synapses. Neuron. 2009;61(4):541–55.
74. Hirokawa N, Sato-Yoshitake R, Kobayashi N, Pfister KK, Bloom GS, Brady ST. Kinesin associates with anterogradely transported membranous organelles in vivo. J Cell Biol. 1991;114(2):295–302.

75. Vale RD, Reese TS, Sheetz MP. Identification of a novel force-generating protein, kinesin, involved in microtubule-based motility. Cell. 1985;42(1):39–50.
76. Hirokawa N, Niwa S, Tanaka Y. Molecular motors in neurons: transport mechanisms and roles in brain function, development, and disease. Neuron. 2010;68(4):610–38.
77. Ross JL, Wallace K, Shuman H, Goldman YE, Holzbaur ELF. Processive bidirectional motion of dynein-dynactin complexes in vitro. Nat Cell Biol. 2006;8(6):562–70.
78. Mallik R, Petrov D, Lex SA, King SJ, Gross SP. Building complexity: an in vitro study of cytoplasmic dynein with in vivo implications. Curr Biol. 2005;15(23):2075–85.
79. Saxton WM, Hollenbeck PJ. The axonal transport of mitochondria. J Cell Sci. 2012;125(Pt 9):2095–104.
80. Cagalinec M, Safiulina D, Liiv M, Liiv J, Choubey V, Wareski P, et al. Principles of the mitochondrial fusion and fission cycle in neurons. J Cell Sci. 2013;126(pt10):2187–97.
81. Ruthel G, Hollenbeck PJ. Response of mitochondrial traffic to axon determination and differential branch growth. J Neurosci. 2003;23(24):8618–24.
82. Kang JS, Tian JH, Pan PY, Zald P, Li C, Deng C, et al. Docking of axonal mitochondria by syntaphilin controls their mobility and affects short-term facilitation. Cell. 2008;132(1):137–48. Synaptic Function Section, The Porter Neuroscience Research Center, National Institute of Neurological Disorders and Stroke, National Institutes of Health, Building 35, Room 3B203, 35 Convent Drive, Bethesda, MD 20892, USA.
83. Acs P, Selak MA, Komoly S, Kalman B. Distribution of oligodendrocyte loss and mitochondrial toxicity in the cuprizone-induced experimental demyelination model. J Neuroimmunol. 2013;262(1–2):128–31.
84. Witte ME, Bø L, Rodenburg RJ, Belien JA, Musters R, Hazes T, et al. Enhanced number and activity of mitochondria in multiple sclerosis lesions. J Pathol. 2009;219(2):193–204.
85. Traugott U, Reinherz EL, Raine CS. Multiple sclerosis. Distribution of T cells, T cell subsets and Ia-positive macrophages in lesions of different ages. J Neuroimmunol. 1983;4(3):201–21.
86. Esiri MM, Reading MC. Macrophage populations associated with multiple sclerosis plaques. Neuropathol Appl Neurobiol. 1987;13(6):451–65.
87. Brück W, Porada P, Poser S, Rieckmann P, Hanefeld F, Kretzschmar HA, et al. Monocyte/macrophage differentiation in early multiple sclerosis lesions. Ann Neurol. 1995;38(5):788–96.
88. Ferguson B. Axonal damage in acute multiple sclerosis lesions. Brain. 1997;120(3):393–9. [Internet].
89. Trapp BD, Peterson J, Ransohoff RM, Rudick R, Mörk S, Bö L. Axonal transection in the lesions of multiple sclerosis. N Engl J Med. 1998;338(5):278–85.
90. Cannella B, Raine CS. The adhesion molecule and cytokine profile of multiple sclerosis lesions. Ann Neurol. 1995;37(4):424–35.
91. Fabriek BO, Van Haastert ES, Galea I, Polfliet MMJ, Döpp ED, Van Den Heuvel MM, et al. CD163-positive perivascular macrophages in the human CNS express molecules for antigen recognition and presentation. Glia. 2005;51(4):297–305.
92. Marik C, Felts PA, Bauer J, Lassmann H, Smith KJ. Lesion genesis in a subset of patients with multiple sclerosis: a role for innate immunity? Brain. 2007;130(Pt 11):2800–15.
93. Cuzner ML, Opdenakker G. Plasminogen activators and matrix metalloproteases, mediators of extracellular proteolysis in inflammatory demyelination of the central nervous system. J Neuroimmunol. 1999;94(1–2):1–14.
94. Lindberg RLP. The expression profile of matrix metalloproteinases (MMPs) and their inhibitors (TIMPs) in lesions and normal appearing white matter of multiple sclerosis. Brain. 2001;124(9):1743–53.
95. Liu JS, Zhao ML, Brosnan CF, Lee SC. Expression of inducible nitric oxide synthase and nitrotyrosine in multiple sclerosis lesions. Am J Pathol. 2001;158(6):2057–66.
96. Aboul-Enein F, Rauschka H, Kornek B, Stadelmann C, Stefferl A, Brück W, et al. Preferential loss of myelin-associated glycoprotein reflects hypoxia-like white matter damage in stroke and inflammatory brain diseases. J Neuropathol Exp Neurol. 2003;62(1):25–33.

97. Cleeter MW, Cooper JM, Darley-Usmar VM, Moncada S, Schapira AH. Reversible inhibition of cytochrome c oxidase, the terminal enzyme of the mitochondrial respiratory chain, by nitric oxide. Implications for neurodegenerative diseases. FEBS Lett. 1994;345(1):50–4.
98. Clementi E, Brown GC, Feelisch M, Moncada S. Persistent inhibition of cell respiration by nitric oxide: crucial role of S-nitrosylation of mitochondrial complex I and protective action of glutathione. Proc Natl Acad Sci U S A. 1998;95(13):7631–6.
99. Granger DL, Lehninger AL. Sites of inhibition of mitochondrial electron transport in macrophage-injured neoplastic cells. J Cell Biol. 1982;95(2 Pt 1):527–35.
100. Bolaños JP, Peuchen S, Heales SJR, Land JM, Clark JB. Nitric oxide-mediated inhibition of the mitochondrial respiratory chain in cultured astrocytes. J Neurochem. 2002;63(3):910–6.
101. Geng Y, Hansson GK, Holme E. Interferon-gamma and tumor necrosis factor synergize to induce nitric oxide production and inhibit mitochondrial respiration in vascular smooth muscle cells. Circ Res. 1992;71(5):1268–76.
102. McCord JM, Keele BB, Fridovich I. An enzyme-based theory of obligate anaerobiosis: the physiological function of superoxide dismutase. Proc Natl Acad Sci. 1971;68(5):1024–7.
103. Qi X, Lewin AS, Sun L, Hauswirth WW, Guy J. Mitochondrial protein nitration primes neurodegeneration in experimental autoimmune encephalomyelitis. J Biol Chem. 2006;281(42):31950–62.
104. Lu F, Selak M, O'Connor J, Croul S, Lorenzana C, Butunoi C, et al. Oxidative damage to mitochondrial DNA and activity of mitochondrial enzymes in chronic active lesions of multiple sclerosis. J Neurol Sci. 2000;177(2):95–103.
105. Lowes DA, Webster NR, Murphy MP, Galley HF. Antioxidants that protect mitochondria reduce interleukin-6 and oxidative stress, improve mitochondrial function, and reduce biochemical markers of organ dysfunction in a rat model of acute sepsis. Br J Anaesth. 2013;110(3):472–80.
106. Vanasco V, Magnani ND, Cimolai MC, Valdez LB, Evelson P, Boveris A, et al. Endotoxemia impairs heart mitochondrial function by decreasing electron transfer, ATP synthesis and ATP content without affecting membrane potential. J Bioenerg Biomembr. 2012;44(2):243–52.
107. Noble F, Rubira E, Boulanouar M, Palmier B, Plotkine M, Warnet J-M, et al. Acute systemic inflammation induces central mitochondrial damage and mnesic deficit in adult Swiss mice. Neurosci Lett. 2007;424(2):106–10.
108. Nikić I, Merkler D, Sorbara C, Brinkoetter M, Kreutzfeldt M, Bareyre FM, et al. A reversible form of axon damage in experimental autoimmune encephalomyelitis and multiple sclerosis. Nat Med. 2011;17(4):495–9. Nature Publishing Group, a division of Macmillan Publishers Limited. All Rights Reserved.
109. Sorbara CD, Wagner NE, Ladwig A, Nikić I, Merkler D, Kleele T, et al. Pervasive axonal transport deficits in multiple sclerosis models. Neuron. 2014;84(6):1183–90.
110. Bö L, Dawson TM, Wesselingh S, Mörk S, Choi S, Kong PA, et al. Induction of nitric oxide synthase in demyelinating regions of multiple sclerosis brains. Ann Neurol. 1994;36:778–86.
111. Martin LJ, Adams NA, Pan Y, Price A, Wong M. The mitochondrial permeability transition pore regulates nitric oxide-mediated apoptosis of neurons induced by target deprivation. J Neurosci. 2011;31(1):359–70.
112. Wang W, Fang H, Groom L, Cheng A, Zhang W, Liu J, et al. Superoxide flashes in single mitochondria. Cell. 2008;134(2):279–90.
113. Haworth RA, Hunter DR. The Ca2+-induced membrane transition in mitochondria. Arch Biochem Biophys. 1979;195(2):460–7.
114. Rizzuto R, Simpson AW, Brini M, Pozzan T. Rapid changes of mitochondrial Ca2+ revealed by specifically targeted recombinant aequorin. Nature. 1992;358(6384):325–7. Department of Biomedical Sciences, University of Padova Italy.
115. Dornonville de la Cour C, Jakobsen J. Residual neuropathy in long-term population-based follow-up of Guillain-Barré syndrome. Neurology. 2005;64(2):246–53.
116. Hughes R, Sanders E, Hall S, Atkinson P, Colchester A, Payan P. Subacute idiopathic demyelinating polyradiculoneuropathy. Arch Neurol. 1992;49:612–6.

117. Miyakawa T, Murayama E, Sumiyoshi S, Deshimaru M, Kamano A. A biopsy case of Landry-Guillain-Barré syndrome. Acta Neuropathol. 1971;17(3):181–7.
118. Wiśniewski H. Landry-Guillain-Barré syndrome. Arch Neurol. 1969;21(3):269.
119. Illes Z. Accumulation of V 7.2-J 33 invariant T cells in human autoimmune inflammatory lesions in the nervous system. Int Immunol. 2004;16(2):223–30.
120. Winer J, Hughes S, Cooper J, Ben-Smith A, Savage C. Gamma delta T cells infiltrating sensory nerve biopsies from patients with inflammatory neuropathy. J Neurol. 2002;249(5):616–21.
121. Wanschitz J, Maier H, Lassmann H, Budka H, Berger T. Distinct time pattern of complement activation and cytotoxic T cell response in Guillain-Barré syndrome. Brain. 2003;126(Pt 9):2034–42.
122. Lehmann HC, Köhne A, Meyer zu Hörste G, Dehmel T, Kiehl O, Hartung H-P, et al. Role of nitric oxide as mediator of nerve injury in inflammatory neuropathies. J Neuropathol Exp Neurol. 2007;66(4):305–12.
123. Cho D-H, Nakamura T, Fang J, Cieplak P, Godzik A, Gu Z, et al. S-nitrosylation of Drp1 mediates beta-amyloid-related mitochondrial fission and neuronal injury. Science. 2009;324(5923):102–5.
124. Barsoum MJ, Yuan H, Gerencser AA, Liot G, Kushnareva Y, Gräber S, et al. Nitric oxide-induced mitochondrial fission is regulated by dynamin-related GTPases in neurons. EMBO J. 2006;25(16):3900–11.
125. De Vos K, Severin F, Van Herreweghe F, Vancompernolle K, Goossens V, Hyman A, et al. Tumor necrosis factor induces hyperphosphorylation of kinesin light chain and inhibits kinesin-mediated transport of mitochondria. J Cell Biol. 2000;149(6):1207–14.
126. Shriver LP, Dittel BN. T-cell-mediated disruption of the neuronal microtubule network: correlation with early reversible axonal dysfunction in acute experimental autoimmune encephalomyelitis. Am J Pathol. 2006;169(3):999–1011.

Chapter 9
Mitochondria, the Synapse, and Neurodegeneration

Alexia Chrysostomou and Doug M. Turnbull

Abstract Mitochondria have a remarkably sophisticated structure and essential function within neurons that ensure normal synaptic transmission and maintained synaptic plasticity and hence normal information processing that can occur within the brain. The importance of mitochondria for synaptic structural preservation and activity is exemplified by a high mitochondrial density at synaptic sites, whereas synaptic mitochondrial dysfunction or loss of mitochondria from synapses leads to synaptic abnormalities. The increasing body of evidence for mitochondrial defects, early synaptic dysfunction, and/or decreased synaptic density in common neurodegenerative disorders places these organelles at centre stage and establishes synaptic mitochondria as good candidates for drug targeting. In this chapter, we describe synaptic mitochondrial functions and present the evidence for synaptic disturbances in common neurodegenerative and mitochondrial diseases. Finally, we discuss the proposed mechanisms of neuronal loss with a focus on early synaptic loss.

Keywords Synapses • Mitochondria • ATP • Calcium buffering • Alzheimer's disease • Parkinson's disease • Huntington's disease • Mitochondrial disease • Synaptopathy • Neurodegeneration

9.1 Neurons and Mitochondria

The highly specialized neuronal morphology allows neurons to form networks that control and facilitate all the information processing that occurs within the brain. Information is conveyed from one neuron to another via synaptic transmission, a process that is highly dependent on mitochondrial function. Hence, mitochondrial

A. Chrysostomou, BSc, MRes, PhD (✉)
Wellcome Trust Centre for Mitochondrial Research, Institute of Neuroscience,
Newcastle upon Tyne, UK
e-mail: Alexia.Chrysostomou2@newcastle.ac.uk

D.M. Turnbull, MD, PhD
Newcastle University Medical School, Wellcome Trust Centre for Mitochondrial Research,
Newcastle upon Tyne, UK

A.K. Reeve et al. (eds.), *Mitochondrial Dysfunction in Neurodegenerative Disorders*, DOI 10.1007/978-3-319-28637-2_9

biogenesis, transport, and quality control are vital to neuronal health and furthermore to brain function.

9.1.1 Mitochondrial Biogenesis and Axonal Transport

Although mitochondria retain a small genome (mitochondrial DNA, or mtDNA) encoding 13 proteins, the majority of mitochondrial proteins are encoded by nuclear DNA. Thus, mitochondrial synthesis and recycling have traditionally been thought to occur in neuronal nuclei [1] and that mitochondria are conveyed to/removed from distal parts of the axon via motor protein-assisted transport. There are two types of axonal mitochondrial transport, anterograde and retrograde. Anterograde transport is facilitated by kinesin motors and involves mitochondrial transport towards the axonal terminal end of microtubules [2]. Vertebrate neuronal mitochondria move anterogradely with an estimated speed of ~0.32 μm/s [3]. Dyneins, on the other hand, facilitate retrograde mitochondrial transport back to the cell body (towards the minus end of microtubules) [4], with a predicted speed of ~0.38 μm/s in vertebrate neurons [3]. Mitochondria constantly undergo cycles of moving, stopping, and restarting. They are capable of moving in both directions, whereas specific neuronal and physiological signals promote mitochondrial anchoring to particular neuronal sites [5–8].

However, some neurons possess long axons, several hundred centimetres in length, which would make it impossible for slow-moving mitochondria to reach distal synapses fast enough to meet the synaptic energetic demands. It is thus of no surprise that there is a growing body of evidence suggesting that there are local protein machineries that account for axonal and presynaptic protein synthesis. More specifically, research using invertebrate systems [9, 10], cultured sympathetic [11, 12] and sensory neurons [13], demonstrated the presence of microRNAs that function locally and affect neuronal development and axonal function [14]. The same was proven to be true for mitochondrial proteins, since short-term inhibition of axonal protein synthesis had adverse effects on mitochondrial function [15].

9.2 Synaptic Mitochondrial Functions

Although the brain only occupies 2 % of our bodies' mass, it consumes 20 % of the organisms' oxygen at rest [16]. Neurons are highly metabolically active and as such are heavily dependent on mitochondria for energy (ATP) production. Nonetheless, mitochondrial function in specific sub-neuronal compartments goes beyond ATP synthesis, placing mitochondria at the centre of neuronal life and death. Synaptic mitochondrial functions are described below; a summary of which is presented in Fig. 9.1.

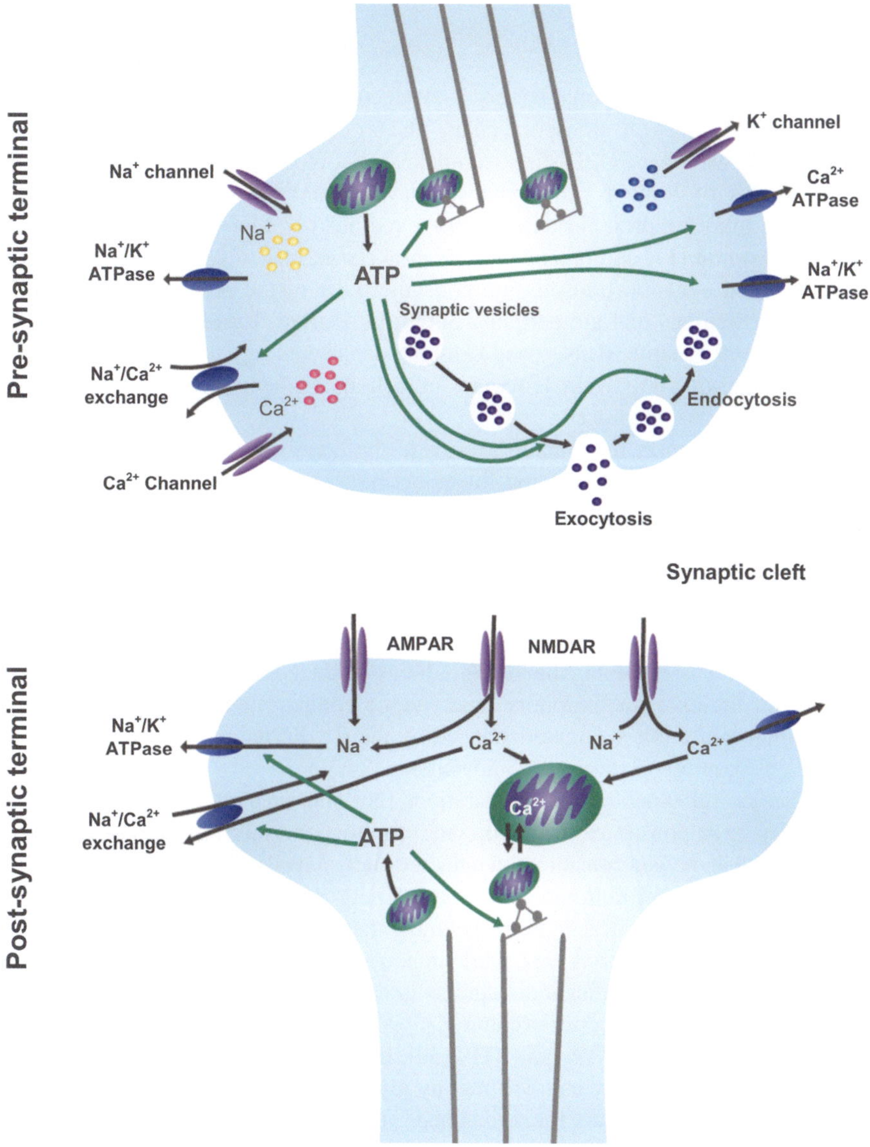

Fig. 9.1 Summary of synaptic mitochondrial functions. Mitochondria are enriched at synaptic sites (axonal terminals and neuronal dendrites), where their function is crucial for synaptic transmission. At the presynaptic terminal, mitochondrial-generated ATP is used to drive synaptic vesicle transport, release, and endocytosis (synaptic vesicle recycling) and regulate ionic fluxes necessary for action potential propagation (via fuelling ATPases) and motor protein-assisted organelle transport to/from the synapse. Postsynaptic mitochondria are essential for synaptic transmission preservation since mitochondrial ATP production and calcium buffering capacities are key for reversing ionic gradients generated following transmitter release and uptake

9.2.1 ATP Production

The vast majority of neuronal energy is produced by mitochondria in the form of ATP, via the electron transport chain (ETC) and oxidative phosphorylation (OXPHOS) (see Chap. 1). Distinct sub-neuronal compartments like active growth cones, nodes of Ranvier, and synapses have considerable metabolic demands explaining mitochondrial clustering in these regions [8]. More specifically, studies performed on rat [17] and human [18] cerebral cortex have revealed that signalling processes and synaptic transmission consume most of the ATP produced in the brain. Brain regions that are particularly rich in neurons, for example, the cortex, have increased synaptic density and exhibit extensive energy usage [19]. Both the pre- and postsynaptic terminals require mitochondrial-generated ATP for normal synaptic transmission to occur.

At the presynapse, mitochondria provide the energy necessary to drive ionic pumps and hence regulate the ion fluxes necessary for action potential initiation [17]. Previous work has suggested that mitochondrial density in nerve terminals is linked to synaptic activity [20], while synaptic vesicle transport [21], exocytosis, and recycling have also been demonstrated to be dependent on ATP supply from mitochondria [22, 23].

The energetic expenditure of postsynaptic compartments is even greater, since ATP is required for reversing the ionic gradients produced following excitation [19]. This is not the case for inhibitory postsynaptic compartments, though, since Cl^- influx upon inhibition is counterbalanced by the Cl^--K^+ uniporter, and thus much less ATP is required to reverse Cl^- movements [24, 25].

Studies on rat synaptosomal preparations [26] and isolated synaptic mitochondria [27] have demonstrated the importance of mitochondrial respiratory chain complex function in oxygen consumption rates and energy availability in nerve terminals. Moreover, studies on cultured hippocampal neurons by Overly and colleagues (1996) were suggestive of less motile but more metabolically active mitochondria in dendrites compared to axons [28], while impaired mitochondrial protein translation in *Drosophila* led to significant disruptions in dendritic morphology [29]. Recently, the use of a fluorescent reporter in cultured hippocampal neurons facilitated the quantification of presynaptic ATP (*syn*-ATP) [30]. Energetic demand at the synapse was driven by synaptic activity and was met by glycolytic and mitochondrial-mediated energy generation, necessary for maintained synaptic function [30].

9.2.2 Calcium Buffering

The primary signal that is sensed by mitochondria in the synapse and causes mitochondrial anchoring to these sites is elevated cytosolic (Ca^{2+}). Mitochondria are important for maintaining neuronal calcium homeostasis, while their buffering capacity is central in establishing neuronal polarity and axonal differentiation [31].

Calcium buffering by mitochondria is key for rapid recovery and therefore synaptic transmission preservation following tetanic stimulation in glutamatergic synapses [32]. Mitochondrial calcium uptake is also important for synaptic plasticity as demonstrated in hippocampal neuronal cultures [33], hippocampal slices [34], and crayfish neuromuscular junctions [35]. Hippocampal neurons cultured from syntaphilin knockout mice were exhibitive of enhanced short-term facilitation following tetanic stimulation (10 and 50 Hz), without however any consequent effects on the frequency, amplitude, or kinetics of excitatory postsynaptic potentials [33]. On the contrary, CA1-CA3 neuronal synaptic plasticity (short- and long-term) was debilitated following treatment with cyclosporin A and subsequent mitochondrial permeability transition pore block [34]. Similarly, impaired mitochondrial Ca^{2+} influx/efflux, due to CCCP, TTP^{+}, or ruthenium red application, abolished post-tetanic potentiation in crayfish neuromuscular junctions [35].

Interestingly, synaptic mitochondria appear to have a lower threshold for calcium uptake before their membrane potential collapses, and mitochondrial function is impaired relative to their non-synaptic counterparts [36]. Overall, synaptic mitochondria appear to be more susceptible to calcium overload and respiratory chain complex dysfunction compared to mitochondria of non-synaptic origin. This inherent susceptibility of synaptic mitochondria could explain the extent and severity of synaptic changes observed in neurodegenerative disorders in which mitochondrial function is implicated in the pathogenesis.

9.3 Evidence for Synaptic Disturbances in Neurodegenerative Diseases

Following evidence that highlights the contribution of synaptic mitochondria to normal synaptic signalling and considering that mitochondrial dysfunction is directly involved in the pathophysiology of neurodegenerative disorders [37, 38], it comes as no surprise that synaptic abnormalities are pronounced in neurodegenerative disorders. Not only are synapses negatively affected in neurodegenerative disease, their dysfunction has actually been demonstrated to be important in the genesis and progression of many neurodegenerative conditions [39, 40].

9.3.1 Alzheimer's Disease

The abnormal deposition of amyloid beta protein (Aβ) in plaques and the presence of hyperphosphorylated *tau* are used to definitively diagnose Alzheimer's disease (AD) [41]. Neuropathological studies have demonstrated that structural and functional synaptic alterations occur well before cortical synaptic and neuronal loss, while decreased synaptic density correlates strongly with the severity of cognitive impairments in patients with this disease [42, 43].

Ultrastructural studies on postmortem brain samples from patients with AD reveal a severe decrease (~45 %) in synaptic bouton density compared to controls [44, 45]. Moreover, transgenic mouse lines that express mutant forms of the amyloid precursor protein (APP) show evidence of decreased dendritic extension and spine density [46], implying that neuronal communication is lost in areas critically affected by the disease. Indeed, electrophysiological recordings from hippocampal slices that originate from mice injected with synthetic forms of Aβ [47] and Aβ oligomers originating from AD patients' brains [48] establish the functional impact of structural synaptic changes on synaptic plasticity. Both synthetic protein motifs and protein oligomers isolated from patients with AD significantly decrease long-term potentiation (LTP), a surrogate experimental model for learning and memory, whereas Aβ oligomers from patients' brains further enhance long-term depression (LTD) [47, 48].

Interestingly, Aβ deposition was evident in synaptic mitochondria isolated from APP transgenic mice early in life (at 4 months) and prior to extracellular Aβ accumulation [49]. Synaptic organelles had decreased respiratory function compared to non-synaptic counterparts, as exemplified by decreased cytochrome *c* oxidase activity, prominent oxidative damage, and increased probability of permeability transition. Additionally, axonal mitochondrial density and trafficking declined in hippocampal neuronal cultures from the same mice [49].

9.3.2 Parkinson's Disease

Parkinson's disease is the second most common age-related neurodegenerative disease (after Alzheimer's), arising mainly due to a combination of genetic and environmental factors (~90 % of PD cases are sporadic). Only a small percentage (~10 %) of PD patients report a family history of the disease, attributed mainly to mutations in genes involved in mitochondrial quality control [50]. These include PTEN-induced putative kinase 1 (PINK1) and parkin, detailed descriptions of which are provided in Chap. 11. For the purpose of this chapter, we will focus on α-synuclein, DJ-1, and leucine-rich repeat serine/threonine-protein kinase 2 (LRRK2), mutations in which are also associated with familial forms of PD [50].

α-Synuclein is enriched at presynaptic sites and is closely associated with synaptic vesicles [51], exerting a direct effect on synaptic transmission. This is exemplified by protein knockout in mice, which caused significantly decreased numbers of undocked synaptic vesicles that were proven to be the limiting factor of neuronal transmission following repetitive depolarization [52]. Moreover, aggregated α-synuclein (the main constituent of Lewy bodies) modifies presynaptic protein interactions with negative downstream effects on synaptic vesicle exocytosis and neurotransmitter release as shown in the human postmortem brain [53].

DJ-1 is believed to be an oxidative stress sensor, since it translocates from the nucleus to mitochondria following oxidative damage [54]. The protein may interact with α-synuclein during both normal and diseased states [55] and is also believed to be important for synaptic transmission. Brain slices from DJ-1 knockout mice demonstrated that the balance between neurotransmitter (dopamine) release and reuptake was disturbed, while LTD failed to be induced. Moreover, neuronal sensitivity to transmitter release was decreased and animals were hypoactive [56].

Goldberg and colleagues have also generated parkin-deficient mice in an attempt to characterize the effects of this protein in nigrostriatal pathway communication. They have noticed that increased currents are required to induce postsynaptic neuronal excitability and spike initiation; however, there was no evidence of neuronal loss in the pathway [57]. Likewise, decreased dopamine release followed by impaired synaptic plasticity (LTP and LTD) was noticed in striatal slices from PINK1-deficient mice [58]. Finally, mouse models for LRRK2 dysfunction are denotive of impaired dopaminergic neurotransmission accompanied with poor motor performance [59].

9.3.3 Huntington's Disease

This trinucleotide repeat expansion disorder has been closely linked with mitochondrial dynamics and function. Key proteins for controlling mitochondrial dynamics have altered expression levels in patients with Huntington's disease (HD) [60]. Postmortem brain samples from patients and huntingtin (htt) transgenic mice displayed increased dynamin-related protein 1 (Drp1) enzymatic activity associated with increased mitochondrial fragmentation. Fragmented organelles were less motile in primary cultures and were mainly localized in neuronal cell bodies. mRNA levels of key synaptic proteins were significantly decreased, while dendritic branching declined [61]. Given that the ubiquitin-proteasome system (UPS) is important in synaptic plasticity and function, sequestration of an E3 ubiquitin ligase (Ube3a) by mutant htt aggregates in the nucleus and its removal from synapses [62] as well as decreased synaptic UPS activity in primary neurons (striatal and cortical) and knock-in mice (expressing full-length htt with extra 150 CAG repeats) may result in synaptic rigidity and dysfunction [63].

Further to mitochondrial structure, mutant huntingtin (htt) affects mitochondrial trafficking in neurons since vesicles and mitochondria are immobilized in the striatum of transgenic mice (expressing full-length human htt with 16 or 72 repeats) and in striatal neurons in vitro [64]. Similarly, axonal htt aggregates in primary cortical neurons impair mitochondrial transport along the axon without, however, affecting the ability of mitochondria to buffer Ca^{2+} [65]. Association of mitochondria with specific N-terminal mutant htt adversely affects the association of mitochondria with proteins involved in its axonal transport, thus explaining for altered distribution and decreased mitochondrial transport rate in cultured cells from knock-in mice (*Hdh* (CAG)150) [66]. Aberrant neuronal distribution of mitochondria results in synaptic ATP shortage with likely downstream effects on synaptic transmission [66].

Mutant htt localisation on the outer mitochondrial membrane is associated with increased mitochondrial vulnerability to calcium, opening of the permeability transition pore, and release of cytochrome *c* [67]. Moreover, the ability of mitochondria to take up calcium is impaired and accompanied by elevated reactive oxygen species (ROS) production, mitochondrial fragmentation, and striatal cellular death in the presence of mutant htt ($STHdh^{Q111/Q111}$) [68]. Thus, mitochondrial dysfunction in the presence of mutant htt impacts on cellular survival.

9.3.4 Amyotrophic Lateral Sclerosis

The preferential loss of lower and upper motor neurons is characteristic of patients with amyotrophic lateral sclerosis (ALS) [69]. This neurodegenerative disease may be familial or sporadic, with the majority (~50 %) of familial cases attributed to a hexanucleotide repeat expansion [70] followed by mutations in the superoxide dismutase 1 (SOD1) gene (~20 %) [71].

Mutant [72, 73] and wild-type [74] forms of SOD1 are localized to mitochondria and therefore may negatively affect mitochondrial structure and function early in disease pathogenesis. Indeed, degenerating mitochondria with disorganized cristae and leaky outer membranes have been detected in motor neuronal processes of presymptomatic SOD1 mouse models, attributed to mutant protein localisation to the mitochondrial membrane [75, 76]. Similarly, mitochondria aggregate within the motor axons of young (2 months prior to disease onset) mutant mice (G39A) [77] and axosomatic synapses, contacting degenerating and normal anterior horn cells, of patients with sporadic ALS [78]. Further to altered organellar morphology, recruitment of mutant SOD1 to mitochondria results in aberrant respiratory chain activity [79, 80], impaired Ca^{2+} buffering capacity [81, 82], and activation of apoptotic cell death [82–85].

Synaptic abnormalities occurring early in disease pathogenesis and prior to motor neuron loss and disease phenotypic manifestation suggest that ALS is a distal axonopathy. Electron microscopic examination of normal anterior horn cells in the lumbar spinal cord of patients with ALS showed atrophy of the cell body and decreased synaptic density and length [86]. Likewise, Fischer and colleagues report on a sporadic case of ALS with preserved motor neuron density but with evidence of acute and chronic changes in muscle fibre innervation [87]. The same group documents early (day 47) endplate denervation and progressive axonal loss (starting at day 80) in mutant SOD1 mice (G93A) with preserved motor neuron cell bodies [87]. Fast-type neuromuscular synapses are preferentially lost, while remaining contact sites lose their ability to undergo plastic alterations [88].

9.4 Synaptic Pathology in Primary Mitochondrial Disease

Mitochondrial disease encompasses a genetically and phenotypically heterogeneous group of disorders due to either primary mtDNA mutations or defects in nuclear DNA-encoded genes involved in mtDNA maintenance, respiratory chain assembly, and function (see also Chaps. 2 and 4). Neurological and non-neurological symptoms may manifest in patients with mitochondrial disease, although neurological disease is the most common feature [89]. Decreased fresh brain weight in patients with mitochondrial disease compared to age-matched controls and abnormal neuroradiological findings is suggestive of severe neurodegeneration and consequent brain atrophy [90] (for a general description of the neuropathology of mitochondrial disease, see Chap. 2).

9.4.1 Human Postmortem Brain Tissue

9.4.1.1 Immunohistochemical Observations

Further to generalized neurodegeneration in the brain of patients with mitochondrial disease, there are sub-neuronal compartment-specific abnormalities. These mainly include the synaptic terminal connections that link Purkinje cells to dentate nucleus neurons within the cerebellum. More specifically, synaptophysin immunoreactivity has indicated a decrease in presynaptic terminal density in the dentate nucleus of patients carrying the m.3243A>G and m.8344A>G point mutations, although Purkinje cell numbers were preserved in the cerebellar cortex. Furthermore, increased synaptophysin deposition has been detected in the absence of apparent postsynaptic dentate nucleus neurons in a patient carrying the m.14709T>C point mutation [91], giving rise to "ghost synapses". Similarly, immunohistochemical staining for synaptophysin (generalized synaptic input) and calbindin (Purkinje cell-specific input) in patients with large-scale mtDNA deletions was exhibitive of sparse immunoreactivity and swollen terminals with empty spaces around dentate nucleus neurons, respectively [92].

In addition to changes in the density and morphology of Purkinje cell presynaptic terminals, neuronal dendrites and axonal regions proximal to Purkinje cell bodies are also abnormal. Multiple axonal swellings, known as torpedoes, were detected in the granular cell layer, and more arborized thickened dendritic trees were found in the molecular layer of a patient with a single large-scale mitochondrial DNA deletion [91].

9.4.1.2 Quadruple Immunofluorescence and Three-Dimensional Reconstruction of Synapses

The frequent involvement of the cerebellum in mitochondrial disease, the relatively simple and well-documented intracerebellar circuitry, and the evidence for disturbed Purkinje cell-dentate nucleus neuronal communication have driven us to focus on these sub-neuronal structures in an attempt to gain further insights into the degenerative process occurring in mitochondrial disease.

A developed quantitative quadruple immunofluorescent technique has allowed the three-dimensional reconstruction of presynaptic terminals and the precise assessment of mitochondrial respiratory chain protein expression within these structures. The majority of patients studied had decreased inhibitory synaptic contact sites to the dentate nucleus neurons, including patients with the m.3243A>G, m.8344A>G, and recessive *POLG* mutations. Interestingly, synaptic loss was associated with increased size (measured as volume) of residual synapses suggesting that the increase in size could reflect a compensatory response in order to maintain the total synaptic contact area constant [93]. Additional to structural changes, the synapses of patients with mitochondrial disease have altered respiratory chain protein expression. Preserved Purkinje cell synapses on dentate nucleus neurons have variably decreased levels of complex I, whereas synaptic complex I expression correlates with the severity of ataxia documented in these patients. Thus, synaptic pathology in patients with mitochondrial disease is likely to contribute to the progression of ataxia [93].

9.4.2 Cell Culture/Animal Models of Synaptic Pathology

Altered mitochondrial dynamics negatively affect synaptic contact and dendritic spine numbers considered to be due to a subsequent decrease in dendritic mitochondrial density. Moreover, mitochondrial distribution throughout the neuron could be a possible determinant of synaptic formation following repetitive excitation, since mitochondrial localization in dendritic protrusions after stimulation favours new spine formation [94].

Mice deficient for the mitochondrial fusion protein optic atrophy 1, OPA1, show a close correlation between mitochondrial localization, dynamics, and synaptic density. Heterozygous rodents aged 10–15 months had decreased retinal ganglion cell synaptic density, preceded by mitochondrial fragmentation and withdrawal back to the soma within the inner plexiform layer [95]. Furthermore, mitochondrial absence from *Drosophila* neuromuscular junctions, as a result of mutations in the mitochondrial fission protein Drp1, severely affected transmitter release. Despite maintained basal synaptic function under resting conditions, application of electrical currents up to 10 Hz impaired normal signal transduction [21]. For more details on the impact of mitochondrial dynamics on neuronal health, refer to Chap. 7.

9.5 Artificial Inhibition of Synaptic Mitochondria

Further to observations made in patients with or models of mitochondrial disease, evidence for the significance of mitochondrial changes to the synapse also arises from artificial inhibition of respiratory chain complexes.

Application of drugs such as rotenone, myxothiazol, and potassium cyanide, known to inhibit complexes I, III, and IV of the electron transport chain, respectively, in isolated synaptic mitochondria suggests that there are activity thresholds above which mitochondrial ATP production is severely affected. Since glycolysis alone cannot compensate for mitochondrial energy depletion in neurons, it is likely that degeneration preferentially occurs in nerve terminals [27].

Further to in vitro synaptic and synaptosomal preparations, some experiments induced deficiencies within the electron transport chain and/or of OXPHOS in animal neuromuscular junctions (NMJs). Blockage of the electron transport chain, oxidative phosphorylation, or depolarization of mitochondria in frog NMJs has highlighted the contribution of these organelles to synaptic transmission. Impairment of mitochondrial calcium buffering capacity and a subsequent increase in cytosolic (Ca^{2+}) leads to calcium influx after stimulation and results in elevated signal outburst frequency [96]. Complex I inhibition (via rotenone application) in *Drosophila* NMJs negatively affects synaptic mitochondrial density and synaptic potentiation capacity [97], whereas partial inhibition of the complex (with amytal) in lizard motor nerve terminals results in mitochondrial membrane depolarisation and cytosolic (Ca^{2+}) elevation following repetitive stimulation [98]. Likewise, the effect of complex III and IV inhibition in fly models was partially rescued by external ATP application, suggesting that mitochondrial-generated energy is central to reserve pool vesicle actuation and thus synaptic strength [21]. Finally, the strength of synaptic signalling was downregulated in snake NMJs following drug (CCCP and/or oligomycin) application, pinpointing the role of mitochondria in neuroplasticity [99].

Given that complex I deficiency is the most common defect noticed within surviving neuronal populations in the brains of patients suffering from mitochondrial disease, it might well be that oxygen consumption is greatly affected in patients' synapses consequently leading to synaptic pathology and eventually to neuronal loss.

9.6 Mechanisms of Neuronal Loss in Primary Mitochondrial Disease

A unifying feature of common neurodegenerative disorders and mitochondrial disease is the far-reaching neuronal loss that occurs at late disease stages and is consequent to upstream abnormalities. The degenerative process is likely to last for years, with destructive mechanisms acting upon the neurites, axons, and eventually neuronal cell bodies [100].

Routine neuroradiological imaging of patients with mitochondrial disease is exhibitive of focal lesions in the cerebral cortex, white matter, basal ganglia, and brainstem, while neuronal loss and consequent structural disorganization are apparent in the cerebellar cortex [92], the olivo-cerebellar pathway [91], and the midbrain [101].

Individuals harbouring the m.3243A>G point mutation, frequently associated with mitochondrial encephalomyopathy with lactic acidosis and stroke-like episodes (MELAS), present with enlarged fourth ventricles, a feature of cerebellar atrophy [102]. Patients with myoclonic epilepsy with ragged red fibre (MERRF) syndrome, due to the m.8344A>G point mutation, show atrophic middle pons, cerebellar hemispheres, and middle and superior cerebellar peduncles [103]. Moreover, patients with the neuropathy, ataxia, and retinitis pigmentosa (NARP) syndrome exhibit generalized pontocerebellar atrophy [104]. Extensive neuronal loss is observed in the cerebellum of patients with Kearns-Sayre syndrome [92], in the olivo-cerebellar pathway of patients with MELAS, and in the posterior inferior cerebellar cortex of patients with the m.3243A>G, m.8344A>G, and autosomal recessive mutations in the sole mtDNA polymerase, polymerase γ (*POLG*) [91]. Neuronal loss in both the midbrain and the cerebellum of a patient suffering from multiple mtDNA deletions, due to *POLG* changes, was also detected [101].

9.6.1 Cell Culture Models

Despite the broad signs of neurodegeneration in patients with mitochondrial disease, the mechanisms responsible for the initiation and progression of the neuronal loss process are yet to be unravelled. Some work has been done towards understanding the impact of mitochondrial dysfunction on neuronal function using *trans*-mitochondrial embryonic cybrids [105, 106], but the synapse and how it contributes to neurodegeneration have not been the point of focus.

Skin fibroblasts extracted from patients with mitochondrial disease are routinely employed in order to investigate the pathophysiological changes in different tissues as a consequence of disease. Fibroblast reprogramming into induced pluripotent stem cells (iPSCs) may alter the heteroplasmy levels of any given mtDNA mutation [107, 108]; however, it provides the opportunity to model specific mitochondrial defects [108] and screen potential therapeutic compounds [109]. A high mutation load of the m.3243A>G point mutation in iPSCs [108], differentiated neurons, and teratomas [107] was demonstrated to selectively impair complex I activity, in accordance with observations in MELAS patients. Interestingly, complex I-positive mitochondria were clustered around neuronal cell bodies and were eliminated by mitophagy with time [107]. Likewise, aberrant complex I activity was detected in fibroblasts extracted from patients with Leigh syndrome, due to mutations in *Ndufs4*, accompanied by complex III deficiencies and subassembled respiratory chain complexes (I and III) [110, 111]. Moreover, altered mitochondrial morphology, membrane potential, calcium signalling, and ROS production levels demonstrate organellar vulnerability to stress factors [112].

9.6.2 Animal Models

Additional to cell lines that carry mtDNA mutations, a small number of mouse models have been developed in an attempt to recapitulate the human disease without providing important information on the degenerative process (see Chap. 13).

Pioneering work from Sligh (2000) and Inoue (2000) involved the introduction of pathogenic mtDNA molecules into embryonic stem cells [113] or fertilized embryos [114], achieving the transmission of pathogenic mutations [113] and deletions [114] into successive generations. A similar approach to directly manipulate mtDNA involved the introduction of severe complex I (*ND6*) and mild complex IV (*COXI*) mutations into the female germ line, though severe defects were selected against during oogenesis [115].

Alternatively, manipulation of nuclear DNA-encoded genes involved in mtDNA maintenance has proved to be an effective method for altering mtDNA integrity. Examples include the "deletor" and the "mutator" mouse due to mutations in mtDNA helicase, *Twinkle* [116], and *POLG* [117], respectively. Accumulation of multiple mtDNA deletions in the deletor mouse induces the muscle features of patients with PEO. Respiratory chain deficiency has mainly been detected in the muscle and hippocampus, though a low percentage of COX-deficient neurons exist in the cerebellum, olfactory bulbs, substantia nigra, and hypothalamus [116]. In contrast, the mutator mouse demonstrates an accelerated ageing phenotype with weight and hair loss, kyphosis, osteoporosis, anaemia, reduced fertility, and cardiomegaly [117]. Accumulation of mtDNA mutations with time results in extensive respiratory chain deficiencies without elevated ROS production [118].

Recent technologies allow inactivation of genes in selective tissues or cell types providing the opportunity to replicate the features of specific mitochondrial disorders. Such an example is the inactivation of *Ndufs4* from Purkinje cells and glia in an effort to model Leigh syndrome [119]. Spongiform degeneration, astrogliosis, and vascular proliferation in the cerebellum constitute the main neuropathological findings, accompanied by abnormal mitochondrial structure within presynaptic terminals contacting Purkinje cell bodies [119]. Phenotypic manifestations of these mice recapitulate those of Leigh syndrome including ataxia, failure to thrive, motor dysfunction, and premature death [119]. However, no basal ganglion pathology is detected, contrary to observations in patients with the disease.

A detailed description of the animal models developed in order to examine mitochondrial dysfunction and its contribution to neurodegenerative diseases are provided in Chap. 13.

9.7 Is Synaptic Pathology a Primary Event in Neurodegeneration?

Although there is still an ongoing debate on the exact mechanisms that contribute to neuronal loss in degenerative disorders, synapses are gaining increased attention and may play a central role in the initiation and progression of neuronal degeneration. A hypothesis on the synaptic-led neuronal death cascade is summarized in Fig. 9.2.

Many neurodegenerative diseases, including Alzheimer's, Parkinson's, and Huntington's, are now being redefined as "synaptopathies". This is mainly due to the fact that abnormalities in synaptic protein gene expression, synaptic

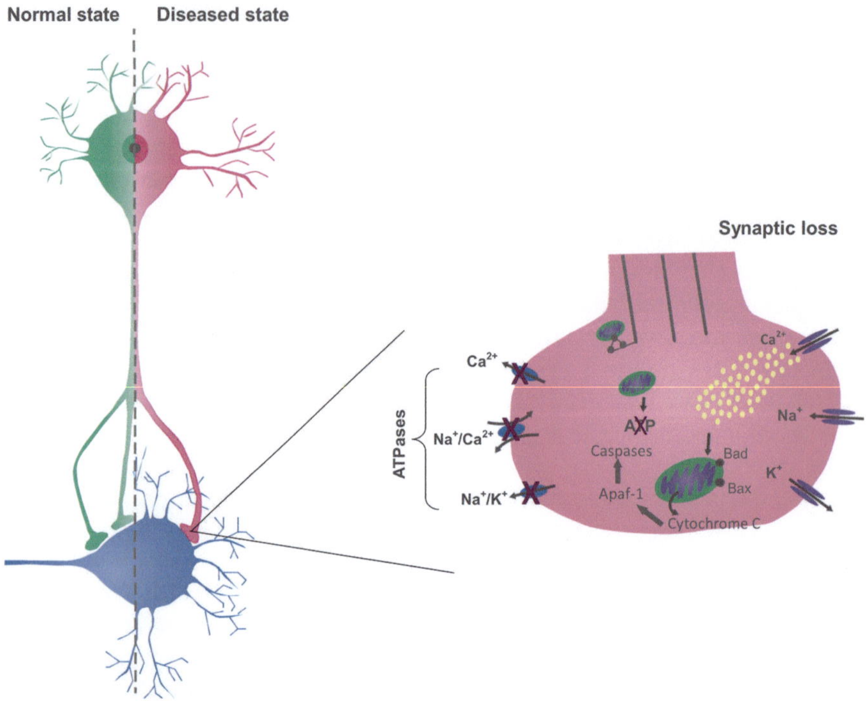

Fig. 9.2 Hypothesis on neuronal death cascade. Disease-related signals and mitochondrial dysfunction at the synapse favour the activation of apoptotic cascades that result in synaptic and eventually neuronal loss [120]. Failure of mitochondria to provide the synapse with ATP combined with impaired mitochondrial calcium buffering capacity causes a build-up of intracellular Ca^{2+} that drives pro-apoptotic factors (e.g. Bax, Bad) to act on mitochondria. Mitochondrial depolarization and opening of permeability transition pores release cytochrome *c* that facilitates apoptosome formation via interaction with the apoptotic protease-activating factor 1 (Apaf-1). Caspases are then activated, causing an "apoptotic-like" synaptic loss and neuronal cell death via dying-back

transmission, and synaptic morphology have been found to occur at early stages [121–124]. This process is similar to the distal axonopathic phenomena that are common in motor neuron disorders and peripheral neuropathies [125].

A novel hybrid protein, the slow Wallerian degeneration (Wld^s) protein, has been demonstrated to protect against axonal and synaptic degeneration [126–128]. In fact, proteomic analysis of striatal synaptic fractions from Wld^s mice revealed altered mitochondrial protein expression, suggesting that differential expression of mitochondrial proteins and mitochondrial function is likely to play a key role in the neuroprotective phenotype [129]. Moreover, mitochondrial motility is not impaired in Wld^s mice following axotomy, while purified mitochondria were shown to be able to buffer more Ca^{2+} [130]. Taking all of the above into consideration, it seems that mitochondrial disease could be a synaptopathy, though evidence in support of this hypothesis has not yet been presented.

9.8 Conclusion

Mitochondrial dysfunction is increasingly being documented in common neurodegenerative disorders, where structural and functional synaptic abnormalities are central to the genesis and progression of the neurodegenerative process. Furthermore, synaptic mitochondria have been proven to be intrinsically more vulnerable to respiratory chain and oxidative stress compared to their non-synaptic counterparts. But is mitochondrial dysfunction at the synapse the driver of synaptic failure and loss? What is the sequence and timescale of events that lead to neuronal degeneration? Work on postmortem tissue from patients with primary mitochondrial disease and reliable cell culture/animal models is likely to help us better understand the consequences of mitochondrial dysfunction at the synapse and shed light into neuronal loss mechanisms. Future work to understand the importance of synaptic plasticity and function to neuronal preservation is crucial if we are to try and prevent or delay neuronal loss.

References

1. Davis AF, Clayton DA. In situ localization of mitochondrial DNA replication in intact mammalian cells. J Cell Biol. 1996;135(4):883–93.
2. Hirokawa N, Noda Y. Intracellular transport and kinesin superfamily proteins, KIFs: structure, function, and dynamics. Physiol Rev. 2008;88(3):1089–118.
3. Morris RL, Hollenbeck PJ. Axonal transport of mitochondria along microtubules and F-actin in living vertebrate neurons. J Cell Biol. 1995;131(5):1315–26.
4. Vallee RB, Williams JC, Varma D, Barnhart LE. Dynein: an ancient motor protein involved in multiple modes of transport. J Neurobiol. 2004;58(2):189–200.
5. MacAskill AF, Kittler JT. Control of mitochondrial transport and localization in neurons. Trends Cell Biol. 2010;20(2):102–12.
6. Sheng ZH. Mitochondrial trafficking and anchoring in neurons: new insight and implications. J Cell Biol. 2014;204(7):1087–98.
7. Schwarz TL. Mitochondrial trafficking in neurons. Cold Spring Harb Perspect Biol. 2013;5(6).
8. Hollenbeck PJ, Saxton WM. The axonal transport of mitochondria. J Cell Sci. 2005;118(Pt 23):5411–9.
9. Lécuyer E, Yoshida H, Parthasarathy N, Alm C, Babak T, Cerovina T, et al. Global analysis of mRNA localization reveals a prominent role in organizing cellular architecture and function. Cell. 2007;131(1):174–87.
10. Gioio AE, Eyman M, Zhang H, Lavina ZS, Giuditta A, Kaplan BB. Local synthesis of nuclear-encoded mitochondrial proteins in the presynaptic nerve terminal. J Neurosci Res. 2001;64(5):447–53.
11. Hillefors M, Gioio AE, Mameza MG, Kaplan BB. Axon viability and mitochondrial function are dependent on local protein synthesis in sympathetic neurons. Cell Mol Neurobiol. 2007;27(6):701–16.
12. Natera-Naranjo O, Aschrafi A, Gioio AE, Kaplan BB. Identification and quantitative analyses of microRNAs located in the distal axons of sympathetic neurons. RNA. 2010;16(8):1516–29.
13. Schacher S, Wu F. Synapse formation in the absence of cell bodies requires protein synthesis. J Neurosci. 2002;22(5):1831–9.

14. Kaplan BB, Kar AN, Gioio AE, Aschrafi A. MicroRNAs in the axon and presynaptic nerve terminal. Front Cell Neurosci. 2013;7:126.
15. Kaplan BB, Gioio AE, Hillefors M, Aschrafi A. Axonal protein synthesis and the regulation of local mitochondrial function. Results Probl Cell Differ. 2009;48:225–42.
16. Mink JW, Blumenschine RJ, Adams DB. Ratio of central nervous system to body metabolism in vertebrates: its constancy and functional basis. Am J Physiol. 1981;241(3):R203–12.
17. Attwell D, Laughlin SB. An energy budget for signaling in the grey matter of the brain. J Cereb Blood Flow Metab. 2001;21(10):1133–45.
18. Lennie P. The cost of cortical computation. Curr Biol CB. 2003;13(6):493–7.
19. Harris Julia J, Jolivet R, Attwell D. Synaptic energy use and supply. Neuron. 2012; 75(5):762–77.
20. Brodin L, Bakeeva L, Shupliakov O. Presynaptic mitochondria and the temporal pattern of neurotransmitter release. Philos Trans R Soc Lond B Biol Sci. 1999;354(1381):365–72.
21. Verstreken P, Ly CV, Venken KJ, Koh TW, Zhou Y, Bellen HJ. Synaptic mitochondria are critical for mobilization of reserve pool vesicles at Drosophila neuromuscular junctions. Neuron. 2005;47(3):365–78.
22. Rowland KC, Irby NK, Spirou GA. Specialized synapse-associated structures within the calyx of Held. J Neurosci. 2000;20(24):9135–44.
23. Perkins GA, Tjong J, Brown JM, Poquiz PH, Scott RT, Kolson DR, et al. The micro-architecture of mitochondria at active zones: electron tomography reveals novel anchoring scaffolds and cristae structured for high-rate metabolism. J Neurosci. 2010;30(3):1015–26.
24. Howarth C, Peppiatt-Wildman CM, Attwell D. The energy use associated with neural computation in the cerebellum. J Cereb Blood Flow Metab. 2010;30(2):403–14.
25. Buzsáki G, Kaila K, Raichle M. Inhibition and brain work. Neuron. 2007;56(5):771–83.
26. Telford JE, Kilbride SM, Davey GP. Complex I is rate-limiting for oxygen consumption in the nerve terminal. J Biol Chem. 2009;284(14):9109–14.
27. Davey GP, Peuchen S, Clark JB. Energy thresholds in brain mitochondria. Potential involvement in neurodegeneration. J Biol Chem. 1998;273(21):12753–7.
28. Overly CC, Rieff HI, Hollenbeck PJ. Organelle motility and metabolism in axons vs dendrites of cultured hippocampal neurons. J Cell Sci. 1996;109(Pt 5):971–80.
29. Chihara T, Luginbuhl D, Luo L. Cytoplasmic and mitochondrial protein translation in axonal and dendritic terminal arborization. Nat Neurosci. 2007;10(7):828–37.
30. Rangaraju V, Calloway N, Ryan TA. Activity-driven local ATP synthesis is required for synaptic function. Cell. 2014;156(4):825–35.
31. Mattson MP, Partin J. Evidence for mitochondrial control of neuronal polarity. J Neurosci Res. 1999;56(1):8–20.
32. Billups B, Forsythe ID. Presynaptic mitochondrial calcium sequestration influences transmission at mammalian central synapses. J Neurosci. 2002;22(14):5840–7.
33. Kang JS, Tian JH, Pan PY, Zald P, Li C, Deng C, et al. Docking of axonal mitochondria by syntaphilin controls their mobility and affects short-term facilitation. Cell. 2008;132 (1):137–48.
34. Levy M, Faas GC, Saggau P, Craigen WJ, Sweatt JD. Mitochondrial regulation of synaptic plasticity in the hippocampus. J Biol Chem. 2003;278(20):17727–34.
35. Y-g T, Zucker RS. Mitochondrial involvement in post-tetanic potentiation of synaptic transmission. Neuron. 1997;18(3):483–91.
36. Yarana C, Sanit J, Chattipakorn N, Chattipakorn S. Synaptic and nonsynaptic mitochondria demonstrate a different degree of calcium-induced mitochondrial dysfunction. Life Sci. 2012;90(19–20):808–14.
37. Gillingwater TH, Wishart TM. Mechanisms underlying synaptic vulnerability and degeneration in neurodegenerative disease. Neuropathol Appl Neurobiol. 2013;7(10):12014.
38. Duchen MR. Mitochondria in health and disease: perspectives on a new mitochondrial biology. Mol Aspects Med. 2004;25(4):365–451.
39. Picconi B, Piccoli G, Calabresi P. Synaptic dysfunction in Parkinson's disease. Adv Exp Med Biol. 2012;970:553–72.

40. Marcello E, Epis R, Saraceno C, Di Luca M. Synaptic dysfunction in Alzheimer's disease. Adv Exp Med Biol. 2012;970:573–601.
41. Glenner GG, Wong CW. Alzheimer's disease and Down's syndrome: sharing of a unique cerebrovascular amyloid fibril protein. Biochem Biophys Res Commun. 1984;122(3): 1131–5.
42. Terry RD, Masliah E, Salmon DP, Butters N, DeTeresa R, Hill R, et al. Physical basis of cognitive alterations in Alzheimer's disease: synapse loss is the major correlate of cognitive impairment. Ann Neurol. 1991;30(4):572–80.
43. Coleman P, Federoff H, Kurlan R. A focus on the synapse for neuroprotection in Alzheimer disease and other dementias. Neurology. 2004;63(7):1155–62.
44. Scheff SW, Price DA. Alzheimer's disease-related alterations in synaptic density: neocortex and hippocampus. J Alzheimers Dis. 2006;9(3 Suppl):101–15.
45. DeKosky ST, Scheff SW. Synapse loss in frontal cortex biopsies in Alzheimer's disease: Correlation with cognitive severity. Ann Neurol. 1990;27(5):457–64.
46. Moolman D, Vitolo O, Vonsattel J-P, Shelanski M. Dendrite and dendritic spine alterations in alzheimer models. J Neurocytol. 2004;33(3):377–87.
47. Chen QS, Kagan BL, Hirakura Y, Xie CW. Impairment of hippocampal long-term potentiation by Alzheimer amyloid beta-peptides. J Neurosci Res. 2000;60(1):65–72.
48. Shankar GM, Li S, Mehta TH, Garcia-Munoz A, Shepardson NE, Smith I, et al. Amyloid-beta protein dimers isolated directly from Alzheimer's brains impair synaptic plasticity and memory. Nat Med. 2008;14(8):837–42.
49. Du H, Guo L, Yan S, Sosunov AA, McKhann GM, ShiDu Yan S. Early deficits in synaptic mitochondria in an Alzheimer's disease mouse model. Proc Natl Acad Sci U S A. 2010;107(43):18670–5.
50. Klein C, Westenberger A. Genetics of Parkinson's disease. Cold Spring Harb Perspect Med. 2012;2(1):a008888.
51. Clayton DF, George JM. The synucleins: a family of proteins involved in synaptic function, plasticity, neurodegeneration and disease. Trends Neurosci. 1998;21(6):249–54.
52. Cabin DE, Shimazu K, Murphy D, Cole NB, Gottschalk W, McIlwain KL, et al. Synaptic vesicle depletion correlates with attenuated synaptic responses to prolonged repetitive stimulation in mice lacking alpha-synuclein. J Neurosci. 2002;22(20):8797–807.
53. Dalfó E, Barrachina M, Rosa JL, Ambrosio S, Ferrer I. Abnormal α-synuclein interactions with rab3a and rabphilin in diffuse Lewy body disease. Neurobiol Dis. 2004;16(1):92–7.
54. Henchcliffe C, Beal MF. Mitochondrial biology and oxidative stress in Parkinson disease pathogenesis. Nat Clin Pract Neurol. 2008;4(11):600–9.
55. Meulener MC, Graves CL, Sampathu DM, Armstrong-Gold CE, Bonini NM, Giasson BI. DJ-1 is present in a large molecular complex in human brain tissue and interacts with alpha-synuclein. J Neurochem. 2005;93(6):1524–32.
56. Goldberg MS, Pisani A, Haburcak M, Vortherms TA, Kitada T, Costa C, et al. Nigrostriatal dopaminergic deficits and hypokinesia caused by inactivation of the familial Parkinsonism-linked gene DJ-1. Neuron. 2005;45(4):489–96.
57. Goldberg MS, Fleming SM, Palacino JJ, Cepeda C, Lam HA, Bhatnagar A, et al. Parkin-deficient mice exhibit nigrostriatal deficits but not loss of dopaminergic neurons. J Biol Chem. 2003;278(44):43628–35.
58. Kitada T, Pisani A, Porter DR, Yamaguchi H, Tscherter A, Martella G, et al. Impaired dopamine release and synaptic plasticity in the striatum of PINK1-deficient mice. Proc Natl Acad Sci U S A. 2007;104(27):11441–6.
59. Tong Y, Pisani A, Martella G, Karouani M, Yamaguchi H, Pothos EN, et al. R1441C mutation in LRRK2 impairs dopaminergic neurotransmission in mice. Proc Natl Acad Sci U S A. 2009;106(34):14622–7.
60. Shirendeb U, Reddy AP, Manczak M, Calkins MJ, Mao P, Tagle DA, et al. Abnormal mitochondrial dynamics, mitochondrial loss and mutant huntingtin oligomers in Huntington's disease: implications for selective neuronal damage. Hum Mol Genet. 2011;20(7): 1438–55.

61. Shirendeb UP, Calkins MJ, Manczak M, Anekonda V, Dufour B, McBride JL, et al. Mutant huntingtin's interaction with mitochondrial protein Drp1 impairs mitochondrial biogenesis and causes defective axonal transport and synaptic degeneration in Huntington's disease. Hum Mol Genet. 2012;21(2):406–20.
62. Maheshwari M, Samanta A, Godavarthi SK, Mukherjee R, Jana NR. Dysfunction of the ubiquitin ligase Ube3a may be associated with synaptic pathophysiology in a mouse model of Huntington disease. J Biol Chem. 2012;287(35):29949–57.
63. Wang J, Wang CE, Orr A, Tydlacka S, Li SH, Li XJ. Impaired ubiquitin-proteasome system activity in the synapses of Huntington's disease mice. J Cell Biol. 2008;180(6):1177–89.
64. Trushina E, Dyer RB, Badger JD, Ure D, Eide L, Tran DD, et al. Mutant huntingtin impairs axonal trafficking in mammalian neurons in vivo and in vitro. Mol Cell Biol. 2004;24(18): 8195–209.
65. Chang DT, Rintoul GL, Pandipati S, Reynolds IJ. Mutant huntingtin aggregates impair mitochondrial movement and trafficking in cortical neurons. Neurobiol Dis. 2006;22(2): 388–400.
66. Orr AL, Li S, Wang CE, Li H, Wang J, Rong J, et al. N-terminal mutant huntingtin associates with mitochondria and impairs mitochondrial trafficking. J Neurosci. 2008;28(11):2783–92.
67. Choo YS, Johnson GVW, MacDonald M, Detloff PJ, Lesort M. Mutant huntingtin directly increases susceptibility of mitochondria to the calcium-induced permeability transition and cytochrome c release. Hum Mol Genet. 2004;13(14):1407–20.
68. Quintanilla RA, Jin YN, von Bernhardi R, Johnson GVW. Mitochondrial permeability transition pore induces mitochondria injury in Huntington disease. Mol Neurodegener. 2013;8:45.
69. Kiernan MC, Vucic S, Cheah BC, Turner MR, Eisen A, Hardiman O, et al. Amyotrophic lateral sclerosis. Lancet. 2011;377(9769):942–55.
70. Renton AE, Majounie E, Waite A, Simon-Sanchez J, Rollinson S, Gibbs JR, et al. A hexanucleotide repeat expansion in C9ORF72 is the cause of chromosome 9p21-linked ALS-FTD. Neuron. 2011;72(2):257–68.
71. Rosen DR, Siddique T, Patterson D, Figlewicz DA, Sapp P, Hentati A, et al. Mutations in Cu/Zn superoxide dismutase gene are associated with familial amyotrophic lateral sclerosis. Nature. 1993;362(6415):59–62.
72. Higgins CM, Jung C, Ding H, Xu Z, Mutant C. Zn superoxide dismutase that causes motoneuron degeneration is present in mitochondria in the CNS. J Neurosci. 2002;22(6):Rc215.
73. Vijayvergiya C, Beal MF, Buck J, Manfredi G. Mutant superoxide dismutase 1 forms aggregates in the brain mitochondrial matrix of amyotrophic lateral sclerosis mice. J Neurosci. 2005;25(10):2463–70.
74. Okado-Matsumoto A, Fridovich I. Subcellular distribution of superoxide dismutases (SOD) in rat liver: Cu, Zn-SOD IN MITOCHONDRIA. J Biol Chem. 2001;276(42):38388–93.
75. Kong J, Xu Z. Massive mitochondrial degeneration in motor neurons triggers the onset of amyotrophic lateral sclerosis in mice expressing a mutant SOD1. J Neurosci. 1998;18(9):3241–50.
76. Wong PC, Pardo CA, Borchelt DR, Lee MK, Copeland NG, Jenkins NA, et al. An adverse property of a familial ALS-linked SOD1 mutation causes motor neuron disease characterized by vacuolar degeneration of mitochondria. Neuron. 1995;14(6):1105–16.
77. Sotelo-Silveira JR, Lepanto P, Elizondo V, Horjales S, Palacios F, Martinez-Palma L, et al. Axonal mitochondrial clusters containing mutant SOD1 in transgenic models of ALS. Antiox Redox Signal. 2009;11(7):1535–45.
78. Sasaki S, Iwata M. Ultrastructural study of synapses in the anterior horn neurons of patients with amyotrophic lateral sclerosis. Neurosci Lett. 1996;204(1–2):53–6.
79. Ferri A, Cozzolino M, Crosio C, Nencini M, Casciati A, Gralla EB, et al. Familial ALS-superoxide dismutases associate with mitochondria and shift their redox potentials. Proc Natl Acad Sci U S A. 2006;103(37):13860–5.
80. Jung C, Higgins CMJ, Xu Z. A quantitative histochemical assay for activities of mitochondrial electron transport chain complexes in mouse spinal cord sections. J Neurosci Methods. 2002;114(2):165–72.

81. Damiano M, Starkov AA, Petri S, Kipiani K, Kiaei M, Mattiazzi M, et al. Neural mitochondrial Ca2+ capacity impairment precedes the onset of motor symptoms in G93A Cu/Zn-superoxide dismutase mutant mice. J Neurochem. 2006;96(5):1349–61.
82. Nguyen KT, Barrett JN, Garcia-Chacon L, David G, Barrett EF. Repetitive nerve stimulation transiently opens the mitochondrial permeability transition pore in motor nerve terminals of symptomatic mutant SOD1 mice. Neurobiol Dis. 2011;42(3):381–90.
83. Takeuchi H, Kobayashi Y, Ishigaki S, Doyu M, Sobue G. Mitochondrial localization of mutant superoxide dismutase 1 triggers caspase-dependent cell death in a cellular model of familial amyotrophic lateral sclerosis. J Biol Chem. 2002;277(52):50966–72.
84. Kirkinezos IG, Bacman SR, Hernandez D, Oca-Cossio J, Arias LJ, Perez-Pinzon MA, et al. Cytochrome c association with the inner mitochondrial membrane is impaired in the CNS of G93A-SOD1 mice. J Neurosci. 2005;25(1):164–72.
85. Pasinelli P, Belford ME, Lennon N, Bacskai BJ, Hyman BT, Trotti D, et al. Amyotrophic lateral sclerosis-associated SOD1 mutant proteins bind and aggregate with Bcl-2 in spinal cord mitochondria. Neuron. 2004;43(1):19–30.
86. Sasaki S, Maruyama S. Synapse loss in anterior horn neurons in amyotrophic lateral sclerosis. Acta Neuropathol. 1994;88(3):222–7.
87. Fischer LR, Culver DG, Tennant P, Davis AA, Wang M, Castellano-Sanchez A, et al. Amyotrophic lateral sclerosis is a distal axonopathy: evidence in mice and man. Exp Neurol. 2004;185(2):232–40.
88. Frey D, Schneider C, Xu L, Borg J, Spooren W, Caroni P. Early and selective loss of neuromuscular synapse subtypes with low sprouting competence in motoneuron diseases. J Neurosci. 2000;20(7):2534–42.
89. McFarland R, Taylor RW, Turnbull DM. A neurological perspective on mitochondrial disease. Lancet Neurol. 2010;9(8):829–40.
90. Lax N, Jaros E. Neurodegeneration in primary mitochondrial disorders. In: Reeve AK, Krishnan KJ, Duchen MR, Turnbull DM, editors. Mitochondrial dysfunction in neurodegenerative disorders. London: Springer; 2012. p. 21–41.
91. Lax NZ, Hepplewhite PD, Reeve AK, Nesbitt V, McFarland R, Jaros E, et al. Cerebellar ataxia in patients with mitochondrial DNA disease: a molecular clinicopathological study. J Neuropathol Exp Neurol. 2012;71(2):148–61.
92. Tanji K, DiMauro S, Bonilla E. Disconnection of cerebellar Purkinje cells in Kearns-Sayre syndrome. J Neurol Sci. 1999;166(1):64–70.
93. Chrysostomou A, Grady JP, Laude A, Taylor RW, Turnbull DM, Lax NZ. Investigating complex I deficiency in Purkinje cells and synapses in patients with mitochondrial disease. Neuropathol Appl Neurobiol. 2015.
94. Timmis JN, Ayliffe MA, Huang CY, Martin W. Endosymbiotic gene transfer: organelle genomes forge eukaryotic chromosomes. Nat Rev Genet. 2004;5(2):123–35.
95. Williams PA, Piechota M, von Ruhland C, Taylor E, Morgan JE, Votruba M. Opa1 is essential for retinal ganglion cell synaptic architecture and connectivity. Brain. 2012;135(Pt 2): 493–505.
96. Alnaes E, Rahamimoff R. On the role of mitochondria in transmitter release from motor nerve terminals. J Physiol. 1975;248(2):285–306.
97. Tong JJ. Mitochondrial delivery is essential for synaptic potentiation. Biol Bull. 2007;212(2):169–75.
98. Talbot J, Barrett JN, Barrett EF, David G. Stimulation-induced changes in NADH fluorescence and mitochondrial membrane potential in lizard motor nerve terminals. J Physiol. 2007;579(Pt 3):783–98.
99. Calupca MA, Prior C, Merriam LA, Hendricks GM, Parsons RL. Presynaptic function is altered in snake K+-depolarized motor nerve terminals containing compromised mitochondria. J Physiol. 2001;532(Pt 1):217–27.
100. Koike T, Yang Y, Suzuki K, Zheng X. Axon & dendrite degeneration: its mechanisms and protective experimental paradigms. Neurochem Int. 2008;52(4–5):751–60.

101. Betts-Henderson J, Jaros E, Krishnan KJ, Perry RH, Reeve AK, Schaefer AM, et al. Alpha-synuclein pathology and Parkinsonism associated with POLG1 mutations and multiple mitochondrial DNA deletions. Neuropathol Appl Neurobiol. 2009;35(1):120–4.
102. Sue CM, Crimmins DS, Soo YS, Pamphlett R, Presgrave CM, Kotsimbos N, et al. Neuroradiological features of six kindreds with MELAS tRNA(Leu) A2343G point mutation: implications for pathogenesis. J Neurol Neurosurg Psychiatry. 1998;65(2):233–40.
103. Ito S, Shirai W, Asahina M, Hattori T. Clinical and brain MR imaging features focusing on the brain stem and cerebellum in patients with myoclonic epilepsy with ragged-red fibers due to mitochondrial A8344G mutation. Am J Neuroradiol. 2008;29(2):392–5.
104. Renard D, Labauge P. Posterior leukoencephalopathy in NARP syndrome. Acta Neurol Belg. 2012;112(4):417–8.
105. Trevelyan AJ, Kirby DM, Smulders-Srinivasan TK, Nooteboom M, Acin-Perez R, Enriquez JA, et al. Mitochondrial DNA mutations affect calcium handling in differentiated neurons. Brain. 2010;133(3):787–96.
106. Kirby DM, Rennie KJ, Smulders-Srinivasan TK, Acin-Perez R, Whittington M, Enriquez JA, et al. Transmitochondrial embryonic stem cells containing pathogenic mtDNA mutations are compromised in neuronal differentiation. Cell Prolif. 2009;42(4):413–24.
107. Hamalainen RH, Manninen T, Koivumaki H, Kislin M, Otonkoski T, Suomalainen A. Tissue- and cell-type-specific manifestations of heteroplasmic mtDNA 3243A>G mutation in human induced pluripotent stem cell-derived disease model. Proc Natl Acad Sci U S A. 2013;110(38):E3622–30.
108. Kodaira M, Hatakeyama H, Yuasa S, Seki T, Egashira T, Tohyama S, et al. Impaired respiratory function in MELAS-induced pluripotent stem cells with high heteroplasmy levels. FEBS Open Bio. 2015;5:219–25.
109. Breuer ME, Willems PH, Smeitink JA, Koopman WJ, Nooteboom M. Cellular and animal models for mitochondrial complex I deficiency: a focus on the NDUFS4 subunit. IUBMB Life. 2013;65(3):202–8.
110. Ugalde C, Janssen RJRJ, van den Heuvel LP, Smeitink JAM, Nijtmans LGJ. Differences in assembly or stability of complex I and other mitochondrial OXPHOS complexes in inherited complex I deficiency. Hum Mol Genet. 2004;13(6):659–67.
111. Scacco S, Petruzzella V, Budde S, Vergari R, Tamborra R, Panelli D, et al. Pathological mutations of the human NDUFS4 gene of the 18-kDa (AQDQ) subunit of complex I affect the expression of the protein and the assembly and function of the complex. J Biol Chem. 2003;278(45):44161–7.
112. Valsecchi F, Grefte S, Roestenberg P, Joosten-Wagenaars J, Smeitink JAM, Willems PHGM, et al. Primary fibroblasts of NDUFS4–/– mice display increased ROS levels and aberrant mitochondrial morphology. Mitochondrion. 2013;13(5):436–43.
113. Sligh JE, Levy SE, Waymire KG, Allard P, Dillehay DL, Nusinowitz S, et al. Maternal germ-line transmission of mutant mtDNAs from embryonic stem cell-derived chimeric mice. Proc Natl Acad Sci U S A. 2000;97(26):14461–6.
114. Inoue K, Nakada K, Ogura A, Isobe K, Goto Y, Nonaka I, et al. Generation of mice with mitochondrial dysfunction by introducing mouse mtDNA carrying a deletion into zygotes. Nat Genet. 2000;26(2):176–81.
115. Fan W, Waymire KG, Narula N, Li P, Rocher C, Coskun PE, et al. Mouse model of mitochondrial disease reveals germline selection against severe mtDNA mutations. Science (New York, NY). 2008;319(5865):958–62.
116. Tyynismaa H, Mjosund KP, Wanrooij S, Lappalainen I, Ylikallio E, Jalanko A, et al. Mutant mitochondrial helicase twinkle causes multiple mtDNA deletions and a late-onset mitochondrial disease in mice. Proc Natl Acad Sci U S A. 2005;102(49):17687–92.
117. Trifunovic A, Wredenberg A, Falkenberg M, Spelbrink JN, Rovio AT, Bruder CE, et al. Premature ageing in mice expressing defective mitochondrial DNA polymerase. Nature. 2004;429(6990):417–23.
118. Trifunovic A, Hansson A, Wredenberg A, Rovio AT, Dufour E, Khvorostov I, et al. Somatic mtDNA mutations cause aging phenotypes without affecting reactive oxygen species production. Proc Natl Acad Sci U S A. 2005;102(50):17993–8.

119. Quintana A, Kruse SE, Kapur RP, Sanz E, Palmiter RD. Complex I deficiency due to loss of Ndufs4 in the brain results in progressive encephalopathy resembling Leigh syndrome. Proc Natl Acad Sci U S A. 2010;107(24):10996–1001.
120. Mattson MP. Apoptosis in neurodegenerative disorders. Nat Rev Mol Cell Biol. 2000;1(2):120–9.
121. Li JY, Plomann M, Brundin P. Huntington's disease: a synaptopathy? Trends Mol Med. 2003;9(10):414–20.
122. Janezic S, Threlfell S, Dodson PD, Dowie MJ, Taylor TN, Potgieter D, et al. Deficits in dopaminergic transmission precede neuron loss and dysfunction in a new Parkinson model. Proc Natl Acad Sci U S A. 2013;110(42):E4016–25.
123. Selkoe DJ. Alzheimer's disease is a synaptic failure. Science. 2002;298(5594):789–91.
124. Dauer W, Przedborski S. Parkinson's disease: mechanisms and models. Neuron. 2003;39(6):889–909.
125. Moloney EB, de Winter F, Verhaagen J. ALS as a distal axonopathy: molecular mechanisms affecting neuromuscular junction stability in the presymptomatic stages of the disease. Front Neurosci. 2014;8:252.
126. Wang MS, Fang G, Culver DG, Davis AA, Rich MM, Glass JD. The WldS protein protects against axonal degeneration: a model of gene therapy for peripheral neuropathy. Ann Neurol. 2001;50(6):773–9.
127. Ferri A, Sanes JR, Coleman MP, Cunningham JM, Kato AC. Inhibiting axon degeneration and synapse loss attenuates apoptosis and disease progression in a mouse model of motoneuron disease. Curr Biol. 2003;13(8):669–73.
128. Sajadi A, Schneider BL, Aebischer P. Wlds-mediated protection of dopaminergic fibers in an animal model of Parkinson disease. Curr Biol. 2004;14(4):326–30.
129. Wishart TM, Paterson JM, Short DM, Meredith S, Robertson KA, Sutherland C, et al. Differential proteomics analysis of synaptic proteins identifies potential cellular targets and protein mediators of synaptic neuroprotection conferred by the slow Wallerian degeneration (Wlds) gene. Mol Cell Proteomics. 2007;6(8):1318–30.
130. Avery MA, Rooney TM, Pandya JD, Wishart TM, Gillingwater TH, Geddes JW, et al. WldS prevents axon degeneration through increased mitochondrial flux and enhanced mitochondrial Ca2+ buffering. Curr Biol. 2012;22(7):596–600.

Chapter 10
Protein Misfolding and Aggregation: Implications for Mitochondrial Dysfunction and Neurodegeneration

Marthe H.R. Ludtmann and Andrey Y. Abramov

Abstract In the early twentieth century, histopathological staining of Parkinson's and Alzheimer's disease patient brains and later other neurodegenerative disorders revealed large intracellular and extracellular protein aggregates. Despite these pathological findings, 100 years later, scientists are yet to fully unravel disease pathogenesis sufficiently enough to allow early disease detection and development of therapeutic agents. While genetic and environmental factors are known to cause protein self-aggregation, recent discoveries have shown that protein intermediates are more toxic than the protein aggregates, serving as invaluable biomarkers for early disease detection. Mutations in alpha-synuclein, amyloid precursor, tau, and huntingtin proteins are known to cause self-aggregation and, via a series of conformational changes, produce oligomers and fibrils before being deposited extra- or intracellularly. In Alzheimer's and Parkinson's diseases, intermediate oligomeric forms of alpha-synuclein and beta-amyloid have been described not only to alter calcium homeostasis, which is vital to neuronal signalling, but also to interact with and alter mitochondrial function and therefore the energetic status of neurons. Research into the effects on the energy status of brain cells is vital to understanding the process of neurotoxicity observed in neurodegenerative diseases.

Keywords Alzheimer's disease • Parkinson's disease • Huntington's disease • Mitochondria • Oligomers • Protein aggregation • Tau • Alpha-synuclein • Huntingtin

M.H.R. Ludtmann, PhD • A.Y. Abramov, PhD, DSc (✉)
Department of Molecular Neuroscience, University College London Institute of Neurology, London, UK
e-mail: m.ludtmann@ucl.ac.uk; a.abramov@ucl.ac.uk

A.K. Reeve et al. (eds.), *Mitochondrial Dysfunction in Neurodegenerative Disorders*, DOI 10.1007/978-3-319-28637-2_10

10.1 Introduction

Most neurodegenerative diseases have been linked to the deposition of misfolded proteins in the brain. These deposits can be either extra- or intracellularly located and are associated with an altered biochemistry, often leading to neuronal loss. Extracellular deposits of beta-amyloid in the form of amyloid plaques and intracellular tau fibrils are histopathological hallmarks in Alzheimer's disease (AD) [1, 2]. In Parkinson's disease (PD), misfolded alpha-synuclein aggregates into intracellular Lewy bodies, while intracellular aggregates of nuclear-encoded mutant huntingtin protein (Htt) are found in the brain of Huntington's disease patients. These deposits consist predominantly of protein fibrils and have long been thought to be the trigger of cellular dysregulations leading to the pathologies seen in these neurodegenerative diseases. However, insights gained more recently have revealed that in fact oligomeric forms of these proteins are more toxic than monomeric or fibrillary forms [3, 4]. Oligomers are cell permeable, allowing them to pass through plasma membranes and propagate throughout the brain. The propagation hypothesis by which proteins can spread progressively between cells and interconnecting brain regions in neurodegenerative diseases found support when oligomeric forms of alpha-synuclein were found in the cerebrospinal fluid (CSF) of affected individuals and in neuronal graft studies [5–7]. Furthermore, beta-amyloid, tau, mutant Htt, and alpha-synuclein have been shown to directly interact with mitochondria which makes bioenergetic disturbances in the cell likely and increases neuronal vulnerability [3, 8, 9].

10.2 Alpha-Synuclein

The exact physiological function of alpha-synuclein is yet to be unravelled despite a presynaptic localisation long being established [10]. However, studies of alpha-synuclein knockout mice and alpha-synuclein overexpressing models have proposed a physiological function related to neurotransmitter release [11, 12]. The fact that genome-wide association studies looking for risk factors for the development of idiopathic PD revealed the alpha-synuclein gene as the biggest risk locus makes research into this protein and its cellular function invaluable [13].

Alpha-synuclein is an intrinsically disordered protein and contains seven 11-amino acid repeats in the N-terminal that are predicted to form amphipathic alpha-helices which are highly conserved in vertebrates. Normal, soluble alpha-synuclein aggregates to form insoluble fibrils via a series of conformational changes and oligomeric intermediates. Native monomeric forms of this protein are α-helical structures, whereas misfolded polymers form more highly ordered β-sheet structures. Mounting evidence suggests that the soluble oligomeric forms generated in this misfolding process are the most toxic, and these have been directly linked to neurodegeneration [3, 14]. Interestingly, alpha-synuclein mutations associated with early-onset, autosomal dominant PD such as A30P (age of onset 54–79 years) and

A53T (average age of onset 46 years) are all located in the N-terminus of this protein and are suggested to increase conformational destabilisation and oligomerisation which in turn favours protein aggregation and deposition [15–18]. However, studies have also revealed that these mutations not only increase the risk of alpha-synuclein misfolding and aggregating but also increase the overall abundance of wild-type alpha-synuclein. Elevated levels of wild-type alpha-synuclein were found to be sufficient to cause protein aggregation and development of early-onset PD (age of onset 35 years) in patients with a duplication of the alpha-synuclein locus [19, 20].

Studies into the effects of alpha-synuclein on cells in culture have revealed a multitude of adverse effects. It has been shown that alpha-synuclein can pass through membranes and associate with mitochondria within 8 min of exposure highlighting potentially devastating effects to cellular bioenergetics and supporting the propagation hypothesis [3, 21].

10.2.1 Cellular Effects of Alpha-Synuclein and Its Role in Neurodegeneration

In vivo and *in vitro* studies have shown that overexpression of alpha-synuclein alters plasma membrane ion permeability, and exogenous oligomeric alpha-synuclein can easily pass across plasma membranes. This altered plasma membrane ion permeability has been shown to increase calcium influx from the extracellular space with detrimental consequences [3, 21, 22]. Intracellular calcium levels are tightly controlled by the endoplasmic reticulum (ER; main intracellular calcium store) and mitochondria (responsible for "fine-tuning" calcium transients) (for details see Chap. 6). Chronic calcium overload or altered handling of calcium fluxes may ultimately damage mitochondria, causing their dysfunction and ultimately the cellular bioenergetic status (e.g., reducing ATP production). This is particularly important for dopaminergic neurons which have a particularly high energy demand (for details see Chap. 6).

In addition to being the major cellular calcium store, the ER plays an important role in protein synthesis, folding, post-translational modification, and transport. Disturbances in ER function (ER stress) can have detrimental effects for the neuron and its survival. In PD, the increased levels of (misfolded) alpha-synuclein are reported to induce ER stress which in turn triggers downstream activation of the unfolded protein response [23]. This response can trigger attenuation of protein translation, altered expression of ER chaperones, and ER-associated degradation which counteracts the accumulation of misfolded proteins within the ER to protect against stress and potentially cell death [24]. ER-associated degradation allows misfolded proteins to be translocated into the cytosol where they are degraded by cytosolic proteasomes as part of the ubiquitin-proteasome system (UPS). In PD, a downregulation of the UPS and autophagy-lysosome pathways (ALP) has been proposed (see Chaps. 11 and 12), which impacts on the cellular clearance of misfolded alpha-synuclein. For example, expression of mutant alpha-synuclein (A30P and A53T) in rat PC12 and neuroblastoma cells was shown to downregulate the UPS [25–27]. The exact mechanism by which alpha-synuclein oligomers interact with and downregulate the proteasome is yet to be fully unravelled, but an interaction of alpha-synuclein with either the 20S β subunit or 19S subunits has been proposed [25, 28]. Alpha-synuclein has also been reported to downregulate ALP via an inhibition of autophagosome formation or a reduction of chaperone-mediated autophagy [29, 30]. Additional support for the importance of the relationship between intracellular protein clearance and alpha-synuclein in PD pathogenesis comes from studies involving the lysosomal enzyme, glucocerebrosidase. Mutations in this gene can cause Gaucher's disease, an autosomal recessive disease, characterised by the accumulation of glucosylceramide (lipid) into fibrils. Gaucher's disease patients may suffer from liver and spleen swelling, osteoporosis, and neurological defects including myoclonus, cognitive impairments, and convulsions (depending on the disease type). Furthermore, Gaucher's disease has been reported to increase the risk of developing PD. It should also be noted that there is a well-documented decrease in the activity of the UPS as well as ALP with age leading to overall slower protein clearance. This reduction in protein clearance and therefore alpha-synuclein may explain partly the late onset of symptoms in sporadic Parkinson's disease (see Chap. 12 for details).

The ER and mitochondria are also known to interact to form the mitochondria-associated endoplasmic reticulum membrane (MAM) which regulates essential calcium and lipid exchange between these organelles [31, 32]. This interaction supports the tricarboxylic acid cycle, electron transport chain, and ultimately ATP production. A recent study by Guardia-Laguarta et al. [33] provided evidence of wild-type alpha-synuclein co-localisation with the MAM, which together with the calcium dysregulation caused by alpha-synuclein is likely to have an effect on energy generation and mitochondrial health. In healthy cells, the ER and mitochondria are distributed throughout axons and dendrites. Considering the multitude of cellular functions supported by MAM interactions, in particular maintenance of the ETC and ATP production, it is not surprising that disruption of MAM will affect cells that have a high energy demand such as dopaminergic neurons. Indeed, reduced MAM interactions have been reported by Guardia-Laguarta et al. in cells that have been exposed to alpha-synuclein [33]. This loss of MAM interactions leads not only to a reduction in MAM function but also in mitochondrial fragmentation, suggesting an important role not only in mitochondrial functioning but also in morphology. These findings warrant further studies into the interaction of alpha-synuclein and its effect on these organelles.

Alpha-synuclein has also been shown to co-localise with the inner mitochondrial membrane which harbours the ETC complexes (Fig. 10.1) [34]. In fact, it has long been known that alpha-synuclein can interact with complex I of the ETC which in turn induces elevated reactive oxygen species (ROS) production leading to oxidative stress [35, 36]. ROS can induce the formation of the permeability transition pore (PTP) which increases the permeability of the inner mitochondrial membrane

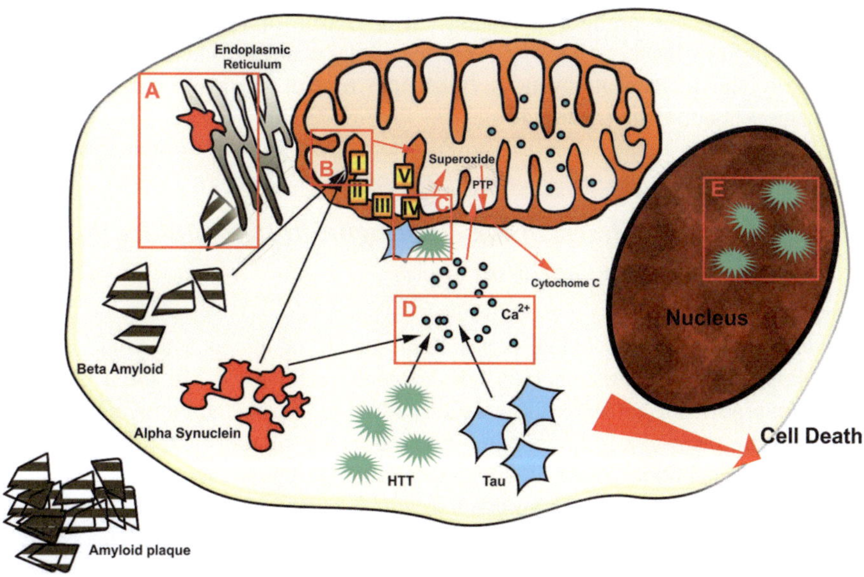

Fig. 10.1 Neurotoxic effects of alpha-synuclein, tau, β-amyloid, and huntingtin proteins. Many of the proteins which aggregate in neurodegenerative diseases are known to interact with various intracellular components and different complexes within the mitochondria. (**a**) Both β-amyloid (*βA*) and alpha-synuclein interact with the endoplasmic reticulum (*ER*) resulting in ER stress and altered mitochondria-associated membrane interaction. (**b**) Alpha-synuclein (in PD) and beta-amyloid (in AD) are known to interact with complex I of the electron transport chain. (**c**) Mitochondrial health can also be compromised by β-amyloid and tau interacting with complex IV. Similarly huntingtin (Htt (in HD)) has been shown to interact with mitochondria through complex IV. (**d**) Dysregulation of calcium homeostasis can be induced by protein inclusions. (**e**) Htt can also interact with the nucleus itself. These proteins have all been shown to interact with mitochondria and induce superoxide production which can ultimately lead to PTP opening and neuronal cell death

allowing ions and solutes to pass more freely. This in turn leads to vast bioenergetic disturbances, ATP depletion and eventually cell death [37]. Furthermore, increased ROS can induce mitochondrial DNA mutations leading to mitochondrial damage and may ultimately lead to cell death [38] (see Fig. 10.1).

10.3 Huntingtin Protein

Huntington's disease (HD) is an autosomal dominant neurodegenerative disease caused by an abnormal expansion of a CAG repeat located in exon 1 of the huntingtin gene. Wild-type huntingtin protein (Htt) localises to the nucleus as well as the cytoplasm, whereas mutant Htt has been found to localise to the mitochondria [39, 40] (see Fig. 10.1). Studies have shown that wild-type Htt plays a vital role in apoptosis, vesicle trafficking, and secretion pathways [41, 42]. It has been found that the

N-terminal cleavage product of mutant Htt is vital to disease progression as the cleavage product has been shown to induce cellular changes, whereas inhibition of caspase-6-dependent proteolysis of full-length mutant Htt protected mutant Htt mice against neurodegeneration [43].

10.3.1 *Cellular Effects of the Huntingtin Protein and Its Role in Neurodegeneration*

The *htt* knockout mouse model is not viable (death occurs at embryonic day 8), providing evidence of the importance of this gene product [44]. Mutations in the *htt* locus are reported to either result in a gain or loss of function affecting a multitude of signalling pathways [45]. Vulnerability and degeneration of excitatory cortical and striatal connections is the first obvious sign of early grade HD.

Mutant Htt co-localises to mitochondria and studies have reported a negative effect on electron transport chain activity and mitochondrial function. Expression of an N-terminal fragment of mutant Htt (82 CAG repeats) in mice resulted in a decreased neuronal mitochondrial membrane potential, suggesting altered mitochondrial health. A lower mitochondrial membrane potential may indicate that one or more of the electron transport chain complexes are not working correctly. In fact, it was found that both the expression and activity of complex II and complex IV were reduced [46–48] (see Fig. 10.1). The lower mitochondrial membrane potential reported in HD results in lowered mitochondrial calcium buffering and efflux. This leads to a prolonged calcium exposure which increases cellular stress and therefore susceptibility to excitotoxic (excitatory-mediated cell death) insults. Whether the mitochondrial calcium dysregulation is a primary or secondary effect caused by the malfunctioning electron transport chain is yet to be unravelled [49].

Interestingly, the total peroxisome proliferator-activated receptor gamma coactivator-1 α RNA level (PGC-1α) is decreased in postmortem brain tissue from HD patients. PGC-1α is a transcriptional coactivator that regulates mitochondrial biogenesis and respiration, and mutant Htt (111 CAG repeats) was found to interact with the PGC-1α promoter in striatal cells. Furthermore, overexpression of PHC-1α rescued not only the mutant Htt-associated mitochondrial phenotypes but also protected mutant Htt primary striatal cultures from cell death making PGC-1α a key player in early HD pathogenesis [50].

10.4 Beta-Amyloid

The amyloid precursor protein (APP) is cleaved by β- and γ-secretase generating a range of β-amyloid (βA) peptides between 39 and 43 amino acid residues long. It is the hydrophobic nature of βA1-40 and βA1-42 which promotes self-aggregation and neurotoxicity revealing the importance of these forms in neurodegenerative disease. A series of conformational changes may occur causing βA aggregation via dimers, oligomers, protofibrils, and fibrils leading ultimately to the formation and deposition of amyloid into characteristic plaques. This deposition of βA is thought to play a central role in the development of Alzheimer's disease (AD). Accumulation of βA in extracellular neuronal plaques is the defining feature for a diagnosis of Alzheimer's disease in postmortem brain tissue [51].

10.4.1 Cellular Effects of β-Amyloid and Its Role in Neurodegeneration

Recently, it has been shown that APP/βA, in addition to its known interactions with the plasma membrane and ER, may also be targeted to mitochondria [52] again suggesting an effect on mitochondrial health (see Fig. 10.1).

Dysregulation of calcium homeostasis has been demonstrated in AD, with βA causing increased cytoplasmic calcium levels, mitochondrial calcium overload, and dysfunction. Mitochondrial dysfunction has been shown to be one of the earliest pathological signs, appearing before neurofibrillary tangles can be detected [53]. Studies have described mitochondrial deficiency in cultured cells overexpressing APP [54] or the spliced form APP_{751} [55]. A reduction in complex IV activity has been demonstrated in mitochondria from the hippocampus and platelets of patients with AD, as well as in AD animal models and AD cybrid cells [56]. In cortical neurons, βA causes mitochondrial dysfunction, reducing ATP levels [57], mostly through inhibition of complex I, causing both mitochondrial depolarisation and a loss of mitochondrial mass [58], while in isolated mitochondria, βA induced respiratory inhibition mostly through inhibition of complex IV [59, 60]. It has also been proposed that βA may increase mitochondrial ROS production [61] which in turn causes further mitochondrial impairment [62].

Application of exogenous βA to mixed cultures of neurons and astrocytes induced two types of mitochondrial depolarisation in astrocytes in the first hours of incubation – a slow and progressive loss and a transient loss of mitochondrial membrane potential [63]. Both types of βA-induced mitochondrial membrane potential change induced ROS production/oxidative stress through interaction with NADPH oxidase and can be blocked by inhibitors of this enzyme [63–65].

Further, it has been shown that βA-induced oxidative stress leads to activation of the DNA repair enzyme poly(ADP-ribose) polymerase (PARP). PARP consumes nicotinamide adenine dinucleotide which in turn decreases substrate availability for mitochondrial complex I of the ETC, resulting in a bioenergetic collapse and cell death [66, 67]. Further, provision of mitochondrial substrates (pyruvate and methyl succinate) reversed βA-induced mitochondrial depolarisation and prevented cell death [63, 65].

Studies have linked βA directly with cyclophilin D which is a key regulator of the PTP. Cyclophilin D is essential for PTP formation since inhibition was shown to inhibit PTP opening. PTP opening depletes mitochondrial calcium, elevates the tosolic calcium, and increases ROS levels, ultimately leading to depletion of cellular ATP and cell death [68]. βA was found to reduce the threshold for mPTP opening and molecular inhibition of cyclophilin D was able to rescue this phenotype. In deed, inhibition of cyclophilin D was able to improve mitochondrial function and learning/memory in an ageing Alzheimer's disease mouse model [69].

10.5 Tau Protein

Tau is a microtubule-associated protein (MAP), encoded by the *MAPT* gene, and known to interact with α- and β-tubulin to facilitate microtubule assembly. With advancing age tau becomes enriched and prone to aggregation in axons and dendrites [70], while in the brain of AD patients, tau is hyperphosphorylated. This process of hyperphosphorylation is known to promote self-assembly, leading to the formation of oligomers and fibrils which eventually leads to neurofibrillary tangle (NFT) aggregation. Hyperphosphorylation of tau results also in microtubule destabilisation and compromised axonal transport which can, together with βA, lead to the pathologies observed in AD.

10.5.1 Cellular Effects of Tau and Its Role in Neurodegeneration

Gomez-Ramos et al. (2006) have shown that application of exogenous tau to neuronal cells leads to an increase in intracellular calcium levels via a muscarinic receptor-mediated mechanism which ultimately results in cell death [71]. In addition, limited studies have provided evidence of a tau-mitochondria interaction but it is thought that tau may act on mitochondria synergistically with βA as well as independently. For example, while βA is known to affect mitochondrial complex IV, tau has been linked to complex I inhibition suggestive of an influence on mitochondrial dynamics and health [72] (see Fig. 10.1).

Quintanilla et al. [73] proposed that the N-terminal fragment of tau may cause mitochondrial dysfunction, while truncated tau (Asp-421) induces mitochondrial fragmentation and elevated mitochondrial superoxide production in immortalised cortical neurons. Furthermore, a truncated tau N-terminal fragment (NH_2-26–44) was found to act on the adenine nucleotide translocator (ANT) which is responsible for the translocation of mitochondrial ATP to the cytosol in exchange of ADP. In addition, Atlante et al. [74] found that the interaction of tau and ANT leads to an impairment of oxidative phosphorylation and therefore a decrease in ATP production. It should be noted that these studies established that it is the truncated N-terminal tau fragment which exerts these deleterious effects on mitochondrial health rather than the full-length tau protein. Overall, these studies suggest that through a disruption of mitochondrial function and therefore neuronal health, tau is a major contributor to the pathological changes and neurodegeneration central to the development of AD.

10.6 Conclusion

Recent studies have provided strong evidence that protein intermediates such as alpha-synuclein oligomers are more toxic to the cell than protein aggregates. Toxic protein intermediates can affect mitochondria and the energetic status of the cell. Disturbances in the energetic status are detrimental to neurons which have a high energy demand. This highlights the importance of continued research to develop therapeutic agents that target these early changes and cellular disruptions that occur before neuronal death and other pathological changes. To achieve early disease intervention, studies must provide robust biological marker analysis as well as develop therapeutic agents that are able to cross the blood-brain barrier to prevent the damage to and changes within the structures of the proteins described above. Most of the modern therapeutic strategies aim to prevent peptide aggregation or indeed protect cells with antibodies against oligomeric forms. However, there is an increasing amount of research being undertaken to prevent/modulate the interactions between these proteins and cellular organelles to delay or prevent subsequent disruption of mitochondrial dysfunction and hence neuronal loss. Despite this, very few therapeutic approaches have undergone clinical trials and even fewer have shown evidence of efficacy making further research into disease onset and progression essential.

References

1. Hardy J, Allsop D. Amyloid deposition as the central event in the aetiology of Alzheimer's disease. Trends Pharmacol Sci. 1991;12(10):383–8.
2. Grundke-Iqbal I, Iqbal K, Tung YC, Quinlan M, Wisniewski HM, Binder LI. Abnormal phosphorylation of the microtubule-associated protein tau (tau) in Alzheimer cytoskeletal pathology. Proc Natl Acad Sci U S A. 1986;83(13):4913–7.
3. Cremades N, Cohen SI, Deas E, Abramov AY, Chen AY, Orte A, et al. Direct observation of the interconversion of normal and toxic forms of alpha-synuclein. Cell. 2012;149(5):1048–59.
4. Lashuel HA, Overk CR, Oueslati A, Masliah E. The many faces of alpha-synuclein: from structure and toxicity to therapeutic target. Nat Rev Neurosci. 2013;14(1):38–48.
5. El-Agnaf OM, Salem SA, Paleologou KE, Cooper LJ, Fullwood NJ, Gibson MJ, et al. Alpha-synuclein implicated in Parkinson's disease is present in extracellular biological fluids, including human plasma. FASEB J. 2003;17(13):1945–7.
6. Kordower JH, Dodiya HB, Kordower AM, Terpstra B, Paumier K, Madhavan L, et al. Transfer of host-derived alpha synuclein to grafted dopaminergic neurons in rat. Neurobiol Dis. 2011;43(3):552–7.

7. Recasens A, Dehay B. Alpha-synuclein spreading in Parkinson's disease. Front Neuroanat. 2014;8:159.
8. Caspersen C, Wang N, Yao J, Sosunov A, Chen X, Lustbader JW, et al. Mitochondrial Abeta: a potential focal point for neuronal metabolic dysfunction in Alzheimer's disease. FASEB J. 2005;19(14):2040–1.
9. Labbadia J, Morimoto RI. Huntington's disease: underlying molecular mechanisms and emerging concepts. Trends Biochem Sci. 2013;38(8):378–85.
10. Maroteaux L, Campanelli JT, Scheller RH. Synuclein: a neuron-specific protein localized to the nucleus and presynaptic nerve terminal. J Neurosci. 1988;8(8):2804–15.
11. Liu S, Ninan I, Antonova I, Battaglia F, Trinchese F, Narasanna A, et al. alpha-Synuclein produces a long-lasting increase in neurotransmitter release. EMBO J. 2004;23(22):4506–16.
12. Vargas KJ, Makani S, Davis T, Westphal CH, Castillo PE, Chandra SS. Synucleins regulate the kinetics of synaptic vesicle endocytosis. J Neurosci. 2014;34(28):9364–76.
13. Simon-Sanchez J, Schulte C, Bras JM, Sharma M, Gibbs JR, Berg D, et al. Genome-wide association study reveals genetic risk underlying Parkinson's disease. Nat Genet. 2009;41(12):1308–12.
14. Spillantini MG, Crowther RA, Jakes R, Hasegawa M, Goedert M. Alpha-Synuclein in filamentous inclusions of Lewy bodies from Parkinson's disease and dementia with Lewy bodies. Proc Natl Acad Sci U S A. 1998;95(11):6469–73.
15. Kruger R, Kuhn W, Muller T, Woitalla D, Graeber M, Kosel S, et al. Ala30Pro mutation in the gene encoding alpha-synuclein in Parkinson's disease. Nat Genet. 1998;18(2):106–8.
16. Conway KA, Lee SJ, Rochet JC, Ding TT, Williamson RE, Lansbury Jr PT. Acceleration of oligomerization, not fibrillization, is a shared property of both alpha-synuclein mutations linked to early-onset Parkinson's disease: implications for pathogenesis and therapy. Proc Natl Acad Sci U S A. 2000;97(2):571–6.
17. Polymeropoulos MH, Lavedan C, Leroy E, Ide SE, Dehejia A, Dutra A, et al. Mutation in the alpha-synuclein gene identified in families with Parkinson's disease. Science. 1997;276(5321):2045–7.
18. Kruger R, Kuhn W, Leenders KL, Sprengelmeyer R, Muller T, Woitalla D, et al. Familial parkinsonism with synuclein pathology: clinical and PET studies of A30P mutation carriers. Neurology. 2001;56(10):1355–62.
19. Kojovic M, Sheerin UM, Rubio-Agusti I, Saha A, Bras J, Gibbons V, et al. Young-onset parkinsonism due to homozygous duplication of alpha-synuclein in a consanguineous family. Mov Disord. 2012;27(14):1827–9.
20. Ibanez P, Bonnet AM, Debarges B, Lohmann E, Tison F, Pollak P, et al. Causal relation between alpha-synuclein gene duplication and familial Parkinson's disease. Lancet. 2004;364(9440):1169–71.
21. Danzer KM, Haasen D, Karow AR, Moussaud S, Habeck M, Giese A, et al. Different species of alpha-synuclein oligomers induce calcium influx and seeding. J Neurosci. 2007;27(34):9220–32.
22. Desplats P, Lee HJ, Bae EJ, Patrick C, Rockenstein E, Crews L, et al. Inclusion formation and neuronal cell death through neuron-to-neuron transmission of alpha-synuclein. Proc Natl Acad Sci U S A. 2009;106(31):13010–5.
23. Bellucci A, Navarria L, Zaltieri M, Falarti E, Bodei S, Sigala S, et al. Induction of the unfolded protein response by alpha-synuclein in experimental models of Parkinson's disease. J Neurochem. 2011;116(4):588–605.
24. Rao RV, Ellerby HM, Bredesen DE. Coupling endoplasmic reticulum stress to the cell death program. Cell Death Differ. 2004;11(4):372–80.
25. Xilouri M, Brekk OR, Stefanis L. alpha-Synuclein and protein degradation systems: a reciprocal relationship. Mol Neurobiol. 2013;47(2):537–51.
26. Stefanis L, Larsen KE, Rideout HJ, Sulzer D, Greene LA. Expression of A53T mutant but not wild-type alpha-synuclein in PC12 cells induces alterations of the ubiquitin-dependent degradation system, loss of dopamine release, and autophagic cell death. J Neurosci. 2001;21(24):9549–60.

27. Petrucelli L, O'Farrell C, Lockhart PJ, Baptista M, Kehoe K, Vink L, et al. Parkin protects against the toxicity associated with mutant alpha-synuclein: proteasome dysfunction selectively affects catecholaminergic neurons. Neuron. 2002;36(6):1007–19.
28. Lindersson E, Beedholm R, Hojrup P, Moos T, Gai W, Hendil KB, et al. Proteasomal inhibition by alpha-synuclein filaments and oligomers. J Biol Chem. 2004;279(13):12924–34.
29. Winslow AR, Chen CW, Corrochano S, Acevedo-Arozena A, Gordon DE, Peden AA, et al. alpha-Synuclein impairs macroautophagy: implications for Parkinson's disease. J Cell Biol. 2010;190(6):1023–37.
30. Alvarez-Erviti L, Rodriguez-Oroz MC, Cooper JM, Caballero C, Ferrer I, Obeso JA, et al. Chaperone-mediated autophagy markers in Parkinson disease brains. Arch Neurol. 2010;67(12):1464–72.
31. Flis VV, Daum G. Lipid transport between the endoplasmic reticulum and mitochondria. Cold Spring Harb Perspect Biol. 2013;5(6):a013235.
32. Raturi A, Simmen T. Where the endoplasmic reticulum and the mitochondrion tie the knot: the mitochondria-associated membrane (MAM). Biochim Biophys Acta. 2013;1833(1):213–24.
33. Guardia-Laguarta C, Area-Gomez E, Rub C, Liu Y, Magrane J, Becker D, et al. alpha-Synuclein is localized to mitochondria-associated ER membranes. J Neurosci. 2014;34(1):249–59.
34. Robotta M, Gerding HR, Vogel A, Hauser K, Schildknecht S, Karreman C, et al. Alpha-synuclein binds to the inner membrane of mitochondria in an alpha-helical conformation. Chembiochem. 2014;15(17):2499–502.
35. Devi L, Raghavendran V, Prabhu BM, Avadhani NG, Anandatheerthavarada HK. Mitochondrial import and accumulation of alpha-synuclein impair complex I in human dopaminergic neuronal cultures and Parkinson disease brain. J Biol Chem. 2008;283(14):9089–100.
36. Parihar MS, Parihar A, Fujita M, Hashimoto M, Ghafourifar P. Mitochondrial association of alpha-synuclein causes oxidative stress. Cell Mol Life Sci. 2008;65(7–8):1272–84.
37. Bernardi P. The mitochondrial permeability transition pore: a mystery solved? Front Physiol. 2013;4:95.
38. Gandhi S, Abramov AY. Mechanism of oxidative stress in neurodegeneration. Oxid Med Cell Longev. 2012;2012:428010.
39. Rockabrand E, Slepko N, Pantalone A, Nukala VN, Kazantsev A, Marsh JL, et al. The first 17 amino acids of Huntingtin modulate its sub-cellular localization, aggregation and effects on calcium homeostasis. Hum Mol Genet. 2007;16(1):61–77.
40. Hoogeveen AT, Willemsen R, Meyer N, de Rooij KE, Roos RA, van Ommen GJ, et al. Characterization and localization of the Huntington disease gene product. Hum Mol Genet. 1993;2(12):2069–73.
41. Brandstaetter H, Kruppa AJ, Buss F. Huntingtin is required for ER-to-Golgi transport and for secretory vesicle fusion at the plasma membrane. Dis Model Mech. 2014;7(12):1335–40.
42. Rigamonti D, Bauer JH, De-Fraja C, Conti L, Sipione S, Sciorati C, et al. Wild-type huntingtin protects from apoptosis upstream of caspase-3. J Neurosci. 2000;20(10):3705–13.
43. Graham RK, Deng Y, Slow EJ, Haigh B, Bissada N, Lu G, et al. Cleavage at the caspase-6 site is required for neuronal dysfunction and degeneration due to mutant huntingtin. Cell. 2006;125(6):1179–91.
44. Nasir J, Floresco SB, O'Kusky JR, Diewert VM, Richman JM, Zeisler J, et al. Targeted disruption of the Huntington's disease gene results in embryonic lethality and behavioral and morphological changes in heterozygotes. Cell. 1995;81(5):811–23.
45. Damiano M, Galvan L, Deglon N, Brouillet E. Mitochondria in Huntington's disease. Biochim Biophys Acta. 2010;1802(1):52–61.
46. Benchoua A, Trioulier Y, Zala D, Gaillard MC, Lefort N, Dufour N, et al. Involvement of mitochondrial complex II defects in neuronal death produced by N-terminus fragment of mutated huntingtin. Mol Biol Cell. 2006;17(4):1652–63.
47. Benchoua A, Trioulier Y, Diguet E, Malgorn C, Gaillard MC, Dufour N, et al. Dopamine determines the vulnerability of striatal neurons to the N-terminal fragment of mutant huntingtin through the regulation of mitochondrial complex II. Hum Mol Genet. 2008;17(10):1446–56.

48. Bae BI, Xu H, Igarashi S, Fujimuro M, Agrawal N, Taya Y, et al. p53 mediates cellular dysfunction and behavioral abnormalities in Huntington's disease. Neuron. 2005;47(1):29–41.
49. Panov AV, Gutekunst CA, Leavitt BR, Hayden MR, Burke JR, Strittmatter WJ, et al. Early mitochondrial calcium defects in Huntington's disease are a direct effect of polyglutamines. Nat Neurosci. 2002;5(8):731–6.
50. Cui L, Jeong H, Borovecki F, Parkhurst CN, Tanese N, Krainc D. Transcriptional repression of PGC-1alpha by mutant huntingtin leads to mitochondrial dysfunction and neurodegeneration. Cell. 2006;127(1):59–69.
51. Naslund J, Haroutunian V, Mohs R, Davis KL, Davies P, Greengard P, et al. Correlation between elevated levels of amyloid beta-peptide in the brain and cognitive decline. JAMA. 2000;283(12):1571–7.
52. Anandatheerthavarada HK, Biswas G, Robin MA, Avadhani NG. Mitochondrial targeting and a novel transmembrane arrest of Alzheimer's amyloid precursor protein impairs mitochondrial function in neuronal cells. J Cell Biol. 2003;161(1):41–54.
53. Hirai K, Aliev G, Nunomura A, Fujioka H, Russell RL, Atwood CS, et al. Mitochondrial abnormalities in Alzheimer's disease. J Neurosci. 2001;21(9):3017–23.
54. Askanas V, McFerrin J, Baque S, Alvarez RB, Sarkozi E, Engel WK. Transfer of beta-amyloid precursor protein gene using adenovirus vector causes mitochondrial abnormalities in cultured normal human muscle. Proc Natl Acad Sci U S A. 1996;93(3):1314–9.
55. Grant SM, Shankar SL, Chalmers-Redman RM, Tatton WG, Szyf M, Cuello AC. Mitochondrial abnormalities in neuroectodermal cells stably expressing human amyloid precursor protein (hAPP751). Neuroreport. 1999;10(1):41–6.
56. Du H, Yan SS. Mitochondrial medicine for neurodegenerative diseases. Int J Biochem Cell Biol. 2010;42(5):560–72.
57. Casley CS, Land JM, Sharpe MA, Clark JB, Duchen MR, Canevari L. Beta-amyloid fragment 25–35 causes mitochondrial dysfunction in primary cortical neurons. Neurobiol Dis. 2002;10(3):258–67.
58. Casley CS, Canevari L, Land JM, Clark JB, Sharpe MA. Beta-amyloid inhibits integrated mitochondrial respiration and key enzyme activities. J Neurochem. 2002;80(1):91–100.
59. Canevari L, Clark JB, Bates TE. beta-Amyloid fragment 25–35 selectively decreases complex IV activity in isolated mitochondria. FEBS Lett. 1999;457(1):131–4.
60. Parks JK, Smith TS, Trimmer PA, Bennett Jr JP, Parker Jr WD. Neurotoxic Abeta peptides increase oxidative stress in vivo through NMDA-receptor and nitric-oxide-synthase mechanisms, and inhibit complex IV activity and induce a mitochondrial permeability transition in vitro. J Neurochem. 2001;76(4):1050–6.
61. Sheehan JP, Swerdlow RH, Miller SW, Davis RE, Parks JK, Parker WD, et al. Calcium homeostasis and reactive oxygen species production in cells transformed by mitochondria from individuals with sporadic Alzheimer's disease. J Neurosci. 1997;17(12):4612–22.
62. Arias C, Montiel T, Quiroz-Baez R, Massieu L. beta-Amyloid neurotoxicity is exacerbated during glycolysis inhibition and mitochondrial impairment in the rat hippocampus in vivo and in isolated nerve terminals: implications for Alzheimer's disease. Exp Neurol. 2002;176(1):163–74.
63. Abramov AY, Canevari L, Duchen MR. Beta-amyloid peptides induce mitochondrial dysfunction and oxidative stress in astrocytes and death of neurons through activation of NADPH oxidase. J Neurosci. 2004;24(2):565–75.
64. Abramov AY, Fraley C, Diao CT, Winkfein R, Colicos MA, Duchen MR, et al. Targeted polyphosphatase expression alters mitochondrial metabolism and inhibits calcium-dependent cell death. Proc Natl Acad Sci U S A. 2007;104(46):18091–6.
65. Abramov AY, Duchen MR. The role of an astrocytic NADPH oxidase in the neurotoxicity of amyloid beta peptides. Philos Trans R Soc Lond B Biol Sci. 2005;360(1464):2309–14.
66. Du L, Zhang X, Han YY, Burke NA, Kochanek PM, Watkins SC, et al. Intra-mitochondrial poly(ADP-ribosylation) contributes to NAD+ depletion and cell death induced by oxidative stress. J Biol Chem. 2003;278(20):18426–33.

67. Abeti R, Abramov AY, Duchen MR. Beta-amyloid activates PARP causing astrocytic metabolic failure and neuronal death. Brain. 2011;134(Pt 6):1658–72.
68. Mattson MP. Calcium and neurodegeneration. Aging Cell. 2007;6(3):337–50.
69. Du H, Guo L, Zhang W, Rydzewska M, Yan S. Cyclophilin D deficiency improves mitochondrial function and learning/memory in aging Alzheimer disease mouse model. Neurobiol Aging. 2011;32(3):398–406.
70. Ittner LM, Ke YD, Delerue F, Bi M, Gladbach A, van E J, et al. Dendritic function of tau mediates amyloid-beta toxicity in Alzheimer's disease mouse models. Cell. 2010;142(3):387–97.
71. Gomez-Ramos A, Diaz-Hernandez M, Cuadros R, Hernandez F, Avila J. Extracellular tau is toxic to neuronal cells. FEBS Lett. 2006;580(20):4842–50.
72. Rhein V, Song X, Wiesner A, Ittner LM, Baysang G, Meier F, et al. Amyloid-beta and tau synergistically impair the oxidative phosphorylation system in triple transgenic Alzheimer's disease mice. Proc Natl Acad Sci U S A. 2009;106(47):20057–62.
73. Quintanilla RA, Matthews-Roberson TA, Dolan PJ, Johnson GV. Caspase-cleaved tau expression induces mitochondrial dysfunction in immortalized cortical neurons: implications for the pathogenesis of Alzheimer disease. J Biol Chem. 2009;284(28):18754–66.
74. Atlante A, Amadoro G, Bobba A, de Bari L, Corsetti V, Pappalardo G, et al. A peptide containing residues 26–44 of tau protein impairs mitochondrial oxidative phosphorylation acting at the level of the adenine nucleotide translocator. Biochim Biophys Acta. 2008;1777(10):1289–300.

Chapter 11
Mitochondrial Degradation, Autophagy and Neurodegenerative Disease

Elsje G. Otten, Diego Manni, and Viktor I. Korolchuk

Abstract Mitochondria are essential for cellular and organismal health, such that mitochondrial dysfunction can contribute to ageing and age-related diseases. In particular, impairment of mitochondrial function has been shown to be an underlying cause of several human neurodegenerative disorders, including Alzheimer's disease and Parkinson's disease. In this chapter, we outline the current understanding of various mechanisms of mitochondrial quality control with a specific focus on mitophagy, the process of mitochondrial degradation via the autophagy pathway. We describe the autophagy and mitophagy pathways and highlight the key molecular players controlling these processes. Additionally, we discuss how mutations in the components of the molecular machinery controlling mitophagy can lead to the loss of neuronal function and viability and eventually result in disease. Studies of these pathways not only produce an important insight into the mechanisms of neurodegenerative diseases but also suggest molecular components of mitochondrial quality control which could be used as targets for therapeutic interventions.

Keywords Autophagy • Mitophagy • Mitochondria • Parkin • PINK1 • Neurodegeneration • Alzheimer's disease • Parkinson's disease • Huntington's disease • Amyloid lateral sclerosis

11.1 Introduction

Amongst several cellular pathways involved in mitochondrial quality control, the process of mitochondrial degradation via the autophagosome-lysosome pathway, termed mitophagy, has emerged as a key determinant of cellular, and in particular neuronal, function. The importance of this process became particularly evident when the mutations in the proteins regulating autophagy and mitophagy pathways

E.G. Otten, MSc • D. Manni, MSc • V.I. Korolchuk, PhD (✉)
Institute for Cell and Molecular Biosciences, Newcastle Institute for Ageing, Newcastle University, Newcastle upon Tyne, UK
e-mail: viktor.korolchuk@ncl.ac.uk

A.K. Reeve et al. (eds.), *Mitochondrial Dysfunction in Neurodegenerative Disorders*, DOI 10.1007/978-3-319-28637-2_11

were found to be genetic determinants in an array of devastating human neurodegenerative diseases [1–3]. This chapter will describe the molecular mechanisms of these pathways and their perturbation in age-related neurodegeneration.

11.2 The Autophagy Pathway

The term autophagy refers to the cellular self-digestion process that occurs through the lysosomal degradation pathway. Autophagy can be activated by different stress conditions including starvation, oxidative stress and an accumulation of damaged proteins. Since the discovery of the autophagy degradation pathway almost 50 years ago [4], it has been implicated in ageing, immunity, development of cancer, myopathies and neurodegeneration [5].

Autophagy is the principle mechanism for the turnover of long-lived proteins, as well as the only known pathway for the degradation and recycling of cytoplasmic organelles [6]. A rapid degradation of irreversibly damaged proteins is essential to preserve normal cellular functions and to prevent the formation of abnormal intermolecular interactions. Autophagy plays this important proteostatic role, and any malfunction in this process is likely to lead to the generation of insoluble aggregates and contribute to the pathology of many human diseases [7].

Autophagy can operate in three different patterns with distinct delivery modality, specificity and regulation: microautophagy, chaperone-mediated autophagy (CMA) and macroautophagy. During microautophagy, the lysosomal membrane plays an active role where it invaginates or protrudes to enclose a portion of cytoplasm [8]. In CMA, proteins presenting the targeting KFERQ motif (a pentapeptide) are recognised by the hsc70 (heat-shock cognate of 70 kDa), and the protein-chaperone complex is internalised directly into lysosomes by binding to the LAMP-2a receptor (lysosomal-associated membrane protein type 2a) [9].

Macroautophagy is the most widely studied and best characterised form of autophagy. In the macroautophagy pathway (for simplicity frequently referred to as autophagy), the formation of a double membrane structure called an autophagosome engulfs cytosolic components. The cargoes are then subsequently delivered to lysosomes via a fusion process resulting in a hybrid structure termed the autolysosome, where hydrolases can degrade trapped material (Fig. 11.1) [10]. After degradation of macromolecules, the building blocks (e.g. amino acids) are exported back to the cytoplasm for reutilisation as a source of energy or for the synthesis of new cellular components. This process is orchestrated by a set of Atg (*au*to*phagy*-related) proteins [11]. To date, more than 30 ATG genes have been identified in yeast, and many of them have mammalian orthologues [12]. The induction of the autophagic process is typically triggered by a lack of nutrients and growth-promoting signals. Insulin, growth factors or amino acid deprivation inhibits mTOR (mechanistic target of rapamycin), an evolutionarily conserved serine/threonine, phosphatidylinositol kinase-related protein kinase (PIKK) which is considered the master

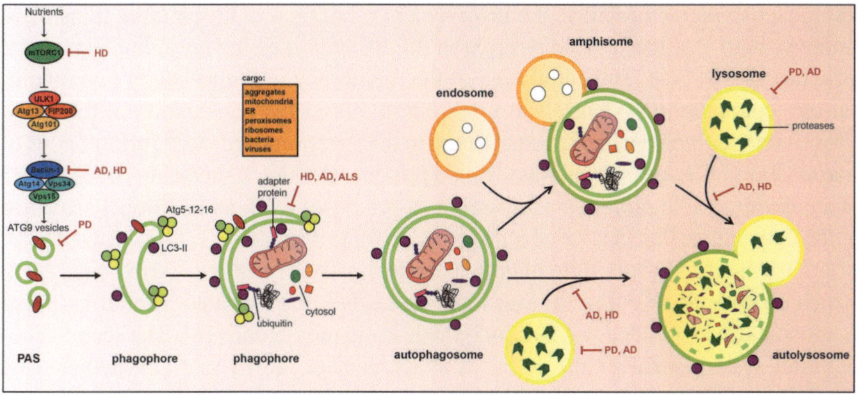

Fig. 11.1 Schematic diagram of autophagy. Mammalian autophagy is typically triggered by nutrient deprivation, which is sensed by mTORC1. The ULK complex (ULK1, Atg13, FIP200 and Atg101) is responsible for autophagy initiation and the PI3K activity of the Vps34 complex (Vps34, Vps15, Atg14 and Beclin-1) is required for the formation of the phagophore. Atg9-positive pre-autophagosomal structures (*PAS*), contribute to the delivery of membrane to the autophagosome. Next, two conjugation systems add the Atg12-Atg5-Atg16 complex and LC3-II to the expansion of the phagophore. The membrane grows and engulfs a portion of the cytosol, and cargo is selectively recruited via receptor proteins and an autophagosome is formed. This is followed by autophagosome maturation by fusion with endosomes (results in the formation of amphisomes) or lysosomes (results in the formation of autolysosomes) and degradation of the autophagosomal cargo through lysosomal proteases, and the degradation products then can be recycled. The steps that are altered in neurodegeneration are indicated in *red*. *HD* Huntington's disease, *PD* Parkinson's disease, *AD* Alzheimer's disease, *ALS* amyotrophic lateral sclerosis

regulator of nutrient signalling [13]. Inhibition of mTOR results in cell growth arrest and in the activation of the autophagic pathway.

11.2.1 The Molecular Machinery of Macroautophagy

The first step in the formation of autophagosomes is the sequestration of cytoplasmic components to a unique membrane called the "phagophore" that expands to enclose the substrate (see Fig. 11.1). This process, called *initiation*, is triggered by ULK1, a serine/threonine kinase that becomes active after mTOR inhibition and forms a complex with Atg13, FIP200 and Atg101 (see Fig. 11.1) [14]. Following initiation, a *nucleation event* takes place driven by Vps34, a phosphatidylinositol 3-kinase (PI3K) that phosphorylates the phosphatidylinositol (PI) generating the phosphatidylinositol 3-phosphate (PI3-P). This lipid then promotes the formation of the autophagosome membrane. The Vps34 forms a complex with Vps15/p150, Vps14 and Vps30/Atg6 (orthologous of mammalian Beclin-1), which starts the nucleation event (see Fig. 11.1) [15]. An important step in the formation of the autophagosome membrane is the involvement of Atg9, a multispanning transmembrane protein that has a principal role in the formation of the *p*re-*a*utophagosomal *s*tructure (PAS) [16]. The *elongation* of the membrane is controlled by two ubiquitin-like conjugation systems. The first system allows the conjugation of the phosphatidylethanolamine (PE) lipid molecule to the C-terminus of Atg8 (orthologous of mammalian LC3, microtubule-associated protein – MAP – light chain 3). Atg8/LC3 is an ubiquitin-like protein synthesised as a precursor, with an additional sequence at the C-terminus that is cleaved by Atg4, a cysteine protease [17]. The cleaved form of LC3 (called LC3-I) is subsequently activated by Atg7 and transferred onto Atg3, which conjugates LC3-I to PE thus forming LC3-II, the lipidated form of LC3 that is specifically associated with the autophagosome membrane [18]. The second conjugation system involves the covalent attachment of Atg12 to Atg5 mediated by Atg7 and Atg10 enzymes [19]. The Atg5-Atg12 complex subsequently associates with Atg16 in a non-covalent way, and then the 350 kDa Atg5-Atg-12-Atg16 complex assists the binding of Atg8/LC3 to PE and the extension of the phagophore (see Fig. 11.1) [20, 21]. Finally, the last step in the autophagosome biogenesis process is termed *maturation*, where autophagosomes fuse with endosomes (thus forming the amphisomes) or directly with lysosomes forming autolysosomes. Amphisomes eventually mature into autolysosomes following their acidification driven by the proton pumps located in the endosomal membranes. Formation of autolysosomes allows the degradation of autophagy substrates (see Fig. 11.1) [22].

11.2.2 Selective Autophagy

For a long time, autophagy was thought to be a non-selective degradation pathway where cytoplasmic components were destroyed in bulk, usually under starvation conditions. However, recently the idea of autophagy as a selective degradation process, where specific substrates are recognised by cargo receptor proteins and targeted for degradation, has been firmly established [12]. Selective autophagy is commonly associated with its housekeeping role promoting the maintenance of cellular homeostasis. An important role of selective autophagy, allowing the clearance of aggregate-prone proteins (aggrephagy) and selective degradation of damaged organelles including mitochondria (mitophagy), is specifically relevant in the context of many neurodegenerative diseases [23, 24]. Other targets of selective autophagy include peroxisomes (pexophagy), endoplasmic reticulum (reticulophagy), ribosomes (ribophagy) and bacteria and viruses (xenophagy) [25–28]. Growing evidence suggests that receptor proteins of selective autophagy, such as p62, NBR1, Nix and NDP52, act to bind ubiquitinated proteins marked for autophagic degradation and shuttle them to the autophagosomal membrane. This process is mediated by the interactions of receptor proteins with both ubiquitin and the LC3-II residing on the autophagic membrane [29].

The signal that leads to the autophagic degradation of substrates appears to be predominantly ubiquitination, which acts as a signalling molecule in the autophagic degradation of both protein aggregates and organelles [30]. It has been reported that Lys63-polyubiquitin chains are predominantly involved in this process [31]; however, other types of ubiquitination have also been found to play role in targeting substrates to autophagic vesicles. Indeed, the lack of autophagy after knocking out the ATG7 or ATG5 genes cause the formation of protein inclusions containing all types of ubiquitin linkages in mice, similar to those observed in neurodegenerative disorders [32].

11.3 Mitochondrial Quality Control

Mitochondria are the cellular power plants responsible for aerobic production of ATP. In addition to energy production, mitochondria are involved in calcium signalling and storage, metabolite synthesis and apoptosis. Mitochondria are also the main cellular source of intracellular reactive oxygen species (ROS). These highly reactive molecules, in particular superoxide anion (O2•-), hydrogen peroxide (H_2O_2) and hydroxyl radical (•HO), are potentially toxic by-products of oxidative phosphorylation (see Chap. 1 for detail). ROS can cause oxidative damage to mitochondrial lipids, DNA and proteins, thus damaging the mitochondrial machinery involved in electrochemical reactions and contributing to further ROS production. Under normal physiological conditions, oxidative damage caused by ROS is buffered by antioxidants like thioredoxin- or glutathione-dependent peroxidases [33]. When the capacity of this antioxidant system is exceeded, cells use several other lines of defence, such as the mitochondrial unfolded protein response (UPR^{mt}), the ubiquitin-proteasome system (UPS) and mitophagy. The UPR^{mt} involves the upregulation of mitochondrial chaperones and proteases that remove misfolded and non-assembled polypeptides in mitochondria [34]. Similarly, the UPS recognises and removes mistargeted and misfolded mitochondrial proteins, including those localised to the outer mitochondrial membrane [35]. The proteases upregulated by the UPR^{mt} include mitochondrial matrix and the intermembrane space (IMS) targeted AAA^+ metalloproteases. These are of central importance for the quality control of the inner mitochondrial membrane (IMM), the site where most ROS are produced. However, the selective removal of damaged/oxidised mitochondrial proteins is not sufficient to remove large numbers of damaged proteins or whole mitochondria. Hence, cells have developed a mechanism to eliminate dysfunctional mitochondria via mitophagy. It has been shown that mitochondria can be completely renewed within 14 days in certain cell types and tissues [36].

11.3.1 Mitophagy

Mitophagy is a type of selective autophagy that mediates clearance of damaged mitochondria, as well as being important in specific cellular contexts such as during maturation of erythrocytes or for the elimination of sperm-derived mitochondria after

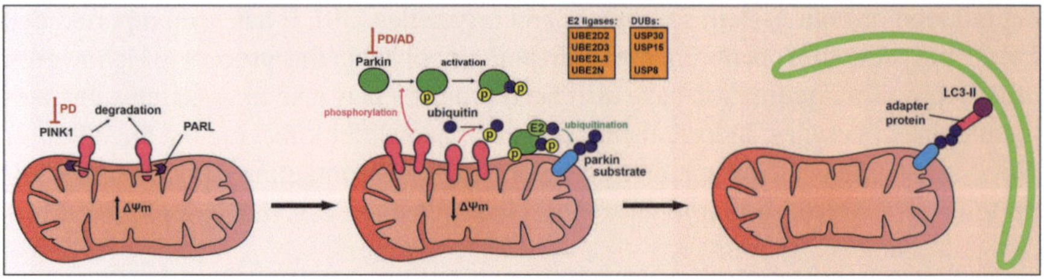

Fig. 11.2 Schematic diagram of PINK1/Parkin-dependent mitophagy. In healthy mitochondria with a high membrane potential (ΔΨm), PINK1 gets proteolytically cleaved by PARL (presenilin-associated rhomboid-like protein), leading to proteasomal degradation. However, upon loss of ΔΨm, PINK1 accumulates at the outer mitochondrial membrane and gets activated. PINK1 phosphorylates Parkin, resulting in its activation. Ubiquitin also gets phosphorylated by PINK1, which is required for the recruitment of Parkin to damaged mitochondria. Several mitochondrial membrane proteins can be ubiquitinated in a Parkin-dependent manner, thereby creating a signal for the removal of the damaged mitochondria. The presence of ubiquitin on the outer mitochondrial membrane results in the recruitment of receptor proteins, which bind to LC3 and target the mitochondria for autophagy-dependent degradation. The steps that are altered in neurodegeneration are indicated in *red*. *PD* Parkinson's disease, *AD* Alzheimer's disease

fertilisation [37–39]. Due to the importance of mitochondria in the development of various degenerative diseases, a lot of interest centres around their selective degradation through mitophagy [40]. Different signals have been shown to result in mitophagy activation, including loss of mitochondrial membrane potential [41], accumulation of misfolded mitochondrial proteins [42], loss of iron from intracellular stores [43] and ROS [44, 45]. Mitochondrial dynamics (fission and fusion (see also Chap. 7) play a critical role in mitochondrial degradation. Specifically, increased fission, resulting in smaller mitochondria that are easier to engulf by autophagosomes, favours degradation of mitochondria, whilst fusion prevents mitophagy [46, 47]. Different mitophagy pathways have been elucidated in the last decade; these will be described in the coming sections. The types of mitophagy taking place during erythrocyte differentiation and oocyte fertilisation are programmed mitochondrial degradation pathways and are relevant to the topic of this chapter due to the mechanistic insight they offer into the molecular events underlying this process, whilst the third pathway, the PINK1/Parkin-dependent mitophagy, is particularly important as it has been extensively implicated in human disease, including neurodegeneration (Fig. 11.2).

11.3.1.1 Mitochondrial Clearance During Erythrocyte Differentiation

During differentiation of reticulocytes to mature erythrocytes, mitochondria and other organelles are eliminated from the cell in a programmed fashion [48]. Nix is a BH3-domain-containing protein and was reported to be a mitochondrial receptor required for the clearance of mitochondria in reticulocytes [49]. Therefore, Nix-deficient mice develop anaemia and severe reticulocytosis caused by inefficient elimination of mitochondria [37]. The basic autophagy machinery is functional in the absence of Nix, but the targeting of mitochondria to autophagosomes is impaired. Nix functions as an adaptor protein and recruits LC3 or *g*amma-*a*mino*b*utyric *a*cid *r*eceptor-*a*ssociated *p*roteins (GABARAP) to damaged mitochondria via a LIR (*L*C3-*i*nteracting *r*egion) motif in its N-terminus [23, 49]. Mitochondrial clearance during reticulocyte differentiation is mainly dependent on ULK1 and to a lesser extent on ATG5 [50]. Even though the molecular mechanism for mitochondrial elimination during this process is well studied, the upstream signal to target the mitochondria for degradation is unknown. The relevance of this mitophagy pathway to human pathophysiology is highlighted by the role of Nix in hypoxia-induced mitochondrial degradation [51].

11.3.1.2 Sperm-Derived Mitochondrial Clearance After Fertilisation

Another example of programmed mitochondrial degradation has been reported during oocyte fertilisation; in sexual reproduction, mitochondria are usually inherited from the mother; thus the paternal mitochondria have to be eliminated when the sperm-derived mitochondria and mitochondrial genome enters the oocyte. In *C. elegans*, mitophagy has been shown to actively target paternal mitochondria for degradation after fertilisation [38, 39]. Immediately after fertilisation, parental mitochondria are surrounded by autophagosomes, which then fuse with lysosomes. In worms with a knock-down or mutation of the homolog of LC3, paternal mitochondria are still present in late-stage embryos [38, 39]. Interestingly, no ubiquitin was detected on parental mitochondria entering the oocyte. This does not rule out that low levels of ubiquitination occur; however, it is possible that another mechanism primes the paternal mitochondria for degradation in worms. In *Drosophila melanogaster* and mice, ubiquitination has been shown to be involved in the degradation of sperm mitochondria. In all studied species, LC3, together with the receptor protein p62, were found on the surface of the ubiquitination mitochondria [39, 52, 53]. Surprisingly, despite recruitment of these autophagy proteins, they are not required for the elimination of paternal mitochondria [53]. Therefore, it remains controversial whether mitophagy is indeed the mechanism of paternal mitochondria and mtDNA elimination after fertilisation.

11.3.1.3 PINK1/Parkin-Dependent Mitophagy

Probably the most studied mitophagy pathway is dependent on the PINK1/Parkin pathway, which is an ubiquitin-dependent process (see Fig. 11.2). When the mitochondrial membrane potential ($\Delta\Psi$m) is high, PINK1, a serine/threonine kinase, binds to PGAM5 (phosphoglycerate mutase family member 5) in the inner mitochondrial membrane, where it gets proteolytically cleaved by PARL (presenilin-associated rhomboid-like protein), leading to proteasomal degradation (see Fig. 11.2) [54–56]. However, upon loss of $\Delta\Psi$m, PINK1 accumulates at the mitochondrial membrane where its activity recruits the E3 ubiquitin ligase, Parkin, from the cytosol to the depolarised mitochondrion [57]. Parkin then ubiquitinates many mitochondrial membrane proteins, thereby creating a signal for the removal of the damaged mitochondrion (see Fig. 11.2) [58]. It is thought that the recruitment and activation of Parkin relies on PINK1 phosphorylation of Parkin itself, but also phosphorylation of mono-ubiquitin and ubiquitin chains, which are required for recruitment of Parkin to damaged mitochondria [59, 60]. Recently, several E2 ubiquitin-conjugating enzymes were identified for Parkin, such a UBE2D2, UBE2D3 and UBE2L3 and UBE2N [61, 62].

The mitochondrial anchored deubiquitylases (DUBs), USP30 and USP15, have been identified as negative regulators of Parkin-mediated mitophagy [63–65]. Another DUB, USP8 selectively deubiquitinates Parkin, but not Parkin substrates [66]. Recently, in a screen to identify positive and negative regulators of mitophagy in yeast, more DUBs were identified. Interestingly, one of the deubiquitination complexes (Ubp3-Bre5) inhibits mitophagy but promotes other types of autophagy [67].

The presence of ubiquitin on the outer mitochondrial membrane results in the recruitment of receptor proteins such as NBR1 and p62 (see Fig. 11.2) [29]. These receptor proteins contain an UBA (*ub*iquitin-*a*ssociated) domain and a LIR (LC3-interacting region) motif, which facilitates simultaneous binding of these adaptors to the ubiquitin-labelled mitochondria and the autophagosomal machinery. As a result, damaged mitochondria are engulfed by phagophores and processed for autophagic degradation.

The importance of these autophagy receptors for the process of Parkin-mediated mitophagy, however, remains controversial. Thus, p62, possibly together with *h*istone *deac*etylase 6 (HDAC6), has been suggested to work as the key mitophagy receptor [68]. Following Parkin-mediated ubiquitination, p62 and HDAC6 bind ubiquitin and drive damaged mitochondria along microtubules to the perinuclear site for degradation [68]. Similarly, Geisler et al. demonstrated that Parkin-dependent mitophagy is dependent on p62 [69]. In support of the role of p62 as a mitophagy receptor protein, studies in *Drosophila melanogaster* demonstrated that ref(2)p, the homolog of mammalian p62, is required for mitophagy [70, 71]. At the same time, Narendra et al. showed that p62 may be required for Parkin-induced mitochondrial clustering, but is not limiting for mitophagy [57, 69]. One possible explanation for these contradictory findings is functional redundancy. In the absence of p62, several intracellular proteins such as NBR1, Alfy (autophagy-linked FYVE protein), BAG3 (Bcl-2-associated athanogene 3), optineurin, NPD52 (nuclear dot protein 52 kDa) or TAX1BP (Tax1-binding protein 1) may be able to link Parkin-mediated ubiquitination to autophagic degradation [30]. Indeed, Lazarou et al. recently demonstrated that optineurin and NPD52, but not p62, are required for mitophagy. HeLa cells with a knock-out of five different autophagy receptors, including optineurin, NPD52, p62, TAX1BP and NBR1, were used, followed by the reintroduction of the different receptors to study their role in mitophagy. Optineurin and NDP52 single knock-out did not affect mitophagy, whilst double knock-out did prevent mitochondrial degradation, suggesting that optineurin and NDP52 are the primary, yet redundant, receptors [72].

Vincow et al. showed that the Pink/Parkin pathway not only promotes mitophagy, but also promotes the selective turnover of mitochondrial respiratory subunits [73]. This is interesting, because the respiratory subunits are prone to oxidation; therefore, quality control of these proteins is essential for mitochondrial function. The mechanism for the selective turnover of respiratory subunits is unknown, and it would be of great interest to further study. Suggested models at the moment are chaperone-mediated extraction of mitochondrial proteins or degradation via mitochondria-derived vesicles [73, 74]. The latter process has been demonstrated to directly deliver parts of the mitochondrion to the lysosomes for degradation in a PINK1/Parkin-dependent manner [75, 76].

The great majority of studies on PINK1/Parkin-dependent mitophagy are done in tissue culture using the acute loss of $\Delta\Psi m$ model, mediated by uncouplers such as CCCP and FCCP [69]. However, depolarisation of the mitochondrial membrane is not sufficient for Parkin recruitment to mitochondria in neurons. Rather, the generation of ROS is required to trigger this mitophagy pathway [44, 77]. The relevance of PINK1/Parkin-dependent mitophagy in neurons remains controversial [78]. Thus, the PINK1/Parkin mitophagy pathway has been robustly demonstrated in immortalised cells and in *Drosophila melanogaster* [79]; however, other models fail to replicate this. For example, in Parkin knock-out mice, no increased neurodegeneration or accumulation of damaged mitochondria was observed [80]. The physiological relevance of mitophagy and the mechanism in neurons remain largely unclear, and future research should focus on mitophagy in neurons in physiologically relevant conditions. This might lead to the discovery of new therapeutic targets for Parkinson's disease and other neurodegenerative diseases.

11.3.1.4 Other Mechanisms of Mitophagy

In addition to the most studied mitophagy pathways as described above, other mechanisms have been observed. One of them involves the translocation of cardiolipin from the inner mitochondrial membrane (IMM) to the mitochondrial surface in neuronal cells, followed by the binding of cardiolipin to the autophagosomal machinery, thereby targeting the mitochondria for degradation [81]. In addition, Melser et al. identified a Rheb-dependent mitophagy pathway that is induced upon stimulation of oxidative phosphorylation [82]. The small GTPase Rheb (*R*as *h*omolog *e*nriched in *b*rain protein) is recruited to the mitochondrial outer membrane upon high oxidative phosphorylation activity, where it promotes mitophagy through an interaction with the mitochondrial adaptor protein Nix and the autophagosomal protein LC3-II. This Rheb-dependent mitophagy pathway is suggested to be important for the renewal of mitochondria and the maintenance of a healthy population of mitochondria. Intriguingly, levels of ROS are also increased upon oxidative conditions, but their role in the initiation of Rheb-dependent mitophagy is not well understood.

In conclusion, mitophagy has been a subject of intense investigations in recent years and significant advances have been made. However, the Pink/Parkin pathway is primarily studied in a very artificial experimental setting requiring overexpression of Parkin and complete depolarisation of mitochondrial membrane; thus, the biological relevance of this pathway in human pathophysiology, specifically in neurodegenerative diseases, requires further clarification. Additionally, whilst the basal mitophagy clearly occurs in neurons [83], mechanistic studies of mitophagy have been largely performed in non-neuronal cells. It will be important to establish whether different mitophagy pathways also occur in neurons and how their impairment can result in neurodegeneration. Some of the existing links between mitochondrial quality control via mitophagy and human age-related neurodegenerative diseases are outlined below.

11.3.2 Impaired Mitophagy and Neurodegeneration

One of the side effects of an extended lifespan in the human population is the increment of age-related neurodegenerative diseases, which are chronic, incurable and terminal conditions affecting the brain, where a structural and functional decline leads to neuronal death. Well-known examples of neurodegenerative diseases are Alzheimer's disease (AD), Parkinson's disease (PD) and Huntington's disease (HD), as well as frontotemporal dementia (FTD), dementia with Lewy bodies (DLB) and many others. Despite identification of some hereditary cases, the overwhelming majority have a sporadic onset, and the causes of which are unknown. Mitochondrial dysfunction as well as autophagy dysregulation is strongly linked to different neurodegenerative diseases [84, 85]. Neurons are postmitotic cells, and therefore, they cannot rely on cell division to dilute the burden of damaged cellular components. It could be predicted that neurons may be particularly dependent on homeostatic processes such as autophagy and specifically mitophagy to prevent accumulation of cellular waste. Indeed, neurons with dysfunctional autophagy accumulate ubiquitinated protein aggregates and damaged mitochondria [86, 87]. In neurodegenerative diseases, autophagy dysfunction can occur at different stages of autophagy, giving rise to different pathologic outcomes. In the following subsections, we will discuss how dysregulation of autophagy and mitophagy might contribute to the pathology of several key neurodegenerative diseases (see Figs. 11.1 and 11.2).

11.3.2.1 Parkinson's Disease

Mitochondrial dysfunction has been widely implicated in Parkinson's disease (PD), which is a disorder characterised by the loss of the dopamine containing neurons of the substantia nigra and the presence of Lewy bodies (intracellular inclusions containing α-synuclein) in remaining cells. Mutations in the genes encoding *PINK1*, *Parkin* and *DJ-1* account for most autosomal-recessive cases of PD and lead to dysfunctional mitophagy and accumulation of damaged mitochondria, which are thought to contribute to neuronal cell death [88, 89].

The first gene identified to be associated with autosomal-recessive PD was *Parkin* (see Fig. 11.2). One hundred twenty-seven pathogenic PD mutations in Parkin have been identified to date, and these are spread throughout the different domains, suggesting that all domains are important for its function [90] (www.molgen.ua.ac.be/PDmutDB/). In the last few years, the structure of the different Parkin domains has become available, thus helping in the understanding of its function. Parkin has a *ub*iquitin-*l*ike domain (Ubl) and four zinc-binding RING domains (RING0, RING1, IBR and RING2) [91–94]. The Ubl is required for substrate binding, whereas RING1 is the binding site for the ubiquitin-conjugated E2 enzyme and RING2 is the catalytic domain [95]. At the same time, the function of the RING0 and IBR domains remains a mystery. In mice, loss-of-function mutations in Parkin cause mitochondrial dysfunction and oxidative damage, and this seems to precede neurodegeneration within the substantia nigra [96]; therefore, mitochondrial dysfunction might be the cause of PD and not a consequence.

The second most frequent gene linked to PD is *PINK1*, which is intimately linked to Parkin. Twenty-eight pathogenic mutations have been described for PINK1, and most of these are located in the kinase domain and result in the loss of kinase activity (www.molgen.ua.ac.be/PDmutDB/). PINK1 deficiency also leads to mitochondrial dysfunction [97], which can be explained by its role in PINK/Parkin-dependent mitophagy (see Fig. 11.2). However, PINK1 also has some Parkin-independent functions, which affect mitochondrial function as well. For example, subunit NDUFA10 of mitochondrial complex I is phosphorylated by PINK, which is important for complex I activity. Expression of a phosphomimetic NDUFA10 in neurons derived from iPS (induced pluripotent stem) cells from PD patients with PINK1 mutations rescued complex I function and mitochondrial membrane potential [98]. In addition, in *Drosophila melanogaster*, expression of the NDUFA10 homolog partially rescues complex I activity and restores locomotor function. Parkin overexpression also rescues the phenotypes of PINK1 mutations in *Drosophila*, but does not restore complex I function [99]. Together, these observations suggest that complex I dysfunction is not sufficient to cause impairment of locomotor function in PD.

The *DJ-1* gene has also been associated with PD with six pathogenic mutations identified to date. The role of DJ-1 in PD is less understood compared to PINK1 and Parkin, but it appears to function in the same pathway [89]. This is supported by the fact that loss of DJ-1 results in mitochondrial dysfunction and can be rescued by Parkin overexpression [100]. DJ-1 is a small oxidation sensing protein, but surprisingly its biochemical function is still not completely clear. What is known is that DJ-1 takes part in the oxidative stress response [101, 102].

Above-mentioned genes all have a significant role in maintaining mitochondrial homeostasis and mitophagy. In contrast, mutations in some other PD-related proteins primarily locate to the cytosol, like LRRK2 (leucine-rich repeat kinase 2) and α-synuclein, and these cause autosomal dominant forms of PD. α-synuclein is an aggregate-prone protein, and mutations or gene duplications/triplications result in the formation protein aggregates, resulting in PD [103]. The mutant form of α-synuclein is degraded via autophagy, causes the accumulation of autophagic-vesicular structures and impairs lysosomal function (see Fig. 11.1) [104, 105]. Moreover, α-synuclein overexpression inhibits autophagy by inhibition of Rab1a, which causes mislocalisation of Atg9, leading to impaired autophagosome formation (see Fig. 11.1) [106]. Also, a PD-related mutation in *VPS35* (*v*acuolar *p*rotein sorting-associated protein 35), disrupts autophagosome formation [107], similar to LRRK2, which localises to autophagosomes and is important for autophagic induction [108, 109]. The interplay between LRRK2 and autophagy appears to be complex; in cultured cells, mutations in LRRK2 resulted in increased autophagy, whilst in knock-out mice, autophagy was reduced [110–112]. Furthermore, two other lysosomal proteins have been linked to PD, ATP13A2 (type 5 P-type ATPase) [113] and glucocerebrosidase [114]. To conclude, mutations in genes resulting in defective autophagy (and particularly mitophagy) can result in PD. Autophagosome maturation, cargo loading and lysosomal function are all implicated and thus will be important to further study how these mutations affect mitochondrial function in neuronal models of PD.

11.3.2.2 Alzheimer's Disease

Alzheimer's disease (AD) is the most common neurodegenerative disorder and cause of dementia, characterised by neuronal loss in the cerebral cortex and accumulation of amyloid plaques and neurofibrillary tangles [115]. Mitochondrial dysfunctions have been implicated in the neuropathogenesis of AD since changes in their appearance occur when neurofibrillary tangles are not yet evident in neurons [116, 117]. Abnormalities in mitochondrial morphology in the neurons of AD brains have been suggested to cause an impairment in the ATP production, which can lead to loss of synaptic activity and cognitive decline [118]. It has been reported that AD patients possess defective mitochondrial cytochrome *c* oxidase in both temporal cortex and hippocampus, which may be the cause of reduced energy generation [119]. Importantly, amyloid β [Aβ] can also directly disrupt mitochondrial function by inhibiting the activity of respiratory enzymes such as cytochrome oxidase, α-ketoglutarate dehydrogenase and pyruvate dehydrogenase [120]. Expression of several proteins implicated in mitochondrial fission/fusion is abnormal in AD, leading to reduced mitochondrial density and increased fragmentation [121]. Thus, together with the observation of increased autophagic degradation of mitochondria in AD [122] and decreased number of mitochondria together with increased amounts of mitochondrial DNA and proteins [116], it suggests that mitophagy may be increased in AD in an attempt to degrade damaged mitochondria or that the proteolytic turnover is impaired. On the other hand, impaired macroautophagy has been also implicated in the pathogenesis of AD, where high amounts of autophagic vacuoles (AVs) accumulate in neurons [123, 124]. The increased amount of AVs in AD brains, compared to the low levels seen in healthy neurons, has been firstly connected with increased autophagic activity [124, 125]. Rather, it has become clear that impairment in the clearance of AVs in AD was instead the case, where the fusion between autophagosomes and lysosomes is not efficient (see Fig. 11.1) [123].

Additionally, the regulatory protein beclin1 which activates autophagy appears to be decreased in AD affected brains (see Fig. 11.1) [126]. Autophagy inhibition also appeared to increase the accumulation of hyperphosphorylated and fragmented tau, responsible for the formation of neurofibrillary tangles [127]. Lysosomal function was also inhibited by deposition of fragmented tau on the outer membrane, leading to lysosomal leakage and further release of toxic tau in the cytoplasm [127]. Due to its role in microtubule stabilisation, dysfunctional tau has been implied to interfere with autophagosome trafficking and clearance, as well as with mitochondrial and lysosomal mobility [128].

Parkin, involved in targeting mitochondria to the autophagosomes, has been shown to be reduced in the frontal and temporal cortex of AD brains together with increased Aβ accumulation (see Fig. 11.2) [129]. Importantly, amyloid precursor protein (APP) and Aβ have been found in AVs, together with components of the γ-secretase complex involved in the cleavage of APP [125, 130]. Thus, macroautophagy itself seems to be linked to the production of Aβ in AD brains. Presenilin1 (PS1), a component of the γ-secretase complex, has been shown to be essential for lysosomal acidification and for correct autophagic degradation [131]. AD-related mutations of PS-1 have shown lysosomal and autophagic dysfunctions in AD patients, leading to abnormal accumulation of autophagic vesicles and early-onset AD pathology [131, 132].

In conclusion, it is clear that autophagy dysfunction plays an important role in the development of AD, where accumulation of insoluble protein aggregates and dysfunctional mitochondria lead to neuronal loss.

11.3.2.3 Huntington's Disease

Huntington's disease (HD) is an autosomal dominant neurodegenerative disorder leading to cognitive and motor disorders, caused by an expanded trinucleotide CAG repeat in exon 1 of the gene encoding for the Huntingtin (Htt) protein. Htt is ubiquitously expressed in the brain, and the mutated version (mHtt) aggregates causing neuronal dysregulation and death [133]. Several lines of study suggest the involvement of respiratory chain defects in mitochondria in HD. A remarkably decreased activity in complex II/III and IV in basal ganglia samples from HD patients has been observed [134]. Also mitochondrial biogenesis dysregulation has been implied in the pathogenesis of HD, due to transcriptional repression of the mitochondrial regulator PGC-1α by mHtt [135]. In addition, mitochondrial mobility and axonal transport are thought to be impaired in HD due to mHtt aggregation [136, 137].

Amongst its numerous functions that have been described in the cell, it has recently emerged that Htt plays an important role in the autophagy process [138]. Whilst a decrease in the autophagic flux often occurs in other neurodegenerative diseases [139], this has been reported to be increased in HD, together with an enhanced number of autophagosomes [140, 141]. The induction of autophagy in HD has been attributed to the sequestration of mTOR by mHtt aggregates, leading to the inactivation of this autophagy-inhibitory kinase [142]. In contrast, recruitment and sequestration of beclin1 by mHtt has been suggested to result in reduction of autophagy levels in HD brains, leading to defective protein turnover [143].

Although autophagosome production and clearance by lysosomes was enhanced, cargo recognition and engulfment failed to occur in HD cells, leading to impaired turnover of cytosolic components and accumulation of damaged mitochondria (see Fig. 11.1) [140]. Hence, this would contribute to the increase in protein aggregates and dysfunctional mitochondria seen in HD cells. However, a defect in the axonal transport of autophagosomes mediated by mHtt has also been connected with mitochondrial accumulation, due to defects in autophagosome/lysosome fusion events [144]. Therefore, it has been proposed that Htt acts as an autophagy scaffold in the cell, also due to its similarities with yeast's autophagy regulatory proteins and the prediction of LC3-interacting repeats in its sequence [138, 145]. Moreover, it has been recently reported that Htt positively modulates selective autophagy by interacting with p62, facilitating its interaction with LC3 and Lys63-poly-ubiquitinated substrates [146]. The interaction between Htt and ULK1, which releases it from the mTOR-negative regulation, has also been connected with the induction of selective autophagy [146].

In summary, Htt has a central role in the regulation of protein trafficking and turnover, where disruption of its function leads to the deficiency of selective autophagy and defective mitochondrial function.

11.3.2.4 Amyotrophic Lateral Sclerosis

Amyotrophic lateral sclerosis (ALS) is a progressive neurodegenerative disorder characterised by motor neuron death and muscle wasting. Approximately 10 % of all the cases are familial (fALS), where 20 % of them are caused by a dominant mutation in the gene SOD1 (superoxide dismutase 1) [147, 148]. Abnormalities in mitochondrial structure and localisation have been shown in both spinal cord and muscle of ALS patients, together with defective respiration and electron transport chain [149, 150]. Interestingly, mutant SOD1 has been demonstrated to localise to the mitochondrial outer membrane, intermembrane space and matrix and appears to be concentrated in vacuolated mitochondria [151, 152]. This interaction would lead to damage of the mitochondrial outer membrane proteins as well as cytochrome *c* release and apoptosis [153, 154].

Overexpression of the G93A-SOD1 mutation in mice leads to mitochondrial degeneration and the formation of mitochondrial vacuoles, possibly triggering the onset of the disease [155]. Additionally, the same transgenic mice have shown increased autophagy levels during the presymptomatic and symptomatic stages, exhibiting increased LC3-II levels and inhibition of mTOR [156, 157]. However, the involvement of autophagy in ALS and its role in the disease progression are controversial [158]. Enhanced autophagy has been linked to increased degradation of mutant SOD1, whereas autophagy inhibition led to drastic accumulation of mutant SOD1 and increased neurotoxicity [159]. It has been also shown that p62 accumulates in the G93A-SOD1 mouse spinal cord and mediates the polyubiquitination and aggregation of mutant SOD1, possibly contributing to the ALS aetiology [160]. The identification of several mutations in p62 sequence related to late-onset alS also suggests the importance of selective autophagy in the disease progression (see Fig. 11.1) [161, 162]. However, an opposite role of autophagy in ALS has been shown after treating G93A-SOD1 mice with rapamycin, an inducer of autophagy [163]. Surprisingly, activation of autophagy by rapamycin in the presymptomatic ALS increased motor neuron death and decreased the lifespan of the animals. Thus, accumulation of autophagosomes and p62 would mean impairment in the autophagic flux rather than enhanced autophagic activity.

Nevertheless, as mitochondrial abnormalities and degeneration seem to contribute with the onset of ALS, mitophagy could play an important role in the disease progression. Degradation of damaged mitochondria would possibly ameliorate the mitochondrial pool and attenuate the disease furtherance.

11.4 Conclusion

Impairment of autophagy (and specifically mitophagy), associated either with age-related decline in the pathway's efficiency or caused by the mutations in the key regulatory components, became evident as an important determinant of many human diseases. Neurons appear to be particularly susceptible to the perturbations in these mitochondrial quality control pathways, possibly due to the damage produced by undegraded toxic cellular components which accumulate over the course of decades in these extremely long-lived cells. Whilst accumulation of general cellular "junk" such as aggregate-prone toxic proteins due to their insufficient degradation via autophagy is undoubtedly contributing to the pathology of neurodegenerative diseases (hence their classification as proteinopathies), it is the perturbation of mitochondrial degradation via the mitophagy pathway which appears to be particularly damaging for the neurons [87, 164]. This, in turn, offers a novel concept for the treatment of these diseases by targeting regulatory components of the mitophagy machinery and upregulating this pathway [165]. Thus, drugs promoting the activity of the general autophagy pathway, and which would also be expected to stimulate mitophagy, have now been shown to protect against a wide range of neurodegenerative diseases with several such drugs having already reached clinical trials [166–168]. Whilst similar therapeutic strategies are yet to be established for the specific activation of the mitophagy pathway, studies in animal models show that overexpression of mitophagy proteins such as Parkin can indeed produce beneficial effect and prevent age-related neuronal decline [169–172]. Further mechanistic studies of mitophagy are likely to lead to the identification and validation of therapeutic targets and eventually identification of specific and potent drugs stimulating mitophagy and protecting neurons against senescence and degeneration.

References

1. Scarffe LA, Stevens DA, Dawson VL, Dawson TM. Parkin and PINK1: much more than mitophagy. Trends Neurosci. 2014;37(6):315–24.
2. Lee JK, Shin JH, Lee JE, Choi EJ. Role of autophagy in the pathogenesis of amyotrophic lateral sclerosis. Biochim Biophys Acta. 2015;1852(11):2517–24.
3. Cortes CJ, La Spada AR. The many faces of autophagy dysfunction in Huntington's disease: from mechanism to therapy. Drug Discov Today. 2014;19(7):963–71.
4. De Duve C, Wattiaux R. Functions of lysosomes. Annu Rev Physiol. 1966;28:435–92.
5. Mizushima N, Levine B, Cuervo AM, Klionsky DJ. Autophagy fights disease through cellular self-digestion. Nature. 2008;451(7182):1069–75.
6. Klionsky DJ, Emr SD. Autophagy as a regulated pathway of cellular degradation. Science. 2000;290(5497):1717–21.
7. Kopito RR. Aggresomes, inclusion bodies and protein aggregation. Trends Cell Biol. 2000;10(12):524–30.
8. Li WW, Li J, Bao JK. Microautophagy: lesser-known self-eating. Cell Mol Life Sci. 2012;69(7):1125–36.
9. Kaushik S, Cuervo AM. Chaperone-mediated autophagy: a unique way to enter the lysosome world. Trends Cell Biol. 2012;22(8):407–17.
10. Ravikumar B, Futter M, Jahreiss L, Korolchuk VI, Lichtenberg M, Luo S, et al. Mammalian macroautophagy at a glance. J Cell Sci. 2009;122(Pt 11):1707–11.
11. Klionsky DJ, Cregg JM, Dunn Jr WA, Emr SD, Sakai Y, Sandoval IV, et al. A unified nomenclature for yeast autophagy-related genes. Dev Cell. 2003;5(4):539–45.
12. van der Vaart A, Mari M, Reggiori F. A picky eater: exploring the mechanisms of selective autophagy in human pathologies. Traffic. 2008;9(3):281–9.
13. Hay N, Sonenberg N. Upstream and downstream of mTOR. Genes Dev. 2004;18(16):1926–45.
14. Mizushima N, Komatsu M. Autophagy: renovation of cells and tissues. Cell. 2011;147(4):728–41.
15. Kang R, Zeh HJ, Lotze MT, Tang D. The Beclin 1 network regulates autophagy and apoptosis. Cell Death Differ. 2011;18(4):571–80.
16. Yamamoto H, Kakuta S, Watanabe TM, Kitamura A, Sekito T, Kondo-Kakuta C, et al. Atg9 vesicles are an important membrane source during early steps of autophagosome formation. J Cell Biol. 2012;198(2):219–33.
17. Kirisako T, Ichimura Y, Okada H, Kabeya Y, Mizushima N, Yoshimori T, et al. The reversible modification regulates the membrane-binding state of Apg8/Aut7 essential for autophagy and the cytoplasm to vacuole targeting pathway. J Cell Biol. 2000;151(2):263–76.
18. Ichimura Y, Kirisako T, Takao T, Satomi Y, Shimonishi Y, Ishihara N, et al. A ubiquitin-like system mediates protein lipidation. Nature. 2000;408(6811):488–92.
19. Mizushima N, Noda T, Yoshimori T, Tanaka Y, Ishii T, George MD, et al. A protein conjugation system essential for autophagy. Nature. 1998;395(6700):395–8.
20. Kuma A, Mizushima N, Ishihara N, Ohsumi Y. Formation of the approximately 350-kDa Apg12-Apg5.Apg16 multimeric complex, mediated by Apg16 oligomerization, is essential for autophagy in yeast. J Biol Chem. 2002;277(21):18619–25.
21. Mizushima N, Noda T, Ohsumi Y. Apg16p is required for the function of the Apg12p-Apg5p conjugate in the yeast autophagy pathway. EMBO J. 1999;18(14):3888–96.
22. Puissant A, Fenouille N, Auberger P. When autophagy meets cancer through p62/SQSTM1. Am J Cancer Res. 2012;2(4):397–413.
23. Novak I, Dikic I. Autophagy receptors in developmental clearance of mitochondria. Autophagy. 2011;7(3):301–3.
24. Lamark T, Johansen T. Aggrephagy: selective disposal of protein aggregates by macroautophagy. Int J Cell Biol. 2012;2012:736905.

25. Bernales S, McDonald KL, Walter P. Autophagy counterbalances endoplasmic reticulum expansion during the unfolded protein response. PLoS Biol. 2006;4(12):e423.
26. Kim PK, Hailey DW, Mullen RT, Lippincott-Schwartz J. Ubiquitin signals autophagic degradation of cytosolic proteins and peroxisomes. Proc Natl Acad Sci U S A. 2008;105(52):20567–74.
27. Gomes LC, Dikic I. Autophagy in antimicrobial immunity. Mol Cell. 2014;54(2):224–33.
28. Kraft C, Deplazes A, Sohrmann M, Peter M. Mature ribosomes are selectively degraded upon starvation by an autophagy pathway requiring the Ubp3p/Bre5p ubiquitin protease. Nat Cell Biol. 2008;10(5):602–10.
29. Johansen T, Lamark T. Selective autophagy mediated by autophagic adapter proteins. Autophagy. 2011;7(3):279–96.
30. Kirkin V, McEwan DG, Novak I, Dikic I. A role for ubiquitin in selective autophagy. Mol Cell. 2009;34(3):259–69.
31. Tan JM, Wong ES, Kirkpatrick DS, Pletnikova O, Ko HS, Tay SP, et al. Lysine 63-linked ubiquitination promotes the formation and autophagic clearance of protein inclusions associated with neurodegenerative diseases. Hum Mol Genet. 2008;17(3):431–9.
32. Komatsu M, Ueno T, Waguri S, Uchiyama Y, Kominami E, Tanaka K. Constitutive autophagy: vital role in clearance of unfavorable proteins in neurons. Cell Death Differ. 2007;14(5):887–94.
33. Lu J, Holmgren A. The thioredoxin antioxidant system. Free Radic Biol Med. 2014;66:75–87.
34. Hu F, Liu F. Mitochondrial stress: a bridge between mitochondrial dysfunction and metabolic diseases? Cell Signal. 2011;23(10):1528–33.
35. Karbowski M, Youle RJ. Regulating mitochondrial outer membrane proteins by ubiquitination and proteasomal degradation. Curr Opin Cell Biol. 2011;23(4):476–82.
36. Menzies RA, Gold PH. The turnover of mitochondria in a variety of tissues of young adult and aged rats. J Biol Chem. 1971;246(8):2425–9.
37. Sandoval H, Thiagarajan P, Dasgupta SK, Schumacher A, Prchal JT, Chen M, et al. Essential role for Nix in autophagic maturation of erythroid cells. Nature. 2008;454(7201):232–5.
38. Sato M, Sato K. Degradation of paternal mitochondria by fertilization-triggered autophagy in C. elegans embryos. Science (New York, NY). 2011;334(6059):1141–4.
39. Al Rawi S, Louvet-Vallee S, Djeddi A, Sachse M, Culetto E, Hajjar C, et al. Postfertilization autophagy of sperm organelles prevents paternal mitochondrial DNA transmission. Science (New York, NY). 2011;334(6059):1144–7.
40. Kubli DA, Gustafsson AB. Mitochondria and mitophagy: the yin and yang of cell death control. Circ Res. 2012;111(9):1208–21.
41. Ye X, Sun X, Starovoytov V, Cai Q. Parkin-mediated mitophagy in mutant hAPP neurons and Alzheimer's disease patient brains. Hum Mol Genet. 2015;24(10):2938–51.
42. Jin SM, Youle RJ. The accumulation of misfolded proteins in the mitochondrial matrix is sensed by PINK1 to induce PARK2/Parkin-mediated mitophagy of polarized mitochondria. Autophagy. 2013;9(11):1750–7.
43. Allen GF, Toth R, James J, Ganley IG. Loss of iron triggers PINK1/Parkin-independent mitophagy. EMBO Rep. 2013;14(12):1127–35.
44. Joselin AP, Hewitt SJ, Callaghan SM, Kim RH, Chung YH, Mak TW, et al. ROS-dependent regulation of Parkin and DJ-1 localization during oxidative stress in neurons. Hum Mol Genet. 2012;21(22):4888–903.
45. Rubio N, Coupienne I, Di Valentin E, Heirman I, Grooten J, Piette J, et al. Spatiotemporal autophagic degradation of oxidatively damaged organelles after photodynamic stress is amplified by mitochondrial reactive oxygen species. Autophagy. 2012;8(9):1312–24.
46. Tanaka A, Cleland MM, Xu S, Narendra DP, Suen DF, Karbowski M, et al. Proteasome and p97 mediate mitophagy and degradation of mitofusins induced by Parkin. J Cell Biol. 2010;191(7):1367–80.
47. Twig G, Shirihai OS. The interplay between mitochondrial dynamics and mitophagy. Antioxid Redox Signal. 2011;14(10):1939–51.

48. Ney PA. Normal and disordered reticulocyte maturation. Curr Opin Hematol. 2011;18(3):152–7.
49. Novak I, Kirkin V, McEwan DG, Zhang J, Wild P, Rozenknop A, et al. Nix is a selective autophagy receptor for mitochondrial clearance. EMBO Rep. 2010;11(1):45–51.
50. Honda S, Arakawa S, Nishida Y, Yamaguchi H, Ishii E, Shimizu S. Ulk1-mediated Atg5-independent macroautophagy mediates elimination of mitochondria from embryonic reticulocytes. Nat Commun. 2014;5:4004.
51. Ding WX, Yin XM. Mitophagy: mechanisms, pathophysiological roles, and analysis. Biol Chem. 2012;393(7):547–64.
52. Politi Y, Gal L, Kalifa Y, Ravid L, Elazar Z, Arama E. Paternal mitochondrial destruction after fertilization is mediated by a common endocytic and autophagic pathway in Drosophila. Dev Cell. 2014;29(3):305–20.
53. Luo SM, Sun QY. Autophagy is not involved in the degradation of sperm mitochondria after fertilization in mice. Autophagy. 2013;9(12):2156–7.
54. Yamano K, Youle RJ. PINK1 is degraded through the N-end rule pathway. Autophagy. 2013;9(11):1758–69.
55. Meissner C, Lorenz H, Weihofen A, Selkoe DJ, Lemberg MK. The mitochondrial intramembrane protease PARL cleaves human Pink1 to regulate Pink1 trafficking. J Neurochem. 2011;117(5):856–67.
56. Fedorowicz MA, de Vries-Schneider RL, Rub C, Becker D, Huang Y, Zhou C, et al. Cytosolic cleaved PINK1 represses Parkin translocation to mitochondria and mitophagy. EMBO Rep. 2014;15(1):86–93.
57. Narendra D, Kane LA, Hauser DN, Fearnley IM, Youle RJ. p62/SQSTM1 is required for Parkin-induced mitochondrial clustering but not mitophagy; VDAC1 is dispensable for both. Autophagy. 2010;6(8):1090–106.
58. Sarraf SA, Raman M, Guarani-Pereira V, Sowa ME, Huttlin EL, Gygi SP, et al. Landscape of the PARKIN-dependent ubiquitylome in response to mitochondrial depolarization. Nature. 2013;496(7445):372–6.
59. Shiba-Fukushima K, Arano T, Matsumoto G, Inoshita T, Yoshida S, Ishihama Y, et al. Phosphorylation of mitochondrial polyubiquitin by PINK1 promotes Parkin mitochondrial tethering. PLoS Genet. 2014;10(12):e1004861.
60. Okatsu K, Koyano F, Kimura M, Kosako H, Saeki Y, Tanaka K, et al. Phosphorylated ubiquitin chain is the genuine Parkin receptor. J Cell Biol. 2015;209(1):111–28.
61. Fiesel FC, Moussaud-Lamodiere EL, Ando M, Springer W. A specific subset of E2 ubiquitin-conjugating enzymes regulate Parkin activation and mitophagy differently. J Cell Sci. 2014;127(Pt 16):3488–504.
62. Geisler S, Vollmer S, Golombek S, Kahle PJ. The ubiquitin-conjugating enzymes UBE2N, UBE2L3 and UBE2D2/3 are essential for Parkin-dependent mitophagy. J Cell Sci. 2014;127(Pt 15):3280–93.
63. Bingol B, Tea JS, Phu L, Reichelt M, Bakalarski CE, Song Q, et al. The mitochondrial deubiquitinase USP30 opposes parkin-mediated mitophagy. Nature. 2014;510(7505):370–5.
64. Cunningham CN, Baughman JM, Phu L, Tea JS, Yu C, Coons M, et al. USP30 and parkin homeostatically regulate atypical ubiquitin chains on mitochondria. Nat Cell Biol. 2015;17(2):160–9.
65. Cornelissen T, Haddad D, Wauters F, Van Humbeeck C, Mandemakers W, Koentjoro B, et al. The deubiquitinase USP15 antagonizes Parkin-mediated mitochondrial ubiquitination and mitophagy. Hum Mol Genet. 2014;23(19):5227–42.
66. Durcan TM, Tang MY, Perusse JR, Dashti EA, Aguileta MA, McLelland GL, et al. USP8 regulates mitophagy by removing K6-linked ubiquitin conjugates from parkin. EMBO J. 2014;33(21):2473–91.
67. Muller M, Kotter P, Behrendt C, Walter E, Scheckhuber CQ, Entian KD, et al. Synthetic quantitative array technology identifies the Ubp3-Bre5 deubiquitinase complex as a negative regulator of mitophagy. Cell Rep. 2015;10(7):1215–25.

68. Lee JY, Nagano Y, Taylor JP, Lim KL, Yao TP. Disease-causing mutations in parkin impair mitochondrial ubiquitination, aggregation, and HDAC6-dependent mitophagy. J Cell Biol. 2010;189(4):671–9.
69. Geisler S, Holmstrom KM, Skujat D, Fiesel FC, Rothfuss OC, Kahle PJ, et al. PINK1/Parkin-mediated mitophagy is dependent on VDAC1 and p62/SQSTM1. Nat Cell Biol. 2010;12(2):119–31.
70. Pimenta de Castro I, Costa AC, Lam D, Tufi R, Fedele V, Moisoi N, et al. Genetic analysis of mitochondrial protein misfolding in Drosophila melanogaster. Cell Death Differ. 2012;19(8):1308–16.
71. de Castro IP, Costa AC, Celardo I, Tufi R, Dinsdale D, Loh SH, et al. Drosophila ref(2)P is required for the parkin-mediated suppression of mitochondrial dysfunction in pink1 mutants. Cell Death Dis. 2013;4:e873.
72. Lazarou M, Sliter DA, Kane LA, Sarraf SA, Wang C, Burman JL, et al. The ubiquitin kinase PINK1 recruits autophagy receptors to induce mitophagy. Nature. 2015;524(7565):309–14.
73. Vincow ES, Merrihew G, Thomas RE, Shulman NJ, Beyer RP, MacCoss MJ, et al. The PINK1-Parkin pathway promotes both mitophagy and selective respiratory chain turnover in vivo. Proc Natl Acad Sci U S A. 2013;110(16):6400–5.
74. Margineantu DH, Emerson CB, Diaz D, Hockenbery DM. Hsp90 inhibition decreases mitochondrial protein turnover. PLoS One. 2007;2(10):e1066.
75. McLelland GL, Soubannier V, Chen CX, McBride HM, Fon EA. Parkin and PINK1 function in a vesicular trafficking pathway regulating mitochondrial quality control. EMBO J. 2014;33(4):282–95.
76. Soubannier V, McLelland GL, Zunino R, Braschi E, Rippstein P, Fon EA, et al. A vesicular transport pathway shuttles cargo from mitochondria to lysosomes. Curr Biol. 2012;22(2):135–41.
77. Ashrafi G, Schlehe JS, LaVoie MJ, Schwarz TL. Mitophagy of damaged mitochondria occurs locally in distal neuronal axons and requires PINK1 and Parkin. J Cell Biol. 2014;206(5):655–70.
78. Grenier K, McLelland GL, Fon EA. Parkin- and PINK1-dependent mitophagy in neurons: will the real pathway please stand up? Front Neurol. 2013;4:100.
79. Park J, Lee SB, Lee S, Kim Y, Song S, Kim S, et al. Mitochondrial dysfunction in Drosophila PINK1 mutants is complemented by Parkin. Nature. 2006;441(7097):1157–61.
80. Sterky FH, Lee S, Wibom R, Olson L, Larsson NG. Impaired mitochondrial transport and Parkin-independent degeneration of respiratory chain-deficient dopamine neurons in vivo. Proc Natl Acad Sci U S A. 2011;108(31):12937–42.
81. Chu CT, Ji J, Dagda RK. Cardiolipin externalization to the outer mitochondrial membrane acts as an elimination signal for mitophagy in neuronal cells. Nat Cell Biol. 2013;15(10):1197–205.
82. Melser S, Chatelain EH, Lavie J, Mahfouf W, Jose C, Obre E, et al. Rheb regulates mitophagy induced by mitochondrial energetic status. Cell Metab. 2013;17(5):719–30.
83. Maday S, Wallace KE, Holzbaur EL. Autophagosomes initiate distally and mature during transport toward the cell soma in primary neurons. J Cell Biol. 2012;196(4):407–17.
84. Wong E, Cuervo AM. Integration of clearance mechanisms: the proteasome and autophagy. Cold Spring Harbor Perspect Biol. 2010;2(12):a006734.
85. Johri A, Beal MF. Mitochondrial dysfunction in neurodegenerative diseases. J Pharmacol Exp Ther. 2012;342(3):619–30.
86. Hara T, Nakamura K, Matsui M, Yamamoto A, Nakahara Y, Suzuki-Migishima R, et al. Suppression of basal autophagy in neural cells causes neurodegenerative disease in mice. Nature. 2006;441(7095):885–9.
87. Son JH, Shim JH, Kim KH, Ha JY, Han JY. Neuronal autophagy and neurodegenerative diseases. Exp Mol Med. 2012;44(2):89–98.
88. Corti O, Lesage S, Brice A. What genetics tells us about the causes and mechanisms of Parkinson's disease. Physiol Rev. 2011;91(4):1161–218.

89. Trempe JF, Fon EA. Structure and function of Parkin, PINK1, and DJ-1, the three musketeers of neuroprotection. Front Neurol. 2013;4:38.
90. Cruts M, Theuns J, Van Broeckhoven C. Locus-specific mutation databases for neurodegenerative brain diseases. Hum Mutat. 2012;33(9):1340–4.
91. Riley BE, Lougheed JC, Callaway K, Velasquez M, Brecht E, Nguyen L, et al. Structure and function of Parkin E3 ubiquitin ligase reveals aspects of RING and HECT ligases. Nat Commun. 2013;4:1982.
92. Trempe JF, Sauve V, Grenier K, Seirafi M, Tang MY, Menade M, et al. Structure of parkin reveals mechanisms for ubiquitin ligase activation. Science (New York, NY). 2013;340(6139):1451–5.
93. Tomoo K, Mukai Y, In Y, Miyagawa H, Kitamura K, Yamano A, et al. Crystal structure and molecular dynamics simulation of ubiquitin-like domain of murine parkin. Biochim Biophys Acta. 2008;1784(7–8):1059–67.
94. Wauer T, Komander D. Structure of the human Parkin ligase domain in an autoinhibited state. EMBO J. 2013;32(15):2099–112.
95. Seirafi M, Kozlov G, Gehring K. Parkin structure and function. FEBS J. 2015;282(11):2076–88.
96. Palacino JJ, Sagi D, Goldberg MS, Krauss S, Motz C, Wacker M, et al. Mitochondrial dysfunction and oxidative damage in parkin-deficient mice. J Biol Chem. 2004;279(18):18614–22.
97. Gandhi S, Wood-Kaczmar A, Yao Z, Plun-Favreau H, Deas E, Klupsch K, et al. PINK1-associated Parkinson's disease is caused by neuronal vulnerability to calcium-induced cell death. Mol Cell. 2009;33(5):627–38.
98. Morais VA, Haddad D, Craessaerts K, De Bock PJ, Swerts J, Vilain S, et al. PINK1 loss-of-function mutations affect mitochondrial complex I activity via NdufA10 ubiquinone uncoupling. Science (New York, NY). 2014;344(6180):203–7.
99. Pogson JH, Ivatt RM, Sanchez-Martinez A, Tufi R, Wilson E, Mortiboys H, et al. The complex I subunit NDUFA10 selectively rescues Drosophila pink1 mutants through a mechanism independent of mitophagy. PLoS Genet. 2014;10(11):e1004815.
100. Thomas KJ, McCoy MK, Blackinton J, Beilina A, van der Brug M, Sandebring A, et al. DJ-1 acts in parallel to the PINK1/parkin pathway to control mitochondrial function and autophagy. Hum Mol Genet. 2011;20(1):40–50.
101. Canet-Aviles RM, Wilson MA, Miller DW, Ahmad R, McLendon C, Bandyopadhyay S, et al. The Parkinson's disease protein DJ-1 is neuroprotective due to cysteine-sulfinic acid-driven mitochondrial localization. Proc Natl Acad Sci U S A. 2004;101(24):9103–8.
102. Taira T, Saito Y, Niki T, Iguchi-Ariga SM, Takahashi K, Ariga H. DJ-1 has a role in antioxidative stress to prevent cell death. EMBO Rep. 2004;5(2):213–8.
103. Cox D, Carver JA, Ecroyd H. Preventing alpha-synuclein aggregation: the role of the small heat-shock molecular chaperone proteins. Biochim Biophys Acta. 2014;1842(9):1830–43.
104. Orenstein SJ, Kuo SH, Tasset I, Arias E, Koga H, Fernandez-Carasa I, et al. Interplay of LRRK2 with chaperone-mediated autophagy. Nat Neurosci. 2013;16(4):394–406.
105. Webb JL, Ravikumar B, Atkins J, Skepper JN, Rubinsztein DC. Alpha-Synuclein is degraded by both autophagy and the proteasome. J Biol Chem. 2003;278(27):25009–13.
106. Winslow AR, Chen CW, Corrochano S, Acevedo-Arozena A, Gordon DE, Peden AA, et al. Alpha-Synuclein impairs macroautophagy: implications for Parkinson's disease. J Cell Biol. 2010;190(6):1023–37.
107. Zavodszky E, Seaman MN, Moreau K, Jimenez-Sanchez M, Breusegem SY, Harbour ME, et al. Mutation in VPS35 associated with Parkinson's disease impairs WASH complex association and inhibits autophagy. Nat Commun. 2014;5:3828.
108. Schapansky J, Nardozzi JD, Felizia F, LaVoie MJ. Membrane recruitment of endogenous LRRK2 precedes its potent regulation of autophagy. Hum Mol Genet. 2014;23(16):4201–14.
109. Schapansky J, Nardozzi JD, LaVoie MJ. The complex relationships between microglia, alpha-synuclein, and LRRK2 in Parkinson's disease. Neuroscience. 2014;302:74–88.

110. Bravo-San Pedro JM, Niso-Santano M, Gomez-Sanchez R, Pizarro-Estrella E, Aiastui-Pujana A, Gorostidi A, et al. The LRRK2 G2019S mutant exacerbates basal autophagy through activation of the MEK/ERK pathway. Cell Mol Life Sci. 2013;70(1):121–36.
111. Tong Y, Yamaguchi H, Giaime E, Boyle S, Kopan R, Kelleher 3rd RJ, et al. Loss of leucine-rich repeat kinase 2 causes impairment of protein degradation pathways, accumulation of alpha-synuclein, and apoptotic cell death in aged mice. Proc Natl Acad Sci U S A. 2010;107(21):9879–84.
112. Alegre-Abarrategui J, Christian H, Lufino MM, Mutihac R, Venda LL, Ansorge O, et al. LRRK2 regulates autophagic activity and localizes to specific membrane microdomains in a novel human genomic reporter cellular model. Hum Mol Genet. 2009;18(21):4022–34.
113. Ramirez A, Heimbach A, Grundemann J, Stiller B, Hampshire D, Cid LP, et al. Hereditary parkinsonism with dementia is caused by mutations in ATP13A2, encoding a lysosomal type 5 P-type ATPase. Nat Genet. 2006;38(10):1184–91.
114. Sidransky E, Nalls MA, Aasly JO, Aharon-Peretz J, Annesi G, Barbosa ER, et al. Multicenter analysis of glucocerebrosidase mutations in Parkinson's disease. N Engl J Med. 2009;361(17):1651–61.
115. Salawu FK, Umar JT, Olokoba AB. Alzheimer's disease: a review of recent developments. Ann Afr Med. 2011;10(2):73–9.
116. Hirai K, Aliev G, Nunomura A, Fujioka H, Russell RL, Atwood CS, et al. Mitochondrial abnormalities in Alzheimer's disease. J Soc Neurosci. 2001;21(9):3017–23.
117. Garcia-Escudero V, Martin-Maestro P, Perry G, Avila J. Deconstructing mitochondrial dysfunction in Alzheimer disease. Oxid Med Cell Longev. 2013;2013:162152.
118. Baloyannis SJ. Mitochondrial alterations in Alzheimer's disease. J Alzheimers Dis. 2006;9(2):119–26.
119. Maurer I, Zierz S, Moller HJ. A selective defect of cytochrome c oxidase is present in brain of Alzheimer disease patients. Neurobiol Aging. 2000;21(3):455–62.
120. Casley CS, Canevari L, Land JM, Clark JB, Sharpe MA. Beta-amyloid inhibits integrated mitochondrial respiration and key enzyme activities. J Neurochem. 2002;80(1):91–100.
121. Wang X, Su B, Lee HG, Li X, Perry G, Smith MA, et al. Impaired balance of mitochondrial fission and fusion in Alzheimer's disease. J Neurosci. 2009;29(28):9090–103.
122. Moreira PI, Siedlak SL, Wang X, Santos MS, Oliveira CR, Tabaton M, et al. Autophagocytosis of mitochondria is prominent in Alzheimer disease. J Neuropathol Exp Neurol. 2007;66(6):525–32.
123. Boland B, Kumar A, Lee S, Platt FM, Wegiel J, Yu WH, et al. Autophagy induction and autophagosome clearance in neurons: relationship to autophagic pathology in Alzheimer's disease. J Neurosci. 2008;28(27):6926–37.
124. Nixon RA, Wegiel J, Kumar A, Yu WH, Peterhoff C, Cataldo A, et al. Extensive involvement of autophagy in Alzheimer disease: an immuno-electron microscopy study. J Neuropathol Exp Neurol. 2005;64(2):113–22.
125. Yu WH, Cuervo AM, Kumar A, Peterhoff CM, Schmidt SD, Lee JH, et al. Macroautophagy – a novel Beta-amyloid peptide-generating pathway activated in Alzheimer's disease. J Cell Biol. 2005;171(1):87–98.
126. Pickford F, Masliah E, Britschgi M, Lucin K, Narasimhan R, Jaeger PA, et al. The autophagy-related protein beclin 1 shows reduced expression in early Alzheimer disease and regulates amyloid beta accumulation in mice. J Clin Invest. 2008;118(6):2190–9.
127. Wang Y, Martinez-Vicente M, Kruger U, Kaushik S, Wong E, Mandelkow EM, et al. Tau fragmentation, aggregation and clearance: the dual role of lysosomal processing. Hum Mol Genet. 2009;18(21):4153–70.
128. Funderburk SF, Marcellino BK, Yue Z. Cell "self-eating" (autophagy) mechanism in Alzheimer's disease. Mt Sinai J Med. 2010;77(1):59–68.
129. Rosen KM, Moussa CE, Lee HK, Kumar P, Kitada T, Qin G, et al. Parkin reverses intracellular beta-amyloid accumulation and its negative effects on proteasome function. J Neurosci Res. 2010;88(1):167–78.

130. Yu WH, Kumar A, Peterhoff C, Shapiro Kulnane L, Uchiyama Y, Lamb BT, et al. Autophagic vacuoles are enriched in amyloid precursor protein-secretase activities: implications for beta-amyloid peptide over-production and localization in Alzheimer's disease. Int J Biochem Cell Biol. 2004;36(12):2531–40.
131. Lee JH, Yu WH, Kumar A, Lee S, Mohan PS, Peterhoff CM, et al. Lysosomal proteolysis and autophagy require presenilin 1 and are disrupted by Alzheimer-related PS1 mutations. Cell. 2010;141(7):1146–58.
132. Theuns J, Del-Favero J, Dermaut B, van Duijn CM, Backhovens H, Van den Broeck MV, et al. Genetic variability in the regulatory region of presenilin 1 associated with risk for Alzheimer's disease and variable expression. Hum Mol Genet. 2000;9(3):325–31.
133. Ross CA, Tabrizi SJ. Huntington's disease: from molecular pathogenesis to clinical treatment. Lancet Neurol. 2011;10(1):83–98.
134. Browne SE, Bowling AC, MacGarvey U, Baik MJ, Berger SC, Muqit MM, et al. Oxidative damage and metabolic dysfunction in Huntington's disease: selective vulnerability of the basal ganglia. Ann Neurol. 1997;41(5):646–53.
135. Cui L, Jeong H, Borovecki F, Parkhurst CN, Tanese N, Krainc D. Transcriptional repression of PGC-1alpha by mutant huntingtin leads to mitochondrial dysfunction and neurodegeneration. Cell. 2006;127(1):59–69.
136. Orr AL, Li S, Wang CE, Li H, Wang J, Rong J, et al. N-terminal mutant huntingtin associates with mitochondria and impairs mitochondrial trafficking. J Neurosci. 2008;28(11):2783–92.
137. Reddy PH, Mao P, Manczak M. Mitochondrial structural and functional dynamics in Huntington's disease. Brain Res Rev. 2009;61(1):33–48.
138. Martin DD, Ladha S, Ehrnhoefer DE, Hayden MR. Autophagy in Huntington disease and huntingtin in autophagy. Trends Neurosci. 2015;38(1):26–35.
139. Ghavami S, Shojaei S, Yeganeh B, Ande SR, Jangamreddy JR, Mehrpour M, et al. Autophagy and apoptosis dysfunction in neurodegenerative disorders. Prog Neurobiol. 2014;112:24–49.
140. Martinez-Vicente M, Talloczy Z, Wong E, Tang G, Koga H, Kaushik S, et al. Cargo recognition failure is responsible for inefficient autophagy in Huntington's disease. Nat Neurosci. 2010;13(5):567–76.
141. Kegel KB, Kim M, Sapp E, McIntyre C, Castano JG, Aronin N, et al. Huntingtin expression stimulates endosomal-lysosomal activity, endosome tubulation, and autophagy. J Neurosci. 2000;20(19):7268–78.
142. Ravikumar B, Vacher C, Berger Z, Davies JE, Luo S, Oroz LG, et al. Inhibition of mTOR induces autophagy and reduces toxicity of polyglutamine expansions in fly and mouse models of Huntington disease. Nat Genet. 2004;36(6):585–95.
143. Shibata M, Lu T, Furuya T, Degterev A, Mizushima N, Yoshimori T, et al. Regulation of intracellular accumulation of mutant Huntingtin by Beclin 1. J Biol Chem. 2006;281(20):14474–85.
144. Wong YC, Holzbaur EL. The regulation of autophagosome dynamics by huntingtin and HAP1 is disrupted by expression of mutant huntingtin, leading to defective cargo degradation. J Neurosci. 2014;34(4):1293–305.
145. Ochaba J, Lukacsovich T, Csikos G, Zheng S, Margulis J, Salazar L, et al. Potential function for the Huntingtin protein as a scaffold for selective autophagy. Proc Natl Acad Sci U S A. 2014;111(47):16889–94.
146. Rui YN, Xu Z, Patel B, Chen Z, Chen D, Tito A, et al. Huntingtin functions as a scaffold for selective macroautophagy. Nat Cell Biol. 2015;17(3):262–75.
147. Rosen DR, Siddique T, Patterson D, Figlewicz DA, Sapp P, Hentati A, et al. Mutations in Cu/Zn superoxide dismutase gene are associated with familial amyotrophic lateral sclerosis. Nature. 1993;362(6415):59–62.
148. Bruijn LI, Miller TM, Cleveland DW. Unraveling the mechanisms involved in motor neuron degeneration in ALS. Ann Rev Neurosci. 2004;27:723–49.
149. Bacman SR, Bradley WG, Moraes CT. Mitochondrial involvement in amyotrophic lateral sclerosis: trigger or target? Mol Neurobiol. 2006;33(2):113–31.

150. Lin MT, Beal MF. Mitochondrial dysfunction and oxidative stress in neurodegenerative diseases. Nature. 2006;443(7113):787–95.
151. Jaarsma D, Rognoni F, van Duijn W, Verspaget HW, Haasdijk ED, Holstege JC. CuZn superoxide dismutase (SOD1) accumulates in vacuolated mitochondria in transgenic mice expressing amyotrophic lateral sclerosis-linked SOD1 mutations. Acta Neuropathol. 2001;102(4):293–305.
152. Vijayvergiya C, Beal MF, Buck J, Manfredi G. Mutant superoxide dismutase 1 forms aggregates in the brain mitochondrial matrix of amyotrophic lateral sclerosis mice. J Neurosci. 2005;25(10):2463–70.
153. Liu J, Lillo C, Jonsson PA, Vande Velde C, Ward CM, Miller TM, et al. Toxicity of familial ALS-linked SOD1 mutants from selective recruitment to spinal mitochondria. Neuron. 2004;43(1):5–17.
154. Takeuchi H, Kobayashi Y, Ishigaki S, Doyu M, Sobue G. Mitochondrial localization of mutant superoxide dismutase 1 triggers caspase-dependent cell death in a cellular model of familial amyotrophic lateral sclerosis. J Biol Chem. 2002;277(52):50966–72.
155. Kong J, Xu Z. Massive mitochondrial degeneration in motor neurons triggers the onset of amyotrophic lateral sclerosis in mice expressing a mutant SOD1. J Neurosci. 1998;18(9):3241–50.
156. Li L, Zhang X, Le W. Altered macroautophagy in the spinal cord of SOD1 mutant mice. Autophagy. 2008;4(3):290–3.
157. Morimoto N, Nagai M, Ohta Y, Miyazaki K, Kurata T, Morimoto M, et al. Increased autophagy in transgenic mice with a G93A mutant SOD1 gene. Brain Res. 2007;1167:112–7.
158. Nassif M, Hetz C. Targeting autophagy in ALS: a complex mission. Autophagy. 2011;7(4):450–3.
159. Kabuta T, Suzuki Y, Wada K. Degradation of amyotrophic lateral sclerosis-linked mutant Cu, Zn-superoxide dismutase proteins by macroautophagy and the proteasome. J Biol Chem. 2006;281(41):30524–33.
160. Gal J, Strom AL, Kilty R, Zhang F, Zhu H. p62 accumulates and enhances aggregate formation in model systems of familial amyotrophic lateral sclerosis. J Biol Chem. 2007;282(15):11068–77.
161. Rea SL, Majcher V, Searle MS, Layfield R. SQSTM1 mutations – bridging Paget disease of bone and ALS/FTLD. Exp Cell Res. 2014;325(1):27–37.
162. Teyssou E, Takeda T, Lebon V, Boillee S, Doukoure B, Bataillon G, et al. Mutations in SQSTM1 encoding p62 in amyotrophic lateral sclerosis: genetics and neuropathology. Acta Neuropathol. 2013;125(4):511–22.
163. Zhang X, Li L, Chen S, Yang D, Wang Y, Zhang X, et al. Rapamycin treatment augments motor neuron degeneration in SOD1(G93A) mouse model of amyotrophic lateral sclerosis. Autophagy. 2011;7(4):412–25.
164. Palikaras K, Tavernarakis N. Mitophagy in neurodegeneration and aging. Front Genet. 2012;3:297.
165. Santos RX, Correia SC, Carvalho C, Cardoso S, Santos MS, Moreira PI. Mitophagy in neurodegeneration: an opportunity for therapy? Curr Drug Targets. 2011;12(6):790–9.
166. Hochfeld WE, Lee S, Rubinsztein DC. Therapeutic induction of autophagy to modulate neurodegenerative disease progression. Acta Pharmacol Sin. 2013;34(5):600–4.
167. Rubinsztein DC, Codogno P, Levine B. Autophagy modulation as a potential therapeutic target for diverse diseases. Nat Rev Drug Discov. 2012;11(9):709–30.
168. Cheng Y, Ren X, Hait WN, Yang JM. Therapeutic targeting of autophagy in disease: biology and pharmacology. Pharmacol Rev. 2013;65(4):1162–97.
169. Rana A, Rera M, Walker DW. Parkin overexpression during aging reduces proteotoxicity, alters mitochondrial dynamics, and extends lifespan. Proc Natl Acad Sci U S A. 2013;110(21):8638–43.
170. Yang Y, Gehrke S, Imai Y, Huang Z, Ouyang Y, Wang JW, et al. Mitochondrial pathology and muscle and dopaminergic neuron degeneration caused by inactivation of Drosophila Pink1 is rescued by Parkin. Proc Natl Acad Sci U S A. 2006;103(28):10793–8.

171. Hong X, Liu J, Zhu G, Zhuang Y, Suo H, Wang P, et al. Parkin overexpression ameliorates hippocampal long-term potentiation and beta-amyloid load in an Alzheimer's disease mouse model. Hum Mol Genet. 2014;23(4):1056–72.
172. Bian M, Liu J, Hong X, Yu M, Huang Y, Sheng Z, et al. Overexpression of parkin ameliorates dopaminergic neurodegeneration induced by 1-methyl-4-phenyl-1,2,3,6-tetrahydropyridine in mice. PLoS One. 2012;7(6):e39953.

Chapter 12
The Deleterious Duo of Neurodegeneration: Lysosomes and Mitochondria

Matthew Nguyen, Ellen Sidransky, and Wendy Westbroek

Abstract Many studies have demonstrated that the accumulation of aggregate-prone proteins due to defects in cellular quality control systems contributes to the development of neurodegenerative diseases. One form of quality control within neurons is autophagy, an intracellular pathway involved in the breakdown of cytosolic constituents. Lysosomes mediate autophagy, and their dysfunction may contribute to perturbations in cellular homeostasis and affect other organelles such as mitochondria. Mitochondrial malfunction may then further perpetuate lysosomal damage and initiate inflammatory responses. Therefore, lysosomes and mitochondria share a reciprocal relationship where dysfunction in one often affects the function of the other. These consequences of lysosome and mitochondrial impairment complete a deleterious feedback loop that concludes not only in neurodegeneration but also neuroinflammation. Herein, we discuss the primary types of autophagy and their underlying mechanisms, the regulation of lysosomal biogenesis and function, and the link between lysosomal and mitochondrial dysfunction. We conclude this chapter by assessing the role of lysosomal dysfunction in neurodegenerative diseases.

Keywords Alpha-synuclein • TFEB • TFE3 • Autophagy • Lysosome • Synucleinopathies • Lysosomal storage disorders • Gaucher disease • Inflammasome • Reactive oxygen species • Mitophagy • Glucocerebrosidase

---

M. Nguyen, BS • W. Westbroek, PhD
National Institutes of Health, Medical Genetics Branch, National Human Genome Research Institute, Bethesda, MD, USA

E. Sidransky, MD (✉)
Section of Molecular Neurogenetics, National Institutes of Health, Medical Genetics Branch, National Human Genome Research Institute, Bethesda, MD, USA
e-mail: sidranse@mail.nih.gov

A.K. Reeve et al. (eds.), *Mitochondrial Dysfunction in Neurodegenerative Disorders*, DOI 10.1007/978-3-319-28637-2_12

12.1 Autophagy

In 1963, Christian de Duve first described autophagy: a quality control system involved in the degradation of unnecessary or nonfunctional cellular components [1]. Lysosomes, with their acidic pH and lytic hydrolases, mediate autophagy in response to perturbations in cellular homeostasis that may occur as a result of organelle dysfunction, genetic mutations, or nutrient deprivation. Autophagy is particularly important in the context of neurodegeneration, as it has been shown that in the absence of any disease-associated gene products, the loss of autophagy alone is sufficient to cause neural dysfunction and eventually neuronal cell death [2, 3]. Depending on the mechanism by which cellular cargo is transported to the lysosome, autophagy can be classified into three types: microautophagy, macroautophagy, and chaperone-mediated autophagy.

12.1.1 Microautophagy

Little is currently known about microautophagy, especially within mammalian cells; as a result, the majority of our knowledge regarding this autophagic process stems from studies in a few species of yeast: *Saccharomyces cerevisiae*, *Pichia pastoris*, and *Hansenula polymorpha* [4–8]. In mammalian microautophagy, the lysosomal membrane directly invaginates and sequesters cytoplasmic constituents into vesicles intended for degradation [9–11]. This invagination may occur through either a concave retreat of the lysosomal membrane or the lysosomal wrapping mechanism (LWM) [12]. During the LWM, the lysosome elongates from a spherical to tubular shape and extends an arm-like protrusion that envelops soluble cytoplasmic components. The tip of this extension then meets and fuses with the lysosomal membrane, sealing these cellular constituents within a vesicle for breakdown inside the lysosome [13, 14]. Microautophagy, as a form of quality of control, has been implicated in the basal turnover of cellular components, but recent literature suggests that this autophagic process may possess additional functions. Microautophagy could be associated with balancing the influx of membrane components introduced by macroautophagy and may be related to multivesicular body formation as a lysosomal microautophagy-like mechanism involved in selectively delivering cytosolic proteins to late endosomes during biogenesis [11, 15–17]. As highlighted in a review article by Mijalica et al. microautophagy is a field in need of further investigation [10].

12.1.2 Macroautophagy

In contrast to microautophagy, macroautophagy is better understood and is described in detail in Chap. 11. In macroautophagy, cytoplasmic cargo is sequestered within double-membrane vesicles called autophagosomes for transportation to the lysosome (Fig. 12.1c) [18, 19]. Initiation of autophagy starts with the formation of a

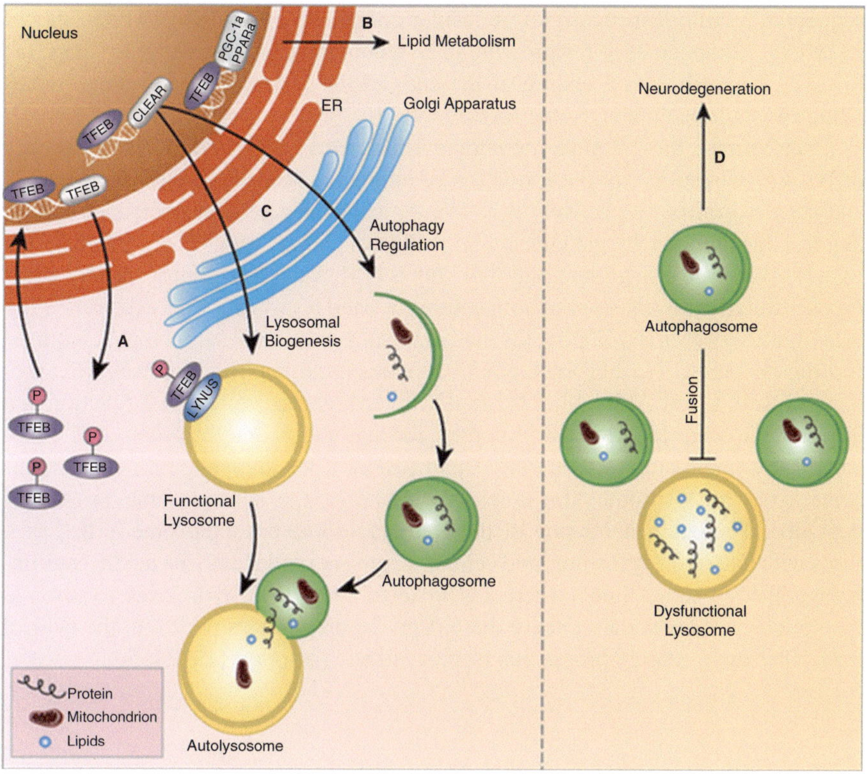

Fig. 12.1 (**a**) TFEB is regulated at both transcriptional and posttranscriptional levels. During nutrient-rich conditions, mTORC1 phosphorylates TFEB, preventing its translocation into the nucleus and sequestering the transcription factor in the cytosol and lysosomal membrane. Under stressful conditions, mTORC1 dissociates from the lysosomal membrane without phosphorylating TFEB, allowing for its translocation to the nucleus. Within the nucleus, TFEB regulates its own transcription as well as genes involved in lipid metabolism and the CLEAR network. (**b**) TFEB contributes to lipid metabolism by activating transcription of PPARα and PGC-1α. These two proteins initiate pathways involved in the degradation of lipids for energy production. (**c**) TFEB activates CLEAR network genes involved in biogenesis of lysosomes and regulation of autophagy, which are both required for proper cellular quality control. (**d**) Defects in the lysosome-autophagy pathway promote accumulation of aggregate-prone proteins such as α-synuclein, huntingtin, and Tau leading to neurodegeneration

double-membrane structure exclusive to macroautophagy termed the phagophore [20–24]. Phagophore assembly is suspected to occur de novo, and within mammalian systems, this process may take place at multiple locations in the cytoplasm [25–29]. The growth of the phagophore is an area of intense debate, as the source of the membrane components used in expansion is currently unknown. Multiple studies have suggested that these building blocks arise from pre-existing membrane compartments, such as those of the mitochondria, endoplasmic reticulum (ER), Golgi apparatus, and plasma membrane. Others speculate that the membrane may originate from a phosphatidylinositol-3-phosphate-enriched portion of the ER named the omegasome [30–36]. Phagophore elongation and expansion involves sequential recruitment of several molecular complexes and subsequent delivery of membrane components to the growing phagophore (discussed in detail in Chap. 11) [37–42]. Apposition and sealing of the ends of the expanding phagophore membrane concludes autophagosome formation [39].

Once formed, these double-membrane vesicles may fuse with either endosomes or lysosomes resulting in the formation of chimeric organelles called amphisomes and autolysosomes, respectively [43, 44]. Amphisomes are intermediate structures that link the autophagic and endosomal pathways and are capable of further fusion with lysosomes to also produce autolysosomes [45]. Autophagosomes are dependent on microtubules for transport to both endosomes and lysosomes where their merging is facilitated by a variety of proteins including both endosomal sort complex required for transport (ESCRT) and soluble NSF attachment protein receptor (SNARE) proteins as well as Rab7 [46–48]. Of peculiar interest, proper lysosomal function and acidification were found to be important for autophagosome-lysosome and endosome-lysosome fusion as well [49]. Upon fusion of the autophagosome or amphisome with the lysosome, the inner membrane of the autolysosome is quickly degraded. The outer membrane of the autophagosome or amphisome is lost as it fuses with that of the lysosome, and cellular cargo is expelled into the acidic environment of the lysosomal lumen where it is broken down by lytic hydrolases. Lysosomal membrane permeases then release these degradation products back into the cytosol for further use in energy production or biosynthetic pathways [19, 24, 50].

12.1.3 Chaperone-Mediated Autophagy

Not all forms of autophagy require the formation of vesicles for transporting cargo to the lysosomal lumen. In chaperone-mediated autophagy (CMA), cellular cargo is individually targeted and directly transported across the lysosomal membrane into the lumen for degradation. This catabolic process, currently only described in mammals, was the first subtype of autophagy for which selectivity was demonstrated. Selectivity occurs through multiple steps [51–53]. The first is the recognition of substrate proteins containing a specific KFERQ pentapeptide motif by heat-shock cognate protein of 70 kDa (hsc70). All known CMA substrate proteins possess this pentapeptide motif, and several studies have demonstrated that it is both necessary and sufficient for lysosomal targeting [51, 54–57]. Recognition of protein substrates by hsc70 is regulated by the accessibility of the pentapeptide motif, as this structure may be concealed by protein folding, protein-protein interactions, and binding to subcellular membranes. Several studies have also demonstrated that posttranslational modifications of KFERQ-like motifs, which are peptide motifs that contain four out of the five required amino acid residues typically found in a KFERQ motif, regulate substrate binding as well. Since the KFERQ motif depends on its charge for proper interaction with hsc70, phosphorylation or acetylation of KFERQ-like motifs may compensate for changes in charge caused by the absence of one of the required amino acid residues typically included in a KFERQ motif. In this manner, a CMA-targeting motif may be formed upon modification of the KFERQ-like motif, allowing for association with hsc70 [52, 56, 58, 59].

Once protein substrates are bound to hsc70, this substrate/chaperone complex is delivered to the lysosomal membrane where it interacts with the cytosolic tail of lysosome-associated membrane protein type 2A (LAMP-2A), a single-span membrane protein [60]. LAMP-2A exists as a monomer at the lysosomal membrane, but, upon binding of its protein substrate, forms a multiprotein complex composed of free LAMP-2A monomers and other proteins [61]. Protein substrates may bind the monomeric form of LAMP-2A in their folded conformations; however, in order to be translocated into the lysosomal lumen, these proteins must first be unfolded. This unfolding process occurs before the assembly of the multiprotein translocation complex and may be mediated by hsc70 and its associated co-chaperones [53, 62, 63]. While numerous factors, such as the protein density and fluidity of the lysosomal membrane, influence assembly of the multiprotein complex, very few proteins have been identified as regulators of this molecular machinery [64]. Recently, work by Bandyopadhyay and colleagues have demonstrated that both GFAP and EF1α modulate the assembly and disassembly of this translocation complex in a GTP-dependent manner [65]. Furthermore, two chaperones, hsc70 and heat-shock protein 90 (hsp90), have also been implicated in this process. These chaperones function not only in the assembly and disassembly of the translocation complex but also contribute to the stabilization of LAMP-2A during multimerization [51, 61, 66].

Translocation of the unfolded protein into the lysosomal lumen occurs one by one through the LAMP-2A multiprotein complex and is dependent on the presence of an intralysosomal isoform of hsc70 (lys-hsc70) [63, 67]. This isoform of hsc70 is located in the lysosomal lumen and may enter the lysosome via fusion with hsc70-containing late endosomes [53, 66]. To date, the exact mechanism by which lys-hsc70 contributes to substrate translocation is unknown; however, it has been proposed that this protein may "pull" the substrate through the LAMP-2A translocation complex or passively "hold" the substrate, preventing its release back into the cytosol [53]. Depending on cellular conditions and cell type, the population of lysosomes containing lys-hsc70 fluctuates between 20 % and 80 %, and therefore, not all lysosomes are capable of CMA [68]. Upon entry of the protein substrate into the lysosomal lumen, substrate degradation occurs by resident hydrolases, and this process is accompanied by the dissociation of the LAMP-2A translocation complex. These resulting monomers of LAMP-2A are then free to further bind substrates and initiate new cycles of CMA [61].

12.2 Transcription Factor EB (TFEB)

12.2.1 The Lysosome

While investigating the mechanism of action of insulin within liver cells, Christian de Duve serendipitously stumbled upon a nonspecific acid phosphatase that possessed a phantasmic enzymatic activity. Tantalized by the "vanishing acts" that this acid phosphatase performed, de Duve abandoned his work on insulin and pursued this accidental finding instead. In 1955, after years of investigating this unexpected observation, de Duve described the lysosome. Roughly 20 years later, he was awarded the Nobel Prize in Physiology and Medicine for this discovery [69, 70]. Traditionally, lysosomes, Greek for "digestive body" and formed from the combination of the words *lysis and soma*, have been considered static cellular organelles that are not influenced by environmental cues. They are primarily implicated in the catabolism of macromolecules obtained from multiple cellular processes such as endocytosis, autophagy, and phagocytosis [10, 52, 71–74]. However, the new concept of lysosomal adaptation has subsequently broadened our perspective from merely its static role in cellular clearance [75]. Recent studies by Ballabio and colleagues have demonstrated that a majority of the 96 lysosomal genes involved in lysosomal biogenesis and function coordinately express and are influenced by environmental factors, both extracellular and intracellular, through a basic helix-loop-helix transcription factor known as transcription factor EB (TFEB), the master regulator of the coordinate lysosomal expression and regulation (CLEAR), which encompasses these lysosomal genes [76, 77]. Lysosomes have now emerged as a critical player involved in nutrient sensing, signaling, and metabolism, in addition to their established duty in cellular macromolecule degradation. At the heart of this adaptive and dynamic lysosome model is the activity of TFEB, the master regulator of lysosomal biogenesis and function that modulates the interplay between lysosome-mediated cellular processes and environmental influences.

12.2.2 Regulation of TFEB

TFEB is located within the cytoplasm and on the surface of the cholesterol sparse lysosomal membrane, where it is regulated through an "inside-out" signaling model initiated by the level of accumulated amino acids in the acidic lumen of the lysosome (see Fig. 12.1a). These amino acid levels are sensed by the lysosome nutrient signaling (LYNUS) machinery, which propagates a signal through a protein complex known as Ragulator to RAG GTPases that subsequently recruit mTORC1 to the lysosomal surface [78–85]. The subcellular localization of TFEB depends largely on the mTORC1-mediated phosphorylation of TFEB at two crucial serine residues: Ser142 and Ser211 [77, 84, 86, 87]. Under favorable conditions, mTORC1 phosphorylates TFEB, sequestering the transcription factor in the cytoplasm and lysosomal surface. However, during adverse cellular circumstances—starvation, stress, and lysosome dysfunction or inhibition—mTORC1 dissociates from the LYNUS complex without phosphorylating TFEB, allowing for TFEB nuclear translocation [76, 77, 81, 84, 86, 87]. Within the nucleus, TFEB activates transcription of the CLEAR network genes pertinent in the lysosomal-autophagy pathway, as well as those involved in lipid metabolism such as peroxisome proliferator-activated receptor α (PPARα), peroxisome proliferator-activated receptor-gamma coactivator-1α (PGC-1α), and their respective target genes. Furthermore, once inside the nucleus TFEB also positively regulates its own function by binding to the CLEAR motif within the promotor region of its associated gene, thus initiating its own transcription. Therefore, the activity of TFEB is regulated both at the transcriptional and posttranscriptional levels (see Fig. 12.1a), and presents a mechanism for which environmental influences, both extracellular and intracellular, can be transmitted from lysosome to nucleus [75–77].

12.2.3 Lipid Catabolism

TFEB is also implicated in lipid catabolism due to the intertwined features of autophagy and lipid metabolism. Through an autophagic process known as macrolipophagy, lipid droplets are transported to lysosomes via autophagosomes, where they are degraded into free fatty acids and glycerol [88, 89]. Mouse liver cells overexpressing TFEB exhibited upregulation of genes implicated in lipid catabolic processes, such as lipophagy and fatty acid oxidation, as well as downregulation of those involved in lipid biogenesis [75]. TFEB exerts its transcriptional control of lipid metabolism by inducing two key modulators of energy metabolism: PPARα and PGC-1α (see Fig. 12.1b) [75–90]. Cells stressed by starvation undergo TFEB-activated transcription of both PPARα and PGC-1α, which subsequently initiates a metabolic response where energy is produced through the breakdown of lipid reserves [75]. Furthermore, studies performed on *Atg7* knockout mice, whose autophagic pathways are suppressed, demonstrate that TFEB mediates lipid metabolism through an autophagy-dependent manner [75]. Therefore, TFEB weaves together the lysosomal-autophagic pathway with lipid metabolism. It is becoming increasingly clear that the lysosome is not only a cellular garbage disposal, but it also serves as an intricate player involved in nutrient sensing, signaling, and metabolism.

12.2.4 TFEB, Autophagy, and Neurodegeneration

In many neurodegenerative proteopathies, the lysosomal-autophagy pathway is disrupted and the pathogenic formation of misfolded protein aggregates occurs (see Fig. 12.1c) [91–93]. Therefore, as the master regulator of the lysosomal-autophagy pathway, the role of TFEB in neurodegeneration is an area of intensive investigation. Studies utilizing a rat model of Parkinson's disease generated by overexpressing human alpha-synuclein (α-syn) in the midbrain demonstrated that elevated α-syn levels induced TFEB retention in the cytoplasm, leading to lysosomal dysfunction, α-syn accumulation in autophagosomes, and a progressive increase in α-syn oligomers [94]. Since α-syn is structurally and functionally similar to 14-3-3 proteins, a group of proteins known to interact with and trap phosphorylated TFEB within the cytoplasm, a possible pathogenic mechanism, has been postulated where aggregates of α-syn bind to phosphorylated TFEB, preventing its nuclear translocation and eventually leading to impairment of autophagic processes [86, 87, 94–96].

Overexpression of TFEB, whether by genetic or pharmacological means, has been shown to mediate the clearance of α-syn and halt the progression of Parkinsonian neurodegeneration in both rat nigral dopaminergic neurons and human neuroglioma cells through an autophagy-dependent pathway [94, 97]. Recent research has also demonstrated that overexpression of TFEB is neuroprotective in both Huntington's and Alzheimer's diseases as well. In a Huntington's disease mouse model, TFEB-mediated induction of PGC-1α was shown to rescue neurotoxicity via cellular clearance of huntingtin protein aggregation and reduction of oxidative stress [98]. Furthermore, TFEB overexpression in a mouse model of Alzheimer's disease showed that TFEB is capable of allaying phosphorylated Tau and neurofibrillary tangle-associated neuropathology by enhancing the clearance of both hyperphosphorylated and misfolded Tau proteins [99].

12.2.5 Transcription Factor E3

Recently, a second member of the MiTF/TFE family was found to be a CLEAR network regulator. Like TFEB, translocation of transcription factor E3 (TFE3) from the cytosol to the nucleus is dependent upon interaction with Rag GTPases, mTORC1-dependent phosphorylation status, and nutrient availability [100]. Under conditions of starvation, TFE3 translocates from the cytosol to the nucleus where it binds to the CLEAR motif of promoters of genes belonging to the CLEAR network. Research indicates that TFE3-induced autophagy and lysosomal biogenesis are independent of TFEB. Relative protein abundance of both transcription factors, which has been shown to be different in various cell types, is speculated to be the decisive factor in taking on the role of master regulator. Just like TFEB, overexpression of TFE3 can promote lysosomal exocytosis and clearage of lysosomal substrate storage in several lysosomal storage disorders [100]. Together, these results suggest that TFEB and TFE3, both master regulators of the lysosomal-autophagy pathway, may be promising therapeutic targets for the development of a broad-spectrum neuroprotective drug.

12.3 Lysosomal and Mitochondrial Dysfunction: Are They Connected?

Mitochondria are eukaryotic organelles involved in a variety of cellular processes ranging from energy production to regulation of cellular calcium concentration and apoptosis. Mitochondria consist of a double-membrane structure that separates the inner-membrane space from the mitochondrial matrix [101]. Located in the inner mitochondrial membrane is the electron transport chain where oxidative phosphorylation occurs. This is also the primary site of both cellular energy and reactive oxygen species (ROS) production [102–105] (see Chap. 1). Growing evidence demonstrates that lysosomes and mitochondria share a mutual relationship, where dysfunction in one organelle often impairs the function of the other. Lysosomal damage not only directly disturbs the lysosomal-autophagy pathway but can also cause lysosomal membrane permeabilization (LMP) resulting in the release of lysosomal luminal contents into the cytosol [74, 106–108]. Whether by insufficient turnover of damaged mitochondria or via direct interaction with lysosomal cathepsins released as a result of LMP, the consequences of lysosomal dysfunction impair mitochondrial function and lead to the loss of mitochondrial membrane potential, increased ROS production, decreased generation of ATP, and eventually the discharge of mitochondrial components into the cytosol via mitochondrial membrane permeabilization [109–112]. Mitochondrial dysfunction may also occur through mutations in gene-encoding proteins involved in mitochondrial quality control and homeostasis such as PTEN-induced putative kinase 1 (*PINK1*/*PARK6*), the E3 ubiquitin ligase parkin (*PARK2*), and DJ-1 (*PARK7*). Mutations in these genes have been implicated in familial forms of Parkinson disease [101, 113–117] (see Chap. 11). While it is unclear whether lysosomal dysfunction precedes mitochondrial damage or vice versa, disruption of the intricate balance between the functions of these two organelles establishes a deleterious feedback loop [107]. A failure to degrade defective mitochondria by the lysosome results in the accumulation of dysfunctional mitochondria and the subsequent leakage of ROS. In turn, these oxidative species may perpetuate lysosome dysfunction and subsequently enhance mitochondrial stress culminating in inflammation and cell death [107, 111, 114, 118]. Therefore, aberrant quality control of both lysosomes and mitochondria has profound consequences on the pathogenesis of a multitude of diseases, especially for those distinguished by neurodegeneration [112, 119–121].

12.3.1 Mitophagy

Neurons are peculiarly susceptible to the subtle sequela of lysosomal and mitochondrial dysfunction due to their reduced capability for glycolysis and reliance on oxidative phosphorylation for energy production [122, 123]. In order to maintain lysosomal and mitochondrial function and homeostasis, the cell employs specialized turnover pathways that target specific organelles (see also Chap. 11). Lysosomes mediate mitochondria-specific autophagy, often referred to as mitophagy, within cells, and this catabolic mechanism is regulated by two key proteins: PINK1 and parkin. In viable mitochondria, PINK1 is continually expressed and recruited to the outer mitochondrial membrane (OMM) where this serine/threonine kinase is imported into the mitochondrial matrix in a mitochondrial membrane potential-dependent manner. Once in the mitochondrial matrix, PINK1 is immediately degraded by proteases, thus regulating its expression [101, 114, 116]. However, upon the accumulation of aberrant proteins or the loss of mitochondrial membrane potential following mitochondrial insult, PINK1 regulation is impaired, allowing for accumulation of PINK1 on the OMM and its activation via autophosphorylation

[124, 125]. Activated PINK1 recruits parkin to the OMM and phosphorylates not only the ubiquitin-like domain of parkin but also ubiquitin itself. Phosphorylation of both protein species is required for full activation of the E3 ubiquitin ligase function of parkin [126–128]. Once relegated to the mitochondrial surface, parkin ubiquitinates various OMM proteins involved in mitochondrial maintenance, and this polyubiquitination labels the damaged mitochondria for turnover [129–131]. Cells with defective lysosomes, and therefore impaired mitophagy, are unable to effectively breakdown defective mitochondria resulting in their accumulation. These damaged mitochondria may leak reactive oxygen species (ROS), such as superoxide (O_2^-) and hydrogen peroxide (H_2O_2), from complex I and III of the respiratory chain. If left unchecked, these oxidative species may ultimately cause neuronal death by not only further perpetuating organelle damage but also by activating the NLRP3 inflammasome and therefore initiating inflammatory responses [111, 112].

12.3.2 Oxidative Stress

ROS are important mediators of the downstream consequences stemming from the combined effects of lysosomal and mitochondrial dysfunction. The mitochondrial respiratory chain is the principal producer of ROS, primarily in the forms of O_2^- and H_2O_2, within the cell and during abnormal lysosomal and mitochondrial function, and contributes to the elevation of oxidative stress [112]. Lysosomes are particularly vulnerable to ROS-induced damage, as these oxidative species peroxidize lysosomal membrane lipids resulting in destabilization of the membrane and, potentially, even LMP. Due to their ability to inherit cargo during fusion with autophagosomes, lysosomes may acquire large amounts of iron during the degradation of macromolecules. This accumulation of iron within lysosomes has been speculated to contribute to lysosomal susceptibility to oxidative damage [107, 132]. Simultaneously, elevated oxidative stress may damage mitochondria as well. Reactive oxygen species may peroxidize lipids in the mitochondrial membrane resulting in loss of membrane potential as well as fragmented mitochondrial morphology [112, 116, 133]. Under high levels of oxidative stress, mitochondrial membrane permeabilization may also occur and release cytochrome c into the cytosol, therefore activating apoptosis [134]. Mitochondrial DNA (mtDNA) neighbors the electron transport chain within the inner mitochondrial membrane and is prone to oxidation and release during mitochondrial damage. Once released into the cytosol, oxidized mtDNA and local ROS trigger assembly and activation of the NLRP3 inflammasome [111, 134–136]. Beyond these organelles, ROS may also regulate the activity of proteins through posttranslational modification. One example is DJ-1, a neuroprotective protein encoded by *PARK7* that has been implicated in familial forms of Parkinson's disease when mutated. DJ-1 is dependent on localization to the mitochondria for proper neuroprotective activity, and this translocation is redox regulated by oxidation of the Cys106 residue into cysteine sulfinic acid by ROS. After translocation to the mitochondria, DJ-1 protects cells from oxidative stress-induced death by modulating mitophagy and, together with other cellular processes, assumes a role in antioxidant response [137, 138].

12.3.3 Inflammasome Activation

Inflammasomes are intracellular multiprotein complexes that initiate an inflammatory response in response to pathogens and intracellular insults. The nucleotide-binding oligomerization domain-like receptor family, pyrin domain containing 3 (NLRP3) inflammasome is the best characterized and is closely associated with lysosomal and mitochondrial dysfunction [139]. This protein complex is formed at the interface of the mitochondria and endoplasmic reticulum in an area known as the mitochondria-associated endoplasmic reticulum membrane. The NLRP3 inflammasome is chiefly composed of three components: a NOD-like receptor (NLRP3), the adaptor protein ASC, and caspase-1. Once assembled and activated, the NLRP3 inflammasome cleaves proIL-1β and proIL-18 into their bioactive forms, where these proinflammatory messenger molecules may subsequently modulate immune and inflammatory pathways [110, 111, 139–141]. Growing evidence identifies both ROS and oxidized mtDNA as activators of the NLRP3 inflammasome. Cells with impaired mitophagy, and consequently prolonged clearance of defective mitochondria, may spontaneously secrete ROS and oxidized mtDNA into the cytosol, resulting in consistent activation of the NLRP3 inflammasome [12, 36]. Recently, elegant work by Shi and colleagues has demonstrated that autophagy may function as a negative regulator of inflammasome activation. Their data suggests that inflammasome activation concomitantly induces autophagosome formation by initiating nucleotide exchange on the G protein RalB. Inflammasomes subsequently undergo ubiquitination and are transported by adaptor proteins p62 and LC3 to autophagosomes for elimination. These results suggest that autophagy may modulate the intensity of inflammation by directly degrading active inflammasomes and therefore may result in uncontrolled inflammation during lysosomal and mitochondrial dysfunction [142]. Taken together, the inflammasome represents a link between lysosomal and mitochondrial dysfunction and inflammation, which contributes to the pathogenesis of not only neurodegenerative but also autoinflammatory diseases [143, 144].

12.4 Lysosomal Storage Disorders and Neurodegeneration

Lysosomal storage disorders (LSDs) are rare inborn metabolic diseases in which lysosomal function is severely compromised due to mutations in gene-encoding enzymes resident in lysosomes involved in the breakdown of specific substrates. The subsequent accumulation of substrate within lysosomes has a variety of consequences such as lysosomal enlargement, altered lysosomal pH, and diminished activity of lysosomal enzymes. Over 50 different LSDs have been described, and mutations in LSD-associated genes in patients, as well as carriers, have been linked to neurodegeneration, more particularly synucleinopathies [145]. Among LSD-associated genes, the molecular link between mutations in the glucocerebrosidase gene (*GBA1*) and Parkinson's disease (PD) is the most established [146].

12.4.1 GBA1 and Synucleinopathies

Pathological mutations in both alleles of the *GBA1* gene cause Gaucher disease (GD), the most common LSD. This disorder is characterized by lysosomal accumulation of the substrate glucosylceramide (GC), due to a deficiency in the lysosome-resident glucocerebrosidase enzyme (GCase) [147]. The cells most affected in GD patients are macrophages, which are involved in breakdown of senescent cells with GC-rich membranes such as erythrocytes. "Gaucher cells," which are the macrophages that have lysosomes engorged with substrate, can infiltrate the spleen, liver, and bone marrow, resulting in inflammation and organomegaly [148]. GD has been historically classified into non-neuronopathic type 1, acute neuronopathic type 2, and chronic neuronopathic type 3. Today, clinicians acknowledge a broad range of clinical manifestations associated with GD and subsequently can have difficulty classifying patients into specific GD subtypes [147]. Over the last 6 years, large cohort studies have established that the presence of mutations in the *GBA1* gene is a risk factor for the development of synucleinopathies including PD [146], dementia with Lewy bodies (DLB) (Fig. 12.2) [149], and, most recently, multiple system atrophy (MSA) [150]. All three synucleinopathies are characterized by the presence of inclusions of aggregated α-syn, a 14 kDa protein that is speculated to be involved in the regulation of synaptic vesicle dynamics and neurotransmitter release [151, 152]. In PD and DLB, the α-syn-positive Lewy bodies and neurites are mainly located in neurons of the substantia nigra, cerebral cortex, and hippocampus, while in MSA, the α-syn inclusions are located in glial oligodendrocytes [146, 149, 150]. The molecular link between mutations in the *GBA1* gene and PD was established by molecular analyses of the *GBA1* gene on a large pan-ethnic cohort comprising 5,691 patients with PD and 4,898 controls. This study revealed a strong association between *GBA1* mutations and the development of PD with an odds ratio of 5.43 and earlier onset of PD symptoms in patients with *GBA1* mutations [146]. These results have been replicated in multiple large cohorts with different ethnic backgrounds [114, 153, 154]. Today, *GBA1* mutations are widely considered the most common genetic risk factor for PD. However, it is important to keep in mind that most patients with GD and mutant *GBA1* carriers never develop synucleinopathies. These observations suggest that *GBA1* mutations and subsequent dysfunctional GCase enzyme are not a direct cause of synucleinopathy development; other cellular processes affecting organelle homeostasis, such as ER-stress and lysosomal and mitochondrial function, might play a more central role in synucleinopathy pathogenesis. The presence of dysfunctional GCase could exacerbate organelle dysfunction and subsequent α-syn accumulation.

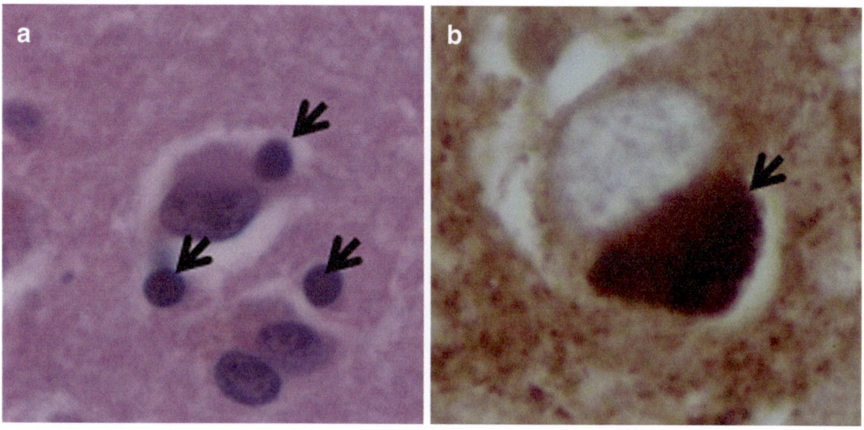

Fig. 12.2 Histology of hippocampal Lewy bodies (*arrows*) from a patient with GD and DLB. (**a**) Hematoxylin-eosin stain and (**b**) anti-α-syn immunostaining of hippocampal tissue from the same individual

12.4.2 GCase and α-syn Homeostasis

Initially, the mechanistic link between dysfunctional GCase enzyme and α-syn aggregation focused on gain-of-function or loss-of function hypotheses, where the former supports the direct involvement of dysfunctional GCase enzyme in the aggregation of α-syn, and the latter supports the role of lysosomal GC substrate accumulation in α-syn aggregation [155]. Currently, in vitro and in vivo research supports a reciprocal relationship between GCase and α-syn where downregulation of GCase protein expression or enzyme activity results in accumulation of α-syn. Increases in α-syn protein expression results in reduced GCase protein expression and enzyme activity (reviewed by [114, 156]). Furthermore, three independent studies support the observation of reduced GCase activity and protein expression in postmortem brains of sporadic PD and DLB patients without *GBA1* mutations, reinforcing the reciprocal relationship in relevant human samples [157–159]. The molecular mechanism of the reciprocal relationship is not fully understood although there is some evidence that an increase in α-syn protein levels inhibits ER-to-Golgi trafficking of GCase, which subsequently results in downregulation of GCase translocation to lysosomes. Less GCase in lysosomes can lead to lysosomal GC substrate accumulation and subsequent lysosomal dysfunction, which in turn may stimulate accumulation and oligomerization of α-syn throughout the cell. Buildup of α-syn aggregates could, in turn, inhibit ER-to-Golgi trafficking of GCase resulting in further decrease of this enzyme within lysosomes [160]. This reciprocal positive feedback loop could eventually lead to neurodegeneration. Evidence for this hypothesis came from a neuronopathic GD type 2 mouse model lacking GCase. Here, autophagy and proteosomal impairment lead to accumulation of fragmented mitochondria and α-syn in cultured neurons and astrocytes of the midbrain [161]. Although this *gba*$^{-/-}$ model is not reflective of PD, it suggested that the lack of GCase expression promotes α-syn accumulation through impairment of cellular turnover pathways [101]. Novel insights into maintenance of α-syn homeostasis by manipulating GCase enzyme levels are promising for the development of new treatments for synucleinopathies. Although GCase enzyme replacement therapy does not improve PD symptoms, as the recombinant enzyme does not cross the blood-brain barrier [162], molecular inhibitors of glucosylceramide synthase for GC substrate reduction therapy and molecular chaperones for enhancing GCase translocation to the lysosomes can cross the blood-brain barrier and therefore show potential as therapeutics [163–166]. Recent research indicates associations similar to that found between mutations in *GBA1*, and the development of synucleinopathies can be expanded to other LSD-associated genes. Large molecular cohort studies suggest that mutations in the sphingomyelin phosphodiesterase (SMPD1) and α-N-acetylglucosaminidase (NAGLU) genes, which are associated with Niemann-Pick disease A and B and mucopolysaccharidosis type III B, respectively, may be implicated in the development PD [167, 168]. These observations suggest that mutations in other lysosomal-resident enzymes might be classified as risk factors for the development of synucleinopathies.

12.5 Conclusion

When first described by Christian de Duve, and for many years after, lysosomes were often considered static organelles primarily involved in the degradation of cellular constituents. However, recent insights into lysosomal function and regulation have demonstrated otherwise. In fact, lysosomes are now considered dynamic organelles capable of not only cellular cleanup but also nutrient sensing and lipid catabolism. As mediators of autophagy, lysosomes also play an important role in the development of neurodegenerative diseases. Lysosomal dysfunction leads to not only the accumulation of aggregate-prone proteins but also impairs other organelles such as mitochondria. Together, dysfunction of this deleterious duo might drive a destructive feedback loop that culminates in the neuropathology often found in Parkinson's, Alzheimer's, and Huntington's diseases. The association between LSDs and neurodegenerative diseases such as PD, LBD, and MSA further highlight the importance of proper lysosomal function in neuronal health. Further investigations exploring the relationship between lysosomal and mitochondrial dysfunction hold promise for the discovery of new potential drug targets.

Acknowledgments We thank Julia Feckes for the assistance in drafting Fig. 12.1 and the Laboratory of Pathology of the National Cancer Institute for the images in Fig. 12.2. This work was supported by the Intramural Research Programs of the National Human Genome Research Institute.

References

1. De Duve C. The lysosome. Sci Am. 1963;208:64–72.
2. Hara T, Nakamura K, Matsui M, Yamamoto A, Nakahara Y, Suzuki-Migishima R, et al. Suppression of basal autophagy in neural cells causes neurodegenerative disease in mice. Nature. 2006;441(7095):885–9.
3. Komatsu M, Waguri S, Chiba T, Murata S, Iwata J, Tanida I, et al. Loss of autophagy in the central nervous system causes neurodegeneration in mice. Nature. 2006;441(7095):880–4.
4. Sakai Y, Koller A, Rangell LK, Keller GA, Subramani S. Peroxisome degradation by microautophagy in Pichia pastoris: identification of specific steps and morphological intermediates. J Cell Biol. 1998;141(3):625–36.
5. Kiel JA, Komduur JA, van der Klei IJ, Veenhuis M. Macropexophagy in Hansenula polymorpha: facts and views. FEBS Lett. 2003;549(1–3):1–6.
6. Roberts P, Moshitch-Moshkovitz S, Kvam E, O'Toole E, Winey M, Goldfarb DS. Piecemeal microautophagy of nucleus in Saccharomyces cerevisiae. Mol Biol Cell. 2003;14(1):129–41.
7. Dunn Jr WA, Cregg JM, Kiel JA, van der Klei IJ, Oku M, Sakai Y, et al. Pexophagy: the selective autophagy of peroxisomes. Autophagy. 2005;1(2):75–83.
8. Uttenweiler A, Mayer A. Microautophagy in the yeast Saccharomyces cerevisiae. Methods Mol Biol. 2008;445:245–59.
9. Ahlberg J, Marzella L, Glaumann H. Uptake and degradation of proteins by isolated rat liver lysosomes. Suggestion of a microautophagic pathway of proteolysis. Lab Invest. 1982;47(6):523–32.
10. Mijaljica D, Prescott M, Devenish RJ. Microautophagy in mammalian cells: revisiting a 40-year-old conundrum. Autophagy. 2011;7(7):673–82.
11. Sahu R, Kaushik S, Clement CC, Cannizzo ES, Scharf B, Follenzi A, et al. Microautophagy of cytosolic proteins by late endosomes. Dev Cell. 2011;20(1):131–9.
12. Marzella L, Ahlberg J, Glaumann H. In vitro uptake of particles by lysosomes. Exp Cell Res. 1980;129(2):460–6.
13. Sakai M, Ogawa K. Energy-dependent lysosomal wrapping mechanism (LWM) during autophagolysosome formation. Histochemistry. 1982;76(4):479–88.
14. Sakai M, Araki N, Ogawa K. Lysosomal movements during heterophagy and autophagy: with special reference to nematolysosome and wrapping lysosome. J Electron Microsc Tech. 1989;12(2):101–31.
15. de Waal EJ, Vreeling-Sindelarova H, Schellens JP, Houtkooper JM, James J. Quantitative changes in the lysosomal vacuolar system of rat hepatocytes during short-term starvation. A morphometric analysis with special reference to macro- and microautophagy. Cell Tissue Res. 1986;243(3):641–8.
16. Mortimore GE, Lardeux BR, Adams CE. Regulation of microautophagy and basal protein turnover in rat liver. Effects of short-term starvation. J Biol Chem. 1988;263(5):2506–12.
17. Muller O, Sattler T, Flotenmeyer M, Schwarz H, Plattner H, Mayer A. Autophagic tubes: vacuolar invaginations involved in lateral membrane sorting and inverse vesicle budding. J Cell Biol. 2000;151(3):519–28.
18. Yorimitsu T, Klionsky DJ. Autophagy: molecular machinery for self-eating. Cell Death Differ. 2005;12 Suppl 2:1542–52.
19. Yang Z, Klionsky DJ. Mammalian autophagy: core molecular machinery and signaling regulation. Curr Opin Cell Biol. 2010;22(2):124–31.
20. Ganley IG, Lam du H, Wang J, Ding X, Chen S, Jiang X. ULK1.ATG13.FIP200 complex mediates mTOR signaling and is essential for autophagy. J Biol Chem. 2009;284(18):12297–305.
21. Hosokawa N, Sasaki T, Iemura S, Natsume T, Hara T, Mizushima N. Atg101, a novel mammalian autophagy protein interacting with Atg13. Autophagy. 2009;5(7):973–9.

22. Jung CH, Jun CB, Ro SH, Kim YM, Otto NM, Cao J, et al. ULK-Atg13-FIP200 complexes mediate mTOR signaling to the autophagy machinery. Mol Biol Cell. 2009;20(7):1992–2003.
23. Mercer CA, Kaliappan A, Dennis PB. A novel, human Atg13 binding protein, Atg101, interacts with ULK1 and is essential for macroautophagy. Autophagy. 2009;5(5):649–62.
24. Parzych KR, Klionsky DJ. An overview of autophagy: morphology, mechanism, and regulation. Antioxid Redox Signal. 2014;20(3):460–73.
25. Noda T, Suzuki K, Ohsumi Y. Yeast autophagosomes: de novo formation of a membrane structure. Trends Cell Biol. 2002;12(5):231–5.
26. Kovacs AL, Palfia Z, Rez G, Vellai T, Kovacs J. Sequestration revisited: integrating traditional electron microscopy, de novo assembly and new results. Autophagy. 2007;3(6):655–62.
27. Itakura E, Mizushima N. Characterization of autophagosome formation site by a hierarchical analysis of mammalian Atg proteins. Autophagy. 2010;6(6):764–76.
28. Chen Y, Klionsky DJ. The regulation of autophagy – unanswered questions. J Cell Sci. 2011;124(Pt 2):161–70.
29. Feng Y, He D, Yao Z, Klionsky DJ. The machinery of macroautophagy. Cell Res. 2014;24(1):24–41.
30. Burman C, Ktistakis NT. Regulation of autophagy by phosphatidylinositol 3-phosphate. FEBS Lett. 2010;584(7):1302–12.
31. Hayashi-Nishino M, Fujita N, Noda T, Yamaguchi A, Yoshimori T, Yamamoto A. A subdomain of the endoplasmic reticulum forms a cradle for autophagosome formation. Nat Cell Biol. 2009;11(12):1433–7.
32. Yla-Anttila P, Vihinen H, Jokitalo E, Eskelinen EL. 3D tomography reveals connections between the phagophore and endoplasmic reticulum. Autophagy. 2009;5(8):1180–5.
33. Hailey DW, Rambold AS, Satpute-Krishnan P, Mitra K, Sougrat R, Kim PK, et al. Mitochondria supply membranes for autophagosome biogenesis during starvation. Cell. 2010;141(4):656–67.
34. Ravikumar B, Moreau K, Jahreiss L, Puri C, Rubinsztein DC. Plasma membrane contributes to the formation of pre-autophagosomal structures. Nat Cell Biol. 2010;12(8):747–57.
35. Ravikumar B, Moreau K, Rubinsztein DC. Plasma membrane helps autophagosomes grow. Autophagy. 2010;6(8):1184–6.
36. Takahashi Y, Meyerkord CL, Hori T, Runkle K, Fox TE, Kester M, et al. Bif-1 regulates Atg9 trafficking by mediating the fission of Golgi membranes during autophagy. Autophagy. 2011;7(1):61–73.
37. Weidberg H, Shvets E, Elazar Z. Biogenesis and cargo selectivity of autophagosomes. Ann Rev Biochem. 2011;80:125–56.
38. Hanada T, Noda NN, Satomi Y, Ichimura Y, Fujioka Y, Takao T, et al. The Atg12-Atg5 conjugate has a novel E3-like activity for protein lipidation in autophagy. J Biol Chem. 2007;282(52):37298–302.
39. Weidberg H, Shvets E, Shpilka T, Shimron F, Shinder V, Elazar Z. LC3 and GATE-16/GABARAP subfamilies are both essential yet act differently in autophagosome biogenesis. EMBO J. 2010;29(11):1792–802.
40. Bjorkoy G, Lamark T, Brech A, Outzen H, Perander M, Overvatn A, et al. p62/SQSTM1 forms protein aggregates degraded by autophagy and has a protective effect on huntingtin-induced cell death. J Cell Biol. 2005;171(4):603–14.
41. Kirkin V, Lamark T, Sou YS, Bjorkoy G, Nunn JL, Bruun JA, et al. A role for NBR1 in autophagosomal degradation of ubiquitinated substrates. Mol Cell. 2009;33(4):505–16.
42. Young AR, Chan EY, Hu XW, Kochl R, Crawshaw SG, High S, et al. Starvation and ULK1-dependent cycling of mammalian Atg9 between the TGN and endosomes. J Cell Sci. 2006;119(Pt 18):3888–900.
43. Gordon PB, Seglen PO. Prelysosomal convergence of autophagic and endocytic pathways. Biochem Biophys Res Commun. 1988;151(1):40–7.

44. Yang Z, Klionsky DJ. Eaten alive: a history of macroautophagy. Nat Cell Biol. 2010;12(9):814–22.
45. Fader CM, Colombo MI. Autophagy and multivesicular bodies: two closely related partners. Cell Death Differ. 2009;16(1):70–8.
46. Kochl R, Hu XW, Chan EY, Tooze SA. Microtubules facilitate autophagosome formation and fusion of autophagosomes with endosomes. Traffic. 2006;7(2):129–45.
47. Monastyrska I, Rieter E, Klionsky DJ, Reggiori F. Multiple roles of the cytoskeleton in autophagy. Biol Rev Camb Philos Soc. 2009;84(3):431–48.
48. Metcalf DJ, Garcia-Arencibia M, Hochfeld WE, Rubinsztein DC. Autophagy and misfolded proteins in neurodegeneration. Exp Neurol. 2012;238(1):22–8.
49. Lu Y, Dong S, Hao B, Li C, Zhu K, Guo W, et al. Vacuolin-1 potently and reversibly inhibits autophagosome-lysosome fusion by activating RAB5A. Autophagy. 2014;10(11):1895–905.
50. Feng Y, Yao Z, Klionsky DJ. How to control self-digestion: transcriptional, post-transcriptional, and post-translational regulation of autophagy. Trends Cell Biol. 2015;25(6):354–63.
51. Kaushik S, Cuervo AM. Chaperones in autophagy. Pharmacol Res. 2012;66(6):484–93.
52. Kaushik S, Cuervo AM. Chaperone-mediated autophagy: a unique way to enter the lysosome world. Trends Cell Biol. 2012;22(8):407–17.
53. Cuervo AM, Wong E. Chaperone-mediated autophagy: roles in disease and aging. Cell Res. 2014;24(1):92–104.
54. Dice JF, Chiang HL, Spencer EP, Backer JM. Regulation of catabolism of microinjected ribonuclease A. Identification of residues 7–11 as the essential pentapeptide. J Biol Chem. 1986;261(15):6853–9.
55. Chiang HL, Terlecky SR, Plant CP, Dice JF. A role for a 70-kilodalton heat shock protein in lysosomal degradation of intracellular proteins. Science. 1989;246(4928):382–5.
56. Dice JF. Peptide sequences that target cytosolic proteins for lysosomal proteolysis. Trends Biochem Sci. 1990;15(8):305–9.
57. Koga H, Martinez-Vicente M, Macian F, Verkhusha VV, Cuervo AM. A photoconvertible fluorescent reporter to track chaperone-mediated autophagy. Nat Commun. 2011;2:386.
58. Thompson LM, Aiken CT, Kaltenbach LS, Agrawal N, Illes K, Khoshnan A, et al. IKK phosphorylates Huntingtin and targets it for degradation by the proteasome and lysosome. J Cell Biol. 2009;187(7):1083–99.
59. Lv L, Li D, Zhao D, Lin R, Chu Y, Zhang H, et al. Acetylation targets the M2 isoform of pyruvate kinase for degradation through chaperone-mediated autophagy and promotes tumor growth. Mol Cell. 2011;42(6):719–30.
60. Cuervo AM, Dice JF. A receptor for the selective uptake and degradation of proteins by lysosomes. Science. 1996;273(5274):501–3.
61. Bandyopadhyay U, Kaushik S, Varticovski L, Cuervo AM. The chaperone-mediated autophagy receptor organizes in dynamic protein complexes at the lysosomal membrane. Mol Cell Biol. 2008;28(18):5747–63.
62. Salvador N, Aguado C, Horst M, Knecht E. Import of a cytosolic protein into lysosomes by chaperone-mediated autophagy depends on its folding state. J Biol Chem. 2000;275(35):27447–56.
63. Agarraberes FA, Dice JF. A molecular chaperone complex at the lysosomal membrane is required for protein translocation. J Cell Sci. 2001;114(Pt 13):2491–9.
64. Kaushik S, Massey AC, Cuervo AM. Lysosome membrane lipid microdomains: novel regulators of chaperone-mediated autophagy. EMBO J. 2006;25(17):3921–33.
65. Bandyopadhyay U, Sridhar S, Kaushik S, Kiffin R, Cuervo AM. Identification of regulators of chaperone-mediated autophagy. Mol Cell. 2010;39(4):535–47.
66. Xilouri M, Stefanis L. Chaperone mediated autophagy to the rescue: a new-fangled target for the treatment of neurodegenerative diseases. Mol Cell Neurosci. 2015;66(Pt A):29–36.
67. Agarraberes FA, Terlecky SR, Dice JF. An intralysosomal hsp70 is required for a selective pathway of lysosomal protein degradation. J Cell Biol. 1997;137(4):825–34.

68. Cuervo AM, Dice JF, Knecht E. A population of rat liver lysosomes responsible for the selective uptake and degradation of cytosolic proteins. J Biol Chem. 1997;272(9):5606–15.
69. de Duve C. The lysosome turns fifty. Nat Cell Biol. 2005;7(9):847–9.
70. De Duve C, Pressman BC, Gianetto R, Wattiaux R, Appelmans F. Tissue fractionation studies. 6. Intracellular distribution patterns of enzymes in rat-liver tissue. Biochem J. 1955;60(4):604–17.
71. Luzio JP, Parkinson MD, Gray SR, Bright NA. The delivery of endocytosed cargo to lysosomes. Biochem Soc Trans. 2009;37(Pt 5):1019–21.
72. Mizushima N, Levine B, Cuervo AM, Klionsky DJ. Autophagy fights disease through cellular self-digestion. Nature. 2008;451(7182):1069–75.
73. Saftig P. Lysosomes. Georgetown/New York: Landes Bioscience/Eurekah.com/Springer Science+Business Media; 2005;1–17.
74. Settembre C, Ballabio A. Lysosomal adaptation: how the lysosome responds to external cues. Cold Spring Harb Perspect Biol. 2014;6(6):1–15.
75. Settembre C, De Cegli R, Mansueto G, Saha PK, Vetrini F, Visvikis O, et al. TFEB controls cellular lipid metabolism through a starvation-induced autoregulatory loop. Nat Cell Biol. 2013;15(6):647–58.
76. Sardiello M, Palmieri M, di Ronza A, Medina DL, Valenza M, Gennarino VA, et al. A gene network regulating lysosomal biogenesis and function. Science. 2009;325(5939):473–7.
77. Settembre C, Di Malta C, Polito VA, Garcia Arencibia M, Vetrini F, Erdin S, et al. TFEB links autophagy to lysosomal biogenesis. Science. 2011;332(6036):1429–33.
78. Bar-Peled L, Schweitzer LD, Zoncu R, Sabatini DM. Ragulator is a GEF for the rag GTPases that signal amino acid levels to mTORC1. Cell. 2012;150(6):1196–208.
79. Sancak Y, Bar-Peled L, Zoncu R, Markhard AL, Nada S, Sabatini DM. Ragulator-Rag complex targets mTORC1 to the lysosomal surface and is necessary for its activation by amino acids. Cell. 2010;141(2):290–303.
80. Sancak Y, Peterson TR, Shaul YD, Lindquist RA, Thoreen CC, Bar-Peled L, et al. The Rag GTPases bind raptor and mediate amino acid signaling to mTORC1. Science. 2008;320(5882):1496–501.
81. Schulze H, Kolter T, Sandhoff K. Principles of lysosomal membrane degradation: cellular topology and biochemistry of lysosomal lipid degradation. Biochim Biophys Acta. 2009;1793(4):674–83.
82. Settembre C, Fraldi A, Medina DL, Ballabio A. Signals from the lysosome: a control centre for cellular clearance and energy metabolism. Nat Rev Mol Cell Biol. 2013;14(5):283–96.
83. Zoncu R, Bar-Peled L, Efeyan A, Wang S, Sancak Y, Sabatini DM. mTORC1 senses lysosomal amino acids through an inside-out mechanism that requires the vacuolar H(+)-ATPase. Science. 2011;334(6056):678–83.
84. Settembre C, Zoncu R, Medina DL, Vetrini F, Erdin S, Erdin S, et al. A lysosome-to-nucleus signalling mechanism senses and regulates the lysosome via mTOR and TFEB. EMBO J. 2012;31(5):1095–108.
85. Martina JA, Puertollano R. Rag GTPases mediate amino acid-dependent recruitment of TFEB and MITF to lysosomes. J Cell Biol. 2013;200(4):475–91.
86. Martina JA, Chen Y, Gucek M, Puertollano R. MTORC1 functions as a transcriptional regulator of autophagy by preventing nuclear transport of TFEB. Autophagy. 2012;8(6):903–14.
87. Roczniak-Ferguson A, Petit CS, Froehlich F, Qian S, Ky J, Angarola B, et al. The transcription factor TFEB links mTORC1 signaling to transcriptional control of lysosome homeostasis. Sci Signal. 2012;5(228):ra42.
88. Singh R, Cuervo AM. Autophagy in the cellular energetic balance. Cell Metab. 2011;13(5):495–504.
89. Singh R, Kaushik S, Wang Y, Xiang Y, Novak I, Komatsu M, et al. Autophagy regulates lipid metabolism. Nature. 2009;458(7242):1131–5.
90. Finck BN, Kelly DP. PGC-1 coactivators: inducible regulators of energy metabolism in health and disease. J Clin Invest. 2006;116(3):615–22.

91. Harris H, Rubinsztein DC. Control of autophagy as a therapy for neurodegenerative disease. Nat Rev Neurol. 2012;8(2):108–17.
92. Wong E, Cuervo AM. Autophagy gone awry in neurodegenerative diseases. Nat Neurosci. 2010;13(7):805–11.
93. Taylor JP, Hardy J, Fischbeck KH. Toxic proteins in neurodegenerative disease. Science. 2002;296(5575):1991–5.
94. Decressac M, Mattsson B, Weikop P, Lundblad M, Jakobsson J, Bjorklund A. TFEB-mediated autophagy rescues midbrain dopamine neurons from alpha-synuclein toxicity. Proc Natl Acad Sci U S A. 2013;110(19):E1817–26.
95. Ostrerova N, Petrucelli L, Farrer M, Mehta N, Choi P, Hardy J, et al. alpha-Synuclein shares physical and functional homology with 14-3-3 proteins. J Neurosci. 1999;19(14): 5782–91.
96. Perez RG, Waymire JC, Lin E, Liu JJ, Guo F, Zigmond MJ. A role for alpha-synuclein in the regulation of dopamine biosynthesis. J Neurosci. 2002;22(8):3090–9.
97. Kilpatrick K, Zeng Y, Hancock T, Segatori L. Genetic and chemical activation of TFEB mediates clearance of aggregated alpha-synuclein. PLoS ONE. 2015;10(3), e0120819.
98. Tsunemi T, Ashe TD, Morrison BE, Soriano KR, Au J, Roque RA, et al. PGC-1alpha rescues Huntington's disease proteotoxicity by preventing oxidative stress and promoting TFEB function. Sci Transl Med. 2012;4(142):142ra97.
99. Polito VA, Li H, Martini-Stoica H, Wang B, Yang L, Xu Y, et al. Selective clearance of aberrant tau proteins and rescue of neurotoxicity by transcription factor EB. EMBO Mol Med. 2014;6(9):1142–60.
100. Martina JA, Diab HI, Lishu L, Jeong AL, Patange S, Raben N, et al. The nutrient-responsive transcription factor TFE3 promotes autophagy, lysosomal biogenesis, and clearance of cellular debris. Sci Signal. 2014;7(309):ra9.
101. Osellame LD, Duchen MR. Quality control gone wrong: mitochondria, lysosomal storage disorders and neurodegeneration. Br J Pharmacol. 2014;171(8):1958–72.
102. Chance B, Sies H, Boveris A. Hydroperoxide metabolism in mammalian organs. Physiol Rev. 1979;59(3):527–605.
103. Mitchell P. Coupling of phosphorylation to electron and hydrogen transfer by a chemiosmotic type of mechanism. Nature. 1961;191:144–8.
104. Mitchell P, Moyle J. Chemiosmotic hypothesis of oxidative phosphorylation. Nature. 1967;213(5072):137–9.
105. Liu Y, Fiskum G, Schubert D. Generation of reactive oxygen species by the mitochondrial electron transport chain. J Neurochem. 2002;80(5):780–7.
106. Appelqvist H, Waster P, Kagedal K, Ollinger K. The lysosome: from waste bag to potential therapeutic target. J Mol Cell Biol. 2013;5(4):214–26.
107. Repnik U, Hafner Cesen M, Turk B. Lysosomal membrane permeabilization in cell death: concepts and challenges. Mitochondrion. 2014;19(Pt A):49–57.
108. Ballabio A, Gieselmann V. Lysosomal disorders: from storage to cellular damage. Biochim Biophys Acta. 2009;1793(4):684–96.
109. Boya P, Andreau K, Poncet D, Zamzami N, Perfettini JL, Metivier D, et al. Lysosomal membrane permeabilization induces cell death in a mitochondrion-dependent fashion. J Exp Med. 2003;197(10):1323–34.
110. Heid ME, Keyel PA, Kamga C, Shiva S, Watkins SC, Salter RD. Mitochondrial reactive oxygen species induces NLRP3-dependent lysosomal damage and inflammasome activation. J Immunol. 2013;191(10):5230–8.
111. Zhou R, Yazdi AS, Menu P, Tschopp J. A role for mitochondria in NLRP3 inflammasome activation. Nature. 2011;469(7329):221–5.
112. Lee J, Giordano S, Zhang J. Autophagy, mitochondria and oxidative stress: cross-talk and redox signalling. Biochem J. 2012;441(2):523–40.
113. Puschmann A. Monogenic Parkinson's disease and parkinsonism: clinical phenotypes and frequencies of known mutations. Parkinsonism Relat Disord. 2013;19(4):407–15.

114. Siebert M, Sidransky E, Westbroek W. Glucocerebrosidase is shaking up the synucleinopathies. Brain. 2014;137(Pt 5):1304–22.
115. Kitada T, Asakawa S, Hattori N, Matsumine H, Yamamura Y, Minoshima S, et al. Mutations in the parkin gene cause autosomal recessive juvenile parkinsonism. Nature. 1998;392(6676):605–8.
116. Ryan BJ, Hoek S, Fon EA, Wade-Martins R. Mitochondrial dysfunction and mitophagy in Parkinson's: from familial to sporadic disease. Trends Biochem Sci. 2015;40(4):200–10.
117. Valente EM, Abou-Sleiman PM, Caputo V, Muqit MM, Harvey K, Gispert S, et al. Hereditary early-onset Parkinson's disease caused by mutations in PINK1. Science. 2004;304(5674):1158–60.
118. van der Burgh R, Boes M. Mitochondria in autoinflammation: cause, mediator or bystander? Trends Endocrinol Metab. 2015;26(5):263–71.
119. Boland B, Kumar A, Lee S, Platt FM, Wegiel J, Yu WH, et al. Autophagy induction and autophagosome clearance in neurons: relationship to autophagic pathology in Alzheimer's disease. J Neurosci. 2008;28(27):6926–37.
120. Narendra D, Tanaka A, Suen DF, Youle RJ. Parkin is recruited selectively to impaired mitochondria and promotes their autophagy. J Cell Biol. 2008;183(5):795–803.
121. Settembre C, Fraldi A, Jahreiss L, Spampanato C, Venturi C, Medina D, et al. A block of autophagy in lysosomal storage disorders. Hum Mol Genet. 2008;17(1):119–29.
122. Bolanos JP, Almeida A, Moncada S. Glycolysis: a bioenergetic or a survival pathway? Trends Biochem Sci. 2010;35(3):145–9.
123. Herrero-Mendez A, Almeida A, Fernandez E, Maestre C, Moncada S, Bolanos JP. The bioenergetic and antioxidant status of neurons is controlled by continuous degradation of a key glycolytic enzyme by APC/C-Cdh1. Nat Cell Biol. 2009;11(6):747–52.
124. Okatsu K, Oka T, Iguchi M, Imamura K, Kosako H, Tani N, et al. PINK1 autophosphorylation upon membrane potential dissipation is essential for Parkin recruitment to damaged mitochondria. Nat Commun. 2012;3:1016.
125. Jin SM, Youle RJ. The accumulation of misfolded proteins in the mitochondrial matrix is sensed by PINK1 to induce PARK2/Parkin-mediated mitophagy of polarized mitochondria. Autophagy. 2013;9(11):1750–7.
126. Kane LA, Lazarou M, Fogel AI, Li Y, Yamano K, Sarraf SA, et al. PINK1 phosphorylates ubiquitin to activate Parkin E3 ubiquitin ligase activity. J Cell Biol. 2014;205(2):143–53.
127. Kazlauskaite A, Kondapalli C, Gourlay R, Campbell DG, Ritorto MS, Hofmann K, et al. Parkin is activated by PINK1-dependent phosphorylation of ubiquitin at Ser65. Biochem J. 2014;460(1):127–39.
128. Koyano F, Okatsu K, Kosako H, Tamura Y, Go E, Kimura M, et al. Ubiquitin is phosphorylated by PINK1 to activate parkin. Nature. 2014;510(7503):162–6.
129. Gegg ME, Cooper JM, Chau KY, Rojo M, Schapira AH, Taanman JW. Mitofusin 1 and mitofusin 2 are ubiquitinated in a PINK1/parkin-dependent manner upon induction of mitophagy. Hum Mol Genet. 2010;19(24):4861–70.
130. Wang X, Winter D, Ashrafi G, Schlehe J, Wong YL, Selkoe D, et al. PINK1 and Parkin target Miro for phosphorylation and degradation to arrest mitochondrial motility. Cell. 2011;147(4):893–906.
131. Sarraf SA, Raman M, Guarani-Pereira V, Sowa ME, Huttlin EL, Gygi SP, et al. Landscape of the PARKIN-dependent ubiquitylome in response to mitochondrial depolarization. Nature. 2013;496(7445):372–6.
132. Kurz T, Terman A, Gustafsson B, Brunk UT. Lysosomes in iron metabolism, ageing and apoptosis. Histochem Cell Biol. 2008;129(4):389–406.
133. Grohm J, Plesnila N, Culmsee C. Bid mediates fission, membrane permeabilization and perinuclear accumulation of mitochondria as a prerequisite for oxidative neuronal cell death. Brain Behav Immun. 2010;24(5):831–8.
134. Giorgio M, Migliaccio E, Orsini F, Paolucci D, Moroni M, Contursi C, et al. Electron transfer between cytochrome c and p66Shc generates reactive oxygen species that trigger mitochondrial apoptosis. Cell. 2005;122(2):221–33.

135. Shimada K, Crother TR, Karlin J, Dagvadorj J, Chiba N, Chen S, et al. Oxidized mitochondrial DNA activates the NLRP3 inflammasome during apoptosis. Immunity. 2012;36(3):401–14.
136. Nakahira K, Haspel JA, Rathinam VA, Lee SJ, Dolinay T, Lam HC, et al. Autophagy proteins regulate innate immune responses by inhibiting the release of mitochondrial DNA mediated by the NALP3 inflammasome. Nat Immunol. 2011;12(3):222–30.
137. Ariga H, Takahashi-Niki K, Kato I, Maita H, Niki T, Iguchi-Ariga SM. Neuroprotective function of DJ-1 in Parkinson's disease. Oxid Med Cell Longev. 2013;2013:683920.
138. Canet-Aviles RM, Wilson MA, Miller DW, Ahmad R, McLendon C, Bandyopadhyay S, et al. The Parkinson's disease protein DJ-1 is neuroprotective due to cysteine-sulfinic acid-driven mitochondrial localization. Proc Natl Acad Sci U S A. 2004;101(24):9103–8.
139. Menu P, Vince JE. The NLRP3 inflammasome in health and disease: the good, the bad and the ugly. Clin Exp Immunol. 2011;166(1):1–15.
140. Martinon F, Mayor A, Tschopp J. The inflammasomes: guardians of the body. Annu Rev Immunol. 2009;27:229–65.
141. Marchi S, Patergnani S, Pinton P. The endoplasmic reticulum-mitochondria connection: one touch, multiple functions. Biochim Biophys Acta. 2014;1837(4):461–9.
142. Shi CS, Shenderov K, Huang NN, Kabat J, Abu-Asab M, Fitzgerald KA, et al. Activation of autophagy by inflammatory signals limits IL-1beta production by targeting ubiquitinated inflammasomes for destruction. Nat Immunol. 2012;13(3):255–63.
143. van der Burgh R, Boes M. Mitochondria in autoinflammation: cause, mediator or bystander? Trends Endocrinol Metab: TEM. 2015;26(5):263–71.
144. Heneka MT, Kummer MP, Latz E. Innate immune activation in neurodegenerative disease. Nat Rev Immunol. 2014;14(7):463–77.
145. Shachar T, Lo Bianco C, Recchia A, Wiessner C, Raas-Rothschild A, Futerman AH. Lysosomal storage disorders and Parkinson's disease: Gaucher disease and beyond. Mov Disord. 2011;26(9):1593–604.
146. Sidransky E, Nalls MA, Aasly JO, Aharon-Peretz J, Annesi G, Barbosa ER, et al. Multicenter analysis of glucocerebrosidase mutations in Parkinson's disease. N Engl J Med. 2009;361(17):1651–61.
147. Beutler E, Grabowski GA. Gaucher disease. In: Scriver CBA, Beaudet AL, Sly WS, Valle D, editors. The metabolic & molecular bases of inherited disease. New York: McGraw-Hill; 2001. p. 3635–68.
148. Sidransky E. Gaucher disease: complexity in a "simple" disorder. Mol Genet Metab. 2004;83(1–2):6–15. Review.
149. Nalls MA, Duran R, Lopez G, Kurzawa-Akanbi M, McKeith IG, Chinnery PF, et al. A multicenter study of glucocerebrosidase mutations in dementia with Lewy bodies. JAMA Neurol. 2013;70(6):727–35.
150. Mitsui J, Matsukawa T, Sasaki H, Yabe I, Matsushima M, Durr A, et al. Variants associated with Gaucher disease in multiple system atrophy. Ann Clin Transl Neurol. 2015;2(4):417–26.
151. Ito H, Nakayama K, Jin C, Suzuki Y, Yazawa I. alpha-Synuclein accumulation reduces GABAergic inhibitory transmission in a model of multiple system atrophy. Biochem Biophys Res Commun. 2012;428(3):348–53 [Research Support, Non-U.S. Gov't].
152. Vargas KJ, Makani S, Davis T, Westphal CH, Castillo PE, Chandra SS. Synucleins regulate the kinetics of synaptic vesicle endocytosis. J Neurosci. 2014;34(28):9364–76.
153. Malec-Litwinowicz M, Rudzinska M, Szubiga M, Michalski M, Tomaszewski T, Szczudlik A. Cognitive impairment in carriers of glucocerebrosidase gene mutation in Parkinson disease patients. Neurol Neurochir Pol. 2014;48(4):258–61.
154. Pulkes T, Choubtum L, Chitphuk S, Thakkinstian A, Pongpakdee S, Kulkantrakorn K, et al. Glucocerebrosidase mutations in Thai patients with Parkinson's disease. Parkinsonism Relat Disord. 2014;20(9):986–91.
155. Westbroek W, Gustafson AM, Sidransky E. Exploring the link between glucocerebrosidase mutations and parkinsonism. Trends Mol Med. 2011;17(9):485–93.

156. Sardi SP, Cheng SH, Shihabuddin LS. Gaucher-related synucleinopathies: the examination of sporadic neurodegeneration from a rare (disease) angle. Prog Neurobiol. 2015;125:47–62.
157. Gegg ME, Burke D, Heales SJ, Cooper JM, Hardy J, Wood NW, et al. Glucocerebrosidase deficiency in substantia nigra of parkinson disease brains. Ann Neurol. 2012;72(3):455–63.
158. Murphy KE, Halliday GM. Glucocerebrosidase deficits in sporadic Parkinson disease. Autophagy. 2014;10(7):1350–1.
159. Chiasserini D, Paciotti S, Eusebi P, Persichetti E, Tasegian A, Kurzawa-Akanbi M, et al. Selective loss of glucocerebrosidase activity in sporadic Parkinson's disease and dementia with Lewy bodies. Mol Neurodegener. 2015;10(1):15.
160. Mazzulli JR, Xu YH, Sun Y, Knight AL, McLean PJ, Caldwell GA, et al. Gaucher disease glucocerebrosidase and alpha-synuclein form a bidirectional pathogenic loop in synucleinopathies. Cell. 2011;146(1):37–52 [Research Support, N.I.H., Extramural Research Support, N.I.H., Intramural].
161. Osellame LD, Rahim AA, Hargreaves IP, Gegg ME, Richard-Londt A, Brandner S, et al. Mitochondria and quality control defects in a mouse model of Gaucher disease – links to Parkinson's disease. Cell Metab. 2013;17(6):941–53.
162. Zimran A, Elstein D. Management of Gaucher disease: enzyme replacement therapy. Pediatr Endocrinol Rev. 2014;12 Suppl 1:82–7.
163. Cabrera-Salazar MA, Deriso M, Bercury SD, Li L, Lydon JT, Weber W, et al. Systemic delivery of a glucosylceramide synthase inhibitor reduces CNS substrates and increases lifespan in a mouse model of type 2 Gaucher disease. PLoS ONE. 2012;7(8):e43310 [Comparative Study Research Support, Non-U.S. Gov't].
164. Bendikov-Bar I, Maor G, Filocamo M, Horowitz M. Ambroxol as a pharmacological chaperone for mutant glucocerebrosidase. Blood Cells Mol Dis. 2013;50(2):141–5 [Research Support, Non-U.S. Gov't].
165. Luan Z, Li L, Higaki K, Nanba E, Suzuki Y, Ohno K. The chaperone activity and toxicity of ambroxol on Gaucher cells and normal mice. Brain Dev. 2013;35(4):317–22 [Research Support, Non-U.S. Gov't].
166. Patnaik S, Zheng W, Choi JH, Motabar O, Southall N, Westbroek W, et al. Discovery, structure-activity relationship, and biological evaluation of noninhibitory small molecule chaperones of glucocerebrosidase. J Med Chem. 2012;55(12):5734–48.
167. Gan-Or Z, Ozelius LJ, Bar-Shira A, Saunders-Pullman R, Mirelman A, Kornreich R, et al. The p.L302P mutation in the lysosomal enzyme gene SMPD1 is a risk factor for Parkinson disease. Neurology. 2013;80(17):1606–10 [Research Support, N.I.H., Extramural Research Support, Non-U.S. Gov't].
168. Winder-Rhodes SE, Garcia-Reitbock P, Ban M, Evans JR, Jacques TS, Kemppinen A, et al. Genetic and pathological links between Parkinson's disease and the lysosomal disorder Sanfilippo syndrome. Mov Disord. 2012;27(2):312–5 [Research Support, Non-U.S. Gov't].

Part IV
Measuring the Contribution of Mitochondrial Dysfunction to Neurodegenerative Disease

Chapter 13
Can We Accurately Model Mitochondrial Dysfunction in Neurodegeneration?

Marijana Aradjanski and Aleksandra Trifunovic

Abstract Mitochondrial diseases are connected to a plethora of clinical phenotypes, with the majority of them connected to pathologies within the nervous system. Mutations in either mtDNA or nDNA genes coding for mitochondrial proteins are known to lead to catastrophic diseases in humans that are the most common cause of inborn errors of metabolism, with a frequency of about 1 in 5000. Therefore, an understanding of mitochondrial roles in normal physiology and pathological conditions is essential for the development of possible treatments for patients suffering from various forms of mitochondrial disease. Many attempts have been made to model mitochondrial dysfunction, but this has proven to be challenging due to unique features of mitochondrial genetics. Despite this, over the last 20 years, a number of very important transgenic mouse models have been developed that in a lesser or higher degree recapitulated changes detected in human patients. Here we discuss some of the most important mouse models generated to mimic mitochondrial encephalomyopathies and the lessons learned from them.

Keywords Mitochondrial encephalomyopathies • Mitochondrial diseases • mtDNA • OXPHOS deficiency • Cybrids • Transgenic mice • CaMKIIα-Cre • mtDNA mutations • Neuronal dysfunction • Ataxia • Leucoencephalopathy

M. Aradjanski, MSc
Cologne Excellence Cluster on Cellular Stress Responses in Aging-Associated Diseases (CECAD), Medical Faculty, Institute for Mitochondrial Diseases and Aging, University of Cologne, Cologne, Germany

A. Trifunovic, PhD (✉)
Cologne Excellence Cluster on Cellular Stress Responses in Aging-Associated Diseases (CECAD), Medical Faculty, Institute for Mitochondrial Diseases and Aging, Cologne University Clinic, University of Cologne, Cologne, Germany
e-mail: aleksandra.trifunovic@uk-koeln.de

A.K. Reeve et al. (eds.), *Mitochondrial Dysfunction in Neurodegenerative Disorders*, DOI 10.1007/978-3-319-28637-2_13

13.1 Introduction

Mitochondria are unique organelles that allowed the development of multicellular life on our planet by providing eukaryotic cells with a highly efficient form of energy production, through the process of oxidative phosphorylation (OXPHOS). Besides this, mitochondria play an important role in versatile cell processes and signalling pathways, including calcium homeostasis, cell cycle regulation, apoptosis, reactive oxygen species (ROS) production, and thermogenesis [1]. As a relic of their alphaproteobacterial origin, they possess their own genome, mitochondrial DNA (mtDNA) that encodes 13 polypeptide subunits of OXPHOS complexes I, III, IV, and V and a full set of RNAs (22 tRNAs and 2 rRNAs) necessary for the effective protein synthesis inside mitochondria. All other polypeptides ($\approx$1500), including other OXPHOS subunits, various metabolic enzymes, and a full complement of proteins required for mtDNA maintenance and expression, are encoded by the nuclear genome (nDNA), translated in the cytoplasm, and imported into mitochondria (see Chap. 1). The dual origin of mitochondrial proteins gives clear ideas about the importance of the constant and regulated communication between nDNA and mtDNA for the maintenance of cell metabolism and homeostasis. Moreover, mitochondrial dynamics and involvement in cellular adaptive responses and quality control through processes of autophagy and apoptosis add additional layers to the integral picture of mitochondria complexity.

The integration of mitochondria into different aspects of cell physiology is reflected by dramatic repercussions linked to mitochondrial dysfunction. Syndromes featuring impaired mitochondrial energy production are commonly referred to as "mitochondrial diseases". Although mitochondrial diseases are usually multisystemic, the brain and muscle are the most commonly affected tissues. This notion became apparent almost four decades ago when paediatric neurologists coined the term "mitochondrial encephalomyopathies" to call attention to the frequent occurrence of brain disease in children with mitochondrial alterations in their muscle biopsies [2]. This term today is reserved for defects of the respiratory chain (OXPHOS). Even within these boundaries, the classification of the mitochondrial encephalomyopathies became quite complicated, including two types of primary mtDNA mutations (impairment of global mitochondrial protein synthesis and of the translation of specific respiratory chain subunits), and a much larger menu of Mendelian disorders (due to mutations in nuclear-encoded mitochondrial respiratory chain proteins). Also, genetic errors in other fundamental mitochondrial functions that do not directly affect the respiratory chain have major deleterious effects on the nervous system, including impaired import of mitochondrial proteins and defects of mitochondrial dynamics, such as motility, fission, fusion, and distribution.

Central nervous system (CNS) involvement in mitochondrial disorders is clinically heterogeneous, manifesting as epilepsy, stroke-like episodes, migraine, ataxia, spasticity, extrapyramidal abnormalities, bulbar dysfunction, psychiatric abnormalities, neuropsychological deficits, or hypophyseal abnormalities [3]. Mutations in mtDNA, the generation and presence of ROS, and environmental factors may contribute to energy failure and lead to neurodegenerative diseases. Furthermore, mitochondrial dysfunction has been associated with the pathogenesis of neurodegeneration in age-associated diseases such as Parkinson's, Huntington's, and Alzheimer's diseases [3, 4].

Taking into consideration the complexity of multifaceted mitochondria, in this chapter we will make an overview of the research being done in order to model mitochondrial dysfunction in neurodegeneration, specifically connected to mitochondrial diseases. Although many different attempts have been made to model mitochondrial diseases in various organisms, this chapter will focus on genetic manipulations in the mammalian model system – transgenic mice. We have classified different mouse models according to whether they targeted the mtDNA integrity and stability or the introduced mutation is in nucleus-encoded proteins affecting activity, levels, and stability of OXPHOS complexes (Table 13.1).

13.2 Modelling the Role of mtDNA Mutations and Loss of mtDNA Integrity in Neurodegeneration

Generation of animal models carrying specific mtDNA mutations has proven to be extremely challenging mainly because stable DNA transfection into mammalian mitochondria is still basically unfeasible. The introduced DNA must cross not one, but three membranes – the plasma membrane plus two mitochondrial membranes, the innermost of which is impermeable to large hydrophilic polyanions such as DNA or RNA. Even if the DNA molecule were able to access the matrix, it would need to be stably recombined into an endogenous copy of mtDNA or maintained independently. As the level of mtDNA recombination is believed to be very low in many species, including mammals, only entire mitochondrial genomes could be used as a potential vector for replication, meaning that an incoming genome needs to rapidly associate with the soluble and membrane-bound factors necessary to promote mtDNA transmission. Finally, as mtDNA is present in many copies in any one mammalian cell, it is unlikely that transfection would introduce many copies of foreign DNA inside mitochondria certifying that only a minute portion of mitochondria will carry the desired mutation. However, a minimum critical proportion of mutated mtDNA is necessary before biochemical defects and tissue dysfunction become apparent. Typically, this threshold value is in the range of 60–90 % mutant to wild-type mtDNA, resulting in a requisite of numerous transformation events to ensure introduced mtDNA reaches meaningful levels [31]. Several attempts have been made to transform mitochondria within cells using: (1) the "ProtoFection" technology that uses recombinant human mitochondrial transcription factor A (TFAM) engineered with an N-terminal protein transduction domain (PTD) followed by the mitochondrial localisation signal (MLS) to deliver mtDNA cargo to the mitochondria of living cells [7] or (2) nanocarriers, including DQAsomes

Table 13.1 Mouse models for neuropathological phenotypes in mitochondrial diseases

Mouse model	Genetic manipulation	Phenotype	Reference
Transmitochondrial mice			
CAP^R mouse	A2379T mutation in mitochondrial 16s rRNA gene	Changes in the optic nerve	Marchington et al. [5]
ΔmtDNA mouse	4696 bp deletion in mtDNA (including 6tRNAs and 7 polypeptides)	Impairment of long-term memory	Inoue et al. [6]
			Tanaka et al. [7]
MT-COI mouse	T6589C mutation in mtDNA	COX deficiency in brain	Kasahara et al. [8]
MT-ND6 mouse	G13997A mutation in mtDNA	Leber's hereditary optic neuropathy (LHON)-like phenotype	Lin et al. [9]
		Axonal swelling	
		Neuronal accumulation of abnormal mitochondria	
		Loss of small retinal fibres and swollen axons	
MT-TK mouse	G7731A mutation in mitochondrial $tRNA^{Lys}$ gene	Mild features of mitochondrial disease	Shimizu et al. [10]
Manipulation of proteins involved in mtDNA maintenance			
mtDNA mutator mouse	Knock-in:	Stochastic brain malformations	Trifunovic et al. [11]
	$Polg^{D257A}$ (lacking proofreading activity)		Ross et al. [12]
Deletor mouse	Overexpression of $Twinkle^{dup352-364}$	Brain respiratory deficiency	Tyynismaa et al. [13]
	$Twinkle^{A359T}$ (ubiquitous)		
	Overexpression of $Twinkle^{dup352-364}$ (in dopaminergic neurons; TH promoter)	Reduction in parkin levels; dopaminergic neurodegeneration	Song et al. [14]
MILON mouse	Knockout:	Late-onset neurodegeneration followed by gliosis	Sorensen et al. [15]
	Forebrain-specific neurons (CaMKIIα-Cre)		
MitoPark mouse	Knockout:	Progressive Parkinsonism	Ekstrand et al. [16]
	Dopaminergic neurons (Dat-Cre)		
TK2-deficient mouse	Knockout:	Encephalopathy and premature death	Zhou et al. [17]
	Whole body (homologous recombination)		
	Knock-in:	Encephalopathy and premature death	Akman et al. [18]
	$Tk2^{H126N}$		

Table 13.1 (continued)

Mouse model	Genetic manipulation	Phenotype	Reference
TYMP-deficient mouse	Knockout:	Leucoencephalopathy	Nishino et al. [19]
	Whole body		Lopez et al. [20]
	Whole body with UP deficiency		Haraguchi et al. [21]
Manipulation of nDNA encoded OXPHOS proteins			
Ndufs4-deficient mouse	Knockout:		
	Whole body (Mox2-Cre)	Leigh-like phenotype	Kruse et al. [22]
	Neurons and glia (Nestin-Cre)	Fatal progressive encephalopathy, ataxia, glial reactivity, and neuronal loss	Quintana et al. [23]
	Dopaminergic neurons (Dat-Cre)	No overt neurodegeneration	Sterky et al. [24]
Ndufa5-deficient mouse	Knockout:	Partial defects of CI in neurons, lethargy, loss of motor skills	Peralta et al. [25]
	Neurons (CaMKIIα-Cre)		
Coq9-deficient mouse	Knock-in:	Encephalopathy with astrogliosis and neuronal death	Garcia-Corzo et al. [26]
	239R>X mutation		
RISP-deficient mouse	Knockout:	Oxidative damage followed by neuronal death	Diaz et al. [27]
	Neurons (CaMKIIα-Cre)		
SURF1-deficient mice	Knockout:	Surprising lack of neurological phenotypes, long lived	Dell'agnello et al. [28]
	Whole body (ubiquitous Cre – not specified)		
COX10-deficient mouse	Knockout:		
	Neurons (CaMKIIα-Cre)	Cortical neurodegeneration, behavioural abnormalities	Diaz et al. [27]
	Oligodendrocytes and Schwann cells (Cnp1-Cre)	Peripheral nervous system neuropathy, dysmyelination, and muscular atrophy	Funfschilling et al. [29]
ETHE1-deficient mouse	Knockout:	Ethylmalonic encephalopathy	Tiranti et al. [30]
	Whole body (ubiquitous Cre – not specified)		

[*dequalinium-based liposome-like vesicles*], cationic, self-assembling vesicles that target the mitochondrion [32], and MITO-Porter, which enters cells by macropinocytosis and mediates mitochondrial membrane fusion [31]. However, these approaches have been heavily criticised for the lack of a mechanism for how mtDNA crossed the three membranes or the size of mtDNA fragments that could be transported, and even more importantly, so far, no researchers outside the laboratories of the inventors have successfully reported the use of any of these systems for mitochondrial transformation inside living cells [9].

Additional attempts have been tried to introduce DNA into isolated mitochondria with a bit more success in the hands of multiple researchers. Successful import of DNA into isolated mitochondria has been reported using the protein import pathway [33, 34], electroporation [35], natural competence [36], or bacterial conjugation [37]. However, none of these methods are commonly used or generally accepted for mitochondrial transformation, not least as it raises a concern of how these mitochondria would be reintroduced into host cells [9].

These technical difficulties have given rise to different methods to manipulate the mtDNA integrity in experimental animals. To date, two different approaches have been successfully tried for introducing genetically distinct mtDNA into mouse models: (1) cybrid technology using fusion of cell cytoplasts bearing mutant mtDNA to undifferentiated mouse stem cells, followed by injection into mouse blastocysts [38], or directly to mouse single-cell embryos [6] and (2) manipulation of proteins essential for mtDNA maintenance, primarily leading to increases in spontaneous mtDNA mutations [11], rearrangements [13], and depletion of mitochondrial genome [39].

13.2.1 Cybrid Models of Mitochondrial Diseases: Transmitochondrial Mice

Cybrids, or "cytoplasmic hybrids", are cultured cells manipulated to contain introduced mitochondrial DNA (mtDNA). They are produced by fusion of the whole cell with a cytoplast, which contains cytoplasm and organelles, including mitochondria, but lacks nucleus. The enucleation is achieved by simultaneous application of centrifugal force and treatment of the cell with an agent that disrupts the cytoskeleton. A number of laboratories have reported use of cybrid methodologies to create transmitochondrial mouse models [5, 6, 38, 40, 41]. Here we will mention the most important examples and specifically focus on results regarding neurological phenotypes.

13.2.1.1 CAP^R Mice

At the turn of the last century, two groups in parallel produced the very first transmitochondrial mice carrying an m.2379A>T mutation in the highly conserved part of mitochondrial 16S rRNA gene, warranting chloramphenicol resistance (CAP^R) to its host, hence the name CAP^R mice [5, 40]. Although using slightly different approaches, both groups have managed to produce cybrid embryonic stem (ES) cells, mostly carrying mutant mtDNA (90–100 %) that was injected into blastocysts, and the obtained embryos are implanted into pseudopregnant females [5, 40]. Analysis of heteroplasmy in different tissues of obtained chimeric mice demonstrated that despite heavy manipulation and chloramphenicol selection, the ES cells with mtDNA m.2379A>T mutation were able to contribute to the development of different tissues, although at fairly low levels (on average, less than 10 %) [5, 40]. Remarkably, the germ-line transmission of the CAP^R mutation even when present in moderate levels (44–52 %) resulted in severe growth retardation, myopathy, dilated cardiomyopathy, and perinatal or in utero lethality [38]. However, these mice did not present any neural phenotypes, besides prominent hamartomatous-like changes (a benign, focal malformation that resembles a neoplasm) of the optic nerve head [38].

13.2.1.2 ΔmtDNA Mice

A slightly different approach was used to create delta-mtDNA (ΔmtDNA) mice [6]. Mitochondria carrying somatic mtDNA mutation (Δmt4696) from aged mouse brain [synaptosomes] were fused to cells lacking mtDNA (ρ0 cells) to create cybrid clones expressing mtDNA with a 4696 bp deletion that included genes encoding for 6 tRNAs (K, G, R, H, S, L) and 7 polypeptides (ND3, ND4, ND4L, ND5, COXIII, ATP6, and ATP8) [6]. These cybrids were then enucleated, again fused to pronuclear-state embryos, and implanted into pseudopregnant females [6]. Although this procedure produced a low level of alive heteroplasmic animals, some founder females were identified, and with subsequent breeding, germ-line transmission of the mtDNA deletion was obtained through three generations [6]. This came as a surprise, as germ-line transmission of large mtDNA deletions rarely occurs in humans. Furthermore, although high tissue-specific differences in the amount of mutant mtDNAs are common in patients suffering from mitochondrial diseases, individual ΔmtDNA mice showed similar proportions of deleted mtDNA in all tissues that varied greatly among individuals (5–90 %) [6]. Furthermore, gene mapping of transmitted mtDNA revealed the presence of a partially duplicated mtDNA molecule consisting of one wild-type and one Δmt4696 mtDNA. The skeletal muscle, heart, and blood contained 0–17 % of the partially duplicated mtDNA, but it was not detected in other tissues [6].

There was a strong correlation between pathological phenotypes and the level of deleted mtDNA. Mice with about 30–50 % of Δmt4696 at birth were healthy, but developed mitochondrial respiratory deficiency and disease phenotypes when the load of mtDNA deletions reached 75–80 %, leading to shortening of the lifespan to about 1.5 years. On the other hand, ΔmtDNA mice carrying more than 70 % mtDNA with the deletion at birth died at around 6 months of age [40]. Most clinical phenotypes, including low body weight, lactic acidosis, systemic ischaemia, hearing loss, male infertility, and a very prominent renal failure, conveyed when the mtDNA with deletions reached the threshold of around 80 % [6, 40, 42, 43]. This came as a bit of a surprise because renal failure is not a common pathology associated with mitochondrial diseases. Nevertheless, phenotypes of ΔmtDNA mice mimic early-onset Pearson syndrome in humans that is caused by a single heteroplasmic mtDNA deletion and leads to anaemia, mitochondrial myopathy, and pancreatic and renal insufficiency [6].

Mice carrying more than 60 % of Δmt4696 already had mitochondrial respiration defects in the visual cortex and dentate gyrus, accompanied by impairment of memory at long retention delays, but not learning and temporal memory [7]. The authors associated this with reduced levels of Ca2+/calmodulin-dependent kinase II-α (CaMKIIα), a protein important for the establishment of spatial remote memory [7]. The mechanism by which moderate respiratory deficiency affects CaMKIIα is not completely clear, but it was proposed that either (1) increased lactic acidosis resulting from respiratory deficiency could deplete calcium and disturb signalling in nerve cells, which would lead to a general reduction of the CaMKIIα protein levels [44], or (2) reduction and depletion of the mitochondrial energy supply could affect the processes of translation and/or targeting of CaMKIIα mRNAs [7]. Although both of these models await experimental confirmation, it seems that the ΔmtDNA mouse could be a valuable model since it represents the only mammalian model with impairment of spatial remote memory caused by mitochondrial respiration deficiency [7]. On the other hand, ΔmtDNA mice do not faithfully recreate the most common phenotypes caused by a single large mtDNA deletion in humans, ranging from mild myopathy to devastating multisystemic syndromes such as Kearns-Sayre syndrome (KSS) [45, 46].

13.2.1.3 *MT-COI* Mice

The first transmitochondrial mouse carrying a point mutation in the mtDNA was created by introducing the missense m.6589T>C mutation in the *COI* gene [8]. These mice are created by fusing respiratory-deficient mouse B82 cell lines to ES cells depleted of mitochondria. As B82 cells lack thymidine kinase, unfused cells could not survive in the selection medium with hypoxanthine/aminopterin/thymidine (HAT) and therefore were effectively eliminated [8]. ES cell clones positive for the mutation were introduced in 8-cell stage embryos and subsequently transferred to a pseudopregnant female. The resulting mice were homoplasmic for the introduced mutation in all analysed tissues [8]. Further analysis was carried in the F6 mice backcrossed to a C57B6 background to show that the observed phenotype does not come from ΔmtDNA nuclear incompatibility. The analysed mice showed growth retardation and 50–70 % decrease in COX activity in the brain, heart, liver, and skeletal muscle (of the control levels) [8]. Although increased blood lactate levels were detected in these mice, there were no signs of epilepsy, as reported in patients with *mt-COI* missense mutation [47]. The authors did not proceed further in their analysis of these mice, and therefore we do not have reports of any additional phenotypes arising in them.

13.2.1.4 *MT-ND6* Mice

Possibly the mouse model that most accurately copies the human pathology was created for Leber's hereditary optic neuropathy (LHON) by introducing the human optic atrophy *ND6 G14600A* (*P25L*) mutation into the mouse [9]. LHON is the first inherited mtDNA disease reported [48] and probably the most prevalent one caused by mtDNA missense mutations, having an estimated frequency of 15 in 100,000 [49]. LHON is typically caused by mutations in one of the three mtDNA-encoded complex I (CI) genes (*MT-ND1*, *MT-ND4*, and *MT-ND6*) [49]. Patients with classic LHON experience loss of central vision, usually in both eyes, between 15 and 35 years of age due to preferential loss of papillomacular bundle nerve fibres [49].

Creation of *MT-ND6* homoplasmic mutant mice started with mutagenesis of murine LMTK cell line, followed by enrichment of introduced mtDNA mutations by ethidium bromide depletion, reamplification, and cloning and subsequent selection of the respiratory-deficient clones using glucose- or galactose-containing media. This approach produced a clone carrying homoplasmic m.13997G>A mutation (*MT-ND6* P25L substitution) with isolated CI deficiency (23 % residual activity), 65 % reduction in ATP synthesis, and increased ROS production [9]. The enucleated clones were fused with mouse ES cells, chimeric females were identified, and the maternal transmission of the homoplasmic *MT-ND6* m.13997G>A mtDNA mutation was confirmed by mtDNA sequencing [9]. Remarkably, despite the systemic reduction of CI activity, like in LHON patients, the phenotype was restricted to the optic nerve. The mice mirrored some of the typical LHON patient phenotypes like: reduction in retinal function by electroretinogram (ERG), age-related decline in central smaller calibre optic nerve fibres with sparing of larger peripheral fibres, neuronal accumulation of abnormal mitochondria, axonal swelling, and demyelination decreased retinal response, preferential loss of small retinal fibres, and swollen axons of retinal ganglion cells, due to the accumulation of morphologically abnormal mitochondria [9]. Reduced CI activity and highly increased ROS levels were detected in mitochondria and synaptosomes isolated from the brain of these mice. Surprisingly, normal ATP levels were detected in synaptosomes of these mice, indicating that the primary cause of the retinal impairment was the chronic oxidative damage rather than the energetic failure [9]. As the aetiology of LHON still requires better understanding, this mouse model could be essential for better characterisation of the relationship between mtDNA and optic nerve dysfunction and could further be used to test antioxidant therapies in vivo.

Surprisingly, mice with the same m.13997G>A mutation that arose independently in C57BL Lewis lung tumour cells on a different mtDNA background had a distinctly different phenotype [50]. They showed moderate complex I defects and increased lactate levels at a young age, but no other phenotypes related to mitochondrial diseases [50]. Later in life, these *MT-ND6* mice developed some age-associated phenotypes, such as development of B lymphoma and diabetes, but no neurological or ophthalmological symptoms were detected [51]. It is currently not clear where this strong discrepancy between the two models is coming from, although, at least in part, it could be ascribed to different genetic backgrounds. Further studies using both models in parallel are clearly needed to address this issue.

13.2.1.5 *MT-TK* Mice

A transmitochondrial mouse with an m.7731G>A mutation in *MT-TK* (mitochondrial $tRNA^{Lys}$) gene was recently generated from a mouse lung carcinoma P29 cell line [10]. Mitochondrial transfer RNA (mt-tRNA) mutations are the most common mtDNA mutations to cause human disease, and this was the very first mouse model with a point mutation in mitochondrial tRNA gene that has a counterpart in human patients affected by mitochondrial disease. The mice were created by fusing ES cells depleted of mitochondria with the cybrid clone having 70 % heteroplasmy for *MT-TK* m.7731G>A [10]. The mutation was transmitted through subsequent generations, if present in ≤85 % heteroplasmy, while oocytes containing a higher mutational load seem to be lost during development [10]. Mice with predominant m.7731G>A heteroplasmy displayed some features of mitochondrial diseases, such as short body length and muscle weakness, but not ragged red fibres (RRFs). However, *MT-TK* m.7731G>A did not reproduce the severe phenotype of patients with matching human *MT-TK* mutation (m.8344G>A) associated with MERRF (myoclonic epilepsy with ragged red fibres) possibly because of the lower heteroplasmy level transmitted to the progeny. At this point, a possibility for a late onset of the disease in mice cannot be excluded. One striking difference between *MT-TK* m.7731G>A and human patients with matching mutation(s) is the loss of oocytes with high mutational load in mice [10]. This aspect underlines once more the difficulties in creating mouse models resembling human defects and, probably, species-specific differences in mtDNA selection during germ-line transmission and segregation in tissues. Nevertheless, we believe that this mouse model will be useful for studying specific aspects of human mitochondrial myopathies and possibly testing therapeutic compounds.

13.2.2 Manipulating Proteins Essential for mtDNA Maintenance

Besides the direct manipulation of mtDNA, manipulation of proteins involved in mitochondrial genome maintenance, primarily mtDNA replication, has been used in order to create animal models for mitochondrial diseases. Replication of the mitochondrial genome requires unique enzymatic machinery, composed of a set of factors encoded by nuclear DNA and recruited to mitochondria. In vitro studies showed that the basic replication fork of human mitochondria can be reconstituted using primed single-stranded DNA substrate and at least five mtDNA maintenance factors: catalytic and accessory subunit of DNA polymerase gamma (POLG and POLG2), mitochondrial RNA polymerase (POLRMT), replicative DNA helicase (TWINKLE), and the mitochondrial single-stranded DNA binding protein (MTSSB) [52]. Although this simple mitochondrial replisome is sufficient to duplicate a DNA substrate in vitro, proper maintenance and effective replication of the mitochondrial genome involve a much longer and still growing list of factors.

13.2.2.1 MtDNA Mutator Mouse

Development of mtDNA mutator mice allowed detailed analysis and understanding of the role of random somatic mtDNA mutations in the development of mitochondrial pathologies [11]. MtDNA mutator mice are created by introducing a single point mutation in the highly conserved exonuclease domain of POLG, resulting in almost complete inactivation of proofreading exonuclease activity in these mice [11]. As the proofreading was abolished, mtDNA mutator mice accumulated three to five times more random mtDNA mutations and developed a range of phenotypes reminiscent of naturally occurring ageing and had a very significantly shortened lifespan [11]. The high number of mtDNA point mutations in mtDNA mutator mice leads to the synthesis of respiratory chain (RC) subunits with many amino acid substitutions that likely cause instability of the RC complexes [53]. Although initial studies indicated that neuronal phenotypes are not predominant in mtDNA mutator mice [11], some studies have specifically addressed changes in the central nervous system (CNS) of these mice [54]. Initially it was reported that mitochondrial dysfunction in the mtDNA mutator mouse brain leads to a metabolic shift from aerobic respiration to glycolytic metabolism [54]. Using proton magnetic resonance spectroscopy, increased brain lactate levels have been detected in these mice, and this was assigned to mitochondrial dysfunction in neurons that, instead of metabolising pyruvate through the tricarboxylic acid (TCA) cycle, supposedly rely on glycolysis and anaerobic metabolism [54]. However, a later study on the CNS phenotypes failed to confirm general mitochondrial dysfunction in mtDNA mutator brains and instead described embryonic-onset dysfunction of somatic stem cells (SCs), including neural progenitors [55]. It has been demonstrated that development of neural and haematopoietic progenitor cells of the mtDNA mutator mice is already affected during foetal development and that neural stem cells showed decreased abundance in vivo as well as reduced self-renewal capacity in vitro [55]. Moreover, the observation that treatment with the antioxidant N-acetylcysteine restores the self-renewal ability of neural progenitor cells has led to the conclusion that an aberrant change in ROS signalling or redox status is sufficient to modify signalling in somatic stem cells and severely disrupt their homeostasis [55]. Recently, another study on mtDNA mutator mice has shown that a combination of inherited and somatic mtDNA mutations causes stochastic brain malformation, which suggests that starting life with healthy mitochondria might be crucial for maintaining brain health during the course of ageing [12]. This study further showed that the initial report on the more general mitochondrial dysfunction in the mtDNA mutator brain is likely a result of the increased, inherited mtDNA mutation load, rather than lifelong somatic mtDNA mutation accumulation. It seems that in mice, unlike humans, spontaneous somatic mtDNA mutations are less likely to cause pathological phenotypes in the brain than other tissues, like the heart, spleen, or skeletal muscle.

Although mtDNA mutator mice were extremely valuable to model and understand the general role of random somatic mtDNA mutations, they cannot be used to model numerous diseases caused by POLG mutations. Mutations in *POLG* have been associated with a number of mitochondrial disorders that affect the stability of mtDNA, primarily leading to mtDNA deletion and depletion syndromes [56, 57]. The most common disease caused by *POLG* mutations is the autosomal dominant (ad) or autosomal recessive (ar) form of familial progressive external ophthalmoplegia (PEO) [57–59]. Additional clinical presentations like autosomal recessive sensory ataxic neuropathy with dysarthria and ophthalmoplegia (SANDO) and Alpers' hepatophatic poliodystrophy have also been associated with *POLG* mutations [60–62]. Furthermore, dominant *POLGA* mutations have shown to cause a severe multisystemic disorder including Parkinsonism and premature menopause, which are not typical of mitochondrial disease [63]. To accurately model these diseases, novel mice that carry specific *POLG* patient mutations should be developed.

13.2.2.2 Deletor Mouse

TWINKLE is a hexameric helicase that together with POLG and MTSSB forms the minimal mtDNA replisome in vitro [52]. Dominant mutations in TWINKLE are associated with PEO [64]. A decade ago, two transgenic mice models overexpressing either in-frame duplication [352–364] or amino acid substitution (A359T) found in human patients were developed [13]. The in-frame duplication is structurally the most severe mutation so far described in TWINKLE patients, and correspondingly Twinkle^{dup} mice showed more severe phenotypes than Twinkle^{A359T} [13]. They are named "deletor mice" as their main feature is the age-dependent accumulation of multiple large mtDNA deletions, along with the respiratory chain dysfunction. A predominant phenotype in deletor mice is mitochondrial myopathy, and the muscles of the mice faithfully replicate all of the key histological, genetic, and biochemical features of PEO patients [13]. Brain respiratory deficiency followed the muscle phenotypes, and COX-deficient (COX-) neurons are primarily found to be Purkinje cells and large pyramidal neurons in the hippocampal CA2 region. A few COX- neurons were also identified from the olfactory bulbs, substantia nigra, and hypothalamus [13]. In contrast to mtDNA mutator mice, deletor mice did not show signs of premature ageing, although they showed progressive respiratory chain deficiency due to the accumulated mtDNA deletions. These mice represent a very good model for the late-onset progressive mitochondrial disease, without affecting the lifespan [13]. More recently, a mouse model expressing mutated TWINKLE (352–364 duplication) specifically in dopaminergic neurons of substantia nigra was created [14]. These mice feature late-onset mild respiratory dysfunction accompanied with increased mtDNA deletions and reduction in parkin levels with alterations in autophagy, which together causes dopaminergic neuron degeneration and movement defects [14]. Although interesting and with results in agreement with previous deletor mice, this specific model conceptually adds novelty in modelling mitochondrial involvement in certain brain regions.

13.2.2.3 MILON and MitoPark Mice

One of the very first mouse models mimicking mitochondrial dysfunction found in patients was the TFAM (mitochondrial transcription factor A)-deficient mouse [39]. *Tfam* encodes a protein with dual function that is on one side important for packaging of mtDNA and on the other is an important transcriptional activator necessary for mitochondrial replication and transcription. TFAM is an essential mitochondrial protein whose constitutive depletion leads to embryonic lethality between E8.5 and E10.5 [39]. Tissue-specific depletion of *Tfam* in forebrain neurons and hippocampus using Cre recombinase under CaMKIIα promoter (CaMKIIα-Cre) leads to creation of the so-called MILON (mitochondrial late-onset neurodegeneration) mouse that showed reduced mtDNA levels already at 2 months and severe respiratory chain deficiency in neurons from 4 months of age [15]. Strong respiratory deficiency led to progressive neurodegeneration and massive apoptosis followed by evident gliosis in the cortex and hippocampus, starting around 5 months of age [15]. MitoPark mouse was another conditional *Tfam* knockout mouse created using Cre recombinase under the dopamine transporter promoter (DAT-Cre) that depleted TFAM in dopaminergic neurons of substantia nigra [16]. MtDNA expression was dramatically reduced at 6 weeks of age, although phenotypic manifestations of Parkinsonism (Parkinson's disease [PD]-like phenotypes) started to develop much later. MitoPark mice showed significantly decreased locomotion and rearing behaviour at 14 weeks of age, and from 20 weeks of age, the mice started to display apparent disease manifestations such as tremor, twitching, and abnormal gait [16]. This slowly progressive impairment of motor function was accompanied by formation of intraneuronal inclusions and dopamine nerve cell death, and like in PD patients, these mice responded to L-DOPA therapy with a differential success depending on disease stage [16]. Therefore, the MitoPark mouse model was the first to recapitulate many of the cardinal clinical features of PD, namely, progressive neurodegeneration and death of neurons, loss of motor function, and therapeutic response to L-DOPA, and is extensively being used in PD research [65–67].

13.2.2.4 Tk2-Deficient Mice

The correct maintenance of mtDNA depends not only on the normal activity of the enzymes involved in replication, but also on a balanced pool of dNTPs. The first step for the salvage pathway synthesis of dNTPs, on which postmitotic tissues primarily rely to maintain their dNTP pools, is provided by the constitutively expressed mitochondrial thymidine kinase (TK2) [68]. Mutations in the *TK2* gene primarily cause mitochondrial DNA depletion syndrome (MDS) with a broad clinical spectrum ranging from severe fatal infantile myopathy with motor regression, spinal muscular atrophy, and rigid spine syndrome to mild mitochondrial myopathy [69, 70]. To model the MDS, two different TK2-deficient mice were developed in parallel: one was created by deleting exons 4 and 5 (*Tk2–/–*) [17] and the other by introducing the H126N mutation (*Tk2*^{H126N}) [18] corresponding to the human pathogenic H121N mutation that drastically reduces TK2 activity [69]. Despite the genetic difference and the presence of some residual TK2 activity in the *Tk2*^{H126N} animals, these two mouse models shared phenotypes of severe isolated encephalopathy with a growth defect, but no major skeletal muscle alterations [17, 18]. Progressive mtDNA depletion was associated with depletion of dNTP pools and a severe mitochondrial defect, characterised by loss of mtDNA-encoded OXPHOS subunits, CI and CIV deficiency, and ATP reduction [17, 18]. As the brain was the most affected tissue, these mice displayed primarily neurological phenotypes like ataxic gait, coarse tremors, impaired motor coordination, abnormal limb clasping, and generalised weakness [17, 18]. This also represents the main difference in comparison to patients that also have a strong myopathic phenotype that was lacking in mice or is possibly masked by a severity of neurological phenotype [17, 18, 69, 70]. However, these mice still represent a very valuable tool to study MDS, but once again show that even mutations homologous to that found in human patients do not necessarily lead to development of the same pathologies.

13.2.2.5 *Tymp*-Deficient Mice

Thymidine phosphorylase (TP), a protein encoded by the *TYMP* gene, is also an enzyme involved in the pyrimidine salvage pathway, required for the reversible reaction catalysing thymidine and phosphate to thymine and deoxyribose-1-phosphate [19]. Mutations in *TYMP* cause mitochondrial neurogastrointestinal encephalopathy (*MNGIE*), an autosomal recessive syndromic disease clinically defined by the occurrence of gastrointestinal dysmotility with intestinal pseudo-obstruction, peripheral neuropathy, ptosis, ophthalmoparesis, and hearing loss [19]. Loss of TP catalytic activity causes an accumulation of thymidine and nucleosides in the blood and tissues of affected patients leading to multiple deletions and depletion of mtDNA [71].

To model MNGIE in mice, both *TYMP* and *UPP1* gene, encoding uridine phosphorylase (UP), had to be knocked down, as in mice, unlike humans, UP can compensate for the loss of TP [21]. Initial study described leucoencephalopathy with enlarged myelinated fibres in *Tymp–/–/Upp1–/–* mice, but no mtDNA abnormalities, leading them to conclude: "encephalopathy in MNGIE patients does not appear to be caused by mtDNA alterations" [21]. However, a consequent study questions these results and conclusion, as they showed that elevated thymidine and uridine levels lead to mtDNA depletion, respiratory chain defects, and histological alterations in the brain of *Tymp–/–/Upp1–/–* mice [20]. However, these mice only partially resembled MNGIE patients, since they presented mtDNA depletion and pathophysiological changes only in the brain, despite ubiquitous TP deficiency [20]. A few possible explanations have been put forward for this discrepancy including: (1) the shorter lifespan of mice may not be sufficient to allow significant accumulation of somatic mtDNA alterations and consequent respiratory dysfunction in most tissues, (2) mitochondrial dNTP pool imbalances in MNGIE patients are more dramatic than in *Tymp–/–/Upp1–/–* mice and therefore could cause mtDNA instability in more tissues, and (3) differential expression and activities of other enzymes involved in the pyrimidine metabolism are likely to influence in the balance of the mitochondrial dNTP pools and contribute to the tissue-specific effects in mice and humans [20]. Despite these differences, this knockout mouse represents an important model to further investigate the pathophysiology of leucoencephalopathy in mitochondrial diseases and is a useful tool to develop therapeutic strategies for MNGIE.

13.3 Manipulating nDNA-Encoded Proteins Involved in Mitochondrial Diseases and Neurodegeneration

The OXPHOS complexes are composed of subunits encoded by both nDNA and mtDNA with the exception of complex II, which is exclusively encoded by nDNA. For fully and correctly assembled OXPHOS complexes, exact coordination of the expression of nuclear and mitochondrial proteins is needed. Defects in OXPHOS result in varied pathologies and metabolic diseases, mostly affecting the organs with high-energy demands, such as the heart, skeletal muscle, and brain. Many attempts to create mouse models for OXPHOS deficiencies have failed because of the essential nature of targeted genes for mitochondrial function, leading to embryonic lethality in generated models [72]. However, in recent years, a number of viable mouse models related to the components of OXPHOS proteins encoded by the nuclear genome were constructed in order to study pathogenic mechanisms underlying these diseases.

13.3.1 Complex I-Deficient Models

Mammalian complex I (CI) or NADH-ubiquinone oxidoreductase is the largest enzyme of the OXPHOS and consists of at least 38 subunits encoded by nDNA and 6 subunits encoded by mtDNA [73]. Isolated complex I deficiency is the most frequently diagnosed mitochondrial defect accounting for almost 23 % of all cases of childhood respiratory chain deficiency [74]. The most common phenotypes associated with isolated complex I deficiency are the Leigh syndrome (LS) and Leigh-like disease which mainly cause psychomotor retardation, brainstem dysfunction, seizures, failure to thrive, muscular hypotonia, dystonia, abnormal eye movements, and lactic acidosis [75].

13.3.1.1 *Ndufs4*-Deficient Mice

NDUFS4 (NADH-ubiquinone oxidoreductase iron-sulphur protein 4) is an 18 kDa protein, inserted at a late stage of CI biogenesis, essential for CI assembly and stability. *Ndufs4* gene is a mutational hot spot in humans leading to development of LS and Leigh-like encephalomyopathy phenotype [76]. The first NDUFS4-deficient mouse model was a whole-body knockout (*Ndufs4–/–*) that rapidly developed a Leigh-like phenotype characterised by ataxia, blindness, retarded growth rate, lethargy, and increased serum lactate, leading to premature death at about 7 weeks of age [22]. These mice developed isolated CI deficiency with a very specific tissue distribution, with liver and CNS being the most affected tissues, whereas CI activity in the skeletal muscle of *Ndufs4–/–* mice was reduced to about 50 % of control levels [22]. It is also quite interesting that these mice lived for about 5 weeks without signs of illness after which they rapidly developed a fatal phenotype. Although it was assumed that NDUFS4 subunit plays an essential role in assembly and stability of the CI, the results from *Ndufs4–/–* mice demonstrated that CI could be formed without it, most probably because of the compensatory activity of other subunits in a tissue-specific manner [22]. Remarkably, mice deficient in NDUFS4, exclusively in neurons and glia (using *Nestin-Cre*), copied the phenotype of full-body knockouts [23]. These mice (*Ndufs4*$^{L/L}$; *Nestin-Cre*) had clear signs of progressive glial activation that promotes neuronal death and ultimately results in mortality. Neurons in the olfactory bulbs (OBs), vestibular nuclei (VN), and posterior lobules of the cerebellar vermis appear to be the most vulnerable [23]. The basic neuropathology featuring mitochondrial disorders consists of nonspecific histological lesions, neuronal loss, necrosis, gliosis, demyelination, and spongiform degeneration [77] that were all apparent when studying the brains of these mice. Therefore, both of these mouse models (*Ndufs4–/–* and *Ndufs4*$^{L/L}$; *Nestin-Cre*) are going to be essential for understanding the progression of symptoms in LS and looking for potential therapeutic interventions [22, 23].

Interestingly, disruption of NDUFS4 in midbrain dopaminergic neurons (*Ndufs4*$^{L/L}$; *DAT-Cre*) did not lead to overt neurodegeneration, loss of striatal innervation, or symptoms of Parkinsonism [24]. However, dopamine (DA) homeostasis was abnormal with impaired DA release and increased levels of DA metabolites allowing the authors to conclude that complex I deficiency still can contribute to the pathophysiology of PD [24].

Another two *Ndufs4*-deficient mouse models were created that did not completely recapitulate observed phenotypes [78, 79]. Firstly, an *Ndufs4* knockout model was created by spontaneous transposable element insertion within the gene that produced a premature stop codon and resulted in an unstable transcript [78]. These *Ndufs4* knockout mice were smaller at birth, lost their fur at 2 weeks of age, displayed some neurological impairment at 5 weeks, and died prematurely at about 7 weeks. However, the authors did not observe significant neuropathological changes in the brain of these mice, in contrast to previous reports [22, 23]. This has been explained by an earlier time point used to analyse these mice, due to ethical permit restrains [78]. In the future, comparative analysis of all three models will be needed to shed more light on the observed phenotypic differences.

A second model was created as a knock-in model, by introducing a premature stop codon in the *Ndufs4* gene [79]. Unexpectedly, homozygous knock-in mice were embryonic lethal, while heterozygous mice showed a mild phenotype, with a reduction of CI activity of about 30 % in the heart, brain, and skeletal muscle mitochondria, accompanied by elevated lactate levels in the brain and heart [79]. It seems that the introduced mutation could have a dominant-negative effect as it creates a stop codon that results in the expression of a truncated NDUFS4 protein of about 14.4 kDa lacking the last 10–15 amino acids. The truncated protein seems to be incorporated at low levels into the fully assembled CI resulting in significant complex deficiency and might explain the differences with other *Ndufs4* knockout models [79].

13.3.1.2 *Ndufa5*-Deficient Mice

Initial attempts to create mice deficient for NADH-ubiquinone oxidoreductase 1 alpha subcomplex subunit 5 (NDUFA5) were done using a gene-trap technology, but without success, as it turned out that *Ndufa5* was an essential mitochondrial protein necessary for embryonic survival [25]. This might be also the reason why there are no human patients with mutations in *NDUFA5* gene. To study the role of NDUFA5 in adult mice, a conditionally targeted *Ndufa5* gene was created and deleted specifically in forebrain neurons and hippocampus using *CaMKIIα-Cre* [25]. These mice were healthy until the age of 10–11 months when they became lethargic and showed significantly reduced motor coordination compared to the controls of the same age. Although levels of the fully assembled CI (75–80 %) and CI activity (40 %) were markedly reduced in the cortex, neither oxidative stress, neuronal loss, nor gliosis was observed in the knockout mice [25]. The mild phenotype of *Ndufa5* knockout mouse could be the result of the compensatory mechanisms such as physiological adjustment to ketogenic environment, since the brain actively metabolises ketone bodies in periods of energy shortage [80]. This was suggested after noting the increased levels of ketogenic enzyme ACAT1 [mitochondrial acetyl-CoA acetyltransferase] in the knockout brains [25]. Another compensatory effect could be the increased electron flow via the electron transfer flavoprotein dehydrogenase to CoQ_{10}, since there was an upregulation of electron-transferring-flavoprotein subunit A (ETFA) in knockout brains [25].

13.3.2 *Coenzyme Q Deficiency*

Coenzyme Q10 (CoQ_{10}) is a mobile lipophilic electron carrier located in the inner mitochondrial membrane that transfers electrons from CI to CII and to complex III [81]. Besides its role in the OXPHOS system, CoQ_{10} is a potent antioxidant molecule and participates in the de novo pyrimidine biosynthesis pathway [81]. CoQ_{10} endogenous biosynthesis occurs ubiquitously in the mitochondria, and several pathogenic mutations in genes encoding proteins involved in this process have been identified [82, 83]. Mutations in CoQ biosynthetic genes produce primary CoQ10 deficiency, a mitochondrial syndrome associated with five major clinical phenotypes: encephalomyopathy, severe infantile multisystemic disease, nephropathy, cerebellar ataxia, and isolated myopathy [83].

13.3.2.1 *Coq9*-Deficient Mice

COQ9 gene encodes 1 of the 11 proteins required for the biosynthesis of CoQ_{10} within mitochondria and is the disease-causing gene reported in a patient with primary CoQ_{10} deficiency [84]. The patient with a *COQ9* homozygous stop mutation developed neonatal lactic acidosis followed by multisystemic disease including intractable seizures, global developmental delay, hypertrophic cardiomyopathy, and renal tubular dysfunction [84]. To better understand the pathophysiological consequences of primary CoQ10 deficiency, a mouse model carrying a homozygous mutation in *Coq9* gene (R239X, $Coq9^{R239X}$), which is homologous to the human R244X mutation, was created [84]. The homozygous $Coq9^{R239X}$ mice were normal at birth, but between 3 and 6 months, they developed a rapid and progressive paralysis followed by death, suggesting an involvement of the central nervous system in the pathogenic mechanism of the disease [26]. Further analysis of $Coq9^{R239X}$ mice showed intense vacuolisation, astrogliosis, and neuronal death in the brain and severe demyelination of peripheral tissues. The lack of CoQ_{10} caused a loss of CI and an increase in free CIII, leading to a decrease in mitochondrial respiration and ATP synthesis only in the brain of the $Coq9^{R239X}$ mice [26]. However, before accepting $Coq9^{R239X}$ mouse as an excellent model of mitochondrial encephalomyopathy associated with CoQ_{10} deficiency, it has to be determined if the observed phenotypes are independent from other metabolic pathways fundamental for proper brain development in foetuses and newborns, like cholesterol metabolism, which shares the biosynthetic pathway of CoQ_{10}.

13.3.3 Complex III-Deficient Models

Complex III (CIII, ubiquinol-cytochrome *c* reductase, or cytochrome bc1 complex) consists of three catalytic and eight structural subunits, yet deficiencies in this complex are very rare, but with a broad range of clinical features and tissue specificity [85]. Although a few mouse models have been created that have deficiency in CIII subunits, only *Uqcrfs1*-deficient mice were analysed for CNS phenotypes.

13.3.3.1 *Uqcrfs1*-Deficient Mice

Rieske iron-sulphur protein (RISP) is one of the catalytic subunits of CIII encoded by nuclear DNA, more specifically, the *UQCRFS1* gene. Similar to the previously described *Ndufa5* gene, *Uqcrfs1* is essential for mouse embryonic development, providing explanation as to why no human patients are described so far. To study the role of RISP in an adult mammalian brain, a conditional strategy was used to create forebrain neurons and hippocampus-specific *Uqcrfs1* knockout mice ($Uqcrfs1^{L/L}$; *CaMKIIα-Cre*) [27]. These mice lived up to 3 months, with CIII deficiency becoming apparent in both cortex and hippocampus already at 1 month of age and progressing rapidly until the end of their lifespan [27]. CIII deficiency was accompanied with an increase in citrate synthase and CIV activities and higher mtDNA/nDNA ratios, all of which indicate increased mitochondrial biogenesis as a compensatory response to respiratory deficiency [27]. *Uqcrfs1* knockout mice also showed extensive oxidative damage followed by neuronal death, mainly in the piriform cortex and hippocampus at 3 months of age and therefore could be a great model for testing antioxidants as therapeutic agents in mitochondrial diseases [27].

13.3.4 Complex IV-Deficient Models

Cytochrome *c* oxidase (COX) or complex IV (CIV) is the terminal enzyme of the electron transport chain in mitochondria that receives electrons from cytochrome c molecules and transfers them to molecular oxygen, thus converting it to two molecules of water. CIV is a very significant factor in the aetiology, progression, and prevalence of numerous human neurodegenerative diseases and represents an important target for developing diagnostic and therapeutic tools against those diseases [86]. COX deficiency may cause clinically heterogeneous disorders ranging from isolated myopathy to severe multisystemic disease [87]. Mutations in genes encoding both the CIV structural subunits (COI and COII) and assembly factors (SURF1, SCO2) have been implicated in human diseases.

13.3.4.1 SURF1-Deficient Mice

Remarkably, mutations in the COX assembly genes are a major cause of isolated COX deficiency and LS [87]. It has been estimated that about one-third of all LS cases are caused by *SURF1* mutations [88]. Therefore, it came as a great surprise that mice lacking SURF1 (*Surf1–/–*) did not show any signs of neurodegeneration [28]. In fact, these mice showed protection from Ca^{2+}-dependent neurotoxicity caused by kainic acid and had significantly increased median lifespan [28]. The biochemical and assembly COX defect was present in *Surf1–/–*mice, but much milder than in human patients, suggesting alternative CIV assembly factors are present in mouse mitochondria [28]. Therefore, despite the fact that SURF1 is evolutionarily conserved, its function in COX assembly in mice seems to be redundant and therefore could be partly taken over by other unknown factors. One possible explanation is that mild mitochondrial dysfunction in *Surf1–/–* mice could lead to activation of compensatory genetic and epigenetic mechanisms that in turn can lead to increased longevity through hormetic responses. Indeed, the results of a most recent study suggested that impaired mitochondrial function as observed in *Surf1–/–* mice can lead to induction of mitochondrial stress pathways like mitochondrial biogenesis and mitochondrial unfolded protein response, to confer protective effects on cellular homoeostasis [89]. It was also reported that loss of SURF1 leads to key metabolic changes resulting in reduced adiposity, improved insulin sensitivity, and induction of mitochondrial biogenesis in white adipose tissue [90]. These novel findings may also contribute to the prolonged lifespan of *Surf1–/–* mice and suggest that although these mice are not a good model for human LS pathologies, they might be invaluable to study other aspects of mitochondrial function, like mitohormesis and the role of mitochondrial dysfunction in mammalian ageing.

13.3.4.2 *Cox10*-Deficient Mice

COX10 is a farnesyltransferase, necessary for the assembly of CIV. In humans, mutations in *COX10* have been associated with leucodystrophy, hypotonia, lactic acidaemia, deafness, etc. [91]. To gain better understanding of the role of COX10 in the mammalian CNS, two different conditional *Cox10* knockout mice were generated: one with a primary deficiency in neurons [27] and the other in oligodendrocytes and Schwann cells [29]. The neuron-specific *Cox10* knockout (*Cox10$^{L/L}$*; *CaMKIIα-Cre*) represents an encephalopathy mouse model with pronounced behavioural abnormalities [27]. Although COX-deficient neurons were detected at 2 months of age in the cortex and hippocampus, the neurodegenerative phenotype was slowly progressing until the age of 8 and 12 months when the mice die [27]. This was accompanied with a delayed onset of oxidative stress and gliosis [27]. The myelin-producing cell-specific ablation of COX10 (*Cox10$^{L/L}$*; *Cnp1-Cre*) showed a severe neuropathy phenotype with dysmyelination, muscle paralysis, and atrophy in the peripheral nervous system. On the contrary, in the adult CNS, no signs of demyelination nor axonal degradation were found, suggesting that postmyelination oligodendrocytes survive well in the absence of COX activity, relying on glycolytic metabolism [29]. Moreover, brain lactate was increased in *Cox10$^{L/L}$/PLPErt2-Cre* mice, which can be effectively used by myelinated axons when energy deprived. This model indicates that an increased rate of oligodendroglial glycolysis can supply aerobic glycolysis products to support axonal energy needs suggesting a model in which axon-glia metabolic coupling serves a physiological function [29].

13.3.4.3 Ethe1-Deficient Mice

ETHE1 gene encodes a *b*-lactamase-like, iron-coordinating metalloprotein that is mutated in ethylmalonic encephalopathy, a devastating infantile metabolic disorder affecting the brain, gastrointestinal tract, and peripheral vessels connected with COX deficiency [92]. In the mouse model deficient in ETHE1 (*Ethe1–/–*), sulphur dioxygenase activity was absent, disabling sulphide detoxification in mitochondria. Sulphide is a powerful inhibitor of COX and short-chain fatty acid oxidation, with vasoactive and vasotoxic effects that explain the microangiopathy in ethylmalonic encephalopathy patients [30]. This mouse model of ethylmalonic encephalopathy is the first example of an inherited mitochondrial disease resulting from toxic inhibition of aerobic energy metabolism [30]. The toxic mechanism underpinning ethylmalonic encephalopathy makes effective therapy a realistic goal. Even though this mouse model isn't mimicking the direct deficiency in OXPHOS, it shows how deficiency in other genes can affect particular parts of OXPHOS, and that is why it represents a valuable model [30].

13.4 Conclusion

Despite the great advance being made in constructing mouse models with OXPHOS defects, we still have a gap in knowledge in explaining tissue specificity and the difference in severity of the phenotypes obtained from these models. Another problem that we are dealing with is the difference between human patients' clinical phenotypes and what we observe in our mouse models, with the same deficiency. The latter complicates finding suitable therapeutic approaches.

The aim of this review was to summarise the mouse models displaying the neurodegenerative phenotype and compare it with the known human patients' diseases. Mitochondrial diseases are extraordinarily diverse, not only in their phenotypic presentation but also in their genetic background. The multifaceted nature of mitochondria is shown in versatile causes of neurodegenerative disease. From disruptions in mtDNA expression, over energy production problems, to defected mitochondrial dynamics and quality control, we can clearly see how every aspect of mitochondrial well-being is important for health maintenance. On the other hand, the mechanisms of many diseases have yet to be elucidated in order to get closer to effective therapeutic approaches. More mouse models depicting human neurodegenerative diseases need to be constructed. Moreover, the inability to genetically transform mammalian mitochondria is a serious impediment to research on mitochondrial disorders and limits gene therapeutic approaches to mtDNA diseases.

References

1. Wallace DC. Diseases of the mitochondrial DNA. Annu Rev Biochem. 1992;61:1175–212.
2. Howard MA, Volkov IO, Noh MD, Granner MA, Mirsky R, Garell PC. Chronic microelectrode investigations of normal human brain physiology using a hybrid depth electrode. Stereotact Funct Neurosurg. 1997;68(1–4 Pt 1):236–42.
3. DiMauro S, Schon EA. Mitochondrial disorders in the nervous system. Annu Rev Neurosci. 2008;31:91–123.
4. Wallace DC. Mitochondrial genetics: a paradigm for aging and degenerative diseases? Science. 1992;256:628–32.
5. Marchington DR, Barlow D, Poulton J. Transmitochondrial mice carrying resistance to chloramphenicol on mitochondrial DNA: developing the first mouse model of mitochondrial DNA disease. Nat Med. 1999;5(8):957–60.
6. Inoue K, Nakada K, Ogura A, Isobe K, Goto Y, Nonaka I, et al. Generation of mice with mitochondrial dysfunction by introducing mouse mtDNA carrying a deletion into zygotes. Nat Genet. 2000;26(2):176–81.
7. Tanaka D, Nakada K, Takao K, Ogasawara E, Kasahara A, Sato A, et al. Normal mitochondrial respiratory function is essential for spatial remote memory in mice. Mol Brain. 2008;1:21.

8. Kasahara A, Ishikawa K, Yamaoka M, Ito M, Watanabe N, Akimoto M, et al. Generation of trans-mitochondrial mice carrying homoplasmic mtDNAs with a missense mutation in a structural gene using ES cells. Hum Mol Genet. 2006;15(6):871–81.
9. Lin CS, Sharpley MS, Fan W, Waymire KG, Sadun AA, Carelli V, et al. Mouse mtDNA mutant model of Leber hereditary optic neuropathy. Proc Natl Acad Sci U S A. 2012;109(49):20065–70.
10. Shimizu A, Mito T, Hayashi C, Ogasawara E, Koba R, Negishi I, et al. Transmitochondrial mice as models for primary prevention of diseases caused by mutation in the tRNA(Lys) gene. Proc Natl Acad Sci U S A. 2014;111(8):3104–9.
11. Trifunovic A, Wredenberg A, Falkenberg M, Spelbrink JN, Rovio AT, Bruder CE, et al. Premature ageing in mice expressing defective mitochondrial DNA polymerase. Nature. 2004;429:417–23.
12. Ross JM, Coppotelli G, Hoffer BJ, Olson L. Maternally transmitted mitochondrial DNA mutations can reduce lifespan. Sci Rep. 2014;4:6569.
13. Tyynismaa H, Mjosund KP, Wanrooij S, Lappalainen I, Ylikallio E, Jalanko A, et al. Mutant mitochondrial helicase Twinkle causes multiple mtDNA deletions and a late-onset mitochondrial disease in mice. Proc Natl Acad Sci U S A. 2005;102(49):17687–92.
14. Song L, Shan Y, Lloyd KC, Cortopassi GA. Mutant Twinkle increases dopaminergic neurodegeneration, mtDNA deletions and modulates Parkin expression. Hum Mol Genet. 2012;21(23):5147–58.
15. Sorensen L, Ekstrand M, Silva JP, Lindqvist E, Xu B, Rustin P, et al. Late-onset corticohippocampal neurodepletion attributable to catastrophic failure of oxidative phosphorylation in MILON mice. J Neurosci. 2001;21(20):8082–90.
16. Ekstrand MI, Terzioglu M, Galter D, Zhu S, Hofstetter C, Lindqvist E, et al. Progressive parkinsonism in mice with respiratory-chain-deficient dopamine neurons. Proc Natl Acad Sci U S A. 2007;104(4):1325–30.
17. Zhou X, Solaroli N, Bjerke M, Stewart JB, Rozell B, Johansson M, et al. Progressive loss of mitochondrial DNA in thymidine kinase 2-deficient mice. Hum Mol Genet. 2008;17(15):2329–35.
18. Akman HO, Dorado B, Lopez LC, Garcia-Cazorla A, Vila MR, Tanabe LM, et al. Thymidine kinase 2 (H126N) knockin mice show the essential role of balanced deoxynucleotide pools for mitochondrial DNA maintenance. Hum Mol Genet. 2008;17(16):2433–40.
19. Nishino I, Spinazzola A, Hirano M. Thymidine phosphorylase gene mutations in MNGIE, a human mitochondrial disorder. Science. 1999;283(5402):689–92.
20. Lopez LC, Akman HO, Garcia-Cazorla A, Dorado B, Marti R, Nishino I, et al. Unbalanced deoxynucleotide pools cause mitochondrial DNA instability in thymidine phosphorylase-deficient mice. Hum Mol Genet. 2009;18(4):714–22.
21. Haraguchi M, Tsujimoto H, Fukushima M, Higuchi I, Kuribayashi H, Utsumi H, et al. Targeted deletion of both thymidine phosphorylase and uridine phosphorylase and consequent disorders in mice. Mol Cell Biol. 2002;22(14):5212–21.
22. Kruse SE, Watt WC, Marcinek DJ, Kapur RP, Schenkman KA, Palmiter RD. Mice with mitochondrial complex I deficiency develop a fatal encephalomyopathy. Cell Metab. 2008;7(4):312–20.
23. Quintana A, Kruse SE, Kapur RP, Sanz E, Palmiter RD. Complex I deficiency due to loss of Ndufs4 in the brain results in progressive encephalopathy resembling Leigh syndrome. Proc Natl Acad Sci U S A. 2010;107(24):10996–1001.
24. Sterky FH, Hoffman AF, Milenkovic D, Bao B, Paganelli A, Edgar D, et al. Altered dopamine metabolism and increased vulnerability to MPTP in mice with partial deficiency of mitochondrial complex I in dopamine neurons. Hum Mol Genet. 2012;21(5):1078–89.
25. Peralta S, Torraco A, Wenz T, Garcia S, Diaz F, Moraes CT. Partial complex I deficiency due to the CNS conditional ablation of Ndufa5 results in a mild chronic encephalopathy but no increase in oxidative damage. Hum Mol Genet. 2014;23(6):1399–412.
26. Garcia-Corzo L, Luna-Sanchez M, Doerrier C, Garcia JA, Guaras A, Acin-Perez R, et al. Dysfunctional Coq9 protein causes predominant encephalomyopathy associated with CoQ deficiency. Hum Mol Genet. 2013;22(6):1233–48.

27. Diaz F, Garcia S, Padgett KR, Moraes CT. A defect in the mitochondrial complex III, but not complex IV, triggers early ROS-dependent damage in defined brain regions. Hum Mol Genet. 2012;21(23):5066–77.
28. Dell'agnello C, Leo S, Agostino A, Szabadkai G, Tiveron C, Zulian A, et al. Increased longevity and refractoriness to Ca(2+)-dependent neurodegeneration in Surf1 knockout mice. Hum Mol Genet. 2007;16(4):431–44.
29. Funfschilling U, Supplie LM, Mahad D, Boretius S, Saab AS, Edgar J, et al. Glycolytic oligodendrocytes maintain myelin and long-term axonal integrity. Nature. 2012;485(7399):517–21.
30. Tiranti V, Viscomi C, Hildebrandt T, Di Meo I, Mineri R, Tiveron C, et al. Loss of ETHE1, a mitochondrial dioxygenase, causes fatal sulfide toxicity in ethylmalonic encephalopathy. Nat Med. 2009;15(2):200–5.
31. Tuppen HA, Blakely EL, Turnbull DM, Taylor RW. Mitochondrial DNA mutations and human disease. Biochim Biophys Acta. 2010;1797(2):113–28.
32. Nakada K, Sato A, Hayashi J. Reverse genetic studies of mitochondrial DNA-based diseases using a mouse model. Proc Jpn Acad Ser B Phys Biol Sci. 2008;84(5):155–65.
33. Seibel P, Trappe J, Villani G, Klopstock T, Papa S, Reichmann H. Transfection of mitochondria: strategy towards a gene therapy of mitochondrial DNA diseases. Nucleic Acids Res. 1995;23(1):10–7.
34. Vestweber D, Schatz G. DNA-protein conjugates can enter mitochondria via the protein import pathway. Nature. 1989;338:170–2.
35. Moraes CT, Bacman SR, Williams SL. Manipulating mitochondrial genomes in the clinic: playing by different rules. Trends Cell Biol. 2014;24(4):209–11.
36. Moraes CT. A magic bullet to specifically eliminate mutated mitochondrial genomes from patients' cells. EMBO Mol Med. 2014;6(4):434–5.
37. Pinto M, Moraes CT. Mitochondrial genome changes and neurodegenerative diseases. Biochim Biophys Acta. 2014;1842(8):1198–207.
38. Sligh JE, Levy SE, Waymire KG, Allard P, Dillehay DL, Nusinowitz S, et al. Maternal germline transmission of mutant mtDNAs from embryonic stem cell-derived chimeric mice. Proc Natl Acad Sci U S A. 2000;97(26):14461–6.
39. Larsson NG, Wang J, Wilhelmsson H, Oldfors A, Rustin P, Lewandoski M, et al. Mitochondrial transcription factor A is necessary for mtDNA maintenance and embryogenesis in mice. Nat Genet. 1998;18(3):231–6.
40. Nakada K, Sato A, Sone H, Kasahara A, Ikeda K, Kagawa Y, et al. Accumulation of pathogenic DeltamtDNA induced deafness but not diabetic phenotypes in mito-mice. Biochem Biophys Res Commun. 2004;323(1):175–84.
41. Irwin MH, Johnson LW, Pinkert CA. Isolation and microinjection of somatic cell-derived mitochondria and germline heteroplasmy in transmitochondrial mice. Transgenic Res. 1999;8(2):119–23.
42. Nakada K, Inoue K, Ono T, Isobe K, Ogura A, Goto YI, et al. Inter-mitochondrial complementation: mitochondria-specific system preventing mice from expression of disease phenotypes by mutant mtDNA. Nat Med. 2001;7(8):934–9.
43. Nakada K, Sato A, Yoshida K, Morita T, Tanaka H, Inoue S, et al. Mitochondria-related male infertility. Proc Natl Acad Sci U S A. 2006;103(41):15148–53.
44. Jackson DC. Surviving extreme lactic acidosis: the role of calcium lactate formation in the anoxic turtle. Respir Physiol Neurobiol. 2004;144(2–3):173–8.
45. Grady JP, Campbell G, Ratnaike T, Blakely EL, Falkous G, Nesbitt V, et al. Disease progression in patients with single, large-scale mitochondrial DNA deletions. Brain. 2014;137(Pt 2):323–34.
46. Berio A, Piazzi A. Multiple endocrinopathies (growth hormone deficiency, autoimmune hypothyroidism and diabetes mellitus) in Kearns-Sayre syndrome. Pediatr Med Chir. 2013;35(3):137–40.
47. Varlamov DA, Kudin AP, Vielhaber S, Schroder R, Sassen R, Becker A, et al. Metabolic consequences of a novel missense mutation of the mtDNA CO I gene. Hum Mol Genet. 2002;11(16):1797–805.

48. Wallace DC, Singh G, Lott MT, Hodge JA, Schurr TG, Lezza AMS, et al. Mitochondrial DNA mutation associated with Leber's hereditary optic neuropathy. Science. 1988;242:1427–30.
49. Sadun AA, La Morgia C, Carelli V. Leber's hereditary optic neuropathy. Curr Treat Options Neurol. 2011;13(1):109–17.
50. Yokota M, Shitara H, Hashizume O, Ishikawa K, Nakada K, Ishii R, et al. Generation of trans-mitochondrial mito-mice by the introduction of a pathogenic G13997A mtDNA from highly metastatic lung carcinoma cells. FEBS Lett. 2010;584(18):3943–8.
51. Hashizume O, Shimizu A, Yokota M, Sugiyama A, Nakada K, Miyoshi H, et al. Specific mitochondrial DNA mutation in mice regulates diabetes and lymphoma development. Proc Natl Acad Sci U S A. 2012;109(26):10528–33.
52. Korhonen JA, Pham XH, Pellegrini M, Falkenberg M. Reconstitution of a minimal mtDNA replisome in vitro. EMBO J. 2004;23(12):2423–9.
53. Edgar D, Trifunovic A. The mtDNA mutator mouse: dissecting mitochondrial involvement in aging. Aging (Albany NY). 2009;1(12):1028–32.
54. Ross JM, Oberg J, Brene S, Coppotelli G, Terzioglu M, Pernold K, et al. High brain lactate is a hallmark of aging and caused by a shift in the lactate dehydrogenase A/B ratio. Proc Natl Acad Sci U S A. 2010;107(46):20087–92.
55. Ahlqvist KJ, Hamalainen RH, Yatsuga S, Uutela M, Terzioglu M, Gotz A, et al. Somatic progenitor cell vulnerability to mitochondrial DNA mutagenesis underlies progeroid phenotypes in Polg mutator mice. Cell Metab. 2012;15(1):100–9.
56. Suomalainen A, Kaukonen J. Diseases caused by nuclear genes affecting mtDNA stability. Am J Med Genet. 2001;106(1):53–61.
57. Van Goethem G, Dermaut B, Lofgren A, Martin JJ, Van Broeckhoven C. Mutation of POLG is associated with progressive external ophthalmoplegia characterized by mtDNA deletions. Nat Genet. 2001;28(3):211–2.
58. Lamantea E, Tiranti V, Bordoni A, Toscano A, Bono F, Servidei S, et al. Mutations of mitochondrial DNA polymerase gammaA are a frequent cause of autosomal dominant or recessive progressive external ophthalmoplegia. Ann Neurol. 2002;52(2):211–9.
59. Agostino A, Valletta L, Chinnery PF, Ferrari G, Carrara F, Taylor RW, et al. Mutations of ANT1, Twinkle, and POLG1 in sporadic progressive external ophthalmoplegia (PEO). Neurology. 2003;60(8):1354–6.
60. Van Goethem G, Schwartz M, Lofgren A, Dermaut B, Van Broeckhoven C, Vissing J. Novel POLG mutations in progressive external ophthalmoplegia mimicking mitochondrial neurogastrointestinal encephalomyopathy. Eur J Hum Genet. 2003;11(7):547–9.
61. Naviaux RK, Nguyen KV. POLG mutations associated with Alpers' syndrome and mitochondrial DNA depletion. Ann Neurol. 2004;55(5):706–12.
62. Ferrari G, Lamantea E, Donati A, Filosto M, Briem E, Carrara F, et al. Infantile hepatocerebral syndromes associated with mutations in the mitochondrial DNA polymerase-{gamma} A. Brain. 2005;128(Pt 4):723–31. Epub 2005 Feb 2.
63. Luoma P, Melberg A, Rinne JO, Kaukonen JA, Nupponen NN, Chalmers RM, et al. Parkinsonism, premature menopause, and mitochondrial DNA polymerase gamma mutations: clinical and molecular genetic study. Lancet. 2004;364(9437):875–82.
64. Suomalainen A. Mitochondrial DNA, and disease. Ann Med. 1997;29(3):235–46.
65. Gellhaar S, Marcellino D, Abrams MB, Galter D. Chronic L-DOPA induces hyperactivity, normalization of gait and dyskinetic behavior in MitoPark mice. Genes Brain Behav. 2015;14(3):260–70.
66. Grauer SM, Hodgson R, Hyde LA. MitoPark mice, an animal model of Parkinson's disease, show enhanced prepulse inhibition of acoustic startle and no loss of gating in response to the adenosine A(2A) antagonist SCH 412348. Psychopharmacology (Berl). 2014;231(7):1325–37.
67. Li X, Redus L, Chen C, Martinez PA, Strong R, Li S, et al. Cognitive dysfunction precedes the onset of motor symptoms in the MitoPark mouse model of Parkinson's disease. PLoS One. 2013;8(8):e71341.

68. Ferraro P, Pontarin G, Crocco L, Fabris S, Reichard P, Bianchi V. Mitochondrial deoxynucleotide pools in quiescent fibroblasts: a possible model for mitochondrial neurogastrointestinal encephalomyopathy (MNGIE). J Biol Chem. 2005;280(26):24472–80.
69. Saada A, Shaag A, Mandel H, Nevo Y, Eriksson S, Elpeleg O. Mutant mitochondrial thymidine kinase in mitochondrial DNA depletion myopathy. Nat Genet. 2001;29(3):342–4.
70. Oskoui M, Davidzon G, Pascual J, Erazo R, Gurgel-Giannetti J, Krishna S, et al. Clinical spectrum of mitochondrial DNA depletion due to mutations in the thymidine kinase 2 gene. Arch Neurol. 2006;63(8):1122–6.
71. Hirano M, Lagier-Tourenne C, Valentino ML, Marti R, Nishigaki Y. Thymidine phosphorylase mutations cause instability of mitochondrial DNA. Gene. 2005;354:152–6.
72. Joza N, Oudit GY, Brown D, Benit P, Kassiri Z, Vahsen N, et al. Muscle-specific loss of apoptosis-inducing factor leads to mitochondrial dysfunction, skeletal muscle atrophy, and dilated cardiomyopathy. Mol Cell Biol. 2005;25(23):10261–72.
73. Hirst J, Carroll J, Fearnley IM, Shannon RJ, Walker JE. The nuclear encoded subunits of complex I from bovine heart mitochondria. Biochim Biophys Acta. 2003;1604(3):135–50.
74. Loeffen JL, Smeitink JA, Trijbels JM, Janssen AJ, Triepels RH, Sengers RC, et al. Isolated complex I deficiency in children: clinical, biochemical and genetic aspects. Hum Mutat. 2000;15(2):123–34.
75. Morris AAM, Leonard IV, Brown GK, Bidouki SK, Bindoff LA, Woodward CE, et al. Deficiency of respiratory chain complex I is a common cause of Leigh disease. Ann Neurol. 1996;40(1):25–30.
76. Anderson SL, Chung WK, Frezzo J, Papp JC, Ekstein J, DiMauro S, et al. A novel mutation in NDUFS4 causes Leigh syndrome in an Ashkenazi Jewish family. J Inherit Metab Dis. 2008;31 Suppl 2:S461–7.
77. Betts J, Lightowlers RN, Turnbull DM. Neuropathological aspects of mitochondrial DNA disease. Neurochem Res. 2004;29(3):505–11.
78. Leong DW, Komen JC, Hewitt CA, Arnaud E, McKenzie M, Phipson B, et al. Proteomic and metabolomic analyses of mitochondrial complex I-deficient mouse model generated by spontaneous B2 short interspersed nuclear element (SINE) insertion into NADH dehydrogenase (ubiquinone) Fe-S protein 4 (Ndufs4) gene. J Biol Chem. 2012;287(24):20652–63.
79. Ingraham CA, Burwell LS, Skalska J, Brookes PS, Howell RL, Sheu SS, et al. NDUFS4: creation of a mouse model mimicking a Complex I disorder. Mitochondrion. 2009;9(3):204–10.
80. Guzman M, Blazquez C. Ketone body synthesis in the brain: possible neuroprotective effects. Prostaglandins Leukot Essent Fatty Acids. 2004;70(3):287–92.
81. Turunen M, Olsson J, Dallner G. Metabolism and function of coenzyme Q. Biochim Biophys Acta. 2004;1660(1–2):171–99.
82. Luna-Sanchez M, Diaz-Casado E, Barca E, Tejada MA, Montilla-Garcia A, Cobos EJ, et al. The clinical heterogeneity of coenzyme Q10 deficiency results from genotypic differences in the Coq9 gene. EMBO Mol Med. 2015;7(5):670–87.
83. Emmanuele V, Lopez LC, Berardo A, Naini A, Tadesse S, Wen B, et al. Heterogeneity of coenzyme Q10 deficiency: patient study and literature review. Arch Neurol. 2012;69(8):978–83.
84. Duncan AJ, Bitner-Glindzicz M, Meunier B, Costello H, Hargreaves IP, Lopez LC, et al. A nonsense mutation in COQ9 causes autosomal-recessive neonatal-onset primary coenzyme Q10 deficiency: a potentially treatable form of mitochondrial disease. Am J Hum Genet. 2009;84(5):558–66.
85. Benit P, Lebon S, Rustin P. Respiratory-chain diseases related to complex III deficiency. Biochim Biophys Acta. 2009;1793(1):181–5.
86. Arnold S. Cytochrome c oxidase and its role in neurodegeneration and neuroprotection. Adv Exp Med Biol. 2012;748:305–39.
87. Shoubridge EA. Cytochrome c oxidase deficiency. Am J Med Genet. 2001;106(1):46–52.
88. Sacconi S, Salviati L, Sue CM, Shanske S, Davidson MM, Bonilla E, et al. Mutation screening in patients with isolated cytochrome c oxidase deficiency. Pediatr Res. 2003;53(2):224–30.

89. Pulliam DA, Deepa SS, Liu Y, Hill S, Lin AL, Bhattacharya A, et al. Complex IV-deficient Surf1(–/–) mice initiate mitochondrial stress responses. Biochem J. 2014;462(2):359–71.
90. Deepa SS, Pulliam D, Hill S, Shi Y, Walsh ME, Salmon A, et al. Improved insulin sensitivity associated with reduced mitochondrial complex IV assembly and activity. FASEB J. 2013;27(4):1371–80.
91. Valnot I, von Kleist-Retzow JC, Barrientos A, Gorbatyuk M, Taanman JW, Mehaye B, et al. A mutation in the human heme a: farnesyltransferase gene (COX10) causes cytochrome c oxidase deficiency. Hum Mol Genet. 2000;9(8):1245–9 [In Process Citation].
92. Tiranti V, D'Adamo P, Briem E, Ferrari G, Mineri R, Lamantea E, et al. Ethylmalonic encephalopathy is caused by mutations in ETHE1, a gene encoding a mitochondrial matrix protein. Am J Hum Genet. 2004;74(2):239–52.

Chapter 14
Mitochondrial Function and Dynamics Imaged In Vivo

Kim Chisholm, Fabian Peters, Dimitra G. Schiza, Mona Sadeghian, and Kenneth Smith

Abstract The previous chapters have described the extraordinary depth of knowledge of mitochondrial biology revealed by in vitro observations where the environment can be closely controlled. However, in recent years there has been increased interest in the study of mitochondria in vivo, where their properties can be studied with high spatial and temporal resolution while ensuring that key factors such as the oxygen and glucose concentrations are physiologically accurate. Advances facilitating such in vivo research include improved microscope systems and mitochondrially targeted dyes, as well as a wide range of transgenic animals expressing fluorescent proteins. Such in vivo observations provide a more realistic picture of mitochondrial involvement in health and disease and also offer the potential to reveal novel targets for therapeutic interventions. For example, loss of mitochondrial membrane potential and alterations in mitochondrial morphology and trafficking have been reported in mouse models of multiple sclerosis and Alzheimer's disease, and redox potential changes have been reported during (patho)physiological changes in oxygen supply and demand.

In this chapter we summarise some techniques used in imaging of mitochondria in vivo, followed by a summary of key findings and recent advances in the study of mitochondrial function and dynamics. We aim to provide insight into the benefits and limitations of intravital imaging of mitochondria in the nervous system.

Keywords Intravital • Confocal • Two-photon • Microscopy • Neurodegeneration • Mitochondrial trafficking • Membrane potential • Fluorescent probes • Redox potential

---

K. Chisholm, BSc (✉) • F. Peters, MSc • D.G. Schiza, MSc • K. Smith, BSc, PhD
Department of Neuroinflammation, University College London Institute of Neurology, London, UK
e-mail: k.chisholm@ucl.ac.uk

M. Sadeghian, PhD
Department of Experimental Neuroinflammation, University College London Institute of Neurology, London, UK

A.K. Reeve et al. (eds.), *Mitochondrial Dysfunction in Neurodegenerative Disorders*, DOI 10.1007/978-3-319-28637-2_14

14.1 Introduction

The ability to study mitochondrial properties in vivo provides an additional perspective to our understanding of mitochondrial biology gained from in vitro observations. When studied in their natural environment in vivo, the factors governing mitochondrial functionality, while not known in detail, are at, or close to, their physiological values. This advantage is especially valuable when considering crucial factors such as oxygen, the concentration of which is key to mitochondrial function, but which is also temporally and spatially highly variable within the central nervous system (CNS). The advantages presented by in vivo techniques have allowed observations that are not possible in vitro, including the study of how mitochondria behave with changes in vascular supply, physiological stimulation and response to pathology, such as inflammation [1–6].

In vivo imaging of mitochondria is necessarily a broad topic and many techniques to probe mitochondrial health in vivo have been developed. We aim to provide the reader with an idea of the potentials and pitfalls of in vivo visualisation of mitochondrial function and dynamics by reviewing some of the work employing these techniques and briefly describing some of the findings.

14.1.1 Arrangements for In Vivo Imaging

Confocal and two-photon microscopy are the techniques most commonly used for in vivo imaging [7]. The development of two-photon microscopy has given investigators the opportunity to perform long-term imaging of tissues as deep as 1.5 mm in the brain [8, 9], with a high sampling frequency and negligible risk of phototoxicity or photobleaching outside the area of interest [10].

When performing intravital imaging, movement artefacts due to cardiac and respiratory movements can be a considerable obstacle as even small shifts can dramatically deteriorate the spatial resolution. A number of strategies have been developed to address and minimise these artefacts and the most widely used include: (1) the construction of specific organ holders which allow the exposure and appropriate positioning of the organ of interest with minimal compromise to tissue integrity and physiology [11–13]; (2) the synchronisation of image acquisition (time lapses or z-stacks) with the animal's heartbeat and respiration, via tracheal intubation and ventilation at a desired rate [14, 15]; and (3) the restraint of movement originating from muscle contraction and pulsation through a combination of tissue immobilisation and deeper anaesthesia [16]. The remaining movements can sometimes be corrected using post-acquisition image registration [17, 18]. The aforementioned approaches have made in vivo imaging of cellular and subcellular structures feasible, even in tissues highly susceptible to movement artefacts, such as in the heart, lungs and vessel walls of rodents [14, 15, 19].

Although the use of in vivo techniques provides the best opportunity to make observations under physiological conditions, persisting limitations include the consequences of any surgery and the anaesthetic used. Additionally, despite the depth of imaging that can be achieved with some preparations, deeper tissue structures remain difficult to visualise. However, advancements are continuously being made, including chronic in vivo imaging techniques, which allow longitudinal assessment [20, 21], even without anaesthetic [22, 23], or the use of in vivo microscopy on human subjects, for example, imaging the skin or the retina [24, 25].

14.2 Visualising Mitochondrial Function In Vivo

14.2.1 Mitochondrial Membrane Potential

Healthy mitochondria exhibit an electrochemical membrane potential which is required for ATP production, Ca^{2+} buffering and the production of reactive oxygen species. The dependence of mitochondria on their membrane potential ($\Delta\psi_m$) to fulfil their pivotal functions qualifies this measure as a fundamental indicator of mitochondrial 'health' and metabolic activity [26]. The study of mitochondrial membrane polarisation in vivo has been facilitated by the development of cationic lipophilic fluorescent dyes that reveal the $\Delta\psi_m$ and change over time. The dyes are readily sequestered by mitochondria in a $\Delta\psi_m$-dependent manner, i.e. the more negative the membrane potential, the more dye is accumulated within mitochondria [26] (Fig. 14.1). Two dyes that are commonly used are TMRM and TMRE (tetramethylrhodamine, methyl ester/ethyl ester) and these are not only fast equilibrating and highly permeant potentiometric dyes, but also perhaps the least toxic, and they have been successfully used to assess the mitochondrial membrane potential in intravital imaging studies of the nervous system under different experimental paradigms. For example, in multiple sclerosis models, such as experimental autoimmune encephalomyelitis (EAE), the in vivo application of TMRM has provided much insight into possible disease mechanisms. EAE, usually induced by the injection of myelin antigens, including various myelin proteins, or by the injection of T cells reactive to such antigens, can lead to ascending paralysis, sensory disturbances and axonal damage, mimicking some of the defects observed in multiple sclerosis patients. The use of TMRM together with the transgenic expression of cyan fluorescent protein (CFP) inside a subset of axonal mitochondria (in *Thy1-MitoCFP* mice) allowed Nikic et al. [1] to report swollen and depolarised mitochondria in axons traversing through inflammatory foci in the spinal cord of anaesthetised mice immunised with a purified myelin protein. These changes were observed in intact axons, suggesting mitochondrial deficits as an early factor in the pathophysiology of EAE, preceding structural axonal damage and demyelination [1]. TMRM intensity was determined in individual mitochondria after being normalised to their CFP fluorescence and the mean TMRM signal of morphologically healthy-appearing

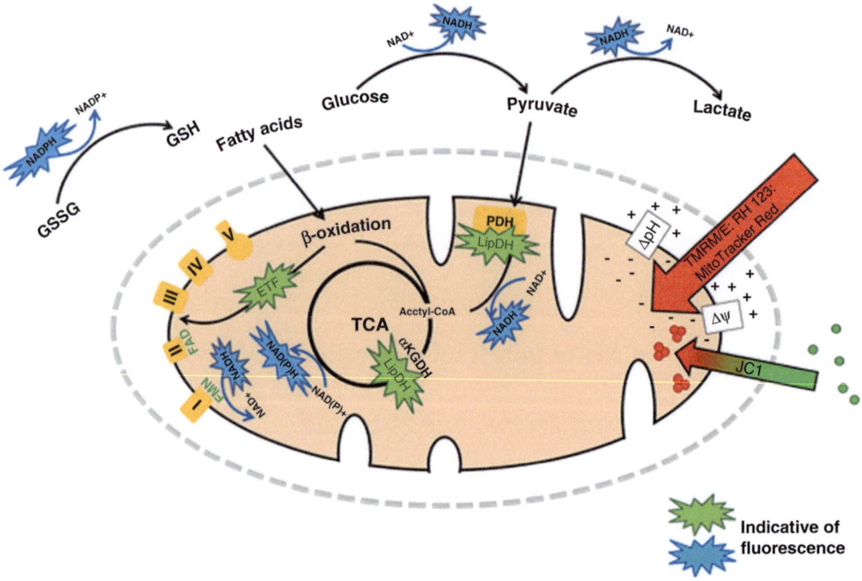

Fig. 14.1 Visualising mitochondrial function. TMRM/E, MitoTracker Red and JC-1 accumulate in mitochondria by virtue of the mitochondrial membrane potential. TMRM/E and MitoTracker Red fluorescence in mitochondria therefore increase when the mitochondrial matrix is more negative while JC-1 exhibits a spectral shift (from *green* to *red* fluorescence) upon accumulation in mitochondria and formation of aggregates. Flavoproteins (*green stars*) and pyridine nucleotides (*blue stars*) are autofluorescent redox carriers involved in mitochondrial and some cytosolic reactions. *Abbreviations*: *GSH* reduced glutathione, *GSSG* oxidised glutathione, *FMN* flavin mononucleotide, *FAD* flavin dinucleotide, *ETF* electron transfer flavoproteins, *LipDH* lipoamide dehydrogenase, *αKGDH* alpha ketoglutarate dehydrogenase, *PDH* pyruvate dehydrogenase complex, *TCA* tricarboxylic acid cycle

mitochondria in each image, to adjust for mitochondrial size and labelling intensity, respectively. This study was further able to suggest a critical role for reactive oxygen and nitrogen species (ROS and NOS) in the mitochondrial and axonal damage observed in this neuroinflammatory model by showing that external application of a nitric oxide donor or H_2O_2 was capable of initiating mitochondrial and axonal damage akin to that seen in EAE, including mitochondrial and axonal swelling and axonal fragmentation, and that scavenging these reactive mediators was capable of attenuating axonal damage in the EAE model.

Whereas Nikic et al. have examined the link between mitochondrial function and axonal degeneration in EAE, we have examined the changes in mitochondrial function and linked them to the expression of neurological disability over the course of this rapidly progressing disease, mapping the mitochondrial changes to the onset, remission and relapse of disability [5]. Confocal imaging of the spinal cord revealed that whereas in naïve mice all axonal mitochondria were polarised as shown by their bright TMRM fluorescence, in the days preceding disease expression in EAE, there was a progressive loss of TMRM fluorescence in approximately half the axonal mitochondria. Despite this accumulating damage, the animals appeared neurologically normal, at least when judged by simple observation of walking behaviour. However, the onset of limb and tail weakness was accompanied by a sharp increase in the magnitude of mitochondrial dysfunction, now affecting 79 % of axonal mitochondria. Mitochondrial function then improved in tandem with remission and worsened again during relapse in chronic disease. The time course of mitochondrial dysfunction therefore mapped quite precisely to the expression of the neurological deficit, providing an adequate explanation for the deficit which traditionally would have been attributed to the concurrent demyelination observed, at least in chronically affected animals [5].

Similar changes in mitochondrial function and morphology, and also mass, have been observed in the brains of mice in a model of Alzheimer's disease [3]. Xie and colleagues employed a battery of mitochondrial dyes to assess mitochondria in vivo in the vicinity of amyloid plaques. The dyes were applied topically or pressure injected into the cortex, and organelle-specific fluorescence was achieved by targeting GFP into the mitochondrial matrix. The authors found that mitochondrial mass was dramatically decreased within 16 μm from the border of the plaques, with the remaining mitochondria in the area characterised by an abnormal, swollen morphology. Membrane potential sensitive dyes (TMRE, JC-1 and MitoTracker Red; see Fig. 14.1) further revealed mitochondrial depolarisation extending as far as 20 μm from the plaque boundaries, which was fully recovered with increasing distance (>50 μm), suggesting that amyloid plaque deposits are associated with mitochondrial dysfunction. These changes were hypothesised to result from the diffusion of toxic molecules from plaques or, possibly more indirectly, through activation of microglia.

In common with TMRM, JC-1 and MitoTracker probes accumulate in a membrane potential-dependent manner. However, in contrast to TMRM, MitoTracker probes can, once sequestered into mitochondria, be retained within them, even after a loss in the membrane potential, by virtue of a thiol-reactive chloromethyl moiety [27]. They could therefore be particularly suitable as stable colabels with, for example, TMRM, when mitochondrial depolarisation is induced during the experimental protocol or when final fixation and/or permeabilisation steps are required. Differently from the other probes, accumulation of JC1 inside mitochondria results in the formation of aggregates and a spectral shift (from green to red), thereby providing a ratiometric measurement. However, as the spectral shift relies on the formation of aggregates, which is a threshold effect, this dye may not be suitable when investigating subtle or graded changes in the membrane potential. Additionally, long equilibration times for the formation of aggregates, which also depends on the surface-to-volume ratio of cellular and subcellular regions [28], might make this probe incompatible with in vivo work.

When applied to the exposed CNS, these dyes nondiscriminately label all surface mitochondria. However it has been possible to focus the accumulation of dyes to a particular subset of CNS cells. In particular, TMRM injected systemically targets astrocytic mitochondria more efficiently [29]. Thus it has been found that following the intravenous injection of TMRM, the dye is capable of crossing the blood-brain barrier and it preferentially labels the mitochondria in cortical astrocytes (confirmed by sulforhodamine 101 co-labelling, which is considered an astrocyte-specific dye [30]) presumably due to the association of astrocyte end feet with cortical blood vessels. Using this technique Zhen et al. were able to demonstrate that their previously observed neuroprotective effect of $P2Y_1R$ activation (a primarily astrocyte-specific G-protein-coupled receptor) in photothrombotic ischaemic brain damage was correlated with enhanced mitochondrial metabolism in astrocytes. The investigators observed astrocytic mitochondrial depolarisation of around 30 mV following ischaemic damage. This depolarisation of astrocytic mitochondria, together with the associated ischaemia-induced dendritic damage, was attenuated after $P2Y_1R$ ligand administration [29]. The findings suggest a neuroprotective role for enhanced astrocytic mitochondrial metabolism during ischaemic brain damage.

It is important to keep in mind, however, that while TMRM is a very versatile mitochondrial dye, it too is not without caveats. The dye's fluorescence depends significantly on its concentration: at low doses, the number of TMRM molecules taken up by mitochondria and their cumulative fluorescence are roughly linearly related. Such concentrations are known as non-quenching [31, 32]. As long as TMRM is used in non-quenching mode, $\Delta\psi_m$ can be estimated from mitochondrial fluorescence relative to the axoplasm and compared across different mitochondrial populations or animals ([33] Chap. 9). If, however, the dye concentration exceeds a certain threshold, the photon emissions of the densely packed TMRM particles within mitochondria begin to interfere and their cumulative fluorescence relative to the axoplasm decreases. In this setting, known as quench mode, the linearity between fluorescence and number of TMRM particles taken up into the mitochondrion is lost which makes the estimation of mitochondrial membrane potential impossible. On the other hand, loading mitochondria with large quantities of TMRM can be useful to monitor acute depolarisations, which lead to very pronounced increases in whole cell fluorescence as previously quenched molecules (i.e. particles which did not add to the mitochondrial and cellular brightness) leave the mitochondrion and increase the fluorescence of the axoplasm [32, 34].

In practice, concentrations of TMRM used in vivo tend to be significantly higher than in vitro (e.g. 7.5 nM in vitro [35] vs 600 nM in vivo [36]). Hence, it is necessary to carefully calibrate the loading of TMRM in order to ensure that the investigator operates in the desired quenching or non-quenching mode. We have found that an established in vitro protocol can be adapted to estimate an appropriate TMRM concentration for in vivo application in axons. The H+ ionophore FCCP rapidly depolarises mitochondria, thus provoking the release of TMRM into the axoplasm. As described above, this increases whole cell fluorescence if the concentration of TMRM used is high enough to induce quenching [31, 33]. No such increase can be seen in non-quench mode, since the mitochondrial depolarisation simply causes a redistribution of unquenched TMRM particles. The quenching threshold can thus be established by applying 10 μM FCCP to preparations with different concentrations of TMRM, starting with a high dose of, e.g. TMRM [31]. Quenching concentrations produce a very obvious spike in whole axon fluorescence upon application of FCCP due to release of TMRM into the axoplasm (Fig. 14.2). The amplitude of

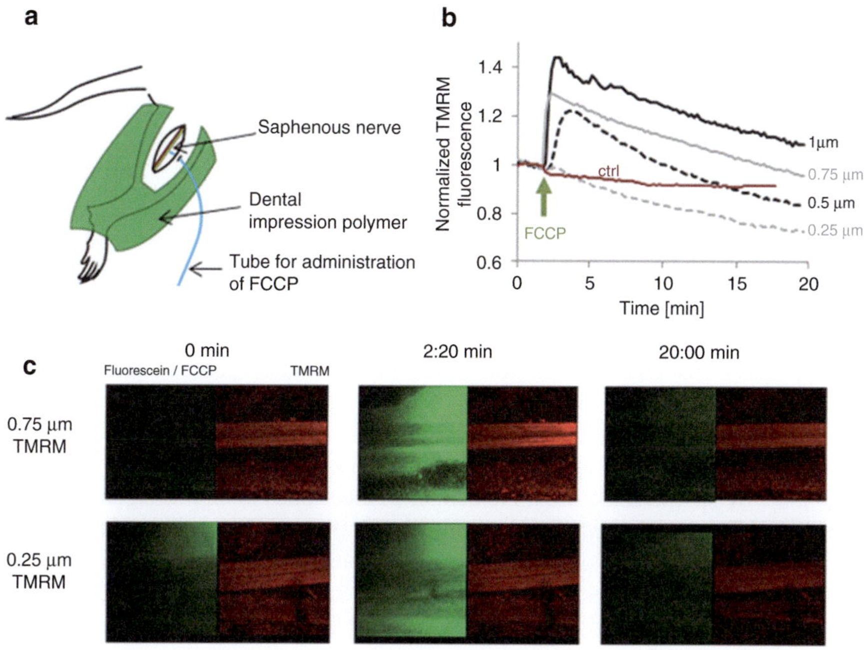

Fig. 14.2 TMRM dose calibration in vivo. (**a**) Setup: The saphenous nerve is exposed, desheathed and imaged under a glass cover slip surrounded by different concentrations of TMRM in sterile saline. After 2 min, 10 μM FCCP together with 0.5 μM fluorescein (for visualisation of the injection) is administered through a thin tube in the skin. (**b**) Whole nerve TMRM fluorescence normalised to baseline. Upon application of FCCP, a rapid spike in TMRM intensity is visible for 1 μM, 0.75 μM and 0.5 μM TMRM which slowly resolves over time. No such spike is seen when using 0.25 μM TMRM. The *red line* indicates a control experiment with 1 μM TMRM (the maximum dose used) and an injection of fluorescein, but no FCCP, demonstrating that there is no significant bleed through between the red and green channel. (**c**) Motion corrected time-lapse images at the beginning of the experiment, just after injection of FCCP and fluorescein and after 20 min. The result suggests a quenching threshold somewhere between 0.25 and 0.5 μM

the spike depends on the TMRM concentration and is absent at non-quenching doses. In the case of the saphenous nerve, we found that TMRM at a 0.25 μM concentration is suitable for non-quenching application (see Fig. 14.2).

14.2.2 Mitochondrial Redox Potential

Since the Britton Chance and coworkers' seminal work in the 1950s and 1960s, it has been known that the redox state of mitochondria can be assessed using the endogenous fluorescence of redox cofactors involved in oxidative phosphorylation [37, 38] (see Fig. 14.1). Within mitochondria, pyridine nucleotides (nicotinamide adenine dinucleotide (phosphate) NAD(P)/NAD(P)H) and flavoproteins act as electron acceptors/donors in various reactions (see Fig. 14.1), thereby undergoing redox transitions depending on mitochondrial function and substrate availability. Reduced pyridine nucleotides fluoresce when excited using UV light (single-photon excitation around 310–370 nm, two-photon excitation around 700–710 nm and emission around 410–510 nm [2, 39, 40]), while oxidised flavoproteins fluoresce when excited with blue light (excitation around 405–500 nm and emission around 520–560 nm [40, 41]). The respective oxidised and reduced counterparts of these molecules do not have such fluorescent properties. These excitation and emission properties provide the opportunity for intravital visualisation of the mitochondrial redox potential using standard fluorescence and laser-based microscopy [e.g. 37, 38] (see Fig. 14.1 for more details).

By virtue of the close coupling of neuronal activity and aerobic metabolism [42–44], in vivo flavoprotein fluorescence has been successfully used as a proxy for oxidative metabolism and neuronal activation [45–50], as confirmed by calcium imaging [48], optical imaging of intrinsic signals [49, 51–53] and recordings of field potentials (which represent the combined electrophysiological signals of neurons located within the recoding range) [51–53], without the need for the addition of dyes. In fact Weber et al. found that flavoprotein fluorescence is spatially and temporally more sensitive for the localisation of cortical activity following peripheral stimulation compared with changes in blood flow [54], which are the basis of functional magnetic resonance imaging.

For example, a series of experiments found that stimulation of parallel fibres in the cerebellar cortex resulted in a transient 'beam-like' increase in flavoprotein fluorescence followed by a delayed reduction in the fluorescence [45–58]. The mitochondrial origin of this signal was confirmed by its response to the mitochondrial inhibitors cyanide (NaCN) and diphenyleneiodonium (DPI), which abolished both phases of the signal without changing neuronal excitability [47].

In addition to the beam-like biphasic signal, Gao et al. also described an 'off-beam' dark phase which was organised into parasagittal bands along and across the on-beam signal [48]. This signal was suggested to originate from molecular layer inhibitory interneurons and was found to have potential clinical significance in a model of spinocerebellar ataxia type 8 [55]. Here the off-beam decrease in signal after parallel fibre stimulation was attenuated, while the on-beam increase in flavoprotein fluorescence was enhanced, suggesting dysfunctional inhibition by cerebellar interneurons and dysfunctional cerebellar circuitry [55].

The use of physiological, in vivo preparations with an intact blood supply and vascular network, together with the spatial and temporal resolution of NAD(P)H/flavoprotein fluorescence, is of particular value when assessing conditions with an oxygen supply/demand imbalance. For example, the marked increase in cerebral oxygen consumption associated with cortical spreading depression [56] was found to result in a distinct pattern of NAD(P)H redox state, characterised by a global drop in fluorescence followed by an increase in fluorescence, only at some distance from vessels [57]. Visualisation of pyridine nucleotides therefore revealed a relationship between redox state and the vascular network in conditions of high oxygen demand

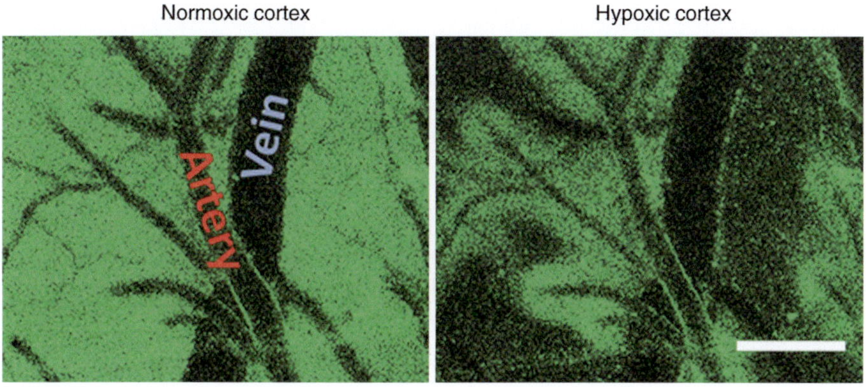

Fig. 14.3 Flavoprotein fluorescence in the hypoxic cerebral cortex. The exposed cortex of an anaesthetised mouse imaged for endogenous flavoprotein fluorescence during normoxic and hypoxic conditions, induced by breathing 5 % oxygen. As hypoxaemia is induced, the oxygen delivered via the arteries is no longer sufficient to supply the entire brain and rather it only supplies a small 'halo' of functioning mitochondria around the arteries, as indicated by preserved flavoprotein fluorescence. Scale bar = 200 μm

which can outstrip the rate of oxygen delivery and reveal hypoxic areas by enhanced pyridine nucleotide fluorescence [57].

A similar relationship between mitochondrial redox potential and cortical microcirculation is seen during reduced oxygen delivery [2, 41]. In vivo imaging of the murine cerebral cortex using NAD(P)H [2] and flavoprotein fluorescence [41] during hypoxaemia has revealed the limits of effective oxygen diffusion from the cortical vasculature. Hypoxia and hypoxaemia, which have been associated with numerous neurological conditions, including traumatic brain injury [58], stroke [59], multiple sclerosis [60] and systemic inflammation [6], resulted in the appearance of hypoxic zones, characterised by an increase in NAD(P)H fluorescence [2] and a decrease in flavoprotein fluorescence [41]. In contrast, areas located close to arteries showed prolonged protection from hypoxaemia, as assessed by preservation of flavoprotein and absence of NAD(P)H fluorescence, due to the proximity to a vascular oxygen supply (e.g. Fig. 14.3). These investigations show that flavoprotein and pyridine nucleotide fluorescence can be used as a proxy for cortical oxygenation by differentiating areas that are sufficiently oxygenated to support functional mitochondria from areas with reduced oxygen supply, unable to support oxidative metabolism.

Interestingly, localised and reversible inactivation of mitochondrial and neuronal function has been achieved in vivo by prolonged exposure to blue light, which can cause reversible inactivation of flavoproteins [61]. This however also has implications on the reliability and effect of prolonged imaging of any autofluorescent molecules, particularly flavoproteins. Laser intensities and duration of imaging should be kept as low as possible and appropriate controls should be included to ensure limited damage.

Despite the relatively weak signal produced by flavoprotein and pyridine nucleotide autofluorescence, such methods provide numerous benefits for the assessment of mitochondrial function, oxidative metabolism and neuronal integrity. For example, in the absence of externally applied dyes, loading problems, including distribution and toxicity, are avoided and the endogenous fluorescence of electron carriers can provide a stable signal over large areas of the CNS.

14.3 Visualising Mitochondrial Dynamics In Vivo

In recent years there has been an increased interest in the dynamics of mitochondria within cells and it has become clear that mitochondria go through a complex life cycle including biogenesis, transport, fusion, fission and mitophagy. This system of continuous turnover is widely believed to be necessary for the maintenance of overall mitochondrial health and to counteract the damage to proteins and mtDNA that accumulates over time, especially in postmitotic cells such as neurons.

Work on cultured cells has been invaluable in revealing the mechanisms underlying mitochondrial movement due to the relative ease that in vitro methods allow for changes in the intracellular and extracellular milieu and for the overexpression and silencing of genes. However, while in vitro work has done much to uncover the molecular basis of mitochondrial movement, the consequences of changes in these dynamics, due, for example, to ageing and pathology, have been harder to assess in culture. Axons in the human sciatic nerve, for example, can extend over 1 m and age over 100 years. While transport from the soma to the axon terminal may take only a few hours in culture, it would take approximately 3 weeks in long axons in vivo assuming uninterrupted transport with an average speed of 0.5 μm/s (a realistic average at least in anaesthetised mice [4, 62]) and so the issues faced in vivo may not arise in vitro. Conversely, properties observed in vitro may be overrepresented when compared to their counterparts in vivo. For example, neurons in culture are typically young and have axons that are usually still growing or regenerating, and such axons are believed to have significantly increased mitochondrial transport [63, 64] and fission and fusion compared to mature axons [65–67]. With regard to ageing and neurodegeneration, fission and fusion are often viewed as critically important to prevent the accumulation of mtDNA mutations, but studies in mice and humans have shown that even with a high preexisting mutation load from birth [68] or a radical reduction of total mitochondrial mass due to antiretroviral therapy [69], it can take weeks or years for any functional deficits to appear – a timescale which far exceeds what is realistically achievable in culture. A possible solution exists in the form of recent induced pluripotent stem cell (iPSC)-based techniques which enable the study of human or murine cells with damaged mtDNA derived from aged fibroblasts [70, 71]. However, while this approach has fascinating applications for the study of genetic diseases such as Leigh syndrome, it is not suitable for investigating some other types of mtDNA mutations, for example, those arising in axons due to local inflammatory processes.

Although observations in vivo can offer advantages to those made in vitro, there are considerable obstacles to overcome. Imaging in live animals is, for example, often hampered by movement artefacts due to breathing and heartbeat, and it can be more difficult to track mitochondria when they are very close to one another [72, 73], a problem that is exacerbated in vivo due to the comparatively higher number of static mitochondria compared with in vitro experiments. Hence, time-lapse sequences must be recorded with a high frame rate and resolution, with the attendant risk of causing damage to the tissue [73]. Furthermore, the majority of publications on mitochondrial dynamics in vitro are based on manual analyses of kymographs [74–78], but due to the size of axons and density of stationary mitochondria in vivo, the kymographs tend to be much more complicated to interpret (see, e.g. the kymographs in [62] versus those in [79]). This complication can affect the choice of tracking algorithm: where only a rough summary measure of anterograde and retrograde transport is needed, a simple nearest neighbour-based algorithm may suffice [4, 80], but details such as starting and stopping behaviour can require more complicated approaches such as single-particle tracking [72, 73, 81] or even manual analysis [82]. A comprehensive comparison of particle tracking algorithms as well as some guidance on how to evaluate them can be found in [83].

An interesting result from the study of mitochondrial dynamics in vivo is that, even in healthy animals, anterograde and retrograde transport are not balanced, and significantly more mitochondrial mass is moved towards the axon terminal in both resting [62, 64, 82, 84] and stimulated/stressed axons [4, 63]. This observation is important in the context of mitophagy. In vitro studies have mostly focused on PARKIN-mediated mitophagy of damaged and depolarised mitochondria and they have produced partially conflicting evidence indicating either transport of damaged mitochondria back to the soma [74, 85] or their degradation in the axon [86, 87]. However, these in vivo observations argue for a combined mechanism with degradation of mitochondria both in the soma and either along the axon or near the axon terminal [84].

An obvious advantage of in vivo observations is the possibility to go beyond naïve animals and investigate mitochondrial dynamics in complex diseases. In one of the earlier examples, Bilsland et al. showed that mitochondrial trafficking in both directions was reduced in a superoxide dismutase 1 mutant model of amyotrophic lateral sclerosis (ALS) [62]. While peak speeds appeared to be identical, mitochondria were found to stop more often and for longer durations. Similarly, in a recent paper investigating mitochondrial trafficking in the inflamed spinal cord in an animal model of multiple sclerosis, transport was found to be significantly reduced in both directions, largely due to an increase in the duration of stops [82]. In both cases the trafficking deficit could be found in normal-appearing axons, suggesting that it preceded neurodegeneration.

In theory, disturbance of mitochondrial transport can either be a consequence of, or contribute to, the damage of mitochondria. This difference was recently demonstrated by a study by Magrane et al. investigating the time course of mitochondrial damage in different ALS mutations [88]. Both the mutation $SOD1^{G93A}$ and $TDP43^{A315T}$ displayed a pattern of reduced retrograde transport, followed by reduced anterograde transport, before the onset of rapid axonal degeneration. However, these changes in trafficking preceded morphological abnormalities of mitochondria in TDP43^{A315T} but followed morphological changes in SOD1^{G93A}, hinting at different patterns of mitochondrial dysfunction.

In addition to pointing out reductions in mitochondrial trafficking, in vivo imaging has also highlighted the controlled increase of transport and relocation of mitochondria due to electrical activity. In 2013, Sajic et al. demonstrated that mitochondrial movement could be significantly increased in peripheral sensory axons by both electrical and chemical stimulation [4]. Impulse conduction at physiological frequencies was found to increase the speed of moving mitochondria by up to 100 % as well as induce fission of stationary particles, thereby supplying additional small mitochondria for transport. Anterograde transport was particularly affected and the result was a relocation of mitochondria towards the peripheral sensory terminals which the authors speculated may be necessary to satisfy a high energy demand at active axon terminals. This finding is in striking contrast to earlier in vitro observations in excised nerves [78] and cultured slices [89] in which stimulation leads to a decrease in transport.

Similar changes in mitochondrial trafficking have also been associated with axonal repair mechanisms. For example, using acute nerve explants, Misgeld et al. demonstrated that anterograde mitochondrial transport was rapidly upregulated 12 h after transection of the intercostal nerve and remained upregulated for weeks during nerve repair [64]. Using the same technique, Mar et al. later found that while peripheral lesions had this effect on mitochondrial transport in both the peripheral and central branches of dorsal root ganglion axons, central lesions did not elicit any such response in the periphery [63]. This intriguing result may be relevant to understanding why central axons generally fail to repair after injury and why peripheral conditioning can restore their capacity to regenerate [90].

Together with the controlled upregulation of axonal transport, anchoring of mitochondria has also been found to be an important mechanism for the maintenance of axonal health. Using knockout mice, Ohno et al. recently identified that syntaphilin is necessary to increase mitochondrial content in demyelinated axons in vivo and that its absence leads to an increase in axonal beading and degeneration [91].

Thus, a picture emerges which highlights the importance of locating mitochondria in axonal compartments with suspected high energy demand. In this context, it is especially interesting that when Sorbara et al. investigated spinal cord dorsal root axons downstream of a site of prolonged trafficking deficit in a chronic EAE model, they found them to be depleted of mitochondria [82]. While most of these axons were not yet undergoing degeneration, it is intriguing to speculate how they might cope if subjected to an increased energy demand.

In conclusion, while the number of publications on mitochondrial dynamics in vivo is still limited, the advantages of in vivo preparations have enabled a new understanding of how mitochondrial dynamics are affected by physiological and pathological influences.

14.4 Conclusion

Despite the complexity of in vivo experimental protocols, recent advances in surgical techniques, microscopy equipment and fluorescent markers have greatly facilitated the assessment of mitochondrial function in vivo. The ability to tag mitochondria using fluorescent markers has allowed the study of mitochondrial dynamics under physiological conditions, revealing new insights into mitochondrial movement and life cycle in the nervous system. Additionally, the ability to visualise mitochondria even without external dye application, using the autofluorescent properties of electron carriers, has provided the opportunity to relate mitochondrial function, neuronal activity and the vasculature in the superficial layers of the CNS in vivo. The benefits of intravital imaging have become particularly valuable in the study of mitochondrial biology under pathological conditions, where the supply of substrates and oxygen can be disturbed in unknown and unpredictable ways.

References

1. Nikić I, Merkler D, Sorbara C, Brinkoetter M, Kreutzfeldt M, Bareyre FM, et al. A reversible form of axon damage in experimental autoimmune encephalomyelitis and multiple sclerosis. Nat Med. 2011;17(4):495–9.
2. Kasischke KA, Lambert EM, Panepento B, Sun A, Gelbard HA, Burgess RW, et al. Two-photon NADH imaging exposes boundaries of oxygen diffusion in cortical vascular supply regions. J Cereb Blood Flow Metab. 2011;31(1):68–81.
3. Xie H, Guan J, Borrelli LA, Xu J, Serrano-Pozo A, Bacskai BJ. Mitochondrial alterations near amyloid plaques in an Alzheimer's disease mouse model. J Neurosci. 2013;33(43):17042–51.
4. Sajic M, Mastrolia V, Lee CY, Trigo D, Sadeghian M, Mosley AJ, et al. Impulse conduction increases mitochondrial transport in adult mammalian peripheral nerves in vivo. Barres BA, editor. PLoS Biol. 2013; 11(12):e1001754.
5. Sadeghian M, Rezaei Haddad A, Mastrolia V, Schiza D, Marija S, Duchen M, et al. Loss of mitochondrial function and impaired trafficking at the onset of neurological deficits in a neuroinflammatory model of multiple sclerosis (MS) studied by in vivo confocal imaging. Mult Scler J. 2013;19:46–7.
6. Chisholm KI, Ida KK, Davies AL, Tachtsidis I, Papkovsky DB, Dyson A, et al. Hypothermia protects brain mitochondrial function from hypoxemia in a murine model of sepsis. J Cereb Blood Flow Metab. 1–10.
7. Weigert R, Porat-Shliom N, Amornphimoltham P. Imaging cell biology in live animals: ready for prime time. J Cell Biol. 2013;201(7):969–79.
8. Theer P, Hasan MT, Denk W. Two-photon imaging to a depth of 1000 microns in living brains by use of a Ti:Al2O3 regenerative amplifier. Opt Lett. 2003;28(12):1022–4.
9. Levene MJ, Dombeck D, Kasischke K, Molloy RP, Webb WW. In vivo multiphoton microscopy of deep brain tissue. J Neurophysiol. 2004;91(4):1908–12.
10. Benninger RKP, Piston DW. Two-photon excitation microscopy for the study of living cells and tissues. Curr Protoc Cell Biol. 2013; Chap. 4: Unit 4.11.1–24.
11. Masedunskas A, Milberg O, Porat-shliom N, Sramkova M, Wigand T, Amornphimoltham P, et al. Intravital microscopy: a practical guide on imaging intracellular structures in live animals. Bioarchitecture. 2012;2(5):143–57.

12. Davalos D, Akassoglou K. In vivo imaging of the mouse spinal cord using two-photon microscopy. J Vis Exp. 2012;59:1–5.
13. Davalos D, Lee JK, Smith WB, Brinkman B, Ellisman MH, Zheng B, et al. Stable in vivo imaging of densely populated glia, axons and blood vessels in the mouse spinal cord using two-photon microscopy. J Neurosci Methods. 2008;169(1):1–7.
14. Li W, Nava RG, Bribriesco AC, Zinselmeyer BH, Spahn JH, Gelman AE, et al. Intravital 2-photon imaging of leukocyte trafficking in beating heart. J Clin Invest. 2012;122(7):2499–508.
15. Presson RG, Brown MB, Fisher AJ, Sandoval RM, Dunn KW, Lorenz KS, et al. Two-photon imaging within the murine thorax without respiratory and cardiac motion artifact. Am J Pathol. 2011;179(1):75–82.
16. Pittet MJ, Weissleder R. Intravital imaging. Cell. 2011;147(5):983–91.
17. Thévenaz P, Ruttimann UE, Unser M. A pyramid approach to subpixel registration based on intensity. IEEE Trans Image Process. 1998;7(1):27–41.
18. Preibisch S, Saalfeld S, Schindelin J, Tomancak P. Software for bead-based registration of selective plane illumination microscopy data. Nat Methods. 2010;7(6):418–9.
19. Megens RT, Reitsma S, Prinzen L, Oude Egbrink MG, Engels W, Leenders PJ, et al. In vivo high-resolution structural imaging of large arteries in small rodents using two-photon laser scanning microscopy. J Biomed Opt. 2010;15(1):011108.
20. Holtmaat A, Bonhoeffer T, Chow DK, Chuckowree J, Paola D, Hofer SB, et al. Long-term, high-resolution imaging in the mouse neocortex through a chronic cranial window. Nat Protoc. 2009;4(8):1128–44.
21. Yang G, Pan F, Parkhurst CN, Grutzendler J, Gan W-B. Thinned-skull cranial window technique for long-term imaging of the cortex in live mice. Nat Protoc. 2010;5(2):201–8.
22. Helmchen F, Denk W, Kerr JND. Miniaturization of two-photon microscopy for imaging in freely moving animals. Cold Spring Harb Protoc. 2013;10:904–13.
23. Dombeck DA, Khabbaz AN, Collman F, Adelman TL, Tank DW. Imaging large-scale neural activity with cellular resolution in awake, mobile mice. Neuron. 2007;56(1):43–57.
24. Masters BR, So PT, Gratton E. Multiphoton excitation fluorescence microscopy and spectroscopy of in vivo human skin. Biophys J. 1997;72(6):2405–12.
25. Webb RH, Hughes GW, Delori FC. Confocal scanning laser ophthalmoscope. Appl Opt. 1987;26(8):1492–9.
26. Perry SW, Norman JP, Barbieri J, Brown EB, Gelbard H. Mitochondrial membrane potential probes and the proton gradient: a practical usage guide. Biotechniques. 2011;50(2):98–115.
27. Poot M, Zhang YZ, Kramer JA, Wells KS, Jones LJ, Hanzel DK, et al. Analysis of mitochondrial morphology and function with novel fixable fluorescent stains. J Histochem Cytochem. 1996;44(12):1363–72.
28. Mathur A. Evaluation of fluorescent dyes for the detection of mitochondrial membrane potential changes in cultured cardiomyocytes. Cardiovasc Res. 2000;46(1):126–38.
29. Zheng W, Talley Watts L, Holstein DM, Wewer J, Lechleiter JD. P2Y1R-initiated, IP3R-dependent stimulation of astrocyte mitochondrial metabolism reduces and partially reverses ischemic neuronal damage in mouse. J Cereb Blood Flow Metab. 2013;33(4):600–11.
30. Nimmerjahn A, Kirchhoff F, Kerr JND, Helmchen F. Sulforhodamine 101 as a specific marker of astroglia in the neocortex in vivo. Nat Methods. 2004;1(1):31–7.
31. Nicholls DG. Fluorescence measurement of mitochondrial membrane potential changes in cultured cells. In: Palmeira CM, Moreno AJ, editors. Methods in molecular biology. Totowa: Humana Press; 2012. p. 119–33.
32. Nicholls DG, Ward MW. Mitochondrial membrane potential and neuronal glutamate excitotoxicity: mortality and millivolts. Trends Neurosci. 2000;23(4):166–74.
33. Nicholls DG, Ferguson S. Bioenergetics. 4th ed. Philadelphia: Elsevier Science; 2013.
34. Brand MD, Nicholls DG. Assessing mitochondrial dysfunction in cells. Biochem J. 2011;435(2):297–312.
35. Gerencser A, Chinopoulos C, Birket MJ, Jastroch M, Vitelli C, Nicholls DG, et al. Quantitative measurement of mitochondrial membrane potential in cultured cells: calcium-induced de- and hyperpolarization of neuronal mitochondria. J Physiol. 2012;590(Pt 12):2845–71.

36. Romanelli E, Sorbara CD, Nikić I, Dagkalis A, Misgeld T, Kerschensteiner M. Cellular, subcellular and functional in vivo labeling of the spinal cord using vital dyes. Nat Protoc. 2013;8(3):481–90.
37. Chance B, Jöbsis F. Changes in fluorescence in a frog sartorius muscle following a twitch. Nature. 1959;184:195–6.
38. Chance B, Schoener B. Fluorometric studies of flavin component of the respiratory chain. Flavins and flavoproteins. 1966. 510 p.
39. Blacker TS, Mann ZF, Gale JE, Ziegler M, Bain AJ, Szabadkai G, et al. Separating NADH and NADPH fluorescence in live cells and tissues using FLIM. Nat Commun. 2014;5:3936.
40. Chance B, Schoener B, Oshino R, Itshak F, Nakase Y. Oxidation-reduction ratio studies of mitochondria in freeze-trapped Samples – NADH and flavoprotein fluorescence signals. J Biol Chem. 1979;254(11):4764–71.
41. Chisholm KI, Ida KK, Davies AL, Papkovsky DB, Singer M, Dyson A, et al. In vivo imaging of flavoprotein fluorescence during hypoxia reveals the importance of direct arterial oxygen supply to cerebral cortex tissue. Oxyg Transp to Tissue XXXVII Adv Exp Med. Chap. 30, 233–41.
42. Fein A, Tsacopoulos M. Activation of mitochondrial oxidative-metabolism by calcium-ions in limulus ventral photoreceptor. Nature. 1988;331(6155):437–40.
43. Shibuki K. Calcium-dependent and ouabain-resistant oxygen-consumption in the rat neurohypophysis. Brain Res. 1989;487(1):96–104.
44. Kann O, Schuchmann S, Buchheim K, Heinemann U. Coupling of neuronal activity and mitochondrial metabolism as revealed by NAD(P)H fluorescence signals in organotypic hippocampal slice cultures of the rat. Neuroscience. 2003;119(1):87–100.
45. Reinert KC, Gao WC, Chen G, Ebner TJ. Flavoprotein autofluorescence imaging in the cerebellar cortex in vivo. J Neurosci Res. 2007;85(15):3221–32.
46. Reinert KC, Gao WC, Chen G, Wang XM, Peng YP, Ebner TJ. Cellular and metabolic origins of flavoprotein autofluorescence in the cerebellar cortex in vivo. Cerebellum. 2011;10(3):585–99.
47. Reinert KC, Dunbar RL, Gao WC, Chen G, Ebner TJ. Flavoprotein autofluorescence imaging of neuronal activation in the cerebellar cortex in vivo. J Neurophysiol. 2004;92(1):199–211.
48. Gao WC, Chen G, Reinert KC, Ebner TJ. Cerebellar cortical molecular layer inhibition is organized in parasagittal zones. J Neurosci. 2006;26(32):8377–87.
49. Husson TR, Mallik AK, Zhang JX, Issa NP. Functional imaging of primary visual cortex using flavoprotein autofluorescence. J Neurosci. 2007;27(32):8665–75.
50. Shibuki K, Hishida R, Murakami H, Kudoh M, Kawaguchi T, Watanabe M, et al. Dynamic imaging of somatosensory cortical activity in the rat visualized by flavoprotein autofluorescence. J Physiol. 2003;549(3):919–27.
51. Takahashi K, Hishida R, Kubota Y, Kudoh M, Takahashi S, Shibuki K. Transcranial fluorescence imaging of auditory cortical plasticity regulated by acoustic environments in mice. Eur J Neurosci. 2006;23(5):1365–76.
52. Tohmi M, Kitaura H, Komagata S, Kudoh M, Shibuki K. Enduring critical period plasticity visualized by transcranial flavoprotein imaging in mouse primary visual cortex. J Neurosci. 2006;26(45):11775–85.
53. Kitaura H, Uozumi N, Tohmi M, Yamazaki M, Sakimura K, Kudoh M, et al. Roles of nitric oxide as a vasodilator in neurovascular coupling of mouse somatosensory cortex. Neurosci Res. 2007;59(2):160–71.
54. Weber B, Burger C, Wyss MT, von Schulthess GK, Scheffold F, Buck A. Optical imaging of the spatiotemporal dynamics of cerebral blood flow and oxidative metabolism in the rat barrel cortex. Eur J Neurosci. 2004;20(10):2664–70.
55. Moseley ML, Zu T, Ikeda Y, Gao WC, Mosemiller AK, Daughters RS, et al. Bidirectional expression of CUG and CAG expansion transcripts and intranuclear polyglutamine inclusions in spinocerebellar ataxia type 8. Nat Genet. 2006;38(7):758–69.
56. Mayevsky A, Weiss HR. Cerebral blood flow and oxygen consumption in cortical spreading depression. J Cereb Blood Flow Metab. 1991;11(5):829–36.

57. Takano T, Tian GF, Peng WG, Lou NH, Lovatt D, Hansen AJ, et al. Cortical spreading depression causes and coincides with tissue hypoxia. Nat Neurosci. 2007;10(6):754–62.
58. Chang JJJ, Youn TS, Benson D, Mattick H, Andrade N, Harper CR, et al. Physiologic and functional outcome correlates of brain tissue hypoxia in traumatic brain injury. Crit Care Med. 2009;37(1):283–90.
59. Read SJ, Hirano T, Abbott DF, Markus R, Sachinidis JI, Tochon-Danguy HJ, et al. The fate of hypoxic tissue on18F-fluoromisonidazole positron emission tomography after ischemic stroke. Ann Neurol. 2000;48(2):228–35.
60. Davies AL, Desai RA, Bloomfield PS, McIntosh PR, Chapple KJ, Linington C, et al. Neurological deficits caused by tissue hypoxia in neuroinflammatory disease. Ann Neurol. 2013;74(6):815–25.
61. Kubota Y, Kamatani D, Tsukano H, Ohshima S, Takahashi K, Hishida R, et al. Transcranial photo-inactivation of neural activities in the mouse auditory cortex. Neurosci Res. 2008;60(4):422–30.
62. Bilsland LG, Sahai E, Kelly G, Golding M, Greensmith L, Schiavo G. Deficits in axonal transport precede ALS symptoms in vivo. Proc Natl Acad Sci U S A. 2010;107(47):20523–8.
63. Mar FM, Simões AR, Leite S, Morgado MM, Santos TE, Rodrigo IS, et al. CNS axons globally increase axonal transport after peripheral conditioning. J Neurosci. 2014;34(17):5965–70.
64. Misgeld T, Kerschensteiner M, Bareyre FM, Burgess RW, Lichtman JW. Imaging axonal transport of mitochondria in vivo. Nat Methods. 2007;4(7):559–61.
65. Jendrach M, Pohl S, Vöth M, Kowald A, Hammerstein P, Bereiter-Hahn J. Morpho-dynamic changes of mitochondria during ageing of human endothelial cells. Mech Ageing Dev. 2005;126:813–21.
66. Arnold B, Cassady SJ, VanLaar VS, Berman SB. Integrating multiple aspects of mitochondrial dynamics in neurons: age-related differences and dynamic changes in a chronic rotenone model. Neurobiol Dis. 2011;41(1):189–200.
67. Figge MT, Reichert AS, Meyer-Hermann M, Osiewacz HD. Deceleration of fusion-fission cycles improves mitochondrial quality control during aging. PLoS Comput Biol. 2012;8(6), e1002576.
68. Ross JM, Stewart JB, Hagström E, Brené S, Mourier A, Coppotelli G, et al. Germline mitochondrial DNA mutations aggravate ageing and can impair brain development. Nature. 2013;501:412–5.
69. Payne BA, Wilson IJ, Hateley C, Horvath R, Santibanez-Koref M, Samuels DC, et al. Mitochondrial aging is accelerated by anti-retroviral therapy through the clonal expansion of mtDNA mutations. Nat Genet. 2011;43(8):806–10.
70. Ma H, Folmes CDL, Wu J, Morey R, Mora-Castilla S, Ocampo A, et al. Metabolic rescue in pluripotent cells from patients with mtDNA disease. Nature. 2015;7564:234–8.
71. Wahlestedt M, Ameur A, Moraghebi R, Norddahl GL, Sten G, Woods N-B, et al. Somatic cells with a heavy mitochondrial DNA mutational load render induced pluripotent stem cells with distinct differentiation defects. Stem Cells. 2014;32(5):1173–82.
72. Jaqaman K, Loerke D, Mettlen M, Kuwata H, Grinstein S, Schmid SL, et al. Robust single-particle tracking in live-cell time-lapse sequences. Nat Methods. 2008;5(8):695–702.
73. Jaqaman K, Danuser G. Computational image analysis of cellular dynamics: a case study based on particle tracking. Cold Spring Harb Protoc. 2009;2009(12):pdb.top65.
74. Miller KE, Sheetz MP. Axonal mitochondrial transport and potential are correlated. J Cell Sci. 2004;117(Pt 13):2791–804.
75. Zambonin JL, Zhao C, Ohno N, Campbell GR, Engeham S, Ziabreva I, et al. Increased mitochondrial content in remyelinated axons: implications for multiple sclerosis. Brain. 2011;134(Pt 7):1901–13.
76. Zala D, Hinckelmann M-V, Yu H, da Cunha MM L, Liot G, Cordelières FP, et al. Vesicular glycolysis provides on-board energy for fast axonal transport. Cell. 2013;152(3):479–91.
77. Obashi K, Okabe S. Regulation of mitochondrial dynamics and distribution by synapse position and neuronal activity in the axon. Eur J Neurosci. 2013;38(3):2350–63.

78. Zhang, Ho PL, Kintner DB, Sun D, Chiu SY. Activity-dependent regulation of mitochondrial motility by calcium and Na/K-ATPase at nodes of Ranvier of myelinated nerves. J Neurosci. 2010;30(10):3555–66.
79. Kiryu-Seo S, Ohno N, Kidd GJ, Komuro H, Trapp BD. Demyelination increases axonal stationary mitochondrial size and the speed of axonal mitochondrial transport. J Neurosci. 2010;30(19):6658–66.
80. Andrews S, Gilley J, Coleman MP. Difference tracker: imageJ plugins for fully automated analysis of multiple axonal transport parameters. J Neurosci Methods. 2010;193(2):281–7.
81. Winter MR, Fang C, Banker G, Roysam B, Cohen AR. Axonal transport analysis using Multitemporal Association Tracking. Int J Comput Biol Drug Des. 2012;5(1):35–48.
82. Sorbara CD, Wagner NE, Ladwig A, Nikić I, Merkler D, Kleele T, et al. Pervasive axonal transport deficits in multiple sclerosis models. Neuron. 2014;84(6):1–8.
83. Chenouard N, Smal I, de Chaumont F, Maška M, Sbalzarini IF, Gong Y, et al. Objective comparison of particle tracking methods. Nat Methods. 2014;11(3):281–90.
84. Plucinska G, Paquet D, Hruscha A, Godinho L, Haass C, Schmid B, et al. In vivo imaging of disease-related mitochondrial dynamics in a vertebrate model system. J Neurosci. 2012;32(46):16203–12.
85. Cai Q, Zakaria HM, Simone A, Sheng Z-H. Spatial parkin translocation and degradation of damaged mitochondria via mitophagy in live cortical neurons. Curr Biol. 2012;22(6):545–52.
86. Ashrafi G, Schlehe JS, LaVoie MJ, Schwarz TL. Mitophagy of damaged mitochondria occurs locally in distal neuronal axons and requires PINK1 and Parkin. J Cell Biol. 2014; 206(5):655–70.
87. Wang X, Winter D, Ashrafi G, Schlehe J, Wong YL, Selkoe D, et al. PINK1 and Parkin target miro for phosphorylation and degradation to arrest mitochondrial motility. Cell. 2011;147(4):893–906.
88. Magrané J, Cortez C, Gan WB, Manfredi G. Abnormal mitochondrial transport and morphology are common pathological denominators in SOD1 and TDP43 ALS mouse models. Hum Mol Genet. 2014;23(6):1413–24.
89. Ohno N, Kidd GJ, Mahad D, Kiryu-Seo S, Avishai A, Komuro H, et al. Myelination and axonal electrical activity modulate the distribution and motility of mitochondria at CNS nodes of Ranvier. J Neurosci. 2011;31(20):7249–58.
90. Neumann S, Woolf CJ. Regeneration of dorsal column fibers into and beyond the lesion site following adult spinal cord injury. Neuron. 1999;23:83–91.
91. Ohno N, Chiang H, Mahad DJ, Kidd GJ, Liu L, Ransohoff RM, et al. Mitochondrial immobilization mediated by syntaphilin facilitates survival of demyelinated axons. Proc Natl Acad Sci U S A. 2014;111(27):1–6.

Part V
The Future

Chapter 15
Development of Treatments and Therapies to Target Mitochondrial Dysfunction

Stephen B. Helliwell

Abstract Mitochondrial diseases affect 1:10,000 people worldwide, with a further 1:6,000 at risk of developing symptoms, and can be caused by mutations in the nuclear genome or by mutations within/deletion of mitochondrial DNA (mtDNA). There is currently no therapy for patients with these diseases. In addition, mitochondrial dysfunction can be considered contributory or causal for other neurological diseases, such that therapies that improve mitochondrial function may have wide-reaching benefits. This chapter summarizes cell- and animal-based attempts to find a therapy by modulating major mitochondrial pathways including mitochondrial biogenesis, fission, fusion and altering metabolic intermediates, mitophagy and the mammalian target of rapamycin (mTOR) pathway. Clinical trials currently in progress include drugs that are direct or indirect antioxidants, a mitogenesis activator, an antiapoptotic compound and a compound that improves electron transport chain efficiency.

Keywords Leber's hereditary optic neuropathy (LHON) • Mitochondrial encephalomyopathy lactic acidosis and stroke-like episodes (MELAS) • Mitochondrial biogenesis • Mitochondrial fission and fusion • Mitophagy • Antioxidant • PGC1-alpha • AICAR • Resveratrol • NAD+ • Rapamycin • Bezafibrate • Clinical trial

15.1 Introduction

Mitochondria are known as the powerhouse of the eukaryotic cell, utilizing nutrients and oxygen to generate ATP, an absolute requirement for life. They are considered endosymbionts, derived originally from proteobacteria and still carrying the remnants of their own genome, the mitochondrial DNA (mtDNA). Human

S.B. Helliwell, BSc, PhD
Development and Molecular Pathways, Novartis Institutes for BioMedical Research, Basel, Switzerland
e-mail: stephen.helliwell@novartis.com

A.K. Reeve et al. (eds.), *Mitochondrial Dysfunction in Neurodegenerative Disorders*, DOI 10.1007/978-3-319-28637-2_15

mtDNA exists as 1,000s of copies per cell and encodes 13 proteins, 22 tRNAs, and 2 rRNAs. Together they encode 13 proteins that are critical components of the ATP-generating electron transport chain (ETC), with the remaining 77 proteins of the ETC being encoded by the nuclear genome [1, 2] (see Chap. 1). An additional 1,000–1,500 nuclear-encoded genes encode the remaining proteins that make up mammalian mitochondria, representing circa 5 % of the protein encoding genome [3–5]. This complexity of mitochondria highlights the fact that they are not just ATP-generating organelles, but perform many other functions including metabolism, iron storage and iron-sulfur cluster formation, calcium homeostasis, apoptosis, hormone signaling/synthesis and thermogenesis, among others [6]. It is therefore not surprising that mitochondrial dysfunction, via inborn or accumulated genetic lesions or via environmental factors, causes a wide variety of diseases affecting many different organs in the body including the central and peripheral nervous system [2, 7–12].

This chapter will briefly describe the causes of mitochondrial defects and then summarize attempts to find therapeutic agents that ameliorate or cure diseases of mitochondrial dysfunction. It will describe molecular pathways and targets most likely to yield therapies when intersected with pharmacological approaches, including a brief summary of those therapeutic agents that have shown promise in cellular or animal models for a given pathway, finishing with a summary of current clinical advances and activities for mitochondrial diseases. The chapter focuses mainly on mitochondrial dysfunction in neurological disorders that represent a clear unmet medical need, with some reference to other disorders. The great hope of everyone in this challenging field is that once we have successfully ameliorated cellular mitochondrial dysfunction with therapeutic agents, those agents will also have broad utility in all mitochondrial disorders as well as in diseases that have been linked to mitochondrial dysfunction, including neurodegenerative diseases such as Parkinson's disease (PD).

15.2 Mitochondrial Disease

Mutations in mtDNA and nuclear DNA (nDNA) can cause mitochondrial disease, generally understood as diseases caused by defective oxidative phosphorylation. Over 300 different point mutations or deletions in mtDNA have been attributed to human disease [13]. Mutations of nuclear mtDNA maintenance genes can also cause mtDNA deletions and depletion (reduction in copy number). mtDNA mutations are primarily linked to a loss of function of the ETC, and the existence of significant copies of mtDNA leads to an additional complication of mtDNA-related disease, the coexistence of mutant and wild-type mtDNA – known as heteroplasmy [14]. One hundred seventy-four nDNA genes have been functionally linked to mitochondrial disease to date [15], and with the advent of inexpensive whole exome/genome sequencing disease, genes are being uncovered at an ever-increasing rate [16–19].

Together the elucidated and as yet unknown mutations account for mitochondrial disease that has an overall clinical prevalence estimated to be 1:10,000 [20]. This would suggest that there are significant numbers of patients such that clinical trial recruitment should be easy – but the major challenge of mitochondrial diseases is the genetic range and clinical complexity, depending on the mutation(s), heteroplasmy (for mtDNA diseases), age, and environment. For this reason there is much interest in targeting the major pathways regulating mitochondrial biogenesis, fission and fusion, and mitochondrial autophagy (mitophagy), in addition to attempts to directly shift heteroplasmy or to modulate metabolism related directly to the electron transport chain.

15.3 Mitochondrial Biogenesis for Improved Metabolic Function

Perhaps the most widely studied *in vitro* and *in vivo* approach, increasing the total mass of mitochondria is an obvious approach to increasing overall function of a partially defective electron transport chain [21–23]. This will only be useful if a negative aspect of mitochondrial (dys)function is not also amplified, e.g., a defective respiratory chain that increases reactive oxygen species (ROS) to damaging levels.

15.3.1 PGC1-α

This approach pivots primarily around the activation of PGC1-alpha(α), the key transcriptional co-activator protein that regulates nuclear mitochondrial genes [23–26]. PGC1-α interacts with multiple transcription factors to bring about the transcriptional program that produces nuclear-encoded mitochondrial genes [27]. There are multiple routes to PGC1-α activation using the low-molecular-weight compounds (LMWs) resveratrol, 5-aminoimidazole-4-carboxamide ribonucleotide (AICAR), and bezafibrate that target sirtuin (SIRT1) and/or AMP-activated kinase (AMPK) and peroxisome proliferator-activated receptor (PPAR), respectively (Fig. 15.1). These three routes to mitochondrial biogenesis exemplify the use of both genetic and pharmacological *in vivo* experiments in relevant mouse models [21, 23].

Muscle-specific overexpression of muscle PGC1α increased mitogenesis in *cox10*, *Surf1-/-*, and *Polg* "mutator" mice [28–30] (for more detail of these models, see Chap. 13), improved muscle function and lifespan in the *cox10* mouse and improved motor function and cardiac performance in the mutator mouse [28, 30]. In addition, PGC1α has been overexpressed in mouse models of Duchenne muscular dystrophy (DMD), amyotrophic lateral sclerosis (ALS), and a 1-methyl-4-phenyl-

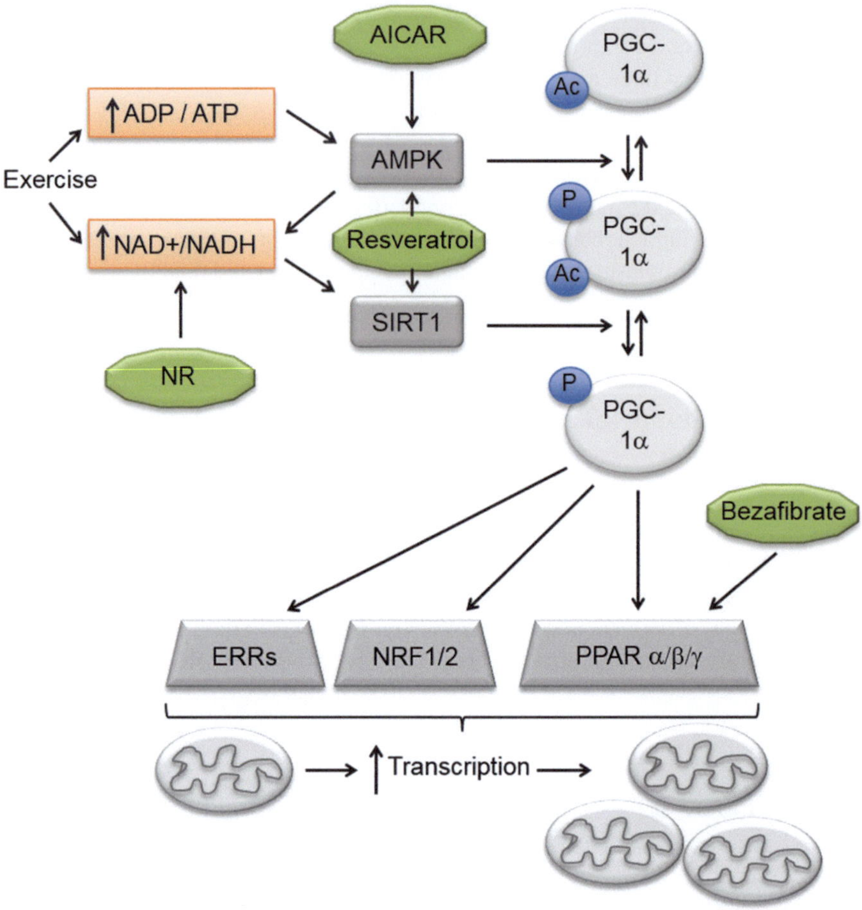

Fig. 15.1 ATP and NADH are depleted during energy expenditure. The subsequent increase in the ADP/ATP ratio is sensed by AMPK that activates PGC-1a. In addition, via NAMPT (not shown), AMPK affects the NAD+/NADH ratio that regulates the deacetylase SIRT1 that fully activates PGC-1a. PGC-1a interacts with groups of transcription factors to increase the transcription of nuclear-encoded and subsequently mitochondrially encoded mitochondrial genes; *ERR* (estrogen-related receptor), *NRF* (nuclear respiratory factor), and *PPAR* (peroxisome proliferator-activating receptor). Compounds that stimulate this pathway are noted in green and are discussed in depth in the text (*NR* nicotinamide riboside)

1,2,3,6-tetrahydropyridine (MPTP)-induced Parkinson's model. Increased mitochondrial biogenesis linked to an improvement in phenotype was shown in some, but not all, cases [31–33]. In particular, in mice treated with MPTP to induce Parkinson's-like phenotypes, PGC1α overexpression through the neuronal-specific thy-1 promoter significantly protected dopaminergic neurons from MPTP-induced degeneration, demonstrating PGC1α's potential utility in ameliorating neuronal phenotypes of mitochondrial disease [33]. Overall there is strong evidence that increased mitochondrial biogenesis via increasing PGC1α activity could have therapeutic utility in mitochondrial disease.

15.3.2 Resveratrol

PGC1α is regulated by AMPK and the deacetylase Sirt1, a member of the sirtuin family (Fig. 15.1). Resveratrol, a natural stilbenoid found in the skin of grapes, was initially proposed as an activator of Sirt1, thereby regulating yeast and metazoan lifespan [34–37]. Sirt1 is a nicotinamide adenine dinucleotide (NAD+)-dependent deacetylase, which is regulated by AMPK activity via altered NAD+/NADH levels. Importantly, NAD+ is also required by the Krebs cycle at multiple steps during oxidative phosphorylation. While resveratrol effects are not necessarily disputed, there is still controversy about the molecular mechanism of action. One view is that resveratrol regulates Sirt1 indirectly via AMPK, regulating NAD+ levels that in turn control Sirt1 activity [38–42]. At the same time resveratrol can act directly on Sirt1 homologues *in vitro* to catalyze a deacetylation reaction on acetyl lysine [34, 43]. Interestingly, it has been proposed that resveratrol may act via tyrosine tRNA synthetase, to activate poly(ADP-ribose) polymerase (PARP-1), a major consumer of NAD+. This appears to contradict the concept that an increase in NAD+ improves mitochondrial function but there is a transient increase in NAD+ levels in mice treated with low doses of resveratrol, before a longer term reduction [44]. This result will require further investigation before the molecular mechanism is truly understood. Nevertheless, resveratrol does increase mitochondrial biogenesis and oxidative respiration in mouse models of obesity and in doing so reduces the development of obesity and retains insulin sensitivity [45, 46]. Based upon this, structurally unrelated compounds have been sought that activate Sirt1 (SRT1720, SRT2183, SRT1460, SRT2104, SRT501) and have been tested in many *in vitro* and *in vivo* situations alongside or independently of resveratrol. In general their effects are consistent with the activation of Sirt1, showing improvement of insulin sensitivity in mouse obesity models [47, 48].

For neurological indications and mitochondrial disease, resveratrol has been tested in mouse models of Friedreich's ataxia, the *mdx* model for DMD, SOD1 ALS models, and models for Alzheimer's disease (AD), Huntington's disease (HD), and PD [21, 49]. Resveratrol was not beneficial in every case. The beneficial mechanisms may or may not be caused solely or partly because of increased mitochondrial function because mitochondrial measurements were not carried out in most of these studies. One exception is the MPTP PD model where PGC1α is a likely efficacy target [33]. Resveratrol has not been tested in any mouse models of direct oxidative phosphorylation defects, but the Friedreich's ataxia study mentioned above does represent an indirect model of mitochondrial function, given that ETC complexes I/II and III contain iron-sulfur clusters and frataxin is required for iron-sulfur cluster synthesis. However in this study there was no data to suggest that resveratrol was having a beneficial effect in terms of the amelioration of disease symptoms [50], but only upon the levels of frataxin. In sum, there is good evidence to suggest that resveratrol could be useful for patients with mitochondrial disorders or neurological dysfunction, although significant analysis of mitochondrial function in the animal models mentioned here is still lacking.

15.3.3 AICAR

AICAR is an adenosine monophosphate (AMP) analogue that stimulates AMP-dependent kinase (AMPK) following intracellular conversion into an intermediate known as 5-amino-4-imidazolecarboxamide ribonucleotide (ZMP) [51, 52]. AMPK acts as the sensor of metabolic status of the cell – the ATP/ADP and ADP/AMP ratios – and has already been implicated in mitochondrial regulation via the use of resveratrol. Thus AICAR is simply another activator of AMPK, but it does so without affecting the ATP/ADP ratios, i.e., it does not affect the energy status of the cell. In *in vitro* experiments, AICAR was shown to improve growth and increase ATP concentration/production better than nine other potential mitochondrial compounds in multiple mitochondrial disease patient fibroblast lines [53].

AICAR is orally effective in mouse models and improved muscle performance in sedentary mice by increasing nuclear mitochondrial gene expression. It also increased COX enzyme expression and function and other mitochondrial gene expressions in multiple muscle-specific COX-deficient mouse models (ACTA-Cox15-/-, Surf1-/-, and Sco2KO/KI). Cox15 is a key enzyme in the heme biosynthesis pathway, and Sco2 is a metallochaperone that regulates cytochrome *c* oxidase (COX); mutations in either of these genes lead to encephalocardiomyopathy in children. Surf1 is a COX assembly factor that when mutated causes severe Leigh syndrome in children [29]. AICAR fully rescued the motor defects of the Sco2 mouse by unknown mechanism(s) but had no effect on the Cox15-/- model, likely reflecting the severity of this model. So, AICAR clearly induces mitochondrial function *in vivo*, but it exhibits poor brain penetrance [54], preventing its obvious utility in neurological situations. Many other direct and indirect LMWs have been proposed to act upon AMPK, mainly indirectly [55], but there are reports of improved direct AMPK activators [56–58]. The driving force behind AMPK activator research is the concept that AMPK is a relevant target for type II diabetes and the metabolic syndrome, as well as certain cancers [59], which may result in better molecules in terms of activity but not necessarily in terms of distribution (i.e., blood-brain barrier penetration).

15.3.4 Bezafibrate

Bezafibrate is a pan-peroxisome proliferator-activated receptor (PPAR) agonist developed initially to treat hyperlipidemia [60]. More recently PPARs were shown to regulate mitochondrial biogenesis via PGC1α [61] and were shown to increase respiratory function in patient-derived cells [62–64]. Parallel to these cell-based activities, bezafibrates were tested in multiple mitochondrial disease model mice with partial success. *ΔCox10* mice receiving 0.5 % w/w dietary bezafibrate exhibited increased muscle mitogenesis, a delay in the onset of myopathy, and an extended lifespan [28]. However, similar dosing in *Surf1-/-* and *ACTA-Cox15-/-* mice resulted in loss of body weight with hepatic toxicity or death, respectively [29]. Further studies in the *Twinkle* mutant "deletor" mouse [65] and the *Polg* "mutator" mouse [66] had varying degrees of success. The "deletor" mouse did exhibit reduced mtDNA deletion load and had fewer COX-negative muscle fibers, but at the expense of body weight and hepatic complications (treatment in this model started significantly later than the other models mentioned here) [67]. The *Polg* "mutator" mouse, despite not displaying an increase in mitochondrial biogenesis during bezafibrate treatment, did show multiple beneficial phenotypes, including a delay in hair loss, reversion of an abnormal spleen seen in this model, and reduced weight loss [68]. Overall bezafibrate has a reasonable success rate in cellular and animal models in terms of increasing mitochondrial function and ameliorating mouse model phenotypes.

15.3.5 NAD+ Precursors

Finally, it is important to consider recent promising developments utilizing NAD+ precursors directly as therapeutic agents, in particular nicotinamide riboside (see Fig. 15.1) [69, 70]. In mitochondrial disease respiratory chain deficiency leads to a decrease in the NAD+/NADH ratio, which sends a "high nutrient" signal to the cell – confounding the original mitochondrial metabolic defect. It makes sense to increase the levels of NAD+ using precursors of NAD+ that can reach the correct cellular compartments. Nicotinamide riboside (NR) is such a precursor and has been shown to increase muscle mitochondrial performance in wild-type and high-fat diet mice [71]. Subsequently, NR was used to treat the "deletor" adult-onset mitochondrial disease model mouse in two dosing schedules representing pre- and post-disease-onset situations (400 mg/kg/day) [69]. NR was beneficial in both dosing situations, increasing intracellular NAD+, increasing mitochondrial biogenesis, and reducing mtDNA deletion accumulation, adding to the hypothesis that increasing NAD+ is likely to be of very significant potential in the treatment of mitochondrial disease.

15.4 Mitochondrial Dynamics and Quality Control: Fission, Fusion, and Mitophagy

Mitochondria are highly dynamic and undergo fission and fusion, as well as being able to move with varying degrees of speed within cells – something of particular relevance in neurons with long cell processes [10, 72–74] (see Chap. 7). Fission and fusion are regulated by GTPases, with Drp1 driving fission and the collaboration of mitofusins 1 and 2 (Mfn1 and Mfn2) with Opa1 and Opa3 facilitating fusion. Fusion allows content mixing, including the mixing of mitochondrial genomes or mitochondrial-encoded proteins, perhaps allowing the dilution of damaged mtDNA or other damaged factors to buffer against stresses [25, 73, 75–77]. Fission allows the regulation of individual mitochondrial size, important for their transport [78] but also for their degradation (see below).

15.4.1 Fusion

Mutations in fusion factors lead to neurological disease. Opa1 defects lead to autosomal dominant optic atrophy (DOA) [79, 80], and Mfn2 mutations cause a peripheral neuropathy, Charcot-Marie-Tooth disease type 2A (CMT2A) [81]. Dominant mutations in Drp1 have been linked to a lethal neurodegenerative condition presenting with multiple phenotypes including microcephaly, optic atrophy, and lactic academia [82]. An extensive review of mouse models generated so far suggests that there are multiple tractable models to study the consequences of these mutations [83]. However, for example, in the case of Mfn2, multiple models have been generated with differing outcomes; some models proved too severe (lethal) [84], while others were too mild in relation to human Charcot-Marie-Tooth neuropathy type 2A [85], This variability in mouse models highlights the challenge of drug discovery for fission-fusion-related diseases.

15.4.2 Fission-Fusion Compounds

There are a limited number of compounds that specifically regulate mitochondrial fission and fusion: Mdivi-1 that inhibits Drp1 and 15-oxospiramilactone (S3) that inhibits USP30, a mitochondrially localized deubiquitinase and a hydrazine (M1) with an unknown target [86–89]. These molecules may represent useful starting points for drugs that are pro-fusion, i.e., they either block fission (mdivi-1) or enhance fusion (M1, S3). While M1 and S3 have never been tested in animal models, mdivi-1 has been tested in multiple studies, mainly concerned with reduction of infarct size and cell death in ischemic-reperfusion models [90–93].

In C57BL/6 mice, mdivi-1 reduced retinal ganglion cell death following pressure-induced ischemia, suggesting potential use for the treatment of glaucoma [91]. More recently inhibition of Drp1 has been tested in two Parkinson's disease models, a PINK1-/- mouse and a 1-methyl-4-phenyl-1,2,3,6-tetrahydropyridine (MPTP) mouse model [94]. This study is important because it was demonstrated that mdivi-1 crosses the blood-brain barrier and in doing so was able to restore a striatal dopamine release deficit and reduce neurotoxicity. This suggests that pushing the balance of fission-fusion toward fusion is beneficial in PD models with altered mitochondrial function. However it should be noted that in this case, the exact relationship between loss of PINK1 function and the pro-fusion function of mdivi-1 is not completely clear, especially given the proposed roles of PINK1 in mitophagy (see below) [94, 95].

15.5 Mitophagy

Mitophagy is the process whereby dysfunctional mitochondria are destroyed by the autophagic machinery (see Chap. 11). This occurs as a developmental process, for example, during the creation of mature erythrocytes [96], or it can occur as a result of loss of function or damage to mitochondria in many different cell types, i.e., a form of mitochondrial quality control [97, 98]. Mutations in *PARK2* (Parkin) and *PARK6* (PINK1) cause juvenile-onset autosomal recessive PD [99–101], and PINK1 and Parkin are now known to be key regulators of mitophagy [95, 102–104]. Briefly, PINK1 accumulates on the surface of damaged mitochondria (those with reduced membrane potential), binds Parkin, and the E3 ubiquitin ligase activity of Parkin converts multiple mitochondrial outer membrane proteins to an ubiquitinated form. These ubiquitinated proteins are recognized by specific receptors of the pre-autophagosome, causing engulfment of whole mitochondria for ultimate destruction in the lysosome. Mitophagy is closely linked to fusion because mitofusin is regulated by PINK1, acts as a receptor for Parkin and is ubiquitinated by Parkin to inactivate it. Presumably mitochondria above a certain size cannot fit within the pre-autophagosomal membrane so linking loss of fusion and mitophagy improves the efficiency of mitophagy. In addition, PINK1 and Parkin have been shown to regulate mitochondrial fission and fusion [105–107].

Thus, one hypothesis for PD is that the poorest functioning mitochondria are normally removed from the cellular pool through mitophagy, but failure of mitophagy due to PINK1 and Parkin mutations permits the accumulation of damaged mitochondria, leading to neuronal death in the *substantia nigra* [97, 98, 108]. It is likely that the *substantia nigra* neurons are more susceptible to loss of mitophagy than other neurons because there is a more dynamic redox environment, due to dopamine metabolism, that increases mtDNA mutations (see Chap. 6) [109]. Indeed, high levels of somatic mtDNA deletions are detected in *substantia nigra* neurons of healthy subjects and PD patients [110–112].

Conceptually then, it makes sense to activate the mitophagy pathway in PINK1/Parkin PD patients – as long as the fidelity and selectivity of damage-dependent mitophagy is maintained – potentially not possible with all alleles of PINK1 and Parkin responsible for disease [97]. In addition, activation of this pathway could preferentially remove mutant mtDNA from heteroplasmic tissue in mitochondrial disease patients. This concept has been genetically validated in several ways. Overexpression of wild-type Parkin in a cybrid cellular model of mtDNA disease (see later for details) was able to improve the heteroplasmic ratio of the wild-type/mutant mtDNA and improve respiratory function [113]. Furthermore, USP30, a de-ubiquitinating enzyme, has been shown to oppose the action of Parkin on mitochondrial outer membrane proteins (MIRO, TOM20) and to inhibit mitophagy. Knockdown of USP30 restores mitophagy in pathogenic Parkin mutant cells and improves mitochondrial function in paraquat-treated flies [97, 114, 115]. In addition, USP8, USP15, and USP35 have also been implicated in the regulation of mitophagy [116–118], These are interesting targets, mainly because the USP proteases are considered "druggable" [119, 120], and an inhibitor of USP30 has already been reported to be involved in fission-fusion regulation of mitochondria via the regulation of mitofusin ubiquitination [88].

15.5.1 Compounds Increasing PINK/Parkin Function

To find compounds that increase Parkin function, a chemical screen was performed using a cell line carrying multiple reporters linked to the PARK2 promoter. This approach yielded several classes of compounds that appear to increase Parkin function *in vitro*, including epigenetic agents, cholesterol synthesis modulators, and Jun kinase (JNK) inhibitors. These all represent potential compounds for further studies in PD and mtDNA disorders [121], given that one of the JNK inhibitors (SR-3066) is known to have neuroprotective effects in an MPTP rodent model [122]. A novel-directed approach to activating PINK1 yielded kinetin triphosphate, a PINK1 neo-substrate, which was able to increase the activity of both wild-type and disease-mutant PINK1^{G309D} in cellular models, resulting in increased Parkin recruitment to depolarized mitochondria [123]. Thus, there are promising compounds that likely influence damage-dependent mitophagy, and future experiments in relevant animal models will tell us more about their therapeutic potential.

15.5.2 mTOR Inhibition

Rapamycin inhibits mammalian target of rapamycin (mTOR), which regulates cell growth via multiple downstream effectors. mTOR is also a negative regulator of autophagy and mitophagy, and as such rapamycin could be beneficial based upon the arguments above [124–126]. However, mTOR positively regulates mitochondrial function in skeletal muscle but does the opposite in adipose tissue, confounding the issue when one thinks of the whole organism [126]. Perhaps more importantly, mTOR regulates aging, with mTOR inhibition increasing the lifespan of eukaryotes across evolution [127, 128].

Mouse models of mitochondrial disorders are often considered useful models of aging, with reduced survival times being common [129, 130]. Based upon this, *Ndufs4-/-* mice that have a lifespan median of 50 days have been treated with different dosing schedules of rapamycin at relatively high doses, such that daily treatment significantly reduced multiple neurological parameters and extended lifespan dramatically [131]. Interestingly mitochondrial mass and function were not altered, but a shift in central metabolism observed in the mutant compared to wild-type mouse was reversed upon rapamycin treatment of the mutant mouse. Importantly, low doses of a rapamycin derivative, RAD001, improve age-dependent immune function in healthy aged subjects, with few side effects [132]. Thus, there is a rationale for testing rapamycin in mitochondrial disease patients, but achieving therapeutic benefit while limiting side effects may be challenging.

15.6 Unbiased Screening in Cellular Systems

One can consider three conceptual approaches to obtaining novel bioactives using large libraries of compounds: First, biochemical screening of a known (protein) target such that modulation of that target should rescue a disease phenotype. Second, screening a cellular model of a known disease [133]. For this phenotypic approach, there must be a detectable phenotype, related to the genotype of interest, and the detection method must be amenable to miniaturization to facilitate large-scale screens. A third approach, a chemical biology approach, is to perform chemical screens on otherwise "normal" cells, organs, or organisms in culture to identify compounds that alter phenotype of choice, yielding tool or probe molecules that can perturb different biological functions [134]. These tools can then be assessed as potential leads for drugs based upon the knowledge surrounding their target.

Several labs have performed medium-throughput phenotypic screens or chemical biology or to identify compounds that alter mitochondrial function in cell lines in culture, using assays that measure cellular viability, cellular dehydrogenase activity, mitochondrial membrane potential, intracellular ATP, NAD+/NADH ratio, reactive oxygen species, and oxygen consumption as primary approaches [135–139]. Additional approaches for LMW altering some or all of these readouts included transcriptomics or assessing compound effects on *C. elegans*. Mining of these datasets for therapeutic opportunities yielded several compounds of interest: first, microtubule inhibitors that drive PGC1α transcription [135]; second, meclizine, an FDA-approved antiemetic, which protects from ischemic-reperfusion injury in animal models if used as a pretreatment and Huntington's disease models [137, 140]; third, a group of FDA-approved drugs that increased oxygen consumption (acarbose, metaraminol, gallamine triethiodide, and acamprosate) [139]; and fourth, a compound (BRD6897) that increases mitochondrial mass and function in endothelial cells [138]. However, given the inherent toxicity issues of microtubule inhibitors [141], it is unlikely that they can be relevant for mitochondrial-dependent neurodegenerative diseases. Previously, FDA-approved drugs that have a positive effect on mitochondrial function in normal and diseased cells and/or fly and rodent models may represent a worthwhile approach, but there is to date no further literature supporting treatment of patients with mitochondrial dysfunction with these drugs.

15.7 Screening in More Relevant Cellular Systems

15.7.1 Patient-Derived Cybrids

By definition, the nature of heteroplasmic mutations makes their study in derived cellular systems challenging. Initial attempts to recreate heteroplasmy *in vitro* led to the production of a cytoplasmic hybrid – known as a cybrid, created by fusion of a rho-zero (lacking mtDNA) transformed cell line with an enucleated cell line carrying donor mtDNA(s) [142]. Creation of cybrids carrying varying point mutations, deletions, and haplotypes have since been created to address a range of questions surrounding mitochondrial disease and the role of mtDNA in other diseases such as Parkinson's or diabetes [143].

Since cybrids can be heteroplasmic, they represent cellular models to test for nutrients or compounds that alter heteroplasmy, increasing the ratio of wild-type to mutant mtDNA. Indeed, several approaches have been described to alter cybrid heteroplasmy, reducing the proportion of mutant mtDNA, including altered nutrient regimes and treatment with compounds. For instance, coenzyme Q [144], glucose/arginine [145], and rapamycin [146] have all been shown to shift the wild-type/mutant mtDNA ratio positively. However, cybrid lines are by definition transformed and may thus reflect abnormal metabolism, and these cells are exclusively mitotic, preventing any examination of the effects of mtDNA mutations in postmitotic cells, which are most often affected in mtDNA disease.

15.7.2 Induced Pluripotent Stem Cells (iPSCs) and Differentiated Neurons

Some of the limitations of using cybrids are now being overcome with the advent of induced pluripotent stem cell (iPSC) technology, enabling the creation of many different mitotic and postmitotic cell types following specific differentiation processes [147]. Since their initial discovery several labs have produced iPSCs derived from patients with mitochondrial defects [148]. In two studies, fibroblasts from patients carrying the MELAS mutation m.3243A>G have been reprogrammed into stem cells where the level of heteroplasmy has a bimodal distribution that remains stable during serial outgrowth of a given clone [149, 150]. In a third study, fibroblasts carrying the MELAS-associated mtDNA mutation, m.13513G>A, were made into iPSCs in which the heteroplasmy remained stable or was significantly reduced [151]. iPSCs carrying mtDNA deletions have also been produced [152] from three different patients with Kearns-Sayre syndrome (KSS)/Pearson's syndrome (PS). During both reprogramming and further outgrowth of clones, a variability of heteroplasmy was observed. Overall it appears that iPSCs can be derived with stable mtDNA mutations and can be differentiated into heteroplasmic neuron-like and hematopoietic cells that demonstrate significant mitochondrial dysfunction [149, 150], and this will aid research and screening for novel therapeutics for these diseases.

To date there have been no reports demonstrating effective rescue of mitochondrial defects in postmitotic-derived mitochondrial disease cells (nuclear or mtDNA). However, several groups have derived neural-like cells from iPSCs carrying PD-causative PINK1 or LRRK2 (leucine-rich repeat kinase 2) mutations [153, 154] and shown pharmacological [153] or genetic rescue [154] of the associated mitochondria dysfunction. Although the overall effects are small, coenzyme Q10 and rapamycin could both rescue valinomycin-induced lactate dehydrogenase (LDH) leakage in LRRK2 G2019S patient-derived neurons, and Q10 could also rescue LRRK2 R1441C and PINK1 Q456X neurons [153]. These initial results using single doses of compound are promising, but the molecular mechanisms involved are unclear. In a genetic approach, neurons carrying LRRK2 G2019S were corrected to wild type using Zn-finger nucleases to demonstrate that the mutant cells accumulate a higher rate of mtDNA mutation than the corrected cells [154]. Whether these mtDNA mutation rates affect mitochondrial function was not assessed in this study.

15.8 Clinical Approaches

Three recent exhaustive reviews assessed all prior reports of clinical trials in mitochondrial disease patients to assess trial design quality and to determine beneficial outcomes [155–157]. The key message is that there is no attempted treatment for mitochondrial disorders that has a clear statistically relevant positive outcome. There are several reasons for this, but a major reason is the trial design itself, because there is a clear bias toward positive reporting when the trials are less well designed – e.g., open label instead of double blind. This sets a poor precedent for all concerned, because patients will receive drugs that are likely to be ineffective, damaging patient's confidence in the medical and research community [155].

Current clinical trials for pharmacological agents in mitochondrial disease space are shown in Table 15.1 (www.clinicaltrials.gov). For mitochondrial disease patients, mercaptamine is proposed to increase levels of glutathione – which acts an antioxidant to combat reactive oxygen species produced by damaged mitochondria. Vatiquinone is also an antioxidant that has been shown to be beneficial in open-label

Table 15.1 Current pharmacological agents in clinical trials for mitochondrial disease

Chemical entity	Phase (mitochondrial disease), trial ID at clinicaltrials.gov	Notes, trial completion dates
Mercaptamine (RP103, transglutaminase inhibitor)	Ph II/III (Leigh syndrome) [launched for nephropathic cystinosis 2013], NCT02023866	Prevents abnormal lysosomal cysteine crystal accumulation. Expected to complete June 2016
Vatiquinone (EPI-743, NADH, or NADPH oxidoreductase modulator)	Ph II (Pearson syndrome (PS), mitochondrial respiratory chain disease, Leigh syndrome). NCT01721733	Antioxidant. PS completion Apr 2016, respiratory chain disease completed, Leigh syndrome completion Apr 2016
RTA-408 (nuclear erythroid 2-related factor 2 stimulator)	Ph II (mitochondrial myopathy). NCT02255422	Antioxidant, anti-inflammatory agent. Completion Jun 2016
Cyclosporine (calcineurin inhibitor)	Ph II (LHON). NCT02176733	Limits apoptosis in acute phase of disease. Completion expected Oct 2015
Bendavia (MTP-131)	Ph I/II (LHON, mitochondrial myopathy). NCT02367014	Targets cardiolipin in mitochondrial membranes. LHON study expected to complete in 2016. Myopathy study expected to complete Dec 2015
Benzafibrate (lipase stimulator; PPAR agonist)	Ph II (mitochondrial myopathy, mt.3243A>G). NCT02398201	Mitochondrial biogenesis stimulator. Completion 2016
KH176	Ph I (mitochondrial disease, MELAS, LHON, Leigh, mt.3243A>G). NCT02544217	Redox modulator. Completion 2015

trials [158] and is now in a randomized double-blind trial for pediatric Leigh syndrome patients. RTA-308 is also an antioxidant and anti-inflammatory agent that acts by increasing Nrf2 activity that in turn enhances the expression of prooxidant genes and represses inflammatory genes in animal models [159]. It is currently in a safety/efficacy trial to assess workload during exercise training in mitochondrial myopathy patients. Cyclosporine A, an immunosuppressive drug that protects cells from cell death caused by opening of the mitochondrial permeability transition pore, is being open-label-trialed in Leber's hereditary optic neuropathy, a mitochondrial disease that affects the basal retinal ganglia – hopefully reducing cell death in the retinal ganglion cell layer of the eye. Bendavia is a peptide that prevents cardiolipin from converting cytochrome c into a peroxidase but leaves its electron chain function intact [160] and is also proposed to improve electron chain efficiency by improving coupling by reducing proton leak. Also, based upon the preclinical data presented earlier and the fact that it has been used extensively in humans already, a trial using bezafibrate in mitochondrial myopathy caused by mutation in mt.3243A>G is under way, with planned completion in mid 2016. Finally, KH176, an antioxidant trolox derivative, is being trialed in phase I rising dose trials in patients with either MELAS, LHON, or mt.3243G<A or Leigh syndrome patients.

It is interesting to note that six of the seven entities being tested are essentially repurposing or expanded indication approaches. Many of these are also general antioxidant therapies that aim to redress the proposed redox defects seen in mitochondrial patients. This reflects the fact that historically there has not been sufficient interest in this class of patients or sufficient knowledge of the pathways in the cell regulating mitochondrial function. This is changing and the hope is that in the immediate future more therapies will be developed with mitochondrial disease patients as the primary focus. Dedicated approaches to processes such as biogenesis and mitophagy, for example, coupled with better clinical trial design, will be key to improving the chance of success for mitochondrial disease patients.

Acknowledgments I would like to thank Nathalie Vancampenhout for her help with clinical trial database searches.

References

1. Turnbull DM, Rustin P. Neurobiology of disease genetic and biochemical intricacy shapes mitochondrial cytopathies. Neurobiol Dis. [Internet]. 2015;S0969-9961(15)00023-6 Feb [Epub ahead of print].
2. Koopman WJ, Distelmaier F, Smeitink JA, Willems PH. OXPHOS mutations and neurodegeneration. EMBO J. 2013;32(1):9–29 [Internet].
3. Calvo SE, Mootha VK. The mitochondrial proteome and human disease. Annu Rev Genomics Hum Genet. 2010;11:25–44.
4. Pagliarini DJ, Calvo SE, Chang B, Sheth SA, Vafai SB, Ong SE, et al. A mitochondrial protein compendium elucidates complex I disease biology. Cell. 2008;134(1):112–23.
5. Smith AC, Blackshaw JA, Robinson AJ. MitoMiner: a data warehouse for mitochondrial proteomics data. Nucleic Acids Res. 2012;40(D1):1160–7.
6. Alberts B, Johnson A, Lewis J, Raff M, Roberts K, Walter P. Molecular biology of the cell. 4th ed. New York: Garland Science; 2002. 1616 p.
7. Vafai SB, Mootha VK. Mitochondrial disorders as windows into an ancient organelle. Nature. 2012;491(7424):374–83 [Internet].
8. Taylor RW, Turnbull DM. Mitochondrial mutations in human disease. Nat Rev Genet. 2005;6(5):389–402.
9. Russell O, Turnbull D. Mitochondrial DNA disease-molecular insights and potential routes to a cure. Exp Cell Res. 2014;325(1):38–43. [Internet]. Elsevier.
10. Burté F, Carelli V, Chinnery PF, Yu-Wai-Man P. Disturbed mitochondrial dynamics and neurodegenerative disorders. Nat Rev Neurol. 2014;11(1):11–24 [Internet].
11. Yu-wai-man P, Grif PG, Chinnery PF. Progress in retinal and eye research mitochondrial optic neuropathies – disease mechanisms and therapeutic strategies. Prog Retin Eye Res. 2011;30(2):81–114.
12. Koopman W, Willems P, Smeitink J. Monogenic mitochondrial disorders. N Engl J Med. 2012;366:1132–41.
13. Lott MT, Leipzig JN, Derbeneva O, Xie HM, Chalkia D, Sarmady M, et al. mtDNA variation and analysis using MITOMAP and MITOMASTER. Curr Protoc Bioinforma. 2013;1(123):1.23.1–1.23.26 [Internet].

14. Wallace DC, Zheng XX, Lott MT, Shoffner JM, Hodge JA, Kelley RI, et al. Familial mitochondrial encephalomyopathy (MERRF): genetic, pathophysiological, and biochemical characterization of a mitochondrial DNA disease. Cell. 1988;55(4):601–10.
15. Scharfe C, Lu HHS, Neuenburg JK, Allen EA, Li GC, Klopstock T, et al. Mapping gene associations in human mitochondria using clinical disease phenotypes. PLoS Comput Biol. 2009;5(4):e1000374.
16. Calvo SE, Compton AG, Hershman SG, Lim SC, Lieber DS, Tucker EJ, et al. Molecular diagnosis of infantile mitochondrial disease with targeted next-generation sequencing. Sci Transl Med. 2012;4(118):118ra10–0.
17. Wortmann S, Koolen D, Smeitink J, van den Heuvel L, Rodenburg R. Whole exome sequencing of suspected mitochondrial patients in clinical practice. J Inherit Metab Dis. [Internet]. 2015;38(3):437–43.
18. Taylor RW, Pyle A, Griffin H, Blakely EL, Duff J, He L, et al. Use of whole-exome sequencing to determine the genetic basis of multiple mitochondrial respiratory chain complex deficiencies. JAMA. 2014;312(1):68 [Internet].
19. Carroll CJ, Brilhante V, Suomalainen A. Next-generation sequencing for mitochondrial disorders. Br J Pharmacol. 2014;171(8):1837–53.
20. Chinnery PF, Johnson MA, Wardell TM, Singh-Kler R, Hayes C, Brown DT, et al. The epidemiology of pathogenic mitochondrial DNA mutations. Ann Neurol. 2000;48(2):188–93.
21. Komen JC, Thorburn DR. Turn up the power – pharmacological activation of mitochondrial biogenesis in mouse models. Br J Pharmacol. 2014;171(8):1818–36.
22. Uittenbogaard M, Chiaramello A, Street NW, Dc W. Mitochondrial biogenesis: a therapeutic target for neurodevelopmental disorders and neurodegenerative diseases. Curr Pharm Des. 2014;20:5574–93.
23. Andreux PA, Houtkooper RH, Auwerx J. Pharmacological approaches to restore mitochondrial function. Nat Rev Drug Discov. 2013;12(6):465–83 [Internet].
24. Wu Z, Puigserver P, Andersson U, Zhang C, Adelmant G, Mootha V, et al. Mechanisms controlling mitochondrial biogenesis and respiration through the thermogenic coactivator PGC-1. Cell. 1999;98(1):115–24.
25. Nunnari J, Suomalainen A. Mitochondria: in sickness and in health. Cell. 2012;148(6):1145–59. [Internet]. Elsevier Inc.
26. Wenz T. Regulation of mitochondrial biogenesis and PGC-1alpha under cellular stress. Mitochondrion. 2013;13(2):134–42 [Internet].
27. Scarpulla RC, Vega RB, Kelly DP. Transcriptional integration of mitochondrial biogenesis. Trends Endocrinol Metab. 2012;23(9):459–66 [Internet].
28. Wenz T, Diaz F, Spiegelman BM, Moraes CT. Activation of the PPAR/PGC-1 a pathway prevents a bioenergetic deficit and effectively improves a mitochondrial myopathy phenotype. Cell Metab. 2008;8:249–56.
29. Viscomi C, Bottani E, Civiletto G, Cerutti R, Moggio M, Fagiolari G, et al. In vivo correction of COX deficiency by activation of the AMPK/PGC-1 a axis. Cell Metab. 2011;14(1):80–90 [Internet].
30. Dillon LM, Williams SL, Hida A, Peacock JD, Prolla TA, Lincoln J, et al. Increased mitochondrial biogenesis in muscle improves aging phenotypes in the mtDNA mutator mouse. Hum Mol Genet. 2012;21(10):2288–97.
31. Selsby JT, Morine KJ, Pendrak K, Barton ER, Sweeney HL. Rescue of dystrophic skeletal muscle by PGC-1alpha involves a fast to slow fiber type shift in the mdx mouse. PLoS ONE. 2012;7(1):1–10.
32. Zhao W, Varghese M, Yemul S, Pan Y, Cheng A, Marano P, et al. Peroxisome proliferator activator receptor gamma coactivator-1alpha (PGC-1α) improves motor performance and survival in a mouse model of amyotrophic lateral sclerosis. Mol Neurodegener. 2011;6(1):51 [Internet], [cited 2015 Apr 20].

33. Mudò G, Mäkelä J, Di Liberto V, Tselykh TV, Olivieri M, Piepponen P, et al. Transgenic expression and activation of PGC-1α protect dopaminergic neurons in the MPTP mouse model of Parkinson's disease. Cell Mol Life Sci. 2012;69(7):1153–65.
34. Howitz KT, Bitterman KJ, Cohen HY, Lamming DW, Lavu S, Wood JG, et al. Small molecule activators of sirtuins extend Saccharomyces cerevisiae lifespan. Nature. 2003;425(6954):191–6 [Internet].
35. Wood JG, Rogina B, Lavu S, Howitz K, Helfand SL, Tatar M, et al. Sirtuin activators mimic caloric restriction and delay ageing in metazoans. Nature. 2004;430(7000):686–9 [Internet].
36. Borra MT, Smith BC, Denu JM. Mechanism of human SIRT1 activation by resveratrol. J Biol Chem. 2005;280(17):17187–95 [Internet].
37. Kaeberlein M, McDonagh T, Heltweg B, Hixon J, Westman EA, Caldwell SD, et al. Substrate-specific activation of sirtuins by resveratrol. J Biol Chem. 2005;280(17):17038–45 [Internet].
38. Beher D, Wu J, Kim KW, Lu S, Atangan L, Wang M. Resveratrol is not a direct activator of SIRT1 enzyme activity. Chem Biol Drug Des. 2009;74(6):619–24.
39. Canto C, Gerhart-Hines Z, Feige JN, Lagouge M, Noriega L, Milne JC, et al. AMPK regulates energy expenditure by modulating NAD 1 metabolism and SIRT1 activity. Nature. 2009;458:1056–60.
40. Canto C, Jiang LQ, Deshmukh AS, Mataki C, Coste A, Lagouge M, et al. Interdependence of AMPK and SIRT1 for metabolic adaptation to fasting and exercise in skeletal muscle. Cell Metab. 2010;11(3):213–9 [Internet].
41. Um J, Park S, Kang H, Yang S, Foretz M, Mcburney MW, et al. AMP-activated protein kinase – deficient mice are resistant to the metabolic effects of resveratrol. Diabetes. 2010;59(3):554–63.
42. Pacholec M, Bleasdale JE, Chrunyk B, Cunningham D, Flynn D, Garofalo RS, et al. SRT1720, SRT2183, SRT1460, and resveratrol are not direct activators of SIRT1. J Biol Chem. 2010;285(11):8340–51.
43. Baur JA. Biochemical effects of SIRT1 activators. Biochim Biophys Acta Protein Proteomics. 2010;1804(8):1626–34 [Internet].
44. Sajish M, Schimmel P. A human tRNA synthetase is a potent PARP1-activating effector target for resveratrol. Nature. 2015;519(7543):370–3 [Internet].
45. Baur JA, Pearson KJ, Price NL, Jamieson HA, Lerin C, Kalra A, et al. Resveratrol improves health and survival of mice on a high-calorie diet. Nature. 2006;444(7117):337–42 [Internet].
46. Lagouge M, Argmann C, Gerhart-hines Z, Meziane H, Lerin C, Daussin F, et al. Resveratrol improves mitochondrial function and protects against metabolic disease by activating SIRT1 and PGC-1 a. Cell. 2006;127:1109–22.
47. Milne JC, Lambert PD, Schenk S, Carney DP, Smith JJ, Gagne DJ, et al. Small molecule activators of SIRT1 as therapeutics for the treatment of type 2 diabetes. Nature. 2007;450(7170):712–6 [Internet].
48. Feige JN, Lagouge M, Canto C, Strehle A, Houten SM, Milne JC, et al. Specific SIRT1 activation mimics low energy levels and protects against diet-induced metabolic disorders by enhancing fat oxidation. Cell Metab. 2008;8:347–58.
49. Herskovits AZ, Guarente L. SIRT1 in neurodevelopment and brain senescence. Neuron. 2014;81(3):471–83 [Internet].
50. Li L, Voullaire L, Sandi C, Pook MA, Ioannou PA, Delatycki MB, et al. Pharmacological screening using an FXN-EGFP cellular genomic reporter assay for the therapy of friedreich ataxia. PLoS ONE. 2013;8(2):e55940.
51. Sullivan JE, Brocklehurst KJ, Marley AE, Carey F, Carling D, Beri RK. Inhibition of lipolysis and lipogenesis in isolated rat adipocytes with AICAR, a cell-permeable activator of AMP-activated protein kinase. FEBS Lett. 1994;353(1):33–6 [Internet].
52. Corton JM, Gillespie JG, Hawley SA, Hardie DG. 5-aminoimidazole-4-carboxamide ribonucleoside. Eur J Biochem. 1995;229(2):558–65 [Internet].

53. Golubitzky A, Dan P, Weissman S, Link G, Wikstrom JD, Saada A. Screening for active small molecules in mitochondrial complex I deficient patient's fibroblasts, reveals AICAR as the most beneficial compound. PLoS ONE. 2011;6(10):e26883.
54. Marangos P, Loftus T, Wiesner J, Lowe T, Rossi E, Browne C, et al. Adenosinergic modulation of homocysteine-induced seizures in mice. Epilepsia. 1990;31:239–46.
55. Arkwright RT, Deshmukh R, Adapa N, Stevens R, Zonder E, Zhang Z, et al. Lessons from nature: sources and strategies for developing AMPK activators for cancer chemotherapeutics. Anticancer Agents Med Chem. 2015;15(5):657–71.
56. Lai Y-C, Kviklyte S, Vertommen D, Lantier L, Foretz M, Viollet B, et al. A small-molecule benzimidazole derivative that potently activates AMPK to increase glucose transport in skeletal muscle: comparison with effects of contraction and other AMPK activators. Biochem J. 2014;375(460):363–75.
57. Giordanetto F, Karis D. Direct AMP-activated protein kinase activators: a review of evidence from the patent literature. Expert Opin Ther Patents. 2012;22:1467.
58. Cool B, Zinker B, Chiou W, Kifle L, Cao N, Perham M, et al. Identification and characterization of a small molecule AMPK activator that treats key components of type 2 diabetes and the metabolic syndrome. Cell Metab. 2006;3:403–16.
59. Steinberg GR, Kemp BE. AMPK in health and disease. Physiol Rev. 2009;89:1025–78.
60. Tenenbaum A, Motro M, Fisman EZ. Dual and pan-peroxisome proliferator-activated receptors (PPAR) co-agonism: the bezafibrate lessons. Cardiovasc Diabetol. 2005;5(1):1–5 [Internet], [cited 2015 Apr 21].
61. Hondares E, Pineda-Torra I, Iglesias R, Staels B, Villarroya F, Giralt M. PPARδ, but not PPARα, activates PGC-1α gene transcription in muscle. Biochem Biophys Res Commun. 2007;354(4):1021–7 [Internet].
62. Bastin J, Aubey F, Rötig A, Munnich A, Djouadi F. Activation of peroxisome proliferator-activated receptor pathway stimulates the mitochondrial respiratory chain and can correct deficiencies in patients' cells lacking its components. J Clin Endocrinol Metab. 2008;93(4):1433–41 [Internet].
63. Srivastava S, Diaz F, Iommarini L, Aure K, Lombes A, Moraes CT. PGC-1α/β induced expression partially compensates for respiratory chain defects in cells from patients with mitochondrial disorders. Hum Mol Genet. 2009;18(10):1805–12 [Internet].
64. Casarin A, Giorgi G, Pertegato V, Siviero R, Cerqua C, Doimo M, et al. Copper and bezafibrate cooperate to rescue cytochrome c oxidase deficiency in cells of patients with SCO2 mutations. Orphanet J Rare Dis. 2012;7(1):21 [Internet], Jan [cited 2015 Apr 23].
65. Tyynismaa H, Mjosund KP, Wanrooij S, Lappalainen I, Ylikallio E, Jalanko A, et al. Mutant mitochondrial helicase twinkle causes multiple mtDNA deletions and a late-onset mitochondrial disease in mice. Proc Natl Acad Sci U S A. 2005;102(49):17687–92.
66. Trifunovic A, Wredenberg A, Falkenberg M, Spelbrink JN, Rovio AT, Bruder CE, et al. Premature ageing in mice expressing defective mitochondrial DNA polymerase. Nature. 2004;429(6990):417–23 [Internet].
67. Yatsuga S, Suomalainen A. Effect of bezafibrate treatment on late-onset mitochondrial myopathy in mice. Hum Mol Genet. 2012;21(3):526–35 [Internet].
68. Dillon LM, Hida A, Garcia S, Prolla TA, Moraes CT. Long-term bezafibrate treatment improves skin and spleen phenotypes of the mtDNA mutator mouse. PLoS ONE. 2012;7(9):e44335 [Internet].
69. Khan NA, Auranen M, Paetau I, Pirinen E, Euro L, Forsström S, et al. Effective treatment of mitochondrial myopathy by nicotinamide riboside, a vitamin B3. EMBO Mol Med. 2014;6(6):721–31.
70. Lightowlers RN, Chrzanowska-Lightowlers ZM. Salvaging hope: is increasing NAD+ a key to treating mitochondrial myopathy? EMBO Mol Med. 2014;6(6):1–3 [Internet].
71. Cantó C, Houtkooper RH, Pirinen E, Youn DY, Oosterveer MH, Cen Y, et al. The NAD+ precursor nicotinamide riboside enhances oxidative metabolism and protects against high-fat diet-induced obesity. Cell Metab. 2012;15(6):838–47.

72. Westermann B. Mitochondrial fusion and fission in cell life and death. Nat Publ Gr. 2010;11(12):872–84 [Internet].
73. Labbé K, Murley A, Nunnari J. Determinants and functions of mitochondrial behavior. Annu Rev Cell Dev Biol. 2014;30(1):357–91 [Internet].
74. Mishra P, Chan DC. Mitochondrial dynamics and inheritance during cell division, development and disease. Nat Publ Gr. 2014;15(10):634–46 [Internet].
75. Chen H, Chan DC. Mitochondrial dynamics-fusion, fission, movement, and mitophagy-in neurodegenerative diseases. Hum Mol Genet. 2009;18(R2):169–76.
76. Chen H, McCaffery JM, Chan DC. Mitochondrial fusion protects against neurodegeneration in the cerebellum. Cell. 2007;130(3):548–62 [Internet].
77. Chen H, Vermulst M, Wang YE, Chomyn A, Prolla TA, McCaffery JM, et al. Mitochondrial fusion is required for mtDNA stability in skeletal muscle and tolerance of mtDNA mutations. Cell. 2010;141(2):280–9 [Internet].
78. Verstreken P, Ly CV, Venken KJT, Koh T-W, Zhou Y, Bellen HJ. Synaptic mitochondria are critical for mobilization of reserve pool vesicles at drosophila neuromuscular junctions. Neuron. 2005;47(3):365–78 [Internet].
79. Delettre C, Lenaers G, Griffoin JM, Gigarel N, Lorenzo C, Belenguer P, et al. Nuclear gene OPA1, encoding a mitochondrial dynamin-related protein, is mutated in dominant optic atrophy. Nat Genet. 2000;26(2):207–10.
80. Alexander C, Votruba M, Pesch UE, Thiselton DL, Mayer S, Moore A, et al. OPA1, encoding a dynamin-related GTPase, is mutated in autosomal dominant optic atrophy linked to chromosome 3q28. Nat Genet. 2000;26(2):211–5.
81. Züchner S, Mersiyanova IV, Muglia M, Bissar-Tadmouri N, Rochelle J, Dadali EL, et al. Mutations in the mitochondrial GTPase mitofusin 2 cause Charcot-Marie-Tooth neuropathy type 2A. Nat Genet. 2004;36(5):449–51.
82. Waterham HR, Koster J, van Roermund CWT, Mooyer PAW, Wanders RJA, Leonard JV. A lethal defect of mitochondrial and peroxisomal fission. N Engl J Med. 2007;356(17):1736–41 [Internet].
83. Pareyson D, Saveri P, Sagnelli A, Piscosquito G. Mitochondrial dynamics and inherited peripheral nerve diseases. Neurosci Lett. 2015;596:66–77 [Internet].
84. Chen H, Detmer SA, Ewald AJ, Griffin EE, Fraser SE, Chan DC. Mitofusins Mfn1 and Mfn2 coordinately regulate mitochondrial fusion and are essential for embryonic development. J Cell Biol. 2003;160(2):189–200 [Internet].
85. Strickland AV, Rebelo AP, Zhang F, Price J, Bolon B, Silva JP, et al. Characterization of the mitofusin 2 R94W mutation in a knock-in mouse model. J Peripher Nerv Syst. 2014;19(2):152–64 [Internet].
86. Lackner LL, Nunnari J. Small molecule inhibitors of mitochondrial division: tools that translate basic biological research into medicine. Chem Biol. 2010;17(6):578–83 [Internet].
87. Wang D, Wang J, Bonamy GMC, Meeusen S, Brusch RG, Turk C, et al. A small molecule promotes mitochondrial fusion in mammalian. Angew Chem. 2012;51:9302–5.
88. Yue W, Chen Z, Liu H, Yan C, Chen M, Feng D, et al. A small natural molecule promotes mitochondrial fusion through inhibition of the deubiquitinase USP30. Nat Publ Gr. 2014;24(4):482–96 [Internet].
89. Cassidy-Stone A, Chipuk JE, Ingerman E, Song C, Yoo C, Kuwana T, et al. Chemical inhibition of the mitochondrial division dynamin reveals its role in Bax/Bak-dependent mitochondrial outer membrane permeabilization. Dev Cell. 2008;14(2):193–204.
90. Ong SB, Subrayan S, Lim SY, Yellon DM, Davidson SM, Hausenloy DJ. Inhibiting mitochondrial fission protects the heart against ischemia/reperfusion injury. Circulation. 2010;121(18):2012–22.
91. Park SW, Kim K-Y, Lindsey JD, Dai Y, Heo H, Nguyen DH, et al. A selective inhibitor of Drp1, Mdivi-1, increases retinal ganglion cell survival in acute ischemic mouse retina. Invest Ophthalmol Vis Sci. 2011;52(5):2837–43 [Internet].

92. Sharp WW, Fang YH, Han M, Zhang HJ, Hong Z, Banathy A, et al. Dynamin-related protein 1 (Drp1)-mediated diastolic dysfunction in myocardial ischemia-reperfusion injury: therapeutic benefits of Drp1 inhibition to reduce mitochondrial fission. FASEB J. 2014;28(1):316–26 [Internet].
93. Disatnik M-H, Hwang S, Ferreira JCB, Mochly-Rosen D. New therapeutics to modulate mitochondrial dynamics and mitophagy in cardiac diseases. J Mol Med (Berl). 2015;93(3):279–87 [Internet].
94. Rappold PM, Cui M, Grima JC, Fan RZ, de Mesy-Bentley KL, Chen L, et al. Drp1 inhibition attenuates neurotoxicity and dopamine release deficits in vivo. Nat Commun. 2014;5:5244 [Internet].
95. Youle RJ, Narendra DP. Mechanisms of mitophagy. Nat Publ Gr. 2011;12(1):9–14 [Internet].
96. Ney PA. Mitochondrial autophagy: origins, significance, and role of BNIP3 and NIX. Biochim Biophys Acta, Mol Cell Res. 2015;1853(0):2775–83 [Internet].
97. Pickrell AM, Youle RJ. The roles of PINK1, parkin, and mitochondrial fidelity in Parkinson' s disease. Neuron. 2015;85(2):257–73 [Internet].
98. Eiyama A, Okamoto K. PINK1/Parkin-mediated mitophagy in mammalian cells. Curr Opin Cell Biol. 2015;33:95–101 [Internet].
99. Matsumine H, Saito M, Shimoda-Matsubayashi S, Tanaka H, Ishikawa A, Nakagawa-Hattori Y, et al. Localization of a gene for an autosomal recessive form of juvenile Parkinsonism to chromosome 6q25.2-27. Am J Hum Genet. 1997;60(3):588–96 [Internet].
100. Kitada T, Asakawa S, Hattori N, Matsumine H, Yamamura Y, Minoshima S, et al. Mutations in the parkin gene cause autosomal recessive juvenile parkinsonism. Nature. 1998;392(6676):605–8 [Internet].
101. Valente EM, Bentivoglio AR, Dixon PH, Ferraris A, Ialongo T, Frontali M, et al. Localization of a novel locus for autosomal recessive early-onset Parkinsonism, PARK6, on human chromosome 1p35-p36. Am J Hum Genet. 2001;68(4):895–900 [Internet].
102. Narendra D, Tanaka A, Suen DF, Youle RJ. Parkin is recruited selectively to impaired mitochondria and promotes their autophagy. J Cell Biol. 2008;183(5):795–803.
103. Geisler S, Holmström KM, Treis A, Skujat D, Weber SS, Fiesel FC, et al. The PINK1/Parkin-mediated mitophagy is compromised by PD-associated mutations. Autophagy. 2010;6(7):871–8.
104. Cai Q, Zakaria HM, Simone A, Sheng ZH. Spatial parkin translocation and degradation of damaged mitochondria via mitophagy in live cortical neurons. Curr Biol. 2012;22(6):545–52 [Internet].
105. Deng H, Dodson MW, Huang H, Guo M. The Parkinson's disease genes pink1 and parkin promote mitochondrial fission and/or inhibit fusion in Drosophila. Proc Natl Acad Sci. 2008;105(38):14503–8 [Internet].
106. Poole AC, Thomas RE, Yu S, Vincow ES, Pallanck L. The mitochondrial fusion-promoting factor mitofusin is a substrate of the PINK1/Parkin pathway. PLoS ONE. 2010;5(4):e10054.
107. Dorn 2nd GW. Mitochondrial dynamism and heart disease: changing shape and shaping change. EMBO Mol Med. 2015;7(7):865–77.
108. Ryan BJ, Hoek S, Fon EA, Wade-Martins R. Mitochondrial dysfunction and mitophagy in Parkinson's: from familial to sporadic disease. Trends Biochem Sci. 2015;40(4):200–10 [Internet].
109. Sanders L, McCoy J, Hu X, Mastroberardino PG, Dickinson BC, Chang CJ, et al. Mitochondrial DNA damage: molecular marker of vulnerable nigral neurons in Parkinson's disease. Neurobiol Dis. 2014;70:214–23 [Internet].
110. Bender A, Krishnan KJ, Morris CM, Taylor GA, Reeve AK, Perry RH, et al. High levels of mitochondrial DNA deletions in substantia nigra neurons in aging and Parkinson disease. Nat Genet. 2006;38(5):515–7.
111. Kraytsberg Y, Kudryavtseva E, McKee AC, Geula C, Kowall NW, Khrapko K. Mitochondrial DNA deletions are abundant and cause functional impairment in aged human substantia nigra neurons. Nat Genet. 2006;38(5):518–20.

112. Reeve AK, Krishnan KJ, Elson JL, Morris CM, Bender A, Lightowlers RN, et al. Nature of mitochondrial DNA deletions in Substantia Nigra neurons. Am J Hum Genet. 2008;82(1):228–35.
113. Suen D-F, Narendra DP, Tanaka A, Manfredi G, Youle RJ. Parkin overexpression selects against a deleterious mtDNA mutation in heteroplasmic cybrid cells. Proc Natl Acad Sci U S A. 2010;107(26):11835–40.
114. Bingol B, Tea JS, Phu L, Reichelt M, Bakalarski CE, Song Q, et al. The mitochondrial deubiquitinase USP30 opposes parkin-mediated mitophagy. Nature. 2014;509(7505):370–5 [Internet].
115. Liang J-R, Martinez A, Lane JD, Mayor U, Clague MJ, Urbé S. USP30 deubiquitylates mitochondrial Parkin substrates and restricts apoptotic cell death. EMBO Rep. 2015;16:618–27. [Internet].
116. Durcan TM, Tang MY, Pérusse JR, Dashti EA, Aguileta MA, Mclelland L, et al. USP 8 regulates mitophagy by removing K 6 -linked ubiquitin conjugates from parkin. EMBO J. 2014;33(21):2473–91.
117. Cornelissen T, Haddad D, Wauters F, Van Humbeeck C, Mandemakers W, Koentjoro B, et al. The deubiquitinase USP15 antagonizes Parkin-mediated mitochondrial ubiquitination and mitophagy. Hum Mol Genet [Internet]. 2014;23(19):5227–42.
118. Wang Y, Serricchio M, Jauregui M, Shanbhag R, Stoltz T, Paolo T Di, et al. Autophagy deubiquitinating enzymes regulate PARK2-mediated mitophagy. Autophagy. 2015;11(4):595–606.
119. Tian X, Isamiddinova NS, Peroutka RJ, Goldenberg SJ, Mattern MR, Nicholson B, et al. Characterization of selective ubiquitin and ubiquitin-like protease inhibitors using a fluorescence-based multiplex assay format. Assay Drugs Dev Technol. 2011;9(2):165–73 [Internet].
120. Lee B-H, Lee MJ, Park S, Oh D-C, Elsasser S, Chen P-C, et al. Enhancement of proteasome activity by a small-molecule inhibitor of Usp14. Nature. 2010;467(7312):179–84 [Internet].
121. Hasson SA, Fogel AI, Wang C, MacArthur R, Guha R, Heman-Ackah S, et al. Chemogenomic profiling of endogenous PARK2 expression using a genome-edited coincidence reporter. ACS Chem Biol. 2015;10:1188–97. [Internet].
122. Chambers JW, Pachori A, Howard S, Ganno M, Hansen D, Kamenecka T, et al. Small molecule c-jun-N-terminal kinase inhibitors protect dopaminergic neurons in a model of Parkinson's disease. ACS Chem Neurosci. 2011;2(4):198–206 [Internet].
123. Hertz NT, Berthet A, Sos ML, Thorn KS, Burlingame AL, Nakamura K, et al. A neo-substrate that amplifies catalytic activity of parkinson's-disease- related kinase PINK1. Cell. 2013;154(4):737–47 [Internet].
124. Laplante M, Sabatini DM. MTOR signaling in growth control and disease. Cell. 2012;149(2):274–93 [Internet].
125. De Cabo R, Carmona-Gutierrez D, Bernier M, Hall MN, Madeo F. The search for antiaging interventions: from elixirs to fasting regimens. Cell. 2014;157(7):1515–28 [Internet].
126. Albert V, Hall MN. mTOR signaling in cellular and organismal energetics. Curr Opin Cell Biol. 2015;33:55–66 [Internet].
127. Johnson SC, Rabinovitch PS, Kaeberlein M. mTOR is a key modulator of ageing and age-related disease. Nature. 2013;493(7432):338–45 [Internet].
128. Li J, Kim SG, Blenis J. Rapamycin: one drug, many effects. Cell Metab. 2014;19(3):373–9 [Internet].
129. Iommarini L, Peralta S, Torraco A, Diaz F. Mitochondrial diseases part II: mouse models of OXPHOS deficiencies caused by defects in regulatory factors and other components required for mitochondrial function. Mitochondrion. 2015;22:96–118 [Internet].
130. Torraco A, Peralta S, Iommarini L, Diaz F. Mitochondrial diseases part I : mouse models of OXPHOS deficiencies caused by defects in respiratory complex subunits or assembly factors. MITOCH. 2015;21:76–91 [Internet].

131. Johnson SC, Yanos ME, Kayser E-B, Quintana A, Sangesland M, Castanza A, et al. mTOR inhibition alleviates mitochondrial disease in a mouse model of Leigh syndrome. Science. 2013;342(6165):1524–8 [Internet].
132. Mannick JB, Del Giudice G, Lattanzi M, Valiante NM, Praestgaard J, Huang B, et al. mTOR inhibition improves immune function in the elderly. Sci Transl Med. 2014;6(268):1–7.
133. Prior M, Chiruta C, Currais A, Goldberg J, Ramsey J, Dargusch R, et al. Back to the future with phenotypic screening. ACS Chem Neurosci. 2014;5:503–13.
134. Schreiber SL. Small molecules: the missing link in the central dogma. Nat Chem Biol. 2005;1(2):64–6.
135. Wagner BK, Kitami T, Gilbert TJ, Peck D, Ramanathan A, Schreiber SL, et al. Large-scale chemical dissection of mitochondrial function. Nat Biotechnol. 2008;26(3):343–52.
136. Andreux PA, Mouchiroud L, Wang X, Jovaisaite V, Mottis A, Bichet S, et al. A method to identify and validate mitochondrial modulators using mammalian cells and the worm C. elegans. Sci Rep. 2014;4:5285. [Internet], (Figure 1).
137. Gohil VM, Sheth SA, Nilsson R, Wojtovich AP, Lee JH, Perocchi F, et al. Nutrient-sensitized screening for drugs that shift energy metabolism from mitochondrial respiration to glycolysis. Nat Biotechnol. 2010;28(3):249–55 [Internet].
138. Kitami T, Logan DJ, Negri J, Hasaka T, Tolliday NJ, Carpenter AE, et al. A chemical screen probing the relationship between mitochondrial content and cell size. PLoS ONE. 2012;7(3):1–7.
139. Sahdeo S, Tomilov A, Komachi K, Iwahashi C, Datta S, Hughes O, et al. High-throughput screening of FDA-approved drugs using oxygen biosensor plates reveals secondary mitofunctional effects. Mitochondrion. 2014;17:116–25 [Internet].
140. Gohil VM, Offner N, Walker JA, Sheth SA, Fossale E, Gusella JF, et al. Meclizine is neuroprotective in models of Huntington's disease. Hum Mol Genet. 2011;20(2):294–300.
141. Casaluce F, Sgambato A, Maione P, Ciardiello F, Gridelli C. Emerging mitotic inhibitors for non-small cell carcinoma. Expert Opin Emerg Drugs. 2013;18(1):97–107 [Internet].
142. King MP, Attardi G. Human cells lacking mtDNA: repopulation with exogenous mitochondria by complementation. Science. 1989;246(4929):500–3.
143. Wilkins HM, Carl SM, Swerdlow RH. Cytoplasmic hybrid (cybrid) cell lines as a practical model for mitochondriopathies. Redox Biol. 2014;2(1):619–31 [Internet].
144. de la Mata M, Garrido-Maraver J, Cotán D, Cordero MD, Oropesa-Ávila M, Izquierdo LG, et al. Recovery of MERRF fibroblasts and cybrids pathophysiology by coenzyme Q10. Neurotherapeutics. 2012;9(2):446–63.
145. Desquiret-Dumas V, Gueguen N, Barth M, Chevrollier A, Hancock S, Wallace DC, et al. Metabolically induced heteroplasmy shifting and l-arginine treatment reduce the energetic defect in a neuronal-like model of MELAS. Biochim Biophys Acta Mol Basis Dis. 2012;1822(6):1019–29 [Internet].
146. Dai Y, Zheng K, Clark J, Swerdlow RH, Pulst SM, Sutton JP, et al. Rapamycin drives selection against a pathogenic heteroplasmic mitochondrial DNA mutation. Hum Mol Genet. 2014;23(3):637–47.
147. Takahashi K, Tanabe K, Ohnuki M, Narita M, Ichisaka T, Tomoda K, et al. Induction of pluripotent stem cells from adult human fibroblasts by defined factors. Cell. 2007;131(5):861–72.
148. Prigione A. Mitochondrial medicine. In: Weissig V, Edeas M, editors. Mitochondrial medicine: volume II, manipulating mitochondrial function, methods in molecular biology [internet]. New York: Springer; 2015. p. 349–56.
149. Fujikura J, Nakao K, Sone M, Noguchi M. Induced pluripotent stem cells generated from diabetic patients with mitochondrial DNA A3243G mutation. Diabetologia. 2012;55:1689–98.
150. Hämäläinen RH, Manninen T, Koivumäki H, Kislin M, Otonkoski T. Tissue- and cell-type – specific manifestations of heteroplasmic mtDNA 3243A>G mutation in human induced plu-

ripotent stem cell-derived disease model. Proc Natl Acad Sci U S A. 2013;110(38):E3622–30.
151. Folmes C, Martinez-fernandez A, Perales-clemente E, Li X, Mcdonald A, Oglesbee D, et al. Disease-causing mitochondrial heteroplasmy segregated within induced pluripotent stem cell clones derived from a patient with. Stem Cells. 2013;31(7):1298–308.
152. Cherry ABC, Gagne KE, McLoughlin EM, Baccei A, Gorman B, Hartung O, et al. Induced pluripotent stem cells with a mitochondrial DNA deletion. Stem Cells. 2013;31(7):1287–97.
153. Cooper O, Seo H, Andrabi S, Guardia-laguarta C. Pharmacological rescue of mitochondrial deficits in iPSC-derived neural cells from patients with familial Parkinson's disease. Sci Transl Med. 2012;4(141):141ra90.
154. Sanders LH, Laganière J, Cooper O, Mak SK, Vu BJ, Huang YA, et al. Neurobiology of disease LRRK2 mutations cause mitochondrial DNA damage in iPSC-derived neural cells from Parkinson's disease patients : reversal by gene correction. Neurobiol Dis. 2014;62:381–6 [Internet].
155. Pfeffer G, Horvath R, Klopstock T, Mootha VK, Suomalainen A, Koene S, et al. New treatments for mitochondrial disease-no time to drop our standards. Nat Rev Neurol. 2013;9(8):474–81 [Internet].
156. Pfeffer G, Majamaa K, Turnbull D, Thorburn D, Chinnery P. Treatment for mitochondrial disorders. Cochrane Database Syst Rev. 2012;1, CD004426.
157. Kerr DS. Review of clinical trials for mitochondrial disorders: 1997–2012. Neurotherapeutics. 2013;10(2):307–19 [Internet].
158. Martinelli D, Catteruccia M, Piemonte F, Pastore A, Tozzi G, Dionisi-Vici C, et al. EPI-743 reverses the progression of the pediatric mitochondrial disease-genetically defined Leigh syndrome. Mol Genet Metab. 2012;107(3):383–8 [Internet].
159. Reisman SA, Lee CYI, Meyer CJ, Proksch JW, Ward KW. Topical application of the synthetic triterpenoid RTA 408 activates Nrf2 and induces cytoprotective genes in rat skin. Arch Dermatol Res. 2014;306(5):447–54.
160. Szeto HH, Birk AV. Serendipity and the discovery of novel compounds that restore mitochondrial plasticity. Clin Pharmacol Ther. 2014;96(6):672–83 [Internet].

Chapter 16
Summary and Conclusions

Amy Katherine Reeve

Abstract More than a decade of research has pushed forward our understanding of the changes that occur within neurons in neurodegenerative disease. From a mitochondrial point of view, we now know that mitochondrial DNA, reactive oxygen species, calcium handling, mitochondrial turnover, movement, neuronal localisation and dynamics are important for neuronal survival. It would seem feasible, therefore, that dysfunction in this organelle will have huge implications for the pathogenesis of a number of degenerative diseases. While we still have much to discover and understand with regard to the causes of these diseases, research in the field of mitochondrial dysfunction can only add to our understanding and hopes for therapeutic interventions.

Keywords Mitochondria • Mitochondrial function • mtDNA genetics • Neurodegeneration

Over the last few decades, there has been an explosion in the amount of research dedicated to understanding the role of mitochondria in the pathogenesis of neurodegenerative diseases. Since the first edition of *Mitochondrial Dysfunction in Neurodegenerative Disorders* 3 years ago, researchers across the globe have been increasing our understanding of this complex relationship. Much of this research has centred on Parkinson's disease and Alzheimer's disease, but as we investigate further the importance of mitochondrial function for neuronal survival, more diseases emerge for which pathogenic mechanisms centre around the mitochondrion. Before one can comprehend the importance of mitochondrial function for neuronal health, one must first understand the complexities of the mitochondrion. Thus, the first section of this book aimed to introduce mitochondria, discussing their functions and genetics and highlighting their importance for all cells including neurons.

A.K. Reeve, BSc(hons), PhD
Newcastle University Medical School, Wellcome Trust Centre for Mitochondrial Research, Newcastle upon Tyne, UK
e-mail: amy.reeve@ncl.ac.uk

A.K. Reeve et al. (eds.), *Mitochondrial Dysfunction in Neurodegenerative Disorders*, DOI 10.1007/978-3-319-28637-2_16

16.1 Mitochondrial Dysfunction: Causes and Effects

Mitochondrial dysfunction can be caused by mutations in both the mitochondrial and nuclear genomes. While mutations in the mitochondrial genome have been shown to be prevalent in the human brain in both normal ageing and neurodegenerative diseases (Chap. 3), such mutations also directly cause a group of diseases characterised by degenerative changes in a number of tissues, predominantly in the brain. As discussed in Chap. 2 (Phillips et al.), the understanding of cell loss in these mitochondrial disorders has allowed research to become focussed on the mechanisms by which mitochondrial dysfunction can lead to neuronal loss. While the type of neurodegenerative changes seen in mitochondrial disorders can vary between individuals, there are similarities with the pathology of more common diseases, including in the brain regions affected (e.g., the basal ganglia) and in the type of changes seen (e.g., demyelination). Recently, a significant body of research has aimed to uncover how changes within neuronal axons and synapses may contribute to neurodegeneration, and Sajic (Chap. 8) and Chrysostomou (Chap. 9) discuss these concepts in their chapters. Despite overlapping pathological changes, there are obvious differences between these disorders and other diseases; for example, the degeneration of neurons in mitochondrial disorders is not associated with the accumulation of protein aggregates so common in other diseases including Alzheimer's and Parkinson's diseases. The importance of comprehending the mechanisms involved in these disorders was also highlighted by Trifunovic, in Chap. 13, through the description of the plethora of animal models which have been used to further investigate the impact of mitochondrial dysfunction. Mitochondrial dysfunction can be also caused by mutations in nuclear genes which encode proteins essential for mitochondrial maintenance and function. In Chap. 4 of this book, Viscomi et al. have highlighted the importance of such genes in neurodegenerative disease.

16.2 Consequences of Mitochondrial Dysfunction for Neuronal Function

Dysfunction of mitochondria caused by changes in mitochondrial DNA, however, is the tip of a very large iceberg. While these changes will undoubtedly cause changes to the ATP production of the cell, which will have detrimental effects for all cellular processes, mitochondrial involvement in other cellular processes could be equally catastrophic. The majority of this book focussed, therefore, on the consequences of the accumulation of mitochondrial dysfunction on neuronal function and how changes in key neuronal processes may exacerbate this dysfunction.

Of particular importance for the pathogenesis of a number of neurodegenerative disorders is the accumulation of damaged or misfolded proteins. In Chap. 10, Ludtmann and Abramov highlight how the aggregation of four proteins affects neuronal homeostasis and may contribute to the development of neurodegeneration, while in Chap. 11 Otten et al. consider how the degradation of long-lived proteins and organelles through the process of autophagy may become overwhelmed and lose efficiency. To complement these two chapters, Nguyen et al. have contributed a chapter which examines recent research into the relationship between mitochondria and the lysosome. The lysosome provides the final digestive step in the process of autophagy, and it is becoming clear that the relationship between these two organelles is dependent on function in both.

Mitochondria are dynamic organelles that exist in a constantly changing, vast reticular network. This network is maintained through the processes of fission and fusion, which like many mitochondrial functions are reliant on the mitochondrial membrane potential. A loss of this membrane potential disrupts mitochondrial dynamics, which will affect the supply of ATP and calcium buffering throughout neuronal processes. In Chap. 7, Mourier discusses the complexities of mitochondrial dynamics and the impact of changes within this process for neuronal health and survival. The transport of mitochondria was also considered in Chap. 8. In this chapter Sajic has highlighted that transport of mitochondria maintains neuronal function by ensuring delivery of mitochondria to discrete sites within the neuron. Furthermore, disruption of the transport of mitochondria has been shown to occur alongside demyelination in multiple sclerosis. Aside from the provision of cellular ATP, mitochondria are also essential for the buffering of intracellular calcium levels. The importance of calcium has been particularly highlighted in the field of Parkinson's disease research since the neurons of the SN show unique pacemaking activity that is dependent on calcium. This aspect of Parkinson's disease research is reviewed in Chap. 6 by Surmeier and colleagues.

In this second edition we wanted to include a chapter that highlighted the advances that have been made in the imaging of mitochondria within living cells. In Chap. 14 Chisholm et al. discuss some of the practical considerations and techniques that have enabled researchers to examine mitochondria in depth within living cells and tissues.

We finish our book with a look to the future. As we delve further into the mechanisms leading to neurodegenerative disease, and unearth additional causes and consequences of mitochondrial dysfunction within neurons, we discover new therapeutic avenues. In our final chapter, Helliwell guides us through recent trials hoping to alleviate the symptoms of neurodegenerative disease by protecting neurons against mitochondrial dysfunction. It would seem that hope for treating these diseases lies in understanding why neurons are lost, and mounting evidence tells us that the mitochondrion is central in the processes which lead to this loss.

16.3 Conclusion

More than a decade of research has pushed forward our understanding of changes in neurons in neurodegenerative disease. From a mitochondrial point of view, we now know that mitochondrial DNA, reactive oxygen species, calcium handling, mitochondrial turnover, movement, neuronal localisation and dynamics are important for neuronal survival. It would seem feasible, therefore, that dysfunction in this organelle will have huge implications for the pathogenesis of a number of degenerative diseases. While we still have much to discover and understand with regard to the causes of these diseases, research in the field of mitochondrial dysfunction can only add to our understanding and hopes for therapeutic interventions.

Index

A.K. Reeve et al. (eds.), *Mitochondrial Dysfunction in Neurodegenerative Disorders*, DOI 10.1007/978-3-319-28637-2